Electron Correlations and Materials Properties 2

Electron Correlations and Materials Properties 2

Edited by

A. Gonis

Lawrence Livermore National Laboratory
Livermore, California

N. Kioussis

California State University—Northridge
Northridge, California

and

M. Ciftan

U.S. Army Research Office
Research Triangle Park, North Carolina

Kluwer Academic / Plenum Publishers
New York, Boston, Dordrecht, London, Moscow

Proceedings of the Second International Workshop on Electron Correlations and Materials Properties, held June 23–29, 2001, in Rhodes, Greece

ISBN: 0-306-47446-8

233 Spring Street, New York, N.Y. 10013

A C.I.P. record for this book is available from the Library of Congress.

10 9 8 7 6 5 4 3 2 1

ACKNOWLEDGMENTS

The conference was assisted through administrative efforts and/or financial aid from the following organizations:

California State University, Northridge

The U. S. Army Research Office

Materials Research Institute,
Lawrence Livermore National Laboratory

PREFACE

This is the second in a series of "International Workshops on Electron Correlations and Materials Properties." The aim of this series of workshops is to provide a periodic (triennial) and in-depth assessment of advances in the study and understanding of the effects that electron–electron interactions in solids have on the determination of measurable properties of materials. The workshop is structured to include exposure to experimental work, to phenomenology, and to *ab initio* theory. Since correlation effects are pervasive the workshop aims to concentrate on the identification of promising developing methodology, experimental and theoretical, addressing the most critical frontier issues of electron correlations on the properties of materials. This series of workshops is distinguished from other topical meetings and conferences in that it strongly promotes an interdisciplinary approach to the study of correlations, involving the fields of quantum chemistry, physics, and materials science.

The First Workshop was held June 28–July 3, 1998, and a proceedings of the workshop was published by Kluwer/Plenum. The Second Workshop was held June 24–29, 2001, and this volume contains the proceedings of that scientific meeting. Through the publications of proceedings, the workshop attempts to disseminate the information gathered during the discussions held at the Workshop to the wider scientific community, and to establish a record of advances in the field.

As of this writing, a great deal of development has taken place in the study of correlated systems, both experimentally and theoretically, and some of them are discussed in the present volume. The effects of size on the magnetism of thin films, the effects of correlations on the phase diagram of plutonium, and the effects of pressure on the properties of magnetic F- and D-electron materials are some of the experimental topics included here.

On the theoretical front, spectral density functional approaches to electron correlations, the application (and limitations) of density functional theory and some of its new extensions, and theoretical insights into magnetism and superconductivity were presented and discussed at the meeting and are included in this proceedings.

The editors wish to express our gratitude to the following organizations who helped us immeasurably (administratively and/or financially) in setting up the workshop. Special thanks are due to the California State University at Northridge for handling the financial aspects of the workshop. Without the financial aid from the U.S. Army Research Office and Lawrence Livermore National Laboratory the meeting would not have materialized—our sincere thanks to them. Finally, we acknowledge the patience and guidance from Kluwer/Plenum in the preparation of this volume.

A. Gonis
N. Kioussis
M. Ciftan

CONTENTS

Part 3. *Ab initio* Studies of Correlation Effects

Part 1

Experimental Indications of Correlation Effects in Materials

PRESSURE-INDUCED EFFECTS IN MAGNETIC MATERIALS WITH F- AND D-ELECTRONS

Vladimír Sechovský, Martin Diviš, Ladislav Havela, Fuminori Honda, Karel Prokeš, Olexandr Syshchenko

Department of Electronic Structures, Charles University, Ke Karlovu 5, 121 16 Prague 2, The Czech Republic.

sech@mag.mff.cuni.cz

Alexander Andreev, Jiří Kamarád

Institute of Physics, Academy of Sciences of the Czech Republic, Na Slovance 2, 182 21 Prague 8, The Czech Republic

Abstract We review recent experiments focused on pressure effects on magnetism in uranium intermetallics, namely the tetragonal compound UNi_2Si_2 and several representatives of hexagonal UTX compounds (T = transition metal, X = p-metal) in which we studied various aspects of magnetism in $5f$-electron intermetallics. As an example of pressure effects on rare-earth – transition metal compounds, we discuss evolution of Co magnetism in $ErCo_2$ with applying high hydrostatic pressures.

Keywords: Magnetism, high pressure, UTX compounds, $ErCo_2$

1. INTRODUCTION

Magnetic phenomena in materials are closely related to the character of states of d- or f- electron in incomplete shells of transition elements. Relevant features of electronic structure can be tested by experiment and first-principles calculations when studying changes of magnetic parameters with applied hydrostatic or uniaxial pressure on a particular material. To interpret the pressure dependence of magnetic parameters correctly, impact of external pressure on interatomic distances should be carefully analyzed considering values of the linear compressibility that is often anisotropic. [1] In this context it is desirable to investigate also the anisotropic thermal expansion and magnetostriction. We will demonstrate on some examples that single crystals are often indispensable to allow obtaining intrinsic material parameters.

In this paper we review results of our recent experiments on several representative UTX compounds (T = transition metal, X = p-metal) and UNi_2Si_2 in which we studied various aspects of magnetism in $5f$-electron intermetallics. Besides the uranium intermetallics, we discuss evolution of Co magnetism in $ErCo_2$ with applying high hydrostatic pressures.

2. UTX COMPOUNDS

We will focus mainly on isostructural hexagonal UTX compounds with the ZrNiAl-type structure (the $P\bar{6}2m$ space group). Here T stays for a late transition metal and X is a p-electron metal, e.g. Al, Ga, In, Sn and Sb. This structure has a distinct layered character. It is built up of U-T and T-X basal-plane layers alternating along the c-axis. Each uranium atom has four nearest U neighbors within the U-T layer. [2]

The U-T layers are separated by the lattice parameter c. The distance between the nearest U neighbors within the basal plane can be found using the relation:

$$d_{\mathrm{U-U}} = a\sqrt{1 - 3x + 3x^2}, \tag{1}$$

where x is a position parameter of the U atoms.

The U ground-state magnetic moment in compounds of this group varies from 0 to 1.6 μ_B depending on the degree of itinerancy of the uranium $5f$ electrons that is mainly determined by the overlap of $5f$ wave-functions centered on neighboring U atoms and by the hybridization of $5f$ states with valence electron states of ligands. [2] The latter mechanism usually called the $5f$-ligand hybridization has been introduced and discussed in various aspects by Koelling and Dunlap. [3] The $5f$-ligand hybridization causes that the T and X atoms affect magnetic properties considerably by influencing the $5f$-electron states although they do not contribute much to the magnetic moment itself. All the compounds of the group, irrespective of the ground state that can be ferromagnetic (the typical representative is UPtAl), [4] antiferromagnetic (UNiAl, [5] UNiGa [2]) or a paramagnetic spin fluctuator with metamagnetic behavior at low temperatures (UCoAl [6]) exhibit a huge uniaxial magnetic anisotropy. [7] It is a consequence of strong $5f$-electron bonding within the basal plane and existence of orbital $5f$ magnetic moment. The characteristic magnetic response is observed along the c-axis whereas the basal-plane directions exhibit Pauli paramagnetism.

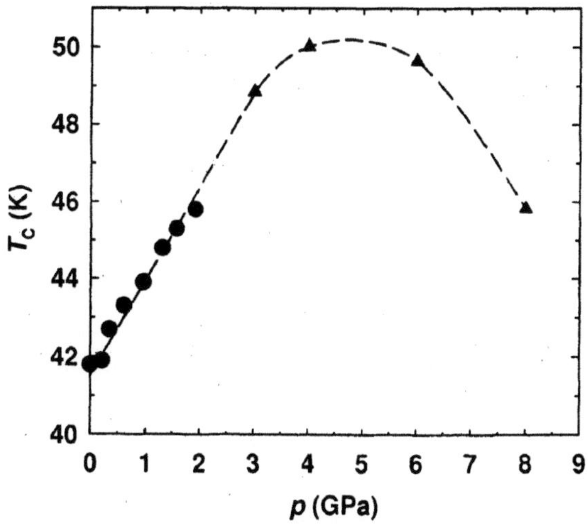

Figure 1. Pressure dependence of Curie temperature of UPtAl; the dashed line is a guide for the eye.

2.1 Collapse of ferromagnetism in UPtAl under hydrostatic pressure

UPtAl exhibits ferromagnetic ordering ($T_C = 42.5$ K) of uranium $5f$-electron magnetic moments. The magnetic phase transition from paramagnetism to ferromagnetism is accompanied by a clear cusp on the temperature dependence of electrical resistivity and by a spontaneous magnetostriction reaching at low temperatures -3×10^{-4} and 2×10^{-4} in the basal plane (λ_a) and along the c-axis (λ_c), respectively. The volume effect $\omega_s = 2\ \lambda_a + \lambda_c$ is $-\ 4 \times 10^{-4}$. Magnetization measurements in pressures up to 0.8 GPa revealed pressure-induced reduction of the spontaneous moment at a rate $\mathrm{d}M_s/\mathrm{d}p = -0.027\ \mu_B$/GPa while T_C increases: $\mathrm{d}T_C/\mathrm{d}p = 2.5$ K/GPa. [8] In experiments with UPtAl single crystals exerted to hydrostatic pressures up to 8 GPa we have observed that the initial increasing tendency of the Curie temperature with pressure persists while the $\mathrm{d}T_C/\mathrm{d}p$ becomes gradually reduced. Finally, a maximum in the T_C vs. p plot (see Figure 1) appears between 4 and 6 GPa followed by a rapid decrease with further increasing pressure.

This result is reminiscent of observations on USe and UTe. [9] To our best knowledge, our result is the first observed on an intermetallics compound that exhibits entirely $5f$-electron band magnetism.

The results on UPtAl can be conceived with a scenario that considers the dual role of the $5f$-ligand hybridization played in magnetism of uranium intermetallics: causing delocalization of $5f$ electrons on one side and mediating exchange interactions between magnetic uranium sites. The hybridization is usually enhanced with increasing pressure because of reduction of interatomic distances and consequently more extended overlaps of wave functions between neighboring atoms in the lattice. When the hydrostatic pressure is increased, the enhancement of $5f$-ligand hybridization causes gradual washout of magnetic moments and the consequent suppression of ordered magnetism. On the other hand, the increased overlaps yield enhancement of exchange integrals that can cause increasing ordering temperature with pressure in a limited pressure interval. In higher pressures, however, the effect of ceasing U magnetic moment dominates and the magnetic ordering finally disappears. This scenario is corroborated by the pressure-induced evolution of the temperature dependence of the resistivity of UPtAl in ordered state that can be approximated by the expression:

$$\rho = \rho_0 + aT^2 + bT(1 + 2T/\Delta)\exp\left(-\frac{\Delta}{T}\right), \qquad (2)$$

where the last term represents scattering of conduction electrons on magnon-like excitations with an energy gap Δ that represents the energy of magnetocrystalline anisotropy.[10] The value of $\Delta \approx 60$ K that is derived from ambient-pressure data does not change within experimental error in pressures up to 4 GPa but decreases rapidly with further increasing pressure to amount only 16 K at 8 GPa. The reduced energy of magnetocrystalline energy is due to the rapidly ceasing U magnetic moment due to more extended delocalization of U $5f$ electrons.

2.2 Pressure induced phenomena in UCoAl

As concerns magnetism, UCoAl has a unique position not only among the UTX (T = transition metal, X = p-electron metal) compounds but also among intermetallics, in general. This material exhibits paramagnetic ground state although the susceptibility has a maximum around $T_{max} \approx 20$ K.[11] The magnetic-field induced ferromagnetic instability qualifies UCoAl to be considered as an itinerant electron metamagnet alike YCo_2.[12] In contrast to the $3d$ itinerant electron metamagnets that show the metamagnetic state in rather high magnetic fields (above 70 T in YCo_2) irrespective to the field direction, the critical field B_c of metamagnetic transition in UCoAl is as low as 0.6 T and it should be applied along the c-axis of the ZrNiAl-type hexagonal structure of UCoAl. The U magnetic moments are then aligned also along c. Fields up to 40 T applied within the basal plane have no effect on the paramagnetic state.[11] The metamagnetic transition is accompanied by a magnetostriction effect[13] of the order of 1×10^{-4}, negative along the c-axis and positive in the basal plane (both almost equal in absolute value) yielding also the volume effect of the same order of magnitude.

Ferromagnetism can be induced in UCoAl by substituting U by small amounts of Y[14] or Lu for U, Fe, Ru, Rh or Ir for Co,[15,16] and Ga, In or Sn for Al.[2,6] This ferromagnetism can be suppressed and the original metamagnetic behavior recovered by applying hydrostatic pressure.[2,6,17] The hydrostatic pressure also causes gradual increase of B_c and suppression of magnetization step ΔM across the metamagnetic transition in UCoAl itself. The estimated critical pressure in which the metamagnetism in UCoAl should disappear is 7 GPa.[11] A qualitatively similar effect of gradually suppressing metamagnetism in UCoAl is observed when Ni, Pt, Pd or Cu substitute the Co atoms.[2,16,18]

To elucidate the phenomena that are characteristic for physics of UCoAl a simple model has been formulated considering the $5f$-ligand hybridization to be the prominent mechanisms that affects the size of U magnetic moments and their exchange coupling. If strong, the hybridization causes considerable delocalization that may lead to total washout of the U moment. On the other hand, the hybridization mediates exchange interactions between U atoms. The UCoAl crystal structure consists of U-Co and Co-Al basal-plane layers alternating along the c-axis. The $5f$-$3d$ hybridization within the U-Co plane mediates a strong ferromagnetic interaction that couples the U moments within the plane to form ferromagnetic basal-plane sheets. The exchange interaction components effective in the perpendicular direction are much weaker. Nevertheless, they are rather important in determining the type of magnetic ground state because they couple U moments located in different layers. Along the c-axis both, the $5f$-$3d$ (U-T) and $5f$-$3p$ (U-Al) hybridization are the sources of exchange interactions in this direction. The analysis of large amount of systematically collected data about the isostructural UTX compounds including UCoAl allows a conclusion that the $5f$-$3d$ hybridization promotes the ferromagnetic coupling of U moments along c, whereas the $5f$-$3p$ (U-Al) hybridization mediates the antiferromagnetic coupling. The non-magnetic ground state of UCoAl is considered to be a consequence of strong delocalization of $5f$-electron due to a strong $5f$-$3d$ hybridization within the basal plane and a delicate balance of the ferromagnetic and antiferromagnetic exchange interaction components along the c-axis. To test this model, experiments studying effects of varying interlayer distances are desirable.

Recently, we have studied evolution of magnetization behavior of UCoAl under uniaxial pressure, as an example of such experiments. [19] The most important finding was that the application of uniaxial pressure along the *c*-axis of UCoAl leads to a gradual decrease of B_c yielding a net spontaneous magnetization for $p > 0.05$ GPa. [19] This result is qualitatively opposite to the effect induced by the hydrostatic pressure.

At this moment one should consider the principal difference the hydrostatic pressure and uniaxial pressure experiments. An important fact should be considered: the linear compressibility of UCoAl is strongly anisotropic. At room temperature $\partial \ln a/\partial p = -4.2 \times 10^{-3}$ GPa^{-1} and $\partial \ln c/\partial p = -1.1 \times 10^{-3}$ GPa^{-1}, i.e. the UCoAl lattice is much softer within the basal plane than along the *c*–axis. [1] Therefore, in the hydrostatic pressure experiment one may expect considerable enhancement of the $5f$-$3d$ hybridization within the basal plane. A consequent effect of delocalization of $5f$-states on the itinerant electron metamagnet UCoAl is the increase of B_c and T_{max} values leading to final suppression of metamagnetism. This was indeed observed in UCoAl. [11] This mechanism probably dominates the effects of varying hybridizations along the *c*-axis due to decreasing interplane distances.

On the other hand, the uniaxial pressure along the *c*-axis primarily causes shrinking interatomic distances in this direction and expansion within the basal plane. Consequently, the increase of $5f$-$3d$ and $5f$-$3p$ wave-function overlap yields enhanced hybridization and the ferromagnetic exchange component mediated by the $5f$-$3d$ hybridization becomes presumably dominating. A natural result is gradual decrease of the B_c value and emerging spontaneous magnetization at higher uniaxial pressures. This tendency is strongly assisted by the reduced $5f$-$3d$ hybridization within the basal plane as a consequence of increasing interatomic distances. The observed uniaxial-pressure effect on the temperature dependence of magnetic susceptibility points to pressure-induced increase of T_C with uniaxial pressure increasing beyond 0.037 GPa. The proposed scenario is plausible also in the light of magnetostriction data in UCoAl across the metamagnetic transition from paramagnetic to ferromagnetic state that is accompanied by shrinkage of the *c*-axis and expansion of the *a*-axis. [13]

2.3 Pressure-induced changes of magnetic propagation in UNiGa

UNiGa belongs to UTX compounds crystallizing in the hexagonal ZrNiAl-type structure that consists of the U-T and T-X basal-plane layers alternating along the *c*-axis. [2] The huge uniaxial magnetocrystalline anisotropy in these materials locks the uranium magnetic moments along the c-axis that leads to Ising-like behavior. The $5f$-d hybridization within the U-T planes mediates a strong ferromagnetic exchange interaction that leads to formation of ferromagnetic basal-plane sheets of U magnetic moments. These ferromagnetic sheets are coupled along the c-axis by exchange interactions mediated by the $5f$-d and $5f$-p hybridization across planes. Consequently, magnetic structures are formed of ferromagnetic sheets with various stacking along c-axis. In UNiGa several types of antiferromagnetic structures (for a magnetic phase diagram see Figure 2) are observed below 40 K. [2, 20] The onset of antiferromagnetic ordering is characterized by an incommensurate antiferromagnetic structure (**ICAF**) with a wave vector $\boldsymbol{q} \approx (0,0,0.36)$. At 37.3 K a first-order transition is observed to a commensurate antiferromagnetic structure with a periodicity $3c$ (we use notation **AF-3c**). This may be conceived with a stacking sequence $+\ 0\ -$, where 0 represents a layer of uranium atoms with the zero ordered magnetic moment (paramagnetic sheet). Such a situation can appear in the case of a balanced antiferromagnetic interaction between nearest neighbor and next-nearest neighbor uranium planes. [20] With further decreasing temperature, UNiGa undergoes gradual transformation to the ground-state antiferromagnetic phase (**AF-6c**) with the $++-+--$ stacking via the "intermediate" $++-+---+-$ phase (**AF-8c**). We note that among the four antiferromagnetic phases and one field-induced phase (there is also a field-forced ferromagnetic phase above 1 T) at ambient pressure there is no such a phase with a simple $+\ -$ stacking. The **ICAF**, **AF-3c** and **AF-8c** phases (and also **AF-6c** above 14 K) are transformed to an uncompensated antiferromagnetic phase **UAF-3c** characterized by the $++-$ stacking already in very small magnetic fields applied parallel to the c-axis. In somewhat higher fields UNiGa exhibits a transition to a ferromagnetic state (**F**). The hatched area in the magnetic phase diagram below 14 K represents a region in which magnetic history phenomena are observed in bulk properties and also on the microscopic scale including the relaxation phenomena. [2, 20] The uranium magnetic moments, which are $\sim 1.4\ \mu_B$ at 2 K, do not decrease substantially with increasing temperature almost up to 35 K, where they retain more than 90 % of their low temperature value. This behavior seems to be more general for the UTX compounds which exhibit high anisotropy yielding an Ising-like behavior.

The combination of relatively high Néel temperature (40 K), very small magnetic fields of metamagnetic transitions (< 1.5 T) and the collection of long-period antiferromagnetic structures depending on temperature indicates that physics of UNiGa is connected with a delicate

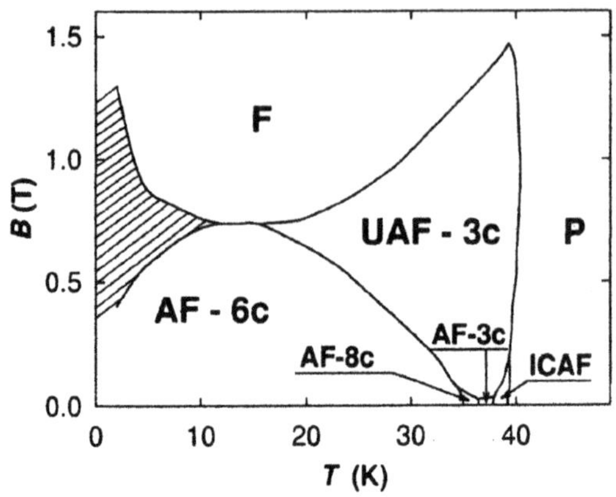

Figure 2. The *B-T* magnetic phase diagram of UNiGa at ambient pressure. Legend: **ICAF** - incommensurate structure, $q = \pm(0, 0, \delta); \delta \approx 0.36$; **AF-3c** – commensurate AF structure with $q = \pm(0, 0, 1/3)$; **AF-8c** - commensurate AF structure with $q = \pm(0, 0, 1/8)$ and $\pm(0, 0, 3/8)$; **AF-6c** - the ground-state - commensurate structure with $q = \pm(0, 0, 1/6), \pm(0, 0, 1/3)$ and $(0, 0, 1/2)$; **UAF-3c** - uncompensated AF structure with $q = \pm(0, 0, 1/3)$; **F** - ferromagnetic state; **P** - paramagnetic phase. All magnetic structures are collinear with U magnetic moments parallel to the *c*-axis and ferromagnetically coupled within the basal plane of the hexagonal ZrNiAl-type structure of UNiGa. In the ground state all uranium atoms carry the same magnetic moment of 1.4 μ_B.

balance of ferromagnetic and antiferromagnetic exchange interactions. To understand the special hierarchy of exchange interactions in UNiGa influence of external pressures that modify interatomic distances is of obvious interest. The magnetic phase transitions in UNiGa are accompanied by distinct magnetotransport [20] and magnetoelastic anomalies that were employed to investigate pressure effects on magnetic phase transitions. [21 – 23]

The influence of hydrostatic pressure on magnetic phase transitions in UNiGa (see the *p-T* magnetic phase diagram in Figure 3) have been studied by measurements of the electrical resistivity at various temperatures, magnetic fields and pressures. [21, 24] Studies in a magnetic field applied along the *c*-axis under external hydrostatic pressures up to 2.2 GPa were continued by the linear thermal expansion and the magnetostriction measurements that indicated appearance of a new magnetic phase above 1.8 GPa. [23]

The linear thermal expansion of UNiGa at ambient pressure exhibits several sharp anomalies between 34 and 40 K [22, 23] that can be associated with magnetic phase transitions separating the **AF-6c**, ... ,

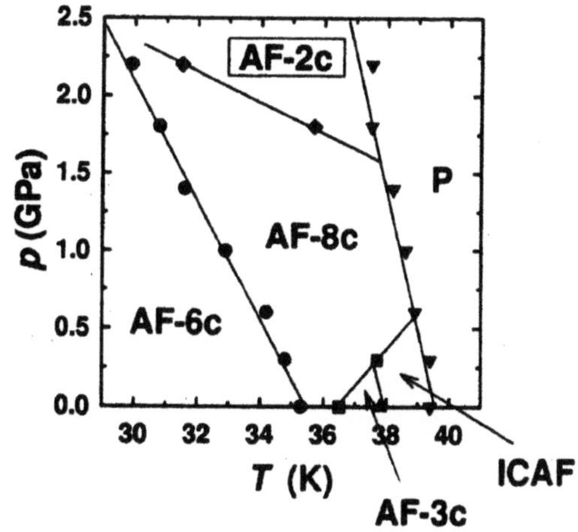

Figure 3. The *p-T* phase diagram of UNiGa. Legend: **AF-2c** – the *"new"* pressure induced antiferromagnetic structure with $q = (0, 0, 1/2)$; other magnetic phases are denoted as in Figure 2.

P as it is evident from the schematic *p-T* magnetic phase diagram in Figure 3. The data sets taken for several pressures between 0.1 and 2.2 GPa reveal that the **ICAF** phase is suppressed already by 0.3 GPa, the **AF-3c** phase ceases in pressures above 0.6 GPa, only the **AF-6c** and **AF-8c** phases remain and the temperature range of the latter phase becomes gradually more extended with increasing pressure on account of the ground state phase. This evolution has been **AF-8c** confirmed by a neutron diffraction experiment in pressures up to 0.9 GPa. [23] The transition temperatures T_N and T_3 (between **AF-6c** and **AF-8c**) become reduced at a rate of -1.1 and -2.6 K·GPa^{-1}, respectively. In pressures $p \geq 1.8$ GPa a new magnetic phase is induced to replace the phase **AF-8c**.

The magnetic phase diagram at hydrostatic pressure of 0.9 GPa is generally similar to that was determined at ambient pressure and no new magnetic phases have been detected. However, some important modifications can be recognized by closer inspection: (i) T_N becomes reduced; (ii) the transition towards the field-induced ferromagnetic phase occurs at higher critical fields, confirming that antiferromagnetic interactions are promoted by hydrostatic pressure. The **AF-2c** phase can be traced, only in increasing field up to about 0.3 T. Negligible reduction of magnetic intensities with application of pressure points to an important conclusion that not only the magnetic periodicity of individual phases, but also the U magnetic moment remain intact pressures up to 0.9 GPa.

The neutron diffraction experiment on UNiGa exerted to pressure of 2.3 GPa reveals that as temperature is lowered below 38 K, magnetic peaks occur which can be indexed by propagation vector $\boldsymbol{q} = (0, 0, 1/2)$. These peaks can be interpreted that at high pressures the simplest antiferromagnetic structure with an stacking $+ -$ along the c-axis is present down to lowest temperatures without indicating any phase transition. An important finding is that at low temperatures only these reflections are present, i.e. that the reflections which are observed in the ground-state **AF-6c** phase (at ambient pressure there are $(h\ k\ l \pm 1/6)$ and $(h\ k\ l \pm 1/3)$ reflections in addition to $(h\ k\ l + 1/2)$ reflections) are missing. Refinement of integrated intensities leads to U magnetic moments of $1.3 \pm 0.1\mu_{\mathrm{B}}$ oriented along the hexagonal axis. This value is only few percent smaller that magnitude found at ambient pressure in agreement with previous experiments. This result can be unambiguously interpreted in terms of the high-pressure induced phase characterized by the simplest $+ -$ antiferromagnetic stacking (we denote it as **AF-2c**). This conclusion is consistent with previously reported pressure effects on fields of metamagnetic transitions that point to enhancement of antiferromagnetic exchange interactions. [23] In order to address different components of strongly anisotropic exchange interactions (ferromagnetic within the basal plane and predominantly antiferromagnetic along the c-axis) experiments on crystals exposed to uniaxial pressure are envisaged.

2.4 Collapse of itinerant 5f-electron antiferromagnetism in UNiAl

UNiAl exhibits attributes of an itinerant 5f-electron antiferromagnet with a strong uniaxial anisotropy of magnetic properties and spin fluctuations phenomena. The reduced U moments in UNiAl order below $T_{\mathrm{N}} = 19.3$ K in a peculiar antiferromagnetic structure, which is characterized by $\boldsymbol{q} = (0.1, 0.1, 0.5)$. [2, 25] The antiferromagnetic ordering in UNiAl is destroyed by a magnetic field $B > B_{\mathrm{c}}$ (≈ 11.35 T below 4 K) applied along the c-axis, but remains intact in fields applied within the basal plane. [5, 26]

Hydrostatic-pressure experiments were focused on the influence on magnetic parameters of UNiAl. Initially, T_{N} is decreasing with increasing pressure at nearly constant rate ($\approx -1.6\mathrm{K} \cdot \mathrm{GPa}^{-1}$) for both current directions. Assuming a linear pressure dependence of T_{N} even at higher pressures, the value of $\partial \ln T_{\mathrm{N}}/\partial p \approx -0.1$ GPa^{-1} can be taken to estimate the critical pressure for suppression of antiferromagnetic ordering yielding the value of $p_c \approx 10$ GPa.

The T_N vs. p dependence becomes strongly non-linear above 1 GPa pointing to antiferromagnetic in UNiAl vanishing around 3 GPa, [27] a value that is much lower than the p_c value estimated by the linear extrapolation of low-pressure data.

The low-temperature $\rho_a(T)$ curves at the pressures between 3 and 8 GPa can be described by the exponential $\rho_0 + aT^{5/3}$ dependence up to 6.5 K at 3 GPa and up to 15 K at 8 GPa. The value of a decreases linearly with increasing pressure from 0.553 $\mu\Omega$ cm $\cdot$ K $^{-5/3}$ at 3 GPa to 0.138 $\mu\Omega$ cm $\cdot$ K $^{-5/3}$ at 8 GPa. Note that the $\rho = \rho_0 + aT^{5/3}$ behavior is predicted by spin-fluctuation theory [28] for a system near to the onset of itinerant antiferromagnetism.

The observed low stability of antiferromagnetic ordering in UNiAl with respect to pressure is one of the principal attributes of itinerant $5f$-electron antiferromagnets. In contrast to UNiAl, the isostructural compound UNiGa with less delocalized $5f$-electrons has the critical pressure higher than 30 GPa. [23] The U moments in UNiAl are considerably suppressed by a pressure, [29] whereas the moments in UNiGa are pressure invariant at least up to 0.9 GPa [23] and only moderately reduced in 2.3 GPa (see Section 2.3.). In this context, we attribute the collapse of antiferromagnetic order in UNiAl under pressure mainly to instability of the U moment with increasing $5f$-ligand hybridization.

3. SUPPRESSION OF SIMPLE ANTIFERROMAGNETISM IN UNi_2Si_2 UNDER MAGNETIC FIELDS AND HYDROSTATIC PRESSURES

UNi_2Si_2 is a member of the group of UT_2X_2 (T = transition metal, X = p-metal) compounds, which crystallize in the body-centered tetragonal $ThCr_2Si_2$-type of structure (space group I4/mmm). In this crystal structure, the U atoms form a square network within the basal plane. The U planes are separated by Si-Ni-Si slabs. [2]

UNi_2Si_2 orders magnetically below $T_N = 124$ K with an incommensurate (IC) modulation of U magnetic moments along the c-axis. The IC phase is characterized by a propagation vector $\boldsymbol{q} \approx (0, 0, 0.745)$. Upon cooling across $T_2 = 103$ K, UNi_2Si_2 undergoes a first-order magnetic phase transition from the IC phase to a body-centered AF-I structure. At temperatures below 53 K, another order-to-order magnetic phase transition to the uncompensated antiferromagnetic (UAF) ground-state phase with $\boldsymbol{q} = (0, 0, 2/3)$ takes place. This transition exhibits a temperature hysteresis of several K. All three ordered phases have basal-plane ferromagnetic sheets of U moments oriented along the c-axis in common and differ mainly in the moment stacking along the c-axis. In particular, the AF-I phase is characterized by $+ -$ coupling and the UAF phase by $+ + -$ coupling of equal U moments. The U moments in the IC phase are supposed to be sinusoidal modulated along c. [30]

Studies of the influence of an external magnetic field [31] revealed that application of the field along the c-axis yields a strong increase of T_1, i.e. an extension of the stability range of the low-temperature phase (UAF) at the expense of the AF-I phase. The high-temperature boundary of the AF-I phase determined by T_2 is initially almost field independent. In fields above 3 T, however, the phase transition at T_2 splits into two transitions, which we denote by the transition temperatures $T_{2A} < T_{2B}$. In the temperature interval between T_{2A} and T_{2B}, the UAF phase is observed. The temperature T_{2A} decreases with increasing field, whereas T_{2B} increases. As a consequence, the interval of the stability of AF-I phase between T_1 and T_{2A} shrinks rapidly with increasing field and the AF-I phase is fully suppressed in fields above 4.4 T. On the other hand, T_N is practically invariable in magnetic fields up to 14.5 T as confirmed by a neutron-diffraction experiment. Owing to the strong easy-axis anisotropy in UNi_2Si_2, no field effects on the magnetic phase transitions are found for $B \perp c$. [2]

It has been found that T_1 (T_2) increases (decreases) also with increasing pressure, whereas T_N seems to be pressure invariant. The AF-I phase disappears at pressures above 3.2 GPa. [32] A possible explanation of this pressure evolution can be given within the Ising model with third-neighbor interaction (A3NNI model). [33]

4. COLLAPSE OF Co METAMAGNETISM IN $ErCo_2$ UNDER HIGH PRESSURE

The Co $3d$-band states in the $RECo_2$ (RE - rare earth element) compounds appear near the critical conditions for Co-moment formation. The compounds with the non-magnetic RE elements, YCo_2 and $LuCo_2$, are exchange-enhanced paramagnets. At sufficiently low temperatures they exhibit a metamagnetic transition with a critical field $\mu_0 H_c \approx 70$ T. [12] The high field state in YCo_2 and $LuCo_2$ characterized by a ferromagnetic ordering of Co $3d$ moments is a consequence of a sudden splitting of the majority- and minority-spin $3d$ subbands at a critical field B_c. This phenomenon called itinerant electron metamagnetism (IEM) is closely related to the instability of itinerant electron ferromagnetism. YCo_2 and $LuCo_2$ are frequently quoted as the archetypes of spin-fluctuation systems with a high characteristic temperature T_{sf}, which correlates with the temperature of susceptibility maximum T_{max}.

The lattice volume of $RECo_2$ compounds is the principal parameter intimately connected with the 3d bandwidth. The role of lattice volume in the appearance of IEM has been demonstrated by the recent total-energy fixed-spin-moment calculations performed for YCo_2 with values of the lattice parameter taken from all the members of the $RECo_2$ series. Results of these calculations indicate that IEM between $TbCo_2$ and $DyCo_2$ that is in agreement with experiment.

The IEM in $ErCo_2$ and some other heavy-RE counterparts ($DyCo_2$ and $HoCo_2$) is, however, observed in zero external magnetic field. In these materials, the Co metamagnetic state is induced in a large effective exchange field, which is mediated from the ferromagnetically ordered Er 4f- moments by the 5d (RE)-3d (Co) hybridization. This field emerges at $T_C = 33$ K when the Er sublattice orders ferromagnetically. [34] Consequently, a first order magnetic phase transition (FOMPT) accompanied by a dramatic drop of the electrical resistivity and a sudden volume expansion is observed at T_C. These phenomena reflect the abrupt change of the density of Co-3d states at E_F and a partial localization of the 3d states as a consequence of the splitting of the majority and minority spin sub-bands and the associated Co moment formation.

The existence of 3d-electron metamagnetism in $ErCo_2$-type compounds has been confirmed by numerous experiments. It has been shown that the value of the Co moment $\mu_{Co} \approx 1\ \mu_B$ when the effective field B_{eff} acting on 3d electrons is higher than a critical value B_c (~ 100 T), but it becomes dramatically reduced as soon as $B_{eff} > B_c$.

In his theoretical work, [35] Yamada has considered the effect of pressure on the itinerant 3d-electron subsystem at finite temperatures. The effect of spin fluctuations has been taken into account in the phenomenological Landau-Ginzburg theory. He has concluded that B_c increases with increasing pressure and the metamagnetism vanishes at a critical pressure p_c. The p_c value decreases with increasing temperature and becomes zero at a critical temperature T_0. Values between 1 and 5 GPa have been estimated for p_c in YCo_2 at 0 K.

In order to see effects of variation of interatomic distances on magnetism in $ErCo_2$ in detail, pressure effects are studied on the resistivity anomalies in the vicinity of T_C. Simultaneously, we have performed *ab initio* electronic structure calculations based on density functional theory (DFT) on $ErCo_2$ for the ambient-pressure lattice parameter a and also for several reduced a-values expected for some applied pressures. [36]

In the pressure dependence of T_C in Figure 4, T_C first decreases linearly with increasing p at a rate $\partial T_C/\partial p = -7.9$ K/GPa ($\partial \ln T_C/\partial p =$

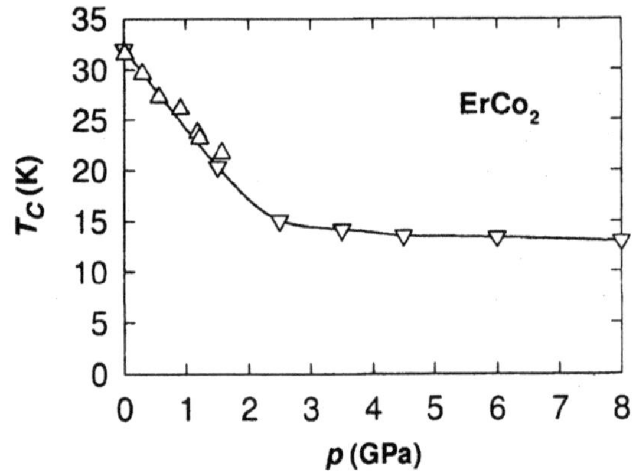

Figure 4. The pressure dependence of T_c of $ErCo_2$. The upright triangles are the data taken from. [37]

-0.24 GPa^{-1}). A linear extrapolation of these data to higher pressures points to a possible critical pressure for disappearance of magnetic ordering $p_c \approx 4.2$ GPa.

Measurements in pressures above 1.5 GPa, however, reveal a gradually decreasing slope of the T_C vs. p dependence as long as the T_C becomes nearly pressure independent for $p > 4$ GPa ($T_C \approx$ 13 K). Also the resistivity drop at T_C becomes rapidly reduced when approaching p_c and vanishes around p_c. The resistivity drop at T_C is closely connected with the stability of Co magnetism. The resistivity above T_C is mainly affected by a spin-disorder scattering on paramagnetic *RE* moments influenced by the crystal field interaction and by a spin-fluctuation scattering depending on the dynamics of spin fluctuations in the Co $3d$-band. [38] When lowering temperature, the ferromagnetically ordered $4f$-moments at T_C assisted by the $5d$ (*RE*)-$3d$ (Co) hybridization produce a strong uniform exchange field acting on the Co $3d$ states. When this field is strong enough to split the $3d$ majority and minority sub-bands abruptly, the $3d$-band metamagnetic state is induced by a first order magnetic phase transition and the spin fluctuations at the Co-sites are quenched. Consequently, the scattering is drastically suppressed, which yields the resistivity drop. The obvious effect of a dramatic reconstruction of Fermi surface on transport properties at the metamagnetic transition should be considered as well. Considering this scenario one may conclude that the loss of Co magnetism in $ErCo_2$ is indicated by the vanishing resistivity drop at T_C for pressures where T_C becomes pressure invariable. For $p > p_c$, the pressure-induced decrease of interatomic distances of Co atoms with neighbors and a consequent increase of the overlap of $3d$ wave functions leads to a critical broadening of the Co $3d$-band. Consequently, the projected Co-$3d$ density of states in the vicinity of E_F decreases and the itinerant Co moment vanishes because the *RE*-Co-*RE* exchange channel becomes ineffective to induce the IEM. Simultaneously, also the magnetic phase transition at T_C becomes second order and the pressure induced decrease of T_C ceases. The *RE* moments then order at a "residual" T_C roughly comparable to the T_C for $ErNi_2$ (that exhibits no $3d$ magnetic moment because of negligible $3d$ density of states at E_F). The T_C value (13 K) is thus determined by the persisting conduction-electron mediated exchange interaction (RKKY-type) coupling the localized *RE* moments for which $\partial T_C/\partial p$ is nearly zero.

This scenario was tested by first principles electronic structure calculations in the framework of DFT. [36] The spin-polarized electronic structure was calculated for the set of ten lattice parameters smaller than the ambient-pressure value $a_{\rm exp} = 714.4$ pm. The Co moment was found to decrease smoothly to the value of 0.78 μ_B for the $a \geq 692$ pm (which confirms the negative pressure effect on the Co moment observed experimentally by magnetovolume measurements). [34] For values $a \leq 688$ pm, however, the moment suddenly drops to the small value smaller than 0.1 μ_B that is in good qualitative agreement with experiment.

References

1 L. Havela, M. Diviš, V. Sechovský, A.V. Andreev, F. Honda, G. Oomi and Y.H.S. Méresse, U ternaries with ZrNiAl structure - lattice properties, *J.Alloys and Compounds* **322**, 7-13 (2001).

2 V. Sechovský and L. Havela, Magnetism of ternary intermetallic compounds of uranium, in: *Handbook of Magnetic Materials* **11**, edited by K. H. J. Buschow (Elsevier Science B.V. Amsterdam, 1998), pp.1-289.

3 D.D. Koelling, B.D. Dunlap and G.W. Crabtree, *f*-electron hybridization and heavy-fermion compounds, *Phys.Rev.B* **31**, 4966-4071 (1985).

4 A.V. Andreev, M. Diviš, P. Javorský, K. Prokeš, V. Sechovský, J. Kuneš and Y. Shiokawa, Electronic structure and magnetism in UPtAl, *Phys.Rev.B*, in press (2001).

5 E. Brück, H. Nakotte, F.R. de Boer, P.F. de Châtel, H.P. van der Meulen, J.J.M. Franse, A.A. Menovsky, N.H. Kim Ngan, L. Havela, V. Sechovský, J.A.A.J. Perenboom, N.C. Tuan and J. Šebek, Electronic properties of UNiAl in high magnetic fields, *Phys.Rev.B* **49**, 8852-8863 (1994).

6 A.V. Andreev, N.V. Mushnikov, T. Goto and V. Sechovský, Alloying and pressure induced transitions between 5*f*-band metamagnetism and ferromagnetism, *Phys.Rev.B* **60**, 1122-1126 (1999).

7 V. Sechovský, L. Havela, F.R. de Boer and E. Brück, Magnetic anisotropy in U*TX* compounds, *J.Alloys and Compounds* **181**, 179-184 (1992).

8 A.V. Andreev, J. Kamarád, F. Honda, G. Oomi, V. Sechovský and Y. Shiokawa, Magnetoelasticity of UPtAl, *J.Alloys and Compounds* **314**, 51-55 (2001).

9 P. Link, U. Benedict, J. Wittig and H. Wuhl, The electrical resistance of USe under high pressure, *Physica B* **190**, 68-71 (1993).

10 K. Prokeš, H. Nakotte, E. Brück, F.R. de Boer, L. Havela, V. Sechovský, P. Svoboda and H. Maletta, Anisotropic magnetic and transport properties of UNiGe, *IEEE Trans.Magn.* **30**, 1214-1216 (1994).

11 N.V. Mushnikov, T. Goto, K. Kamishima, H. Yamada, A.V. Andreev, Y. Shiokawa, A. Iwao and V. Sechovský, Magnetic properties of the $5f$ itinerant electron metamagnet UCoAl under high pressure, *Phys.Rev.B* **59**, 6877-6885 (1999).

12 T. Goto, H. Aruga Katori, T. Sakakibara, H. Mitamura, K. Fukamichi and K. Murata, Itinerant electron metamagnetism and related phenomena in Co-based intermetallic compounds, *J.Appl.Phys.* **76**, 6682-6687 (1994).

13 A.V. Andreev, R.Z. Levitin, Yu.F. Popov and R.Yu. Yumaguzhin, Metamagnetism of UAlCo, *Sov.Phys.Solid State* **27**, 1145-1146 (1985).

14 A.V. Andreev, I.K. Kozlovskaya, N.V. Mushnikov, T. Goto, V. Sechovský, L. Havela, Y. Homma and Y. Shiokawa, Onset of ferromagnetism in the $U_{1-x}Y_x$CoAl system, *J.Magn.Magn.Mater.* **197**, 658-659 (1999).

15 A.V. Andreev, I.K. Kozlovskaya, N.V. Mushnikov, T. Goto, V. Sechovský, Y. Homma and Y. Shiokawa, Magnetic properties of the $UCo_{1-x}T_x$Al solid solutions (T=Rh and Ir), *J.Alloys and Compounds* **284**, 77-81 (1999).

16 A.V. Andreev, H. Aruga Katori and T. Goto, Magnetic properties of $UCo_{0.9}T_{0.1}$Al (T=Fe, Ni, Ru, Pd), *J.Alloys and Compounds* **224** , 117-120 (1995).

17 A.V. Andreev, L. Havela, V. Sechovský, M.I. Bartashevich, T. Goto and K. Kamishima, Effects of alloying and pressure on metamagnetism of UCoAl, *Physica B* **239**, 88-94 (1997).

18 A.V. Andreev, I.K. Kozlovskaya and V. Sechovský, Magnetic properties of the $UCo_{1-x}Pt_x$Al solid solutions, *J.Alloys and Compounds* **165**, 38-41 (1998).

19 V. Sechovský, A.V. Andreev, Y. Ishii, M. Kosaka and Y. Uwatoko, Physics of UCoAl in the light of pressure experiments, in: *Proceeding of 31èmes Journées des Actinides*, edited by H. Noël (Universite de Rennes, Rennes, 2001), p.O2.

20 K. Prokeš, E. Brück, F.R. de Boer, M. Mihalik, A. Menovsky, P. Burlet, J.M. Mignot, L. Havela and V. Sechovský, Giant Magnetoresistance and Magnetic Phase-Diagram of UNiGa, *J.Appl.Phys.* **79**, 6396-6398 (1996).

21 Y. Uwatoko, G. Oomi, V. Sechovský, L. Havela, E. Brück and F.R. de Boer, Effect of pressure on magneic phase transitions of UNiGa, *J.Alloys and Compounds* **213-214**, 281-282 (1994).

22 F. Honda, K. Prokeš, G. Oomi, T. Kagayama, A.V. Andreev, V. Sechovský, L. Havela and E. Brück, Magnetic phase transitions and magnetoelastic phenomena in UNiGa under pressure, *J.Phys.Soc.Japan* **66**, 1904-1907 (1997).

23 F. Honda, K. Prokeš, M. Olšovec, F. Bourdarot, P. Burlet, T. Kagayama, G. Oomi, L. Havela, V. Sechovský, A.V. Andreev, E. Brück, F.R. de Boer, A. Menovsky and M. Mihalik, Magnetic phases and magnetoelastic anomalies in UNiGa under pressure, *J.Alloys and Compounds* **271-273**, 495-498 (1998).

24 V. Sechovský, L. Havela, F.R. de Boer, E. Brück, Y. Uwatoko and G. Oomi, Effect of Pressure on Magnetic Transitions in UNiGa, *Physica B* **186-188**, 697-699 (1993).

25 K. Prokeš, F. Bourdarot, P. Burlet, P. Javorský, M. Olšovec, V. Sechovský, E. Brück, F.R. de Boer and A.A. Menovsky, Antiferromagnetic structure of UNiAl, *Phys.Rev.B* **58**, 2692-2698 (1998).

26 K. Prokeš, F. Bourdarot, P. Burlet, M. Olšovec, V. Sechovský, P. Javorský, E. Brück, A. Goukassov, F.R. de Boer and A.A. Menovsky, Magnetic structures of UNiAl in magnetic fields, *Physica B* **259-261**, 246-247 (1999).

27 O. Mikulina, J. Kamarád, A. Lacerda, O. Syshchenko, T. Fujita, K. Prokeš, V. Sechovský, H. Nakotte and W.P. Beyermann, Pressure effects on antiferromagnetism in UNiAl, *J.Appl.Phys.* **87**, 5152-5154 (2001).

28 K. Ueda, Electrical resistivity of antiferromagnetic metals, *J.Phys.Soc.Japan* **43**, 1497-1508 (1977).

29 K. Prokeš, T. Fujita, N.V. Mushnikov, S. Hane, T. Tomita, T. Goto, V. Sechovský, A.V. Andreev and A.A. Menovsky, Magnetic properties of UNiAl under pressure, *Phys.Rev.B* **59**, 8720-8724 (2001).

30 L. Rebelsky, H. Lin, M.W. McElfresh, M.F. Collins, J.D. Garrett, W.J.L. Buyers and M.S. Torikachvili, The magnetic phase diagram of UNi_2Si_2, *PhysicaB* **180-181**, 43-45 (1992).

31 A. Syshchenko, V. Sechovský, A.V. Andreev, R.V. Dremov, J. Šebek and A.A. Menovsky, Electrical resistivity of UNi_2Si_2 in magnetic fields, *J.Appl.Phys.* **85**, 4554-4555 (1999).

32 F. Honda, M. Matsuda, G. Oomi, N. Mori, N. Takeshita, V. Sechovský, A.V. Andreev and A.A. Menovsky, Suppression of simple antiferromagnetism in UNi_2Si_2 under high pressure, *Phys.Rev.B* **61**, 11267-11269 (2000).

33 M.L. Plumer, Mean-field phase diagrams of UT_2X_2 compounds. *Phys.Rev.B* **50**, 13003-13005 (1994).

34 T.D. Cuong, L. Havela, V. Sechovský, Z. Arnold, J. Kamarád and N.H. Duc, Magnetism and related phenomena in $RE(Co_{1-x}Si_x)_2$, *J.Alloys and Compounds* **262-263**, 141-146 (1997).

35 H. Yamada, Metamagnetic transition and susceptibility maximum in an itinerant-electron system, *Phys.Rev.B* **47**, 11211-11219 (1993).

36 O. Syshchenko, T. Fujita, V. Sechovský, M. Diviš and H. Fujii, Magnetism in $ErCo_2$ under high pressure, *Phys.Rev.B* **63**, 54433 (2001).

37 E.W. Lee and F. Pourarian, Magnetoelastic properties of (rare-earth)-Co_2 compounds. I. Exchange striction *Phys.Stat.Sol.A* **33**, 483 (1976).

38 N.H. Duc, T.D. Hien, R.Z. Levitin, A.S. Markosyan, P.E. Brommer and J.J.M. Franse, The magnetic phase-transitions in $R(Co, Al)_2$ compounds (R = Dy, Ho, Er), *Physica B* **176**, 232 (1992).

Electron Correlation and Plutonium Phase Diagrams

A. C. Lawson*, Joyce A. Roberts*, Barbara Martinez*
and James W. Richardson, Jr†

1. INTRODUCTION

Phase diagrams of plutonium metal are astonishingly complicated. The main feature is a sequence of crystallographically complex phases. Temperature, pressure and alloying shift the stability of these phases rapidly, so that pressure and alloy phase diagrams are also complicated. Another feature of the plutonium phase diagrams is the anomalously low melting point. How these two features, crystallographic complexity and low melting point, work to determine complicated phase diagrams is shown in Fig. 1. This figure, which is a composite of the relevant binary diagrams, shows that crystallographic complexity and low melting point are nearly exactly coincident. Such behavior is unique to the light actinides. It is natural to seek the source of this behavior in the collective properties of the 5f electrons, and in this paper we will trace some of the paths we are following in this search.

Fig. 2 shows the linear thermal expansion of unalloyed Pu metal. As the temperature is raised, there is a progression of phases from low symmetry to high symmetry as the melting point is reached. The α- and β-phases are monoclinic, the γ-phase is orthorhombic, the δ-phase is FCC and the ε-phase is BCC. The δ′-phase is a transitional, tetragonal phase. A number of surprising features are apparent from this plot. First, the crystallographic complexity of the low temperature phases is unmatched by any other element: the α- and β-phases have 16 and 34 atoms per cell, respectively. This crystallographic complexity seems to be connected to the tendency of Pu to form intermetallic compounds with itself, the so-called "self-intermetallic" compounds

* Los Alamos National Laboratory, Los Alamos, NM 87545 USA
† Argonne National Laboratory, Argonne, IL 60439 USA

Electron Correlations and Materials Properties 2
Edited by Gonis *et al.*, Kluwer Academic/Plenum Publishers, 2002

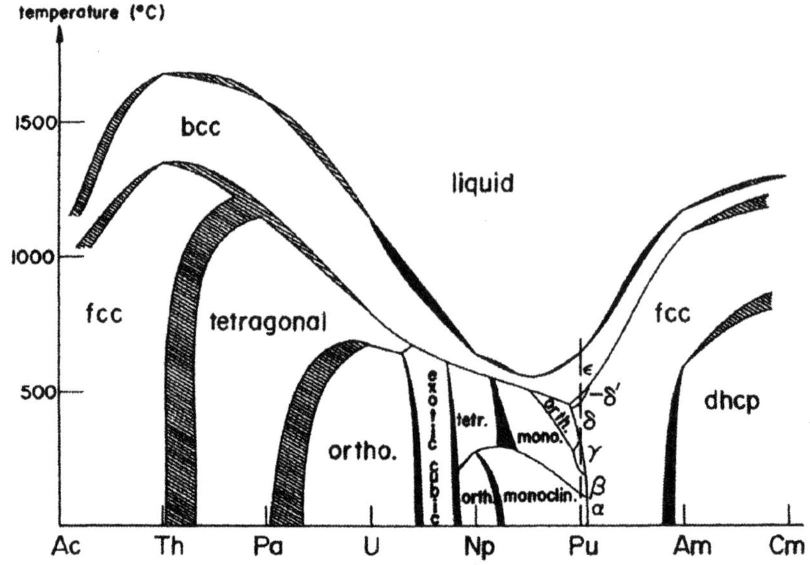

Figure 1. Composite phase diagram of the light actinides. Reprinted from J. Less-Common Metals, Vol. 90, J. L. Smith and E. A. Kmetko, Magnetism or bonding: a nearly periodic table of transition elements, pages 83-88, copyright 1983, with permission from Elsevier Science.

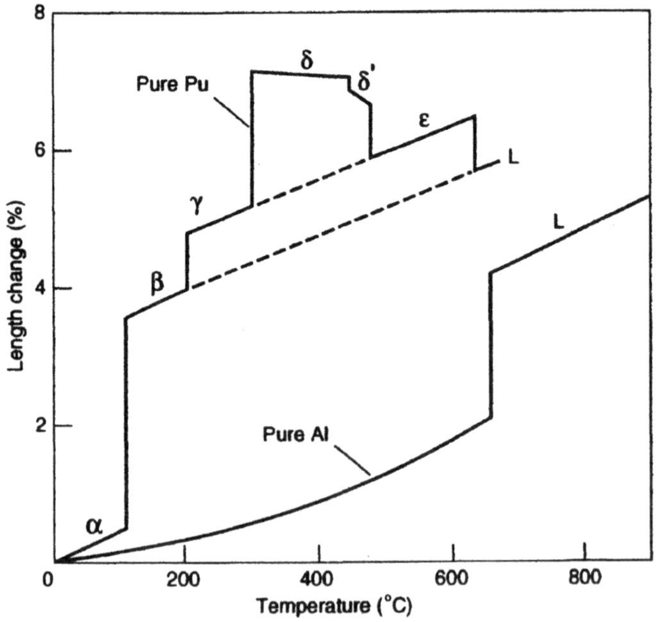

Figure 2. Linear thermal expansion of plutonium metal compared to that of aluminum. (Figure used by permission of Los Alamos Science.)

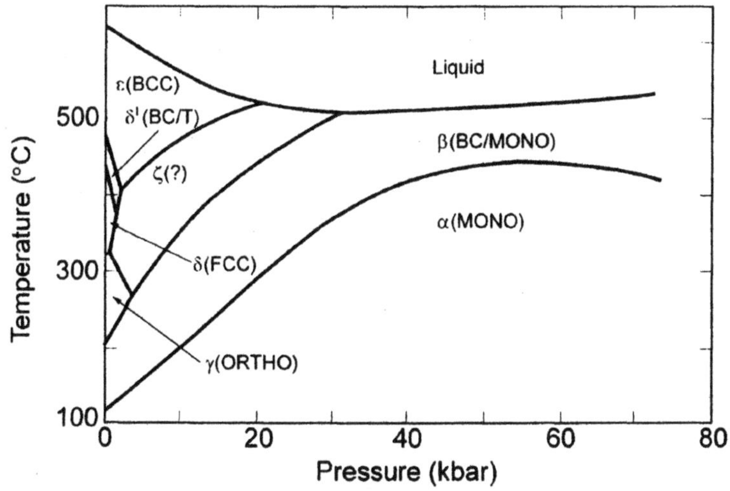

Figure 3. Pressure–temperature phase diagram of unalloyed Pu metal. Reprinted from J. Physics F., Vol. 6, E. A. Kmetko and H. H. Hill, Anomalous melting of f-electron metals (with attention to Pu), pages 1025-1037, copyright 1976, with permission of the Institute of Physics.

(Lawson et al., 1996) or to another, probably equivalent, tendency toward structural complexity in narrow band materials (Söderlind et al., 1995, Söderlind, 1998, Hecker, 2000, and Baskes, 2000). Second, the FCC δ-phase has a surprising *negative* thermal expansion. Third, as noted, the melting point is surprisingly low, with a volume *decrease* at the melting point.

This multiplicity of crystallographically stable phases leads to complex phase diagrams, some of which will be displayed in the next section, and a few features of interest will be pointed out. The following section will discuss some aspects of disorder (or short range order) that may be expected in Pu phases. After that, we will discuss the special case of vibrational disorder and present some measurements of the Debye-Waller factors of the various phases in unalloyed Pu. This leads to a discussion of the melting point of Pu, based on the Lindemann rule. Finally, there is a discussion of the thermodynamics of Pu, which leads us back to the topic of phase diagrams. At this point it is possible to make some contact with the topic of electron correlation.

2. Pu PHASE DIAGRAMS

The pressure-temperature phase diagram of unalloyed Pu, based on the work of Liptai and Friddle (1966) and modified by Kmetko and Hill (1976), is shown in Fig. 3 and reflects the complexity of the first two figures. An additional feature is the ς-phase, stable only under pressure, whose structure is still undetermined. Notice that the melting point is further depressed by pressure to a minimum at ~30 kbar, after which there is a gradual increase. There is an extremely suggestive analogy between the pressure phase diagram and the alloy phase diagrams that was originally put forward by Elliott (ca. 1970). The idea is that alloying with elements that stabilize the δ-phase (Ga, Al, Am, and some others) behaves like a negative pressure. This analogy is shown in Fig. 4. In this plot, the effect of alloying is shown with a phase diagram that is a composite of the Pu-Am (smaller solute) and Pu-Np (larger solute) diagrams. The Pu pressure phase diagram, shown in mirror image, matches the Pu-Np diagram.

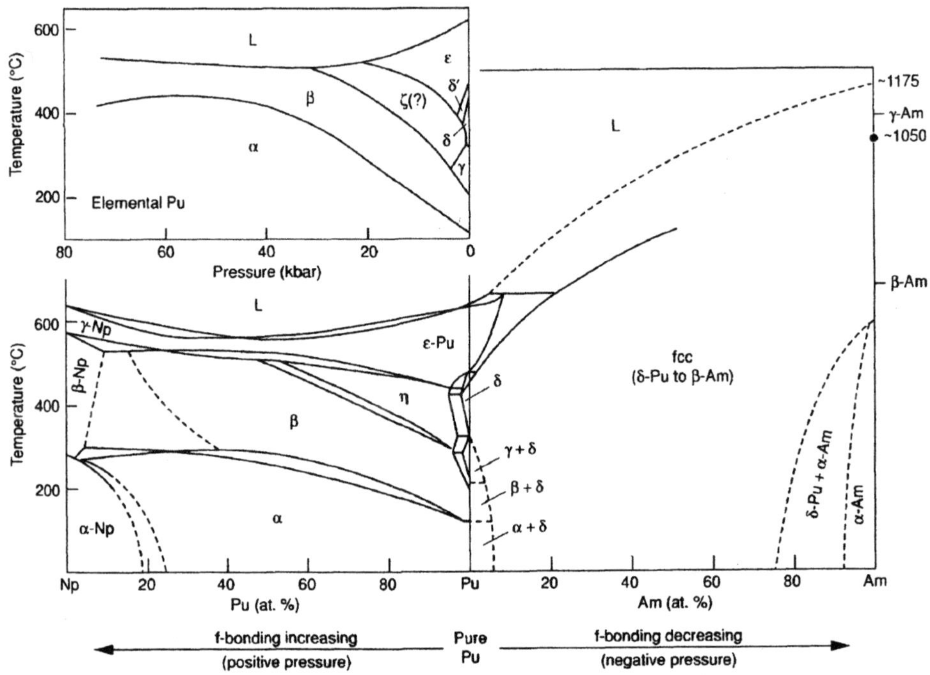

Figure 4. Composite alloy phase diagram for Pu-Np and Pu-Am. The Pu-Np bears a strong resemblance to the pressure-temperature diagram of Fig. 3, which is shown in mirror image. The Pu-Am diagram is similar to the Pu-Ga diagram. (Figure used by permission of Los Alamos Science.)

The Pu-Ga phase diagram is of technological importance, as a few percent Ga is found to stabilize the ductile FCC δ-phase. The version of this diagram that was accepted in the United States for a long time is presented in Fig. 5a and shows how the addition of Ga rapidly suppresses the low-symmetry phases in favor of the ductile δ-phase. Russian workers had believed that the δ-phase alloy is thermodynamically unstable at low temperatures, and their diagram is shown in Fig. 5b. Hecker and Timofeeva (2000) tell the fascinating story of how consensus is being reached in a recent issue of Los Alamos Science. It has been extraordinarily difficult to reach agreement on the true, equilibrium phase diagram, as the equilibration times exceed the time-scale of ordinary laboratory experiments. For example, the eutectoid reaction in Fig. 5b has never been observed in neutron diffraction experiments, which typically have a duration of a few days.

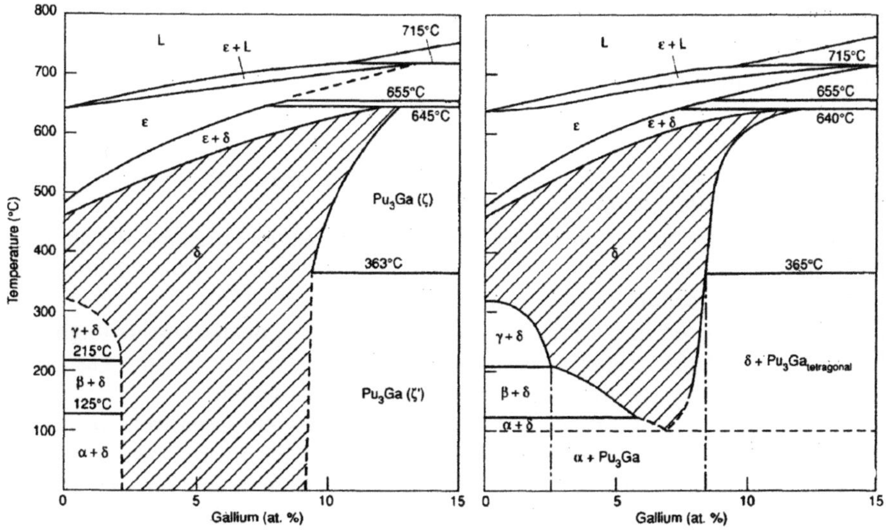

Figure 5. Alternative Pu-Ga alloy phase diagrams. The diagram at left (5a) indicates stability for Pu-2 at. % Ga at low temperatures, while the diagram at right (5b) shows eutectoidal decomposition of the alloy. (Figure used by permission of Los Alamos Science.)

3. Pu DISORDER

The structure of an ordered, crystalline material is correctly given by specifying the space group and atomic positions of the atoms. But perfectly ordered, crystalline materials do not exist in nature. Atomic positions may not be fully occupied, stoichiometric disorder or local atomic displacements may occur in alloys, and even in the purest materials, thermal vibrations will introduce deviations from perfect, periodic symmetry specified by the crystal space group. Deviations from perfect order are in principle observable as diffuse scattering. In practice, however, the diffuse scattering may be too small to be observable, even when the underlying disorder is large enough to be of theoretical significance. In addition one should heed the warning given by Welberry and Butler (1994) that the relationship between observed diffuse scattering and the underlying disorder is complicated and often counterintuitive.

As a specific example, let us consider an alloy Pu-2 at. % Ga. According to the phase diagrams that we have presented, this material contains enough gallium to exhibit the FCC crystal structure over a wide range of temperatures: from 450°C down to either the lowest temperatures (Fig. 5a) or at least to 100°C (Fig 5b). The structure of this alloy is usually presumed to be "disordered" FCC. This means that 2% of the Pu sites in the structure are occupied by gallium. Since the occupation is assumed to be random, the probability of occupation of any site by gallium is 2%, independent of the occupation of neighboring sites. However, this assumption of independence need not be fulfilled exactly, as the Ga occupation of any site may influence the occupation of neighboring sites, and the occupation of gallium is then correlated. This correlation, described by the well-known Cowley order parameters, can be positive or negative, so that the gallium atoms may be thought to attract or repel each other, respectively (Warren, 1969). Whatever the state of correlation in Pu-2 at. % Ga, it is unobservable by x-ray or neutron diffraction, as the signal from only 2% solute is actually invisible in a practical experiment. (The classic example of the observation of such correlation is disordered Cu_3Au.) In fact, for neutron diffraction experiments discussed in this paper, the correlation is unobservable for an additional reason, namely that the neutron scattering cross sections for ^{242}Pu and for Ga are nearly identical.

In a Pu-Ga alloy, the Ga atoms are much smaller than the Pu atoms: the atomic radii are 1.4 and 1.6 Å, respectively. We can therefore anticipate that the structure will be disordered by the random displacement of the Ga atoms. If the Ga occupancies are correlated, as discussed in the previous paragraph, then the displacements might also be correlated, and diffuse scattering will be expected. However, in the case of Pu-2 at. % Ga, the expected scattering is again too small to be observed.

It is clear that crystal diffraction techniques are relatively insensitive to short-range order because the strength of the signal from the periodically ordered crystal swamps the relatively less intense diffuse scattering from the short-range order. The complementary EXAFS technique, which is sensitive to short-range order only, has been producing interesting results (Conradson, 2000) that are so far not completely understood.

4. Pu VIBRATIONS

4.1 Vibrational Correlations

In every solid, the atoms vibrate. The motions of the atoms are given by the lattice dynamics, which can be calculated if the interaction between the atoms is known with sufficient accuracy. In favorable cases, the motions of the atoms can also be determined experimentally by inelastic neutron scattering. A case can be favorable if a single crystal is available; this is not currently the case for Pu-2 at. % Ga.

The existence of the atomic vibrations affects the result of a diffraction experiment in three ways. By far the largest effect is the attenuation of the Bragg reflections according to the Debye-Waller factor

$$I = I_0 e^{-k^2 <u^2>} \qquad (1)$$

where $<u^2>$ is mean-square vibration amplitude of the atom, averaged over all atoms, all frequencies and all directions. $<u^2>$ can be calculated according to the Einstein model of lattice vibrations, which assumes that each atom vibrates independently of all the rest. This is a good model for the behavior of the Bragg intensities, and it is used routinely for the fitting of experimental diffraction data. The determination of $<u^2>$ vs. temperature data can be used to determine the Debye-Waller temperature, Θ_{DW}, as a function of temperature (Lawson et al., 2000a). Θ_{DW} is expected to be much more independent of temperature than is the heat capacity Debye temperature, Θ_D, because its moment of the phonon spectrum shifts between -1 and -2 between low and high temperatures, while that of Θ_D shifts between -3 and +2 (Grimvall, 1999), and the second moment has hardly any temperature dependence at all at high temperatures (Lovesey, 1984). Experimentally, there is a good experimental correspondence between the values of Θ_D and Θ_{DW} measured at low temperatures (Lawson et al., 2000b)

But the atomic motions are in fact correlated, and this gives rise to diffuse scattering. The diffuse scattering can be calculated if a lattice dynamical model or measured phonon spectrum is available. If an exact calculation of the diffuse scattering is not possible, a good approximation is provided by the Debye model, which is based on a continuum approximation for the lattice dynamics. This turns out to be a good approximation to the correlation that is actually observed.

Fig. 6a shows a diffraction pattern of unalloyed δ′-Pu at 465°C. The data have been fit by Rietveld analysis, and the tetragonal δ′ structure is confirmed. The error curve displayed below the data shows very clearly an oscillatory background due to vibrational correlations that have not been accounted for in the analysis. Fig. 6b shows the same data fitted with the δ′ structure but with the background fitted with a series of oscillatory terms

$$I(k) = \sum_{i=1}^{R_{max}} A_i \frac{k^2 \sin(kR_i)}{kR_i}. \quad (2)$$

This function is an approximation to the thermal diffuse scattering for δ′-Pu at 465°C. The A_i's, which are fit by Rietveld analysis, represent the correlation corrections to the Debye-Waller factor (which in turn represents the Einstein model).

The effect of vibrational correlations on the structure of a material can be represented by the pair distribution function, or PDF (Egami, 1998, McQueeney, 1998). The PDF is a one-dimensional plot of the density of atomic pairs in real space. Each peak in the PDF contains N(R) pairs at neighboring distance R, given by the integral

$$N(R) = \int_{peak} 4\pi r^2 \rho(r) dr. \quad (3)$$

A "synthetic" PDF can be constructed from knowledge of the crystal structure. For each neighbor distance, one adds to the PDF a Gaussian term whose normalization gives the correct number of neighbors at that distance and whose width gives the vibrational correlation (including the autocorrelation). For each peak, the correlation can be found in one of three ways. At the lowest level of approximation, based on the Einstein model of independent atomic vibrations, one can use simply the width measured by the Debye-Waller factor. At this level of approximation, all the peaks in the PDF have the same width, which represents the autocorrelation. At the next level of approximation, the Debye model gives

$$< u_0 u_0 >_R = < u^2 >_{DW} \left(1 - \frac{Si(k_D R)}{k_D R}\right) \quad (4)$$

where k_D is the Debye wavenumber, which depends only on the atomic density (Warren, 1967). The Debye model includes interatomic correlations calculated on a continuum model of the solid. A third, and even better level of approximation may be obtained by measuring the correlation corrections individually for each R by Rietveld analysis (Lawson, 2000b), as was done in Fig. 6. The Einstein approximation becomes exact for large enough R, e.g. ~20 Å, as the interatomic correlations fall off with distance. In principle, the correlated atomic motions produce shifts in the peak positions of the PDF, and distortions of the Gaussian peak shape, but these are generally too small to be determined within the limits of resolution set by the Nyquist theorem, π/Q_{max}, where Q_{max} is the maximum scattering wave number of the measurement, say about 25Å^{-1}. The limitation to a Gaussian peak shape centered on fixed crystallographic distance could be relaxed with an extended model for the correlation corrected PDF.

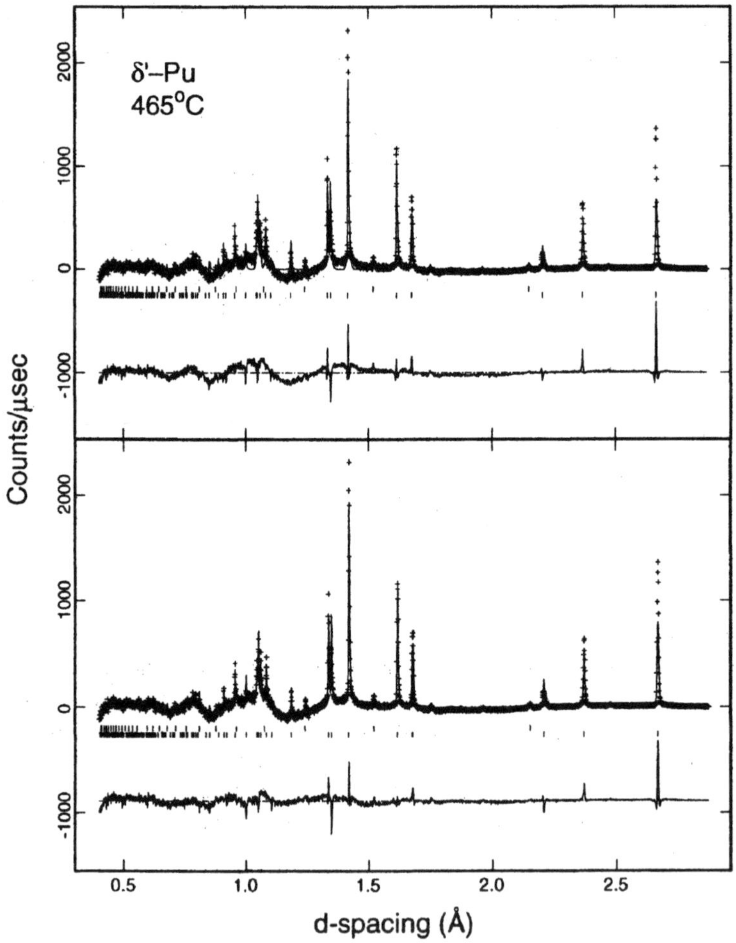

Figure 6. Rietveld fits of diffraction patterns for δ′-Pu. The upper pattern (6a) is fit without correlation correction and the lower pattern (6b) includes only the first two terms of correlation correction. The crosses are the data, the line through the crosses is a fit to the data, and an error curve is shown below each pattern. Two sets of tic-marks indicate the allowed reflections for δ -Pu (below) and for the vanadium radiological containment (above).

An entirely different approach is to use Fourier transformation of the diffraction data to construct the entire PDF (Egami, 1998). This procedure gives the correct vibrational correlations, but places rather high demands on the quality of experimental data. As an example, consider the vibrations of Pb at two different temperatures, 300K and 50K. Fig 7 shows the PDF constructed from the Einstein, Debye and correlation correction methods compared with the experimentally determined PDF obtained by Fourier transformation. As expected, the peaks of the PDFs at 300K are broader, reflecting the larger thermal motion at that temperature. The various PDFs are generally in good agreement, except that the PDF based on the Einstein model overestimates the peak width at small R. The experimental PDF is noisier than usually obtained from this kind of work, because data were collected for only a few hours. Also we have not applied any corrections for sample container scattering, multiple scattering or absorption, or any Placzek corrections. The normalization of this PDF is based on the Rietveld scale factor and had to be adjusted upward by 20% to get the correct number of atoms under the PDF peaks (in other words, to agree with the PDFs obtained by other methods). The plot under the PDF curves shows the difference between the correlation corrected PDF and the Debye PDF. This difference is significant, but small.

Dimitrov et al. have proposed a method for extracting the phonon spectrum from powder diffraction data by matching the PDF obtained by Fourier transformation to that derived from a model phonon spectrum. Good agreement with experimentally determined phonon dispersion curves has been obtained for Ni and for CaF_2. In view of the close agreement with the experimental PDFs for Pb (obtained either by Fourier transformation or by the correlation correction method) with the PDFs estimated by the Debye model, it would appear that experimental PDFs must be *very* accurate for this method to succeed. This is because the difference between a realistic phonon spectrum and one based on the Debye model (very unrealistic) must depend on the difference between experimental PDF and the synthetic PDF based on the Debye model. As we have shown, this latter difference is small.

As a further check on this point, we have compared the PDFs calculated from Monte Carlo simulations of Al and Pb under the assumption of nearest-neighbor forces. The nearest-neighbor model is not expected to produce an accurate prediction of the real metal, but it has the great advantage that the nearest neighbor force constants depend only on the single crystal elastic constants. Fig. 8 shows the anisotropic Young's modulus for Al, Ni, Pb and Ga-stabilized δ-phase Pu. (For the first three, data are from Kittel, 1996, and for Pu, from Ledbetter and Moment, 1976). Pb has a high elastic anisotropy, while Al is nearly isotropic. We can therefore investigate how the elastic anisotropy is reflected in the correlation correction that must be applied to the Einstein PDF. As shown in Fig. 9, the correlation correction for Pb is very similar to that of Al, and both are similar to that expected from the Debye model. One concludes from these results that the PDFs must be measured and analyzed with extraordinary accuracy and that the results of the Dimitrov method may be highly susceptible to experimental error. Very recently (2001), Reichardt and Pintschovius have drawn similarly pessimistic conclusions.

It would be a considerable advantage to know the phonon spectra of the various phases of Pu so that the vibrational free energies can be calculated directly. PDF experiments on δ-phase Pu have been undertaken in an attempt to get the phonon spectrum for this phase without the difficulty of growing a single crystal (Lashley et al., 2000). The discussion given here may argue in favor of growing a single crystal after all.

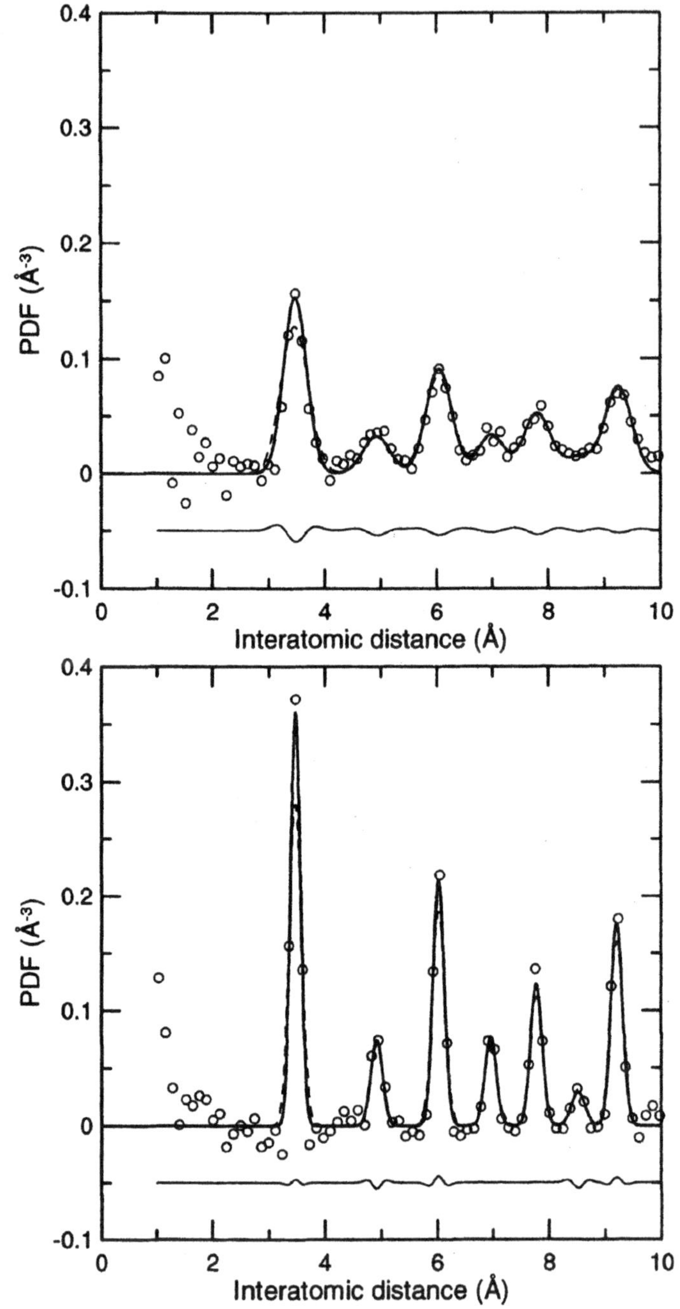

Figure 7. Pair Distribution Functions (PDFs) for lead at 300K (upper panel) and 50K (lower panel). The circles represent the PDF derived from Fourier transformation of the data. The solid line is the correlation-corrected PDF and the dashed line in the Einstein PDF. The curve under the full PDF is the difference between the Debye PDF and the correlation corrected PDF.

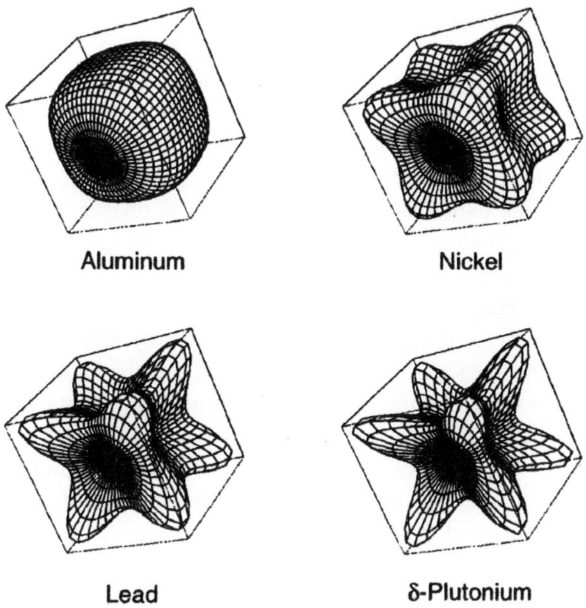

Figure 8. Young's modulus versus crystal direction for Al, Ni, Pb and δ-phase Pu. The scales are arbitrary.

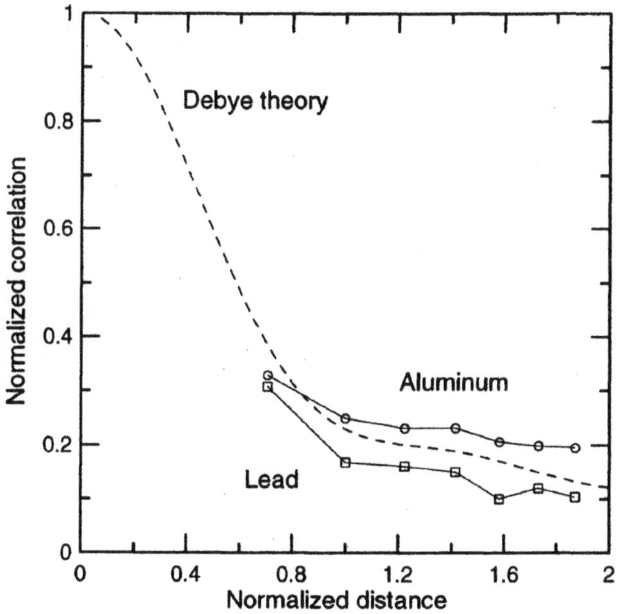

Figure 9. Normalized correlation $\langle u_0 u_n \rangle / \langle u_0^2 \rangle$ versus normalized distance (R_n/a_0) for two Monte Carlo models (Pb and Al) and for the Debye model.

4.2 Debye-Waller Factors of Unalloyed Pu

Fig. 10 shows a plot of $<u^2>$ versus temperature for all the phases of unalloyed Pu. These data were obtained on the General Purpose Powder Diffractometer at IPNS ($T \geq 300K$, Lawson et al., 1998, 1990) or on the High Intensity Powder Diffractometer at LANSCE ($T < 300K$, Lawson et al., 1992, 1993, Kwei et al., 1998). Because the temperature range of the data was small for most of the phases, we fit only Θ_{DW} to the data, except for the α-phase). The errors in the fitted Θ_{DW}s are less than 1K for β- through ε-phases and about 10K for α. (See Lawson et al., 1994, for details of the fitting process.) There is nothing in the $<u^2>$ curve that correlates with the negative thermal expansion of the δ-phase. For each phase, the Θ_{DW} is proportional to the inverse square root of the slope of the $<u^2>$ vs. temperature curves, and the Θ_{DW}s are listed in Table 1 together with some values obtained from heat capacity and ultrasonic methods. The Θ_{DW}s are plotted in Fig. 11 together with a plot of Θ_{DW} for Pu-2 at. % Ga from Lawson et al. (2000a) for comparison. The Θ_{DW}s decrease with increasing temperature: such a trend is required for phase stability and is very similar to that in the alloy, except that there is a big jump between the Θ_{DW}s of α- and β- Pu.

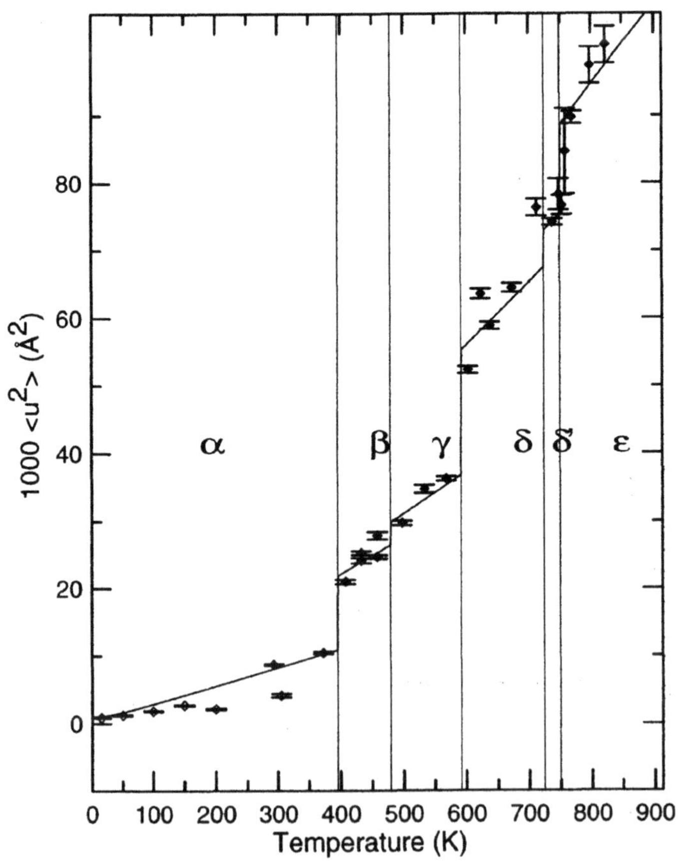

Figure 10. $<u^2>$ versus temperature for all the phases of unalloyed plutonium metal.

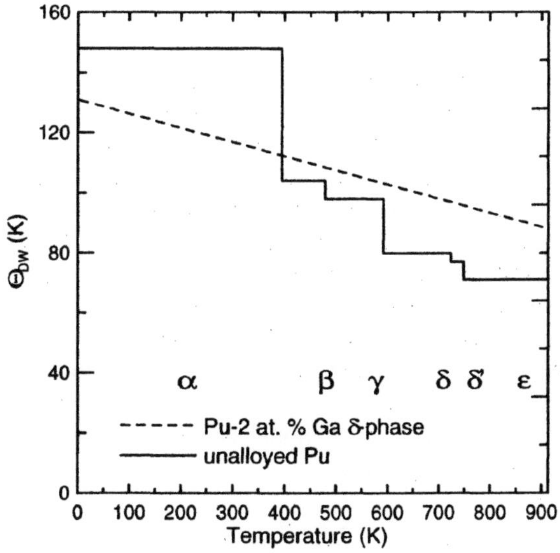

Figure 11. Debye-Waller temperatures for unalloyed Pu compared to those for Pu-2 at. % Ga

TABLE 1. Values of Θ for the phases of Pu.

phase	Θ_{DW} (K) Debye-Waller[a]	Θ_D (K) Heat Capacity	$\Theta_{Ultrasound}$ (K) Ultrasound	Θ_2 (K) Thermodynamic[g]
α	148	165[c]		162
β	104			99.5
γ	98			94.1
δ	80			92.2
δ'	77			84.9
ε	71			53.1
δ-phase Pu-2 at. % Ga	121[b]			
δ-phase Pu-3 at. % Ga Single crystal			115[e]	
δ-phase Pu-5 at. % Al			132[f]	
δ-phase Pu-8 at. % Al		130[d]		

[a] this work
[b] Lawson et al. (2000a)
[c] Lee et al. (1970)
[d] Sandenaw (1960)
[e] Ledbetter and Moment (1976)
[f] Taylor et al. (1965)
[g] Wallace (1998)

5. Pu MELTING

One of the many remarkable physical properties of plutonium metal is its low melting point, as shown in Fig. 1. A few years ago, we showed that this observation could be rationalized with the Lindemann melting rule because the elastic properties (the Debye-Waller temperature, Θ_{DW}) are strongly temperature dependent. The elastic stiffness decreases strongly with temperature as had been shown much earlier by ultrasonic measurements by Taylor et al. (1965). In fact, the Θ_{DW}s obtained by measurement of Debye-Waller factors are in quite good agreement with those obtained by ultrasonic measurement, as shown in Fig. 12.

The Lindemann rule states that a material melts when its vibrational amplitude is roughly equal to 10% of the interatomic distance. The vibrational amplitude at any temperature can be expressed in terms of the Debye temperature, Θ_{DW}. If Θ_{DW} is temperature dependent, as it is for Pu, it should be evaluated at the melting point. A convenient form of the Lindemann rule is

$$T_c = \frac{M\Omega^{2/3} k_B \Theta_{melt}^2 f^2}{3\hbar^2}. \tag{5}$$

which suggests the plot of $T_m/M\Omega^{2/3}$ vs Θ_{melt}^2 that is given in Fig. 13. This plot suggests the validity (with limited accuracy) of the Lindemenn rule. The trick is to evaluate Θ at the melting point, and we have used $\Theta = \Theta_0 + cT$, as suggested by Fig. 12, together with single crystal elastic data or Debye-Waller factor measurements. Then Eq. 5 has the melting point on both sides, but can nevertheless be solved explicitly, as shown by Lawson et al. (2000a, 2000c). The large (negative) value of c shown in Fig 14 gives a calculated value of the melting point that is in good agreement with the observed low melting point. In fact, the value of –c for Pu is one of the largest in the periodic table. (Details are given in Lawson, 2001.)

How can the atomic spring constants of Pu be temperature dependent? The usual explanation has to do with lattice anharmonicity. For an ordinary material, the thermal interatomic potential is asymmetric, and this asymmetry leads to thermal expansion. As this material is heated, the interatomic separation increases, and the vibrational frequencies decrease through the Grüneisen equation of state:

$$\frac{\omega}{\omega_0} = \left[\frac{\Omega}{\Omega_0}\right]^{-\gamma_G} \tag{6}$$

where γ_G is the Grüneisen constant and Ω is the atomic volume. γ_G's are usually in the range of 1-3. A linear decrease of Θ_{DW} with temperature is then a natural consequence of thermal expansion. But for Pu-2 at. % Ga, the thermal expansion is very small, and the Grüneisen effect cannot be important here. Instead, it has been suggested that these materials undergo "harmonic softening" as the temperature increases, and this is assumed to be a purely electronic effect (Lawson et al., 2000a, Manley et al., 2001). According to this suggestion, the large values of –c for the light actinides arise – at least in part – from harmonic softening.

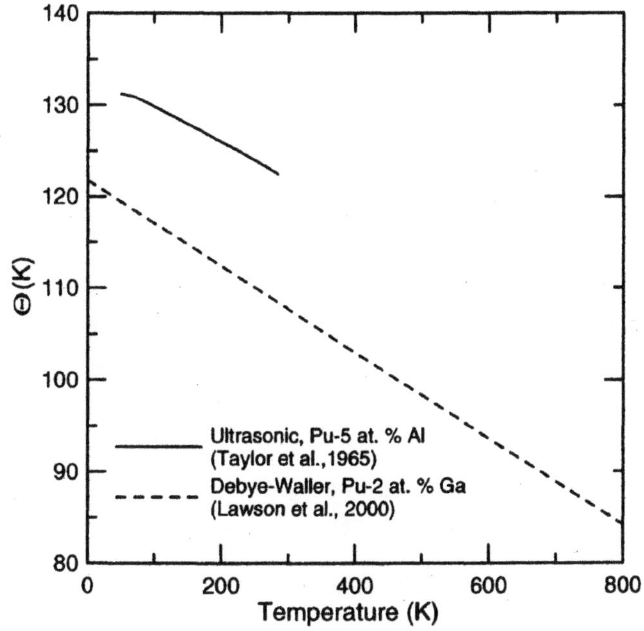

Figure 12. Θ_{DW} and $\Theta_{ultrasonic}$ versus temperature for δ-stabilized Pu alloys.

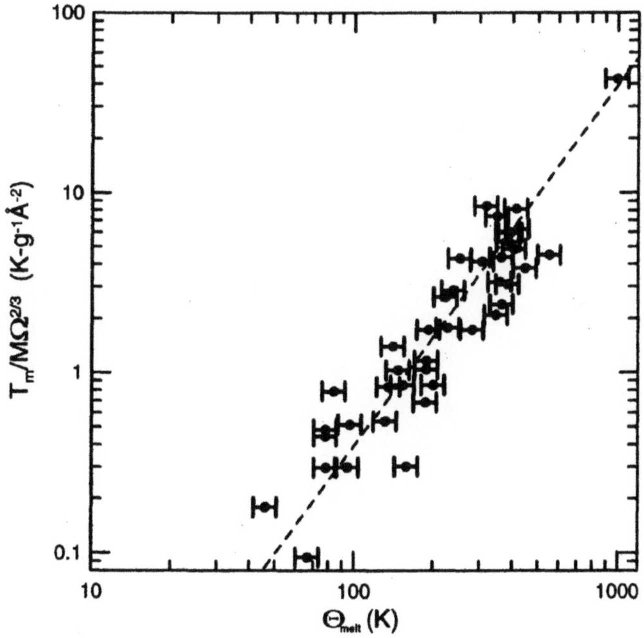

Figure 13. Normalized melting temperature ($T_m/M\Omega^{2/3}$) versus Θ_{melt}, the Debye temperature at the melting point. Ω is the atomic volume and M is the atomic mass. Reprinted from Phil. Mag. B., **81**:255 (2001), A. C. Lawson, An improved Lindemann melting rule, page 255, with permission from Taylor and Francis Limited, http://www.tandf.co.uk.

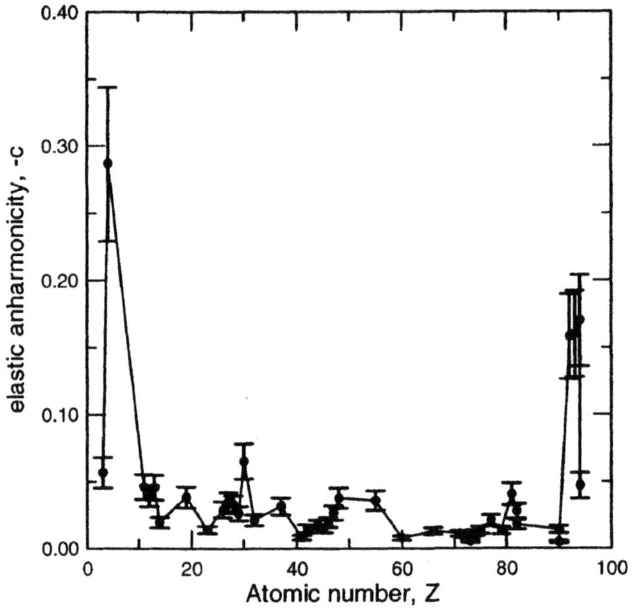

Figure 14. Elastic anharmonicity, -c, defined by Θ = Θ + cT, versus atomic number. Reprinted from Phil. Mag. B., **81**:255 (2001), A. C. Lawson, An improved Lindemann melting rule, page 255, with permission from Taylor and Francis Limited, http://www.tandf.co.uk.

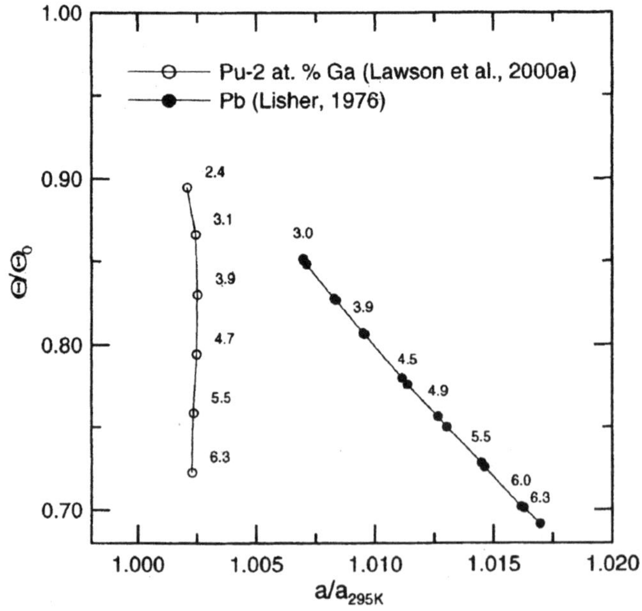

Figure 15. Normalized Debye temperature versus normalized lattice constant for Pb and δ-phase Pu-2 at. % Ga. The parameter is T/Θ_0.

The difference between anharmonic and harmonic softening is illustrated in Fig. 15, which shows plots of normalized Debye temperature, Θ/Θ_0, versus the normalized lattice constant, a/a_{295K} for Pb and for Pu-2 at. % Ga. The data are from Lisher, 1976, and Lawson et al., 2000a, respectively. For Grüneisen solids, such as Pb, the strength of the atomic spring constants is mainly *volume* dependent, but for the light actinides is mainly *temperature* dependent. (The data in this figure could not be used for the evaluation of γ_G, because the temperature is not constant.)

Manley et al. (Manley, 2000 and Manley et al., 2001) have observed harmonic softening in polycrystalline α-U using inelastic neutron scattering over a wide range of temperatures. Their results show a continuous softening of the phonon spectrum in the orthorhombic α-U phase as the temperature is increased.

6. Pu THERMODYNAMICS

For Θ_D independent of temperature, the vibrational heat capacity at constant volume is given by

$$C_{vib} = 3Nk_B\left[12(T/\Theta_D)^3 \int_0^{\Theta_D/T} \frac{u^3du}{e^{\Theta_D/T}-1} + \frac{3\Theta_D/T}{e^{\Theta_D/T}-1}\right] \tag{7}$$

and this formula is modified to

$$C_{anharmonic} = C_{vib}\left(\frac{\Theta_{D,0}}{\Theta_{D,0}+cT}\right)^2 \tag{8}$$

for the case of $\Theta_D = \Theta_{D,0} + cT$. In these equations, Θ_D is the Debye temperature, rather than the Debye-Waller temperature; these are respectively based on the +2 and −2 moments of the phonon spectrum (Grimvall, 1999). To these terms must be added the "dilatation" term

$$C_{dil} = 9\alpha^2\Omega BT \tag{9}$$

and the electronic term, approximated by

$$C_{el} = \gamma_{el}T. \tag{10}$$

These models are evaluated and compared to the measured heat capacities in Fig. 16 for Pb, α-Pu, δ-Pu and α-U, materials for which either good or at least marginal data are available (Pb: Pankratz, 1982; α-Pu: Lee et al., 1970; δ-Pu: Sandenaw, 1960; α-U: Holden, 1958). There is an extra heat capacity contribution for α-U at high temperatures, and this has been seen in three different measurements. This contribution has roughly the shape of a λ-anomaly, and the entropy under the heat capacity curve (C_p-C_{tot}) is Rln(1.6).

This may be a measure of the lattice softening seen by Manley et al. Also unexplained are the enhanced values of C_p in the β- and γ-phases, and the fact that they are constant with temperature, rather than slightly increasing. Data for the heat capacity of δ-phase Pu are marked by some kind of transient phenomenon – perhaps related to radiation damage – that has not received a satisfactory explanation. Better data are required for evaluation.

While the experimental situation is unclear for δ-Pu, it is obvious from the general agreement of C_{tot} with C_{exp} that the temperature dependence of the heat capacity Θ_D for the two phases of Pu is much less than observed in the Debye-Waller factor Θ_{DW}. This apparent disagreement must be resolved by detailed measurements of the frequency dependence of the phonon spectra. Manley's work is a step in this direction.

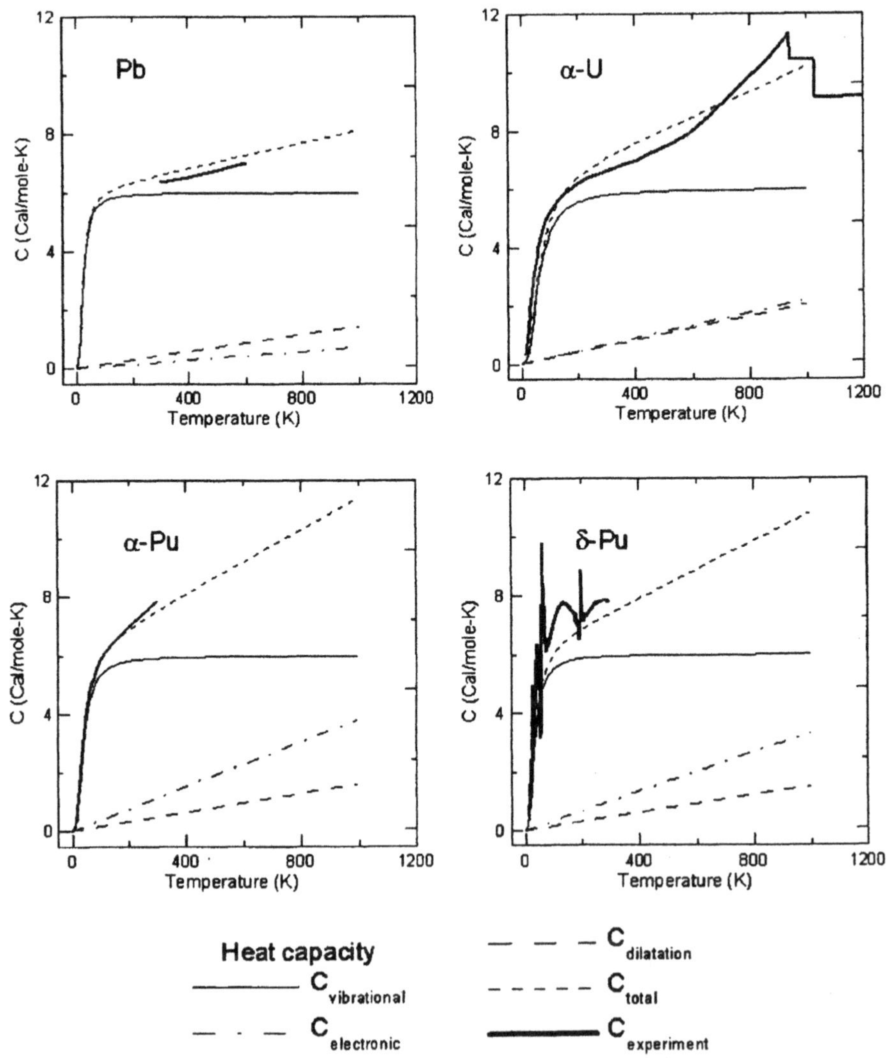

Figure 16. Heat capacities for Pb, α-U, α-Pu and δ-phase Pu-8 at. % Al.

This becomes more than a textbook exercise when one considers all the uncertainties in actinide thermodynamics, and it may be a realistic way of determining the temperature dependence of the electronic term, which is certainly more complicated than Eq. (10) for the f-electron systems we are discussing.

Wallace (1998) has made careful estimates of the thermodynamic functions of all the phases of Pu based on available thermodynamic data. He finds it impossible to separate the "electronic" contributions from the "quasiharmonic" contributions, but the combined contributions from these terms is a significant part of the free energy. This is shown in Fig. 17, which shows the vibrational and electronic plus quasiharmonic contributions estimated by Wallace. It is clear that electron correlation terms could be important for phase stabilization, even if they make only a fractional contribution to the total energy.

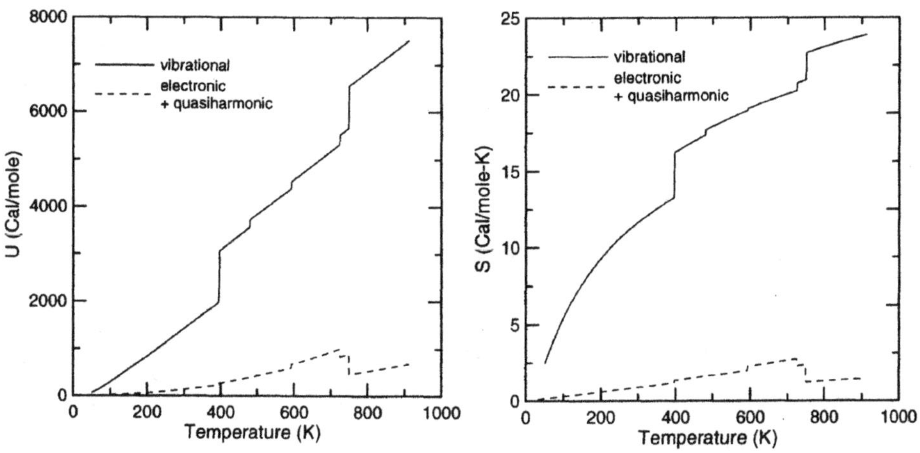

FIGURE 17. Internal energies (left) and entropies (right) for the phases of unalloyed Pu

7. CONCLUSIONS AND CONNECTIONS TO ELECTRON CORRELATION

The energy and entropy curves for unalloyed Pu mark the extent of our progress toward an understanding of the phase diagrams presented at the beginning of the paper; it is clear that there is still a long way to go. It would be desirable to evaluate the vibrational thermodynamic functions from measured phonon spectra. However, this may require the growth of single crystals, if the PDF method using powder diffraction proves impractical for Pu. The measurement of inelastic phonon densities of states from polycrystalline samples might be a practical method for all phases of Pu.

Estimates of the electronic contributions to the thermodynamics have been derived from the work of Wallace (1998). It would be of interest to compare these with theoretical estimates to evaluate the role of electron correlation in the phase stability of plutonium metal. As emphasized by Wallace, this will be an "enormous challenge" to electronic structure theory. In particular, we need a more predictive understanding of how narrow bands in Pu stabilize idiosyncratic crystal structures (Söderlind et al., 1995 and Söderlind, 1998).

Finally we recall the role of electron entropy effects related to magnetism that shape the phase diagram of iron (Zener, 1967, and Kaufmann and Bernstein, 1970). We should be alert for similar effects in plutonium.

ACKNOWLEDGMENTS

We are pleased to thank Al Arko, Mike Baskes, David Becker, Mike Boring, Barry Cooper, Reed Elliott, Matthias Graf, Sig Hecker, Jason Lashley, Mike Manley, Rob McQueeney, Albert Migliori, Thomas Proffen, Mike Stevens, Stuart Trugman, and John Wills for very helpful discussions and we are grateful to Nikki Cooper, Sig Hecker and James L. Smith for use of their figures.

This work was supported under the auspices of the United States Department of Energy. The Manuel Lujan, Jr. Neutron Scattering Center is a National User Facility funded by the United States Department of Energy, Office of Basic Energy Science under contract No. W-7405-ENG-36. The Intense Pulsed Neutron Source is operated as a National User Facility by the United States Department of Energy, Office of Basic Energy Science - Materials Sciences, under contract No. W-31-109-ENG-38.

REFERENCES

Baskes, M. I., 2000, Atomistic model of plutonium, Phys. Rev. B. **62**:15532.

Boring, A. Michael and Smith, James L., 2000, Plutonium condensed matter physics – a survey of theory and experiment, in *Challenges In Plutonium Science*, Necia Grant Cooper, ed., Los Alamos Science, No. 26, pp. 90-127; http://www.lanl.gov/external/science/lascience/.

Conradson, Steven D., 2000, Where Is the Gallium? – Searching the plutonium lattice with XAFS, in *Challenges in Plutonium Science*, op. cit., pp. 356-363.

Dmitrov, D. A., Louca, D., and Röder, H., 1999, Phonons from neutron powder diffraction, Phys. Rev. B **60**:6204.

Egami, T., 1998, PDF analysis applied to crystalline materials, in *Local Structure from Diffraction*, S. L. J. Billinge and M. F. Thorpe, eds, Plenum, New York, pp. 1-21.

Elliott, R. O, 1970, unpublished.

Grimvall, G., 1999, *Thermophysical Properties of Materials, revised edition*, Elsevier, Amsterdam, p. 89.

Hecker, Siegfried S. and Timofeeva, Lidia F., 2000, A tale of two diagrams, in *Challenges in Plutonium Science*, op. cit., pp. 244-251.

Hecker, Siegfried S., 2000, Plutonium and its alloys – from atoms to microstructure, in *Challenges in Plutonium Science*, op. cit., pp. 290-335.

Holden, A. N., 1958, *Physical Metallurgy of Uranium*, Addison-Wesley, Reading MA, p 47.

Kaufman, Larry and Bernstein, Harold, 1970, *Computer Calculations of Phase Diagrams*, Academic Press, New York, pp. 16-26.

Kittel, Charles, 1996, *Introduction to Solid State Physics, 7th Edition*, New York, John Wiley and Sons, p. 91.

Kmetko, E. A., and Hill, H. H., 1976, Anomalous melting of f-electron metals (with attention to Pu), J. Phys. F. **6**:1025.

Kwei, G. H., Louca, D., Billinge, S. J. L. and Rosenfeld, H. D., 1998, Recent local structure studies: metallic alloys, superconductors and proteins, in: *Local Structure from Diffraction*, op. cit., pp. 323-336.

Lashley, Jason C., Blau, Michael S., and Moment, Roger L., 2000, Preparing Single Crystals of Gallium-Stabilized Plutonium, in *Challenges in Plutonium Science*, op. cit., pp. 226-232.

Lawson, A. C., Cort, B., Goldstone, J. A., Wozniak, D., Hitterman, R. L. and Richardson, J. W., 1988, Diffraction studies of plutonium metal, IPNS Progress Report 1986-1988, Argonne National Laboratory, p. 50.

Lawson, A. C., Goldstone, J. A., Vaninaetti, J., Eash, D., Cort, B., Hitteman, R., and Faber, J., 1990, Neutron diffraction study of δ' and ε-plutonium, IPNS Progress Report 1988-1990, Argonne National Laboratory, p 28.

Lawson, A. C., Goldstone, J. A., Cort, B. and Diebolt, L., 1992, Debye-Waller factors of α-plutonium metal, LANSCE Experiment Reports - 1991 Run Cycle, Los Alamos National Laboratory, LA-12407-PR, p 33.

Lawson, A. C., Goldstone, J. A., Cort, B. and Von Dreele, R. B., 1993, Debye-Waller Factors of α-plutonium, LANSCE Experiment Reports – 1992 Run Cycle, Los Alamos National Laboratory, LA-12647-PR, p 46.

Lawson, A. C., Goldstone, J. A., Cort, B., Sheldon, R. I. and Foltyn, E. M., 1994, Debye-Waller factors of the light actinide elements, in: *Actinide Processing: Methods and Materials*, B. Mishra and W. A. Averill, eds., TMS. Warrendale PA, pp. 31-43.

Lawson, A., C., Goldstone, J., A., Cort, Martinez, R., J., Vigil, F., A., Zocco, T. G., Richardson, J. W., Jr., and Mueller, M. H., 1996, Structure of ζ-phase Plutonium-Uranium, Acta Cryst. B **52**:32.

Lawson, A. C., Martinez, B., Roberts, J. A., Bennett, B. I., and Richardson, J. W., Jr., 2000a, Melting of the light actinides, Phil. Mag. B **80**:53.

Lawson, A. C., Martinez, B., Von Dreele, R. B., Roberts, J. A., Sheldon, R. I. and Richardson, J. W., Jr., 2000b, Vibrational order in $Pu_{0.98}Ga_{0.02}$, Phil. Mag. B **80**:1169.

Lawson, Andrew C., Martinez, Barbara, Roberts Joyce A., Richardson, James W., Jr., and Bennett, Bard I., 2000c, Atomic vibrations and melting in plutonium, in *Challenges in Plutonium Science*, op. cit., pp. 190-201.

Lawson, A. C., 2001, An improved Lindemann melting rule, Phil Mag B **81**:255.

Ledbetter, H. M. and Moment, R. L., 1976, Elastic properties of face-centered-cubic plutonium, Acta Met. **24**:891.

Lee, J. A., Mendelssohn, K., and Sutcliffe, P. W., 1970, Specific heats of plutonium and neptunium, Proc. Roy. Soc. Lond. A. **317**:303.

Liptai, R. G., and Friddle, R. J., 1966, The phase diagram of plutonium at pressures up to 75 kbar, J. Less-Common Metals **10**:292.

Lisher, E., J., 1976, The Debye-Waller factors of lead from 296 to 550K, Acta Cryst. A **32**:506.

Lovesey, S. W., 1984, *Theory of Neutron Scattering from Condensed Matter, Vol. I,* Oxford University Press, Oxford, pp. 112-113.

Manley, M. E., Fultz, B., McQueeney, R. J., Brown, C. M., Hults, W. L, Smith, J. L., Thoma, D. J., Osborn, R and Robertson, J. L., 2001, Large Harmonic Softening of the Phonon Density of States of Uranium, Phys. Rev. Lett. **86**:3076.

Manley, Michael E., 2000, Vibrational softening in α-uranium, in *Challenges in Plutonium Science*, op. cit., pp. 202-207.

McQueeney, R. J., 1998, Dynamic radial distribution function from inelastic neutron scattering, Phys. Rev. B **57**:10560.

Pankratz, L. B., 1982, *Thermodynamic Properties of Elements and Oxides*, U.S. Dept. of the Interior, Bureau of Mines, Washington, p. 294.

Reichardt, W., and Pintschovius, L., 2001, Influence of phonons of the pair distribution function deduced from neutron powder diffraction, Phys. Rev. B **63**:174302.

Sandenaw, T. A., 1960, Heat capacity, thermal expansion and electrical resistivity of an 8 a/o aluminum-plutonium (delta-phase stabilized) alloy below 300K, J. Phys. Chem. Solids **16**:329.

Smith, J. L., and Kmetko, E. A., 1983, Magnetism or bonding: a nearly periodic table of transition elements, J. Less-Common Metals **90**:83.

Söderlind, Per, Eriksson, Olle, Johannson, Börje, Willis, J. M. and Boring, A.M. 1995, A unified picture of the crystal structures of metals, Nature **374**:524.

Söderlind, P., 1998, Theory of the crystal structures of cerium and the light actinides, Advances in Physics **47**:959.

Taylor, J. C., Loasby, R., G., Dean, J. D., and Linford, P. F., 1965, Some physical properties of plutonium at low temperatures, in: *Plutonium 1965*, A. E. Kay and M. B. Waldron, eds., Chapman and Hall, London, pp 162-175.

Wallace, D. C., 1998, Electronic and phonon properties of six crystalline phases of Pu metal, Phys. Rev. B **58**:15433.

Warren, B. E., 1969, *X-Ray Diffraction*, Addison Wesley, Reading, MA, reprint:, Dover Publications,. Mineola, NY.

Welberry, T. R., and Butler, B. D., 1994, Interpretation of diffuse x-ray scattering via models of disorder, J. Appl. Cryst. **27**:205.

Wills, John M., and Eriksson, Olle, 2000, Electronic structure of α- and δ-plutonium, in *Challenges in Plutonium Science*, op. cit., pp 129-151.

Zener, Clarence, 1967, Influence of entropy in phase stabilization, in *Phase Stability in Metals an Alloys*, Rudman, Peter S., Stringer, John, and Jaffee, Robert I., eds, McGraw-Hill, New York, pp. 25-38.

SIZE AND INTERFACE EFFECTS IN PATTERNED MAGNETIC AND SUPERCONDUCTING THIN FILMS

Lance E. De Long
Department of Physics and Astronomy,
University of Kentucky, Lexington, KY 40506-0055 U.S.A.
lander@uky.edu

Venkat Chandrasekhar and John B. Ketterson
Department of Physics, Northwestern University
Evanston, IL 60208 U.S.A.
v-chandrasekhar@northwestern.edu
and
j-ketterson@nwu.edu

Vitali V. Metlushko
Department of Electrical Engineering and Computer Science
University of Illinois-Chicago, Chicago, IL 60607-7053 U.S.A.
vmetlush@ece.uic.edu

Abstract An overview is presented of magnetic, electrical transport, and ferromagnetic resonance studies of thin films of superconducting and magnetic materials with patterned features that are laterally confined on the scale of 10 nm to 1 μm. Emphasis is given to finite-size and correlation effects found at the limiting size scale of stability of long-range magnetic order, and the power of superconductivity as a probe of nanoscale magnetism. A summary is given of state-of-art methods of fabricating highly ordered arrays of ferromagnetic features precisely patterned by electron beam and laser interference techniques. Certain directions for future research in the theoretical description of small magnetic features, and the crucial role of mesoscale phenomena in nanotechnology are discussed.

Keywords: superconductivity, ferromagnetism, thin films, film patterning, ferromagnetic resonance, nanotechnology, computer memory

1. FINITE SIZE EFFECTS ON MAGNETISM AND SUPERCONDUCTIVITY

A fundamental understanding of nanoscale magnetic materials and their interfaces with other materials is essential for the design of advanced devices. The properties of thin films and multilayers, and some aspects of magnetic interfaces, have been extensively studied, and film thickness and surface/interface roughness can now be systematically controlled down to near-atomic dimensions. On the other hand, there has been relatively little investigation of *laterally confined magnetic nanostructures.*

Modern device concepts demand further reduction of lateral feature sizes into the nanometer range, a process that necessarily traverses natural length scales such as a minimum ferromagnetic domain or domain wall width $\delta \approx$ 100-1000 nm. Therefore, attaining the penultimate atomic scale in miniaturization necessarily involves crossing a *mesoscopic* regime generally characterized by the strong fluctuations and nonlinearities ("emergent complexity") endemic to finite-size systems. Future device structures also will inevitably require transmission and control of signals from nanoscale architectures to the macroscale of the human environment via interfaces and connections that traverse the mesoscale. Although the *nanoscale* regime is not consistently defined in current literature, we interpret it to be 0.1 - 100 nm. We define the *mesoscale* as the range of feature size that separates bulk thermodynamic (e.g., cooperative states such as ferromagnetism and superconductivity) behavior from the *atomic* scale $\approx$ 1 nm, which may be taken as the rough size threshold for bulk solid-state or many-particle effects.

At and below the mesoscale, the nature and very stability of ferromagnetic and superconducting states, as well as the chemistry, surface properties and interface behavior of material systems, will be strongly modified from typical bulk properties. This statement is based upon two related characteristics of finite systems:

1. Surface energetics are important due to large surface-to-volume ratios.
2. Electronic structure is strongly modified due to finite-size effects that occur in quantum-confined systems, as compared to bulk solids.

Investigation of the unique realm connecting bulk cooperative phenomena to the atomic regime is not only of great fundamental interest, it is simply unavoidable; yet the mesoscale has been largely neglected and is poorly understood at present, mainly due to present limitations in both the experimental and theoretical arenas.

2. PHYSICAL PROPERTIES OF SMALL MAGNETIC FEATURES

Long-range ferromagnetic order is expected to be intrinsically unstable in the nanoscale regime, and therefore difficult to practically control in small device structures. However, this expectation ignores the strong potential for unexpected discoveries of novel nanomagnetic effects and ongoing advances in the theory of finite-size, strongly correlated systems. Already at the submicron scale, the properties of laterally confined magnetic features are governed by significantly different physics compared to continuous films. Feature composition, shape, dimensions and structural defects, can all strongly affect dynamic magnetic response and device performance. For example, recent work indicates that shape anisotropy imposes constraints on magnetization and domain distributions inside an isolated magnetic dot (see Figs. 1 – 4), whereas the direction of the total moment and the dynamics of switching and stability are strongly influenced by the interactions between dots in relatively dense arrays [1–4]. Much more research needs to be done to follow up these results, especially wherever feature sizes can be continuously varied from the micron- to nano-scales.

2.1 Properties of Noninteracting Magnetic Dots

Isolated, micron-sized magnetic dots and dilute dot arrays consisting of relatively independent features display a variety of magnetic structures consisting of only a few domain walls whose arrangements are determined by dot-shape anisotropy. For structures smaller than a characteristic size, the formation of domain walls is energetically unfavorable and a single-domain state is stabilized (see Fig. 1). Changing the shape anisotropy (thickness/diameter or principal-axis aspect ratios) induces variations of polarization across a dot and novel magnetization patterns that depend sensitively on both dot dimensions and applied field direction, as shown in recent magnetic force microscopy (MFM) studies (Figs. 2, 4). These unusual static magnetic dot structures are directly related to dynamic processes that underlie complex magnetic reversals and nonlinear dynamics (see Figs. 5, 6) [7].

The magnetization patterns and local field distributions of single dots and dilute arrays are very different for in-plane and out-of-plane magnetization, suggesting that interesting possibilities exist for developing customized magnetic structures (see Fig. 7 and 8). The strong out-

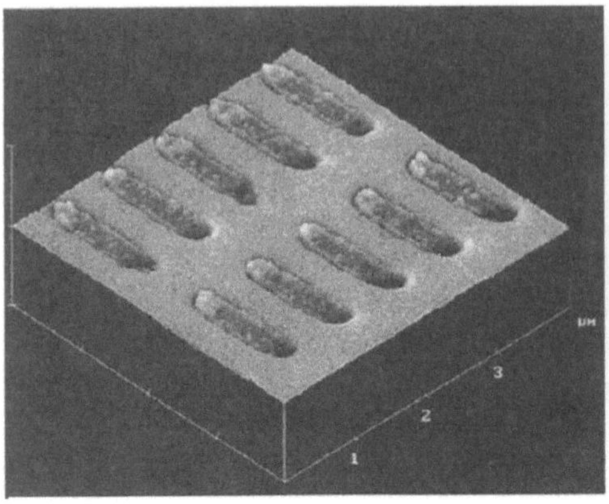

Figure 1. MFM image of the remanent state of "single-domain" Ni nanodots with a long 4:1 elliptical axis ratio. Light and dark correspond to the south and north poles, resp. After [5].

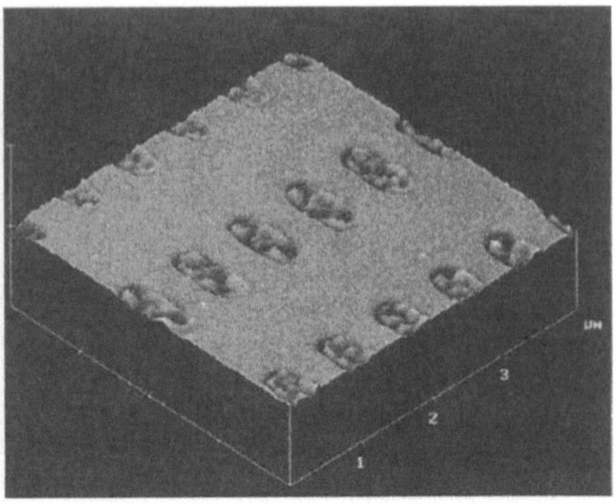

Figure 2. MFM image of 40 nm-thick Ni dots with a short 2:1 axis aspect ratio. The domain wall pattern inside the dots is clearly visible. After [5].

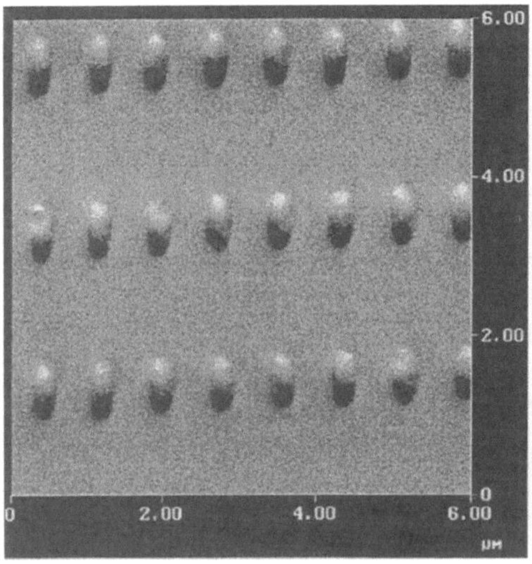

Figure 3. MFM image of the single-domain remanent state of 300 Å Co nanodots after removal of a saturation magnetic field applied along the long elliptical axis. Light and dark shades correspond to south and north poles, respectively. After [6].

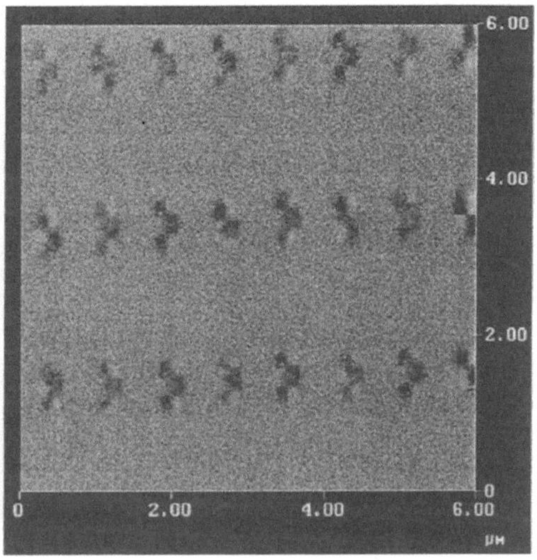

Figure 4. Remanent state of the Co nanodots of Fig. 3 after applying a field along the short elliptical axis. Complex magnetic ("double-vortex") structure is visible. Compare the Co with Ni dots in Figs. 1, 2. Size scales for Figs. 1–4 are in microns. After [6].

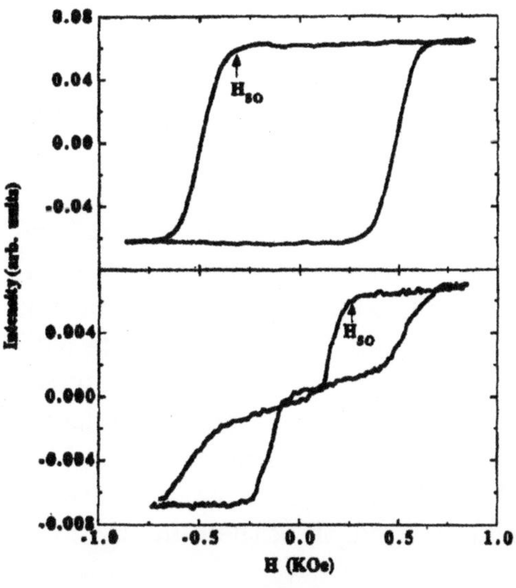

Figure 5. Hysteresis loops for Co nanodots with 6:1 (top) and 1.2:1 (bottom) aspect ratios. The field is applied along easy (long) axis. H_{so} denotes the switching onset field. After [8].

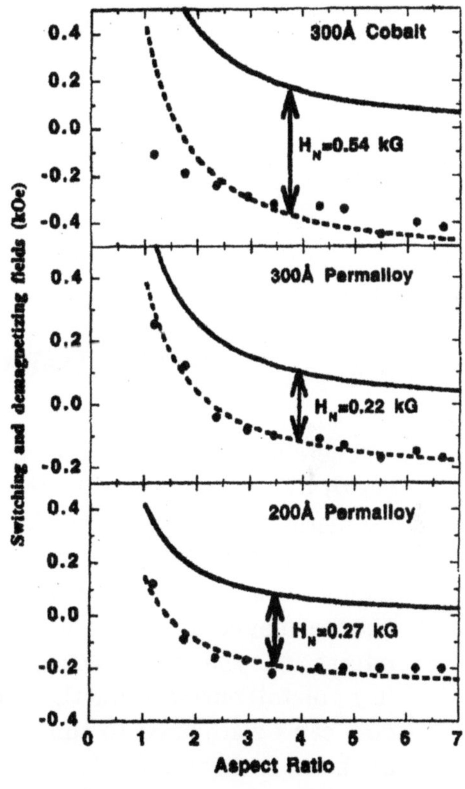

Figure 6. Measured switching fields (filled circles) and demagnetizing fields H_N (dotted line) for 300 Å Co, and 300 and 200 Å permalloy arrays. After [8].

of-plane anisotropy (see Fig. 8) exhibited by Co/Pt multilayers (ML) makes them promising candidates for perpendicular magnetic recording applications [9–12]. Clever patterning of Co/Pt ML may spawn a new generation of magnetic nanostructures with lateral dimensions in the submicron range (see **Sec. 5**), and thickness approaching a monolayer scale (e.g., 4 Å of Co and 10 Å of Pt).

Figure 7. In-plane magnetization for thin 50 – 300 Å polycrystalline dots of Ni, Fe, permalloy and Co. After [6].

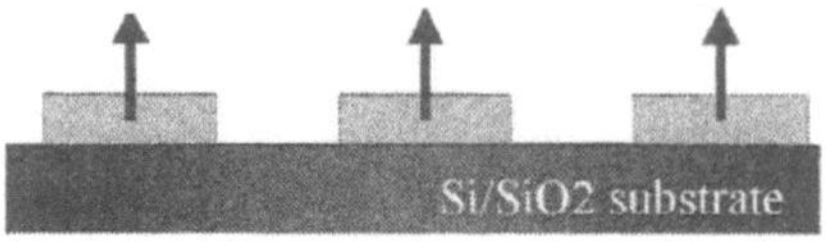

Figure 8. Perpendicular magnetization of an array of [Co/Pt]n ML dots. The thickness of an individual layer is only 4 Å for Co, and 10 Å for Pt. After [6].

2.2 Collective Behavior of Dense Arrays of Magnetic Dots and Antidots

There are many physical variables and fabrication schemes that have yet to be fully exploited to control the magnetic behavior of very small magnetic features. Modern electron beam lithography techniques permit one to place dots so close together that single-dot energies compete with interactions between dots. Dots in high-density periodic arrays interact via the magnetic fields they produce at long range, and via exchange interactions that are relatively short-ranged if there is no active medium (e.g., metal) surrounding the dots. The orientation and distance between dots can be adjusted to tune the relative strength of these interactions, as shown in Figs. 9 & 10.

Recent work [1–3, 7, 8, 13–15] on dense arrays of magnetic dots has revealed novel, complex magnetic structures that are sensitive to array topology, temperature, field and magnetic history. The dynamic processes involved in relaxation and switching of dot magnetization (see Figs. 1, 4) observed under changes in applied field are unconventional, and not well understood. These recent results pose a particularly intriguing set of questions concerning the *collective response of high-density, periodic arrays:* Precisely fabricated magnetic dot lattices (MDL) may exhibit long-range magnetic order and phase transition signatures; and complex structures analogous to ferro-, antiferro-, and ferri-magnetic states may be anticipated, where dots play the analogue to atomic moments in ordinary solids.

Although there is rapidly growing interest in the properties of ferromagnetic dots and their arrays, there have been relatively few studies of the "inverse" geometry of ordered arrays of holes, or "antidot lattices" (ADL). ADL have certain advantages over MDL for magnetic storage applications, such as an absence of any superparamagnetic limit, and the preservation of magnetic continuity of the film around the antidot. The antidot shape and lattice symmetry of the ADL are predicted to influence the magnetic domain patterns and magnetic anisotropy, as shown in Figs. 13 – 15. Three qualitatively different domain patterns were

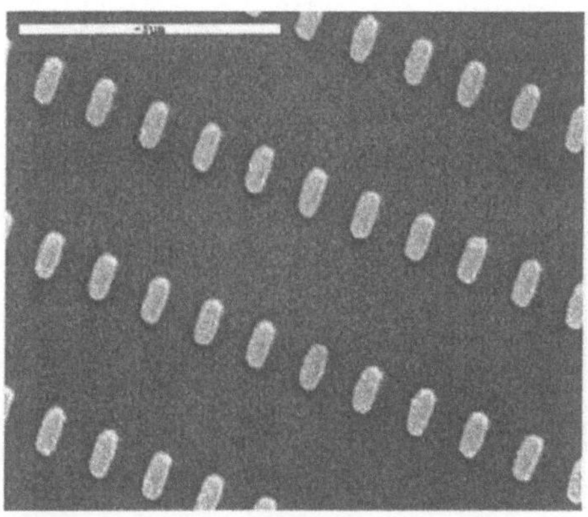

Figure 9. Scanning Electron Microscopy (SEM) image of a rectangular Ni MDL: "Array B". After [8].

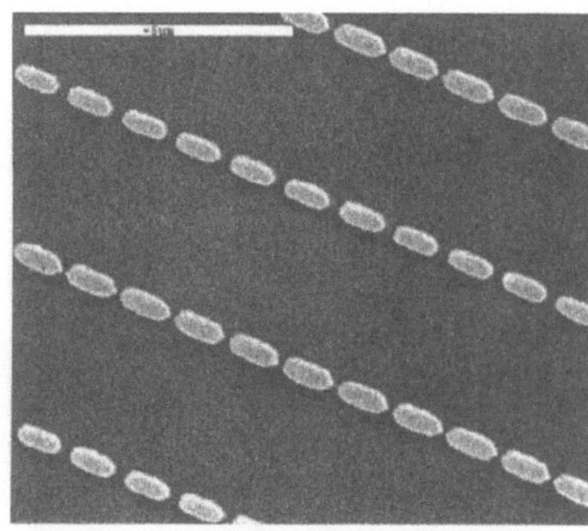

Figure 10. Chains of strongly interacting Ni dots: "Array A". The dot size and period is as in Fig. 9, but the dots are rotated by 90°. After [8].

obtained in micromagnetic simulations by Torres et al. [16], depending upon the film thickness t and the ADL parameters, and the coercivity was found to be approximately triple that expected for an unpatterned film. It was also observed that the demagnetizing effect was independent of antidot size for fixed d/D, but the exchange interaction became relatively more important in domain behavior as either the antidot size or film thickness was decreased.

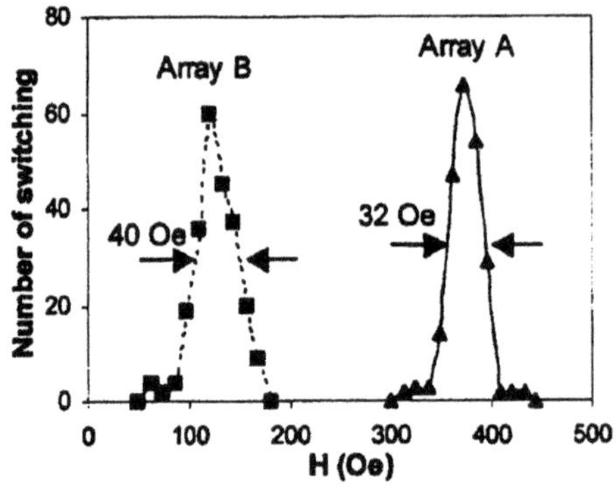

Figure 11. Switching field distribution of arrays "A" and "B" in Figs. 10 and 9, respectively, above. Field was applied along the easy axis in both cases. After [8].

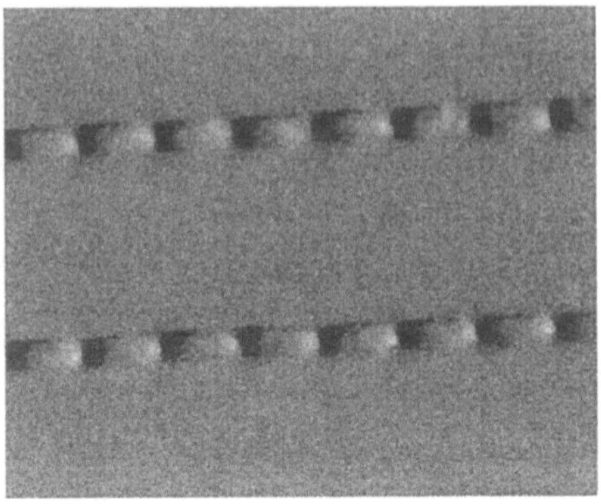

Figure 12. MFM image of chains shown in Fig. 10. A single-domain, polarized state is stabilized by interactions among dots, compared to Fig. 2. After antidot shape and lattice symmetry of the ADL are predicted to influence the magnetic domain patterns and magnetic anisotropy, as shown in Figs. 13 – 15. After [8].

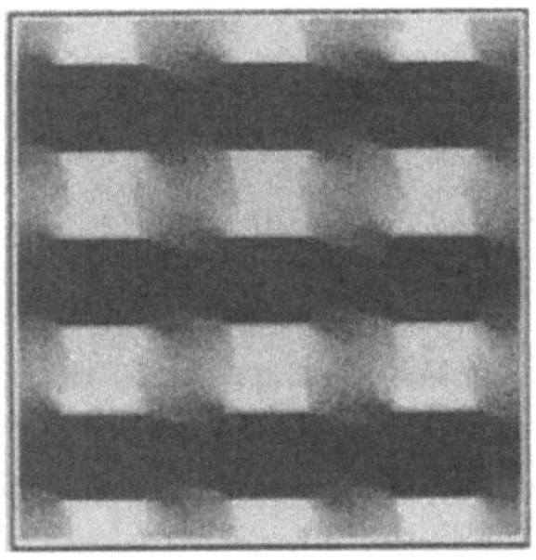

Figure 13. Simulated domain patterns for a permalloy film patterned with a square lattice of square antidots for thickness t = 40 nm, width D = spacing d = 640 nm. After [16].

Figure 14. Simulation for ADL parameters t = 40 nm, D = 640 nm, d = 128 nm. The square antidots are shown as black. After [16].

Figure 15. Simulation for ADL parameters t = 5 nm, D = d = 640 nm. Gray scale measures the horizontal component of the (totally) in-plane magnetization (white is maximal). After [16].

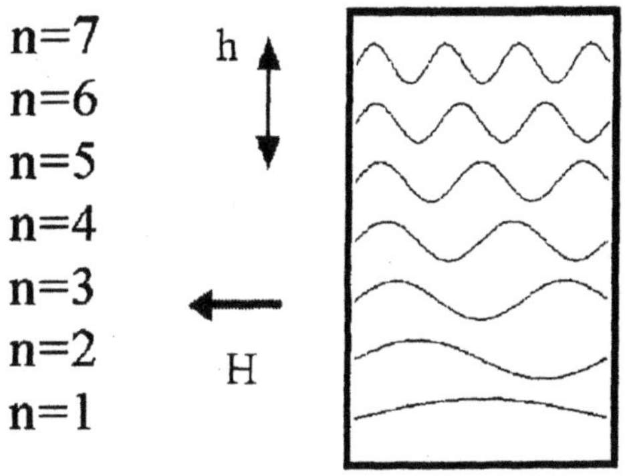

Figure 16. The seven lowest spin wave modes pinned by the upper and lower surfaces of a bulk magnetic thin film, as probed by FMR for an applied magnetic field **H** perpendicular to the film surface and an exciting AC magnetic field **h** parallel to the surface. This geometry is frequently studied in uniform films, but does not probe dot interactions in MDL as effectively as the case with **H** parallel, and **h** perpendicular to the film plane, as in Fig. 18.

3. FERROMAGNETIC RESONANCE STUDIES OF MAGNETIC DOT AND ANTIDOT LATTICES

Ferromagnetic resonance (FMR) is a powerful probe of nanoscale magnetic features that are not easily characterized via conventional magnetometry techniques, and is uniquely suited to investigate thin film samples that present no skin depth limitations. FMR spectra provide direct information about exchange and electromagnetic (dipole) interactions, magnetic shape anisotropy, and dissipation mechanisms that govern magnetic excitations such as spin waves [17–19], which are very sensitive to finite-size and shape effects. Not much is presently known (theoretically or experimentally) about the dynamic magnetic response of laterally confined dots or particles at, or below, the submicron range.

FMR is similar to other magnetic resonance methods, in that a small AC magnetic field **h** is applied perpendicular to a uniform DC applied field **H** in order to tip magnetic moments into a precession mode with reduced local magnetization **M** along **H** (see Fig. 16). The magnetization of a ferromagnetic particle typically follows a time evolution given by the Landau-Lifschitz equation [20]:

$$d\mathbf{M}/dt = -\gamma \mathbf{M} \times \mathbf{H}_{eff} - \gamma\alpha \mathbf{M} \times (\mathbf{M} \times \mathbf{H}_{eff})/M_S \qquad (1)$$

where γ is the gyromagnetic ratio, $\mathbf{H}_{eff}$ is an effective local field, M_S is the saturation magnetization, and α is a damping constant. In the

Figure 17. SEM micrograph of a square lattice of 0.5 mm diameter permalloy dots placed 0.6 mm apart. Size bar corresponds to 1 mm. After [18].

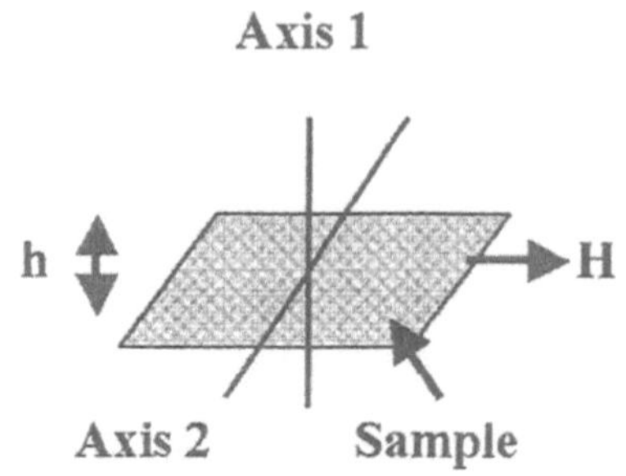

Figure 18. FMR geometry used in [17, 18] with DC magnetic field **H** applied in the film plane and the AC magnetic field **h** applied perpendicular to plane. Two sample rotation axes are shown.

simplest case (an ellipsoidal feature with uniform **M**), only the phase of the spin precession about **H** varies with position, and the associated FMR frequency ω_o is given by the Kittel equation [21]:

$$\omega_o^2 = \gamma^2[H + (N_x - N_z)M][H + (N_y - N_z)M] \tag{2}$$

where N_x, N_y and N_z are demagnetization factors with the **z**-direction defined by the applied field **H**, which is oriented along one of the ellipsoid principal axes. The "uniform precession mode" is essentially one of the extended "magnetostatic" or "Walker" [20] modes (see Fig. 21, below). Such excitations are mediated by long-range dipole interactions that induce relatively small tipping angles and have frequencies that are independent of magnetic feature size if it is large enough.

Shorter wavelength spin-wave modes (see Fig. 16) depend on the linear dimensions and shape of the magnetic feature, as the surface magnetization is pinned by local anisotropies, similar to a vibrating string with fixed ends. Therefore, the mode wavelength is comparable to, or shorter than small sample dimensions. These modes are usually dominated by short-range exchange interactions, and can be observed at lower DC fields and higher frequencies ω_{sw} than the uniform mode [22]:

$$\omega_{sw} = Dk^2 + \omega_o \tag{3}$$

D is a constant that depends upon the exchange interaction between neighboring spins.

3.1 FMR in Dense Magnetic Dot Lattices

A completely new regime should exist over some range of relatively small feature size, such that the magnetic excitations would be determined by both long-range dipolar and short-range exchange interactions of comparable strengths. Numerical simulations [23, 24] predict that such hybrid modes will have resonance frequencies that depend in a complex manner on the size and geometric perfection of the feature, and will lie both above and below the uniform mode.

We have found the first evidence [18] of the existence of hybrid modes having both dipole and exchange couplings in initial FMR experiments on dense permalloy ($Ni_{79}Fe_{21}$) dot lattices (see Fig. 17) patterned by electron beam lithography on Si substrates. Moreover, we observe that the FMR spectra reflect the symmetries of the film patterning and dot (antidot) shape when the external fields generated by neighbors in a dot lattice (or an antidot lattice in an otherwise uniform magnetic film) break the symmetry of an isolated circular feature.

The derivative FMR spectra for several samples having dot thickness t $\approx$ 85 nm, diameters D $\approx$ 500 nm, and various separations 600 nm $\leq$ d $\leq$ 1.50 μm, are shown in Fig. 19. The uniform mode is observed near H = 0.14 T, and does not shift as the direction of the in-plane field is varied (see Fig. 18; rotation about "Axis 1"). A small peak (marked by the arrows in Figs. 19a – 19e) is observed in patterned films just below the field of the uniform mode , but is not seen in the spectrum of an unpatterned reference film, as shown in Fig. 19a. The position of the small peak is shifted slightly as the MDL spacing is reduced, as shown in Figs. 19a – 19e, indicating that this excitation is probably dominated by exchange interactions, but slightly modified by dipolar interactions.

Three summary observations of the data in Fig. 19 indicate that magnetostatic interactions between dots strongly influence the FMR spectra. First, a rich spectrum develops above and below the uniform precession field at the lower values of d investigated. Second, the amplitudes and/or positions of these new peaks depend on the in-plane field orientation. Third, the position and amplitudes of the uniform and new "low-d" modes exhibit a four-fold symmetry of the square MDL, as expected for comparable contributions from both exchange and dipole couplings.

Micromagnetic simulations [18] clarify the relative importance of dipolar coupling in the FMR data, and verify that the lower satellite peak in Fig. 19 is essentially a spin-wave mode, since its simulated counterpart in Fig. 20 disappears for zero exchange stiffness. The time evolution of the magnetization of a dot shown in Fig. 21 verifies that the main resonance is indeed a uniform mode that is only distorted by demagne-

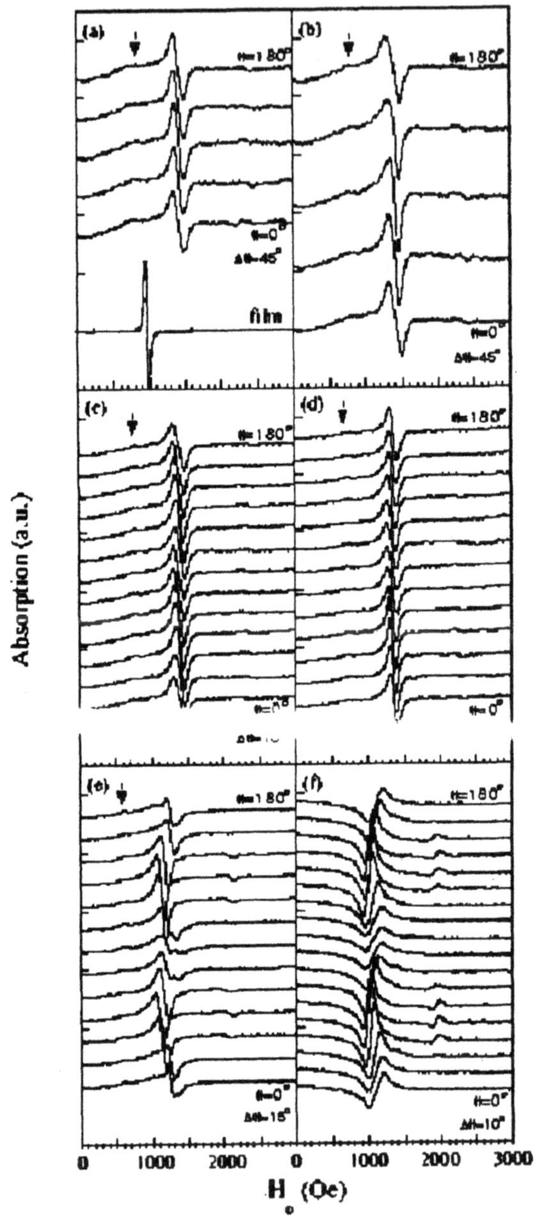

Figure 19. Derivative FMR spectra vs. field for square, permalloy MDL (see Fig. 17) with spacings: a) 1.5 (lowest curve is unpatterned film data), b) 1.1, c) 0.9, d) 0.8, e) 0.6 μm. f) Data for a similar film with D = 0.5 and d = 0.6 μm. After [18].

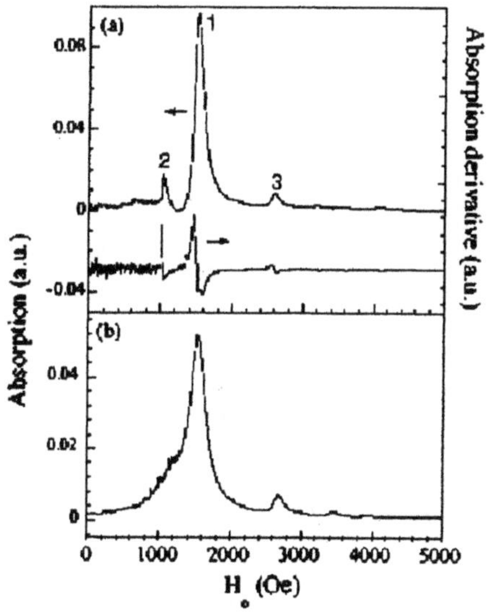

Figure 20. Simulated FMR spectrum of a permalloy dot with D = 0.5 μm, for 9.37 GHz AC drive amplitude h_o = 10 Oe, α = 0.05, DC field saturated ($4\pi M_S = 1.06 \times 10^4$ G) along the x-axis of the film plane, and finite exchange stiffness A = 1.3×10^{-6} ergs/cm (a), or A = 0 (b). After [18].

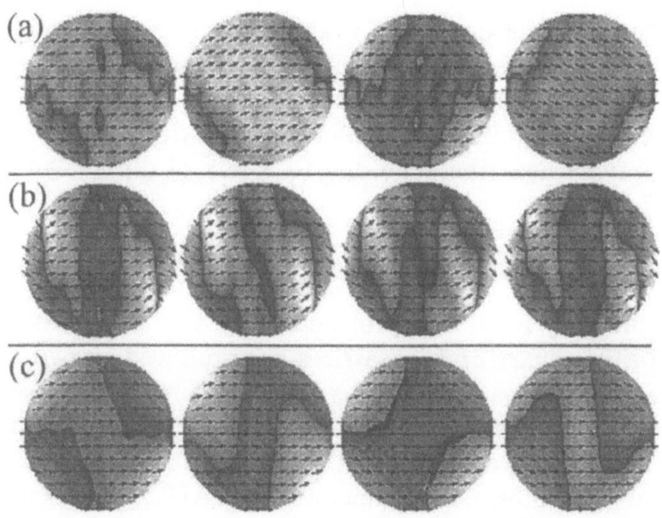

Figure 21. Simulated magnetization of a dot at four equally-spaced times of one period of the perpendicular AC field. Gray scale gives the angle between the magnetization and the x-axis. The top panel (a) is the uniform mode (peak 1, above), the middle panel (b) is a very nonuniform (essentially spin-wave) mode (peak 2), and the panel (c) is a nearly uniform mode (peak 3) with edge variations due to demagnetization (dipolar) effects. After [18].

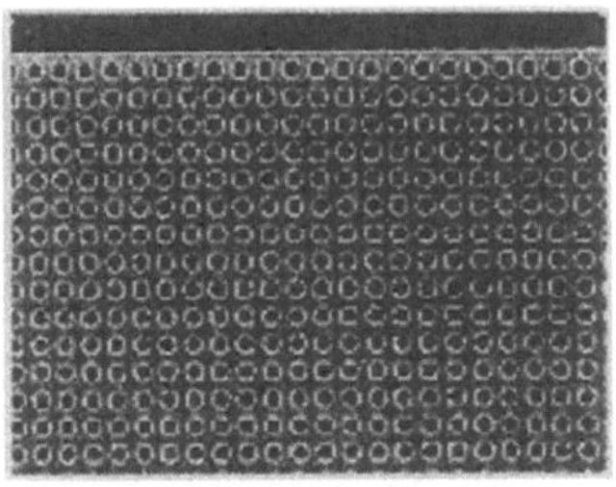

Figure 22. SEM micrograph of a permalloy film of thickness t = 30 nm with a square ADL of diameter D = 450 nm and separation d = 1 μm. After [17]

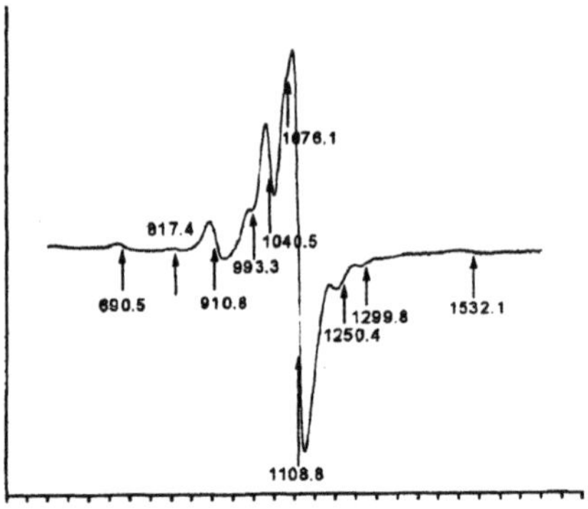

Figure 23. FMR spectrum of the film of Fig. 22, showing mode frequencies. After [17].

tization effects at the dot edge, whereas the spin-wave mode below it exhibits very strong variations in the magnetization over the entire dot area (with small dipolar effects at the dot edge). The higher-field satellite mode has a magnification uniform at the dot center, but exhibits strong variations near the edge due to a mix of dipolar and exchange couplings.

3.2 FMR Studies of Magnetic Antidot Lattices

Our preliminary FMR study of permalloy thin films patterned with various types of square *antidot* lattices has produced intriguing but ambiguous results. All patterned films had structure in their FMR spectra not present in unpatterned films, as shown in Fig. 24. Not all samples exhibited angular dependent FMR spectra, but those that did exhibited the four-fold rotational symmetry of the ADL, as shown in Fig. 27. In particular, there was no systematic relationship between the angular dependences and the ADL spacing. AFM investigations indicated that shadowing effects during deposition induced occasional thickness modulations (see Fig. 28) that could affect the magnetic anisotropy and FMR data. Further studies will be necessary to fully understand the roles of antidot shape and spacing on the FMR spectra.

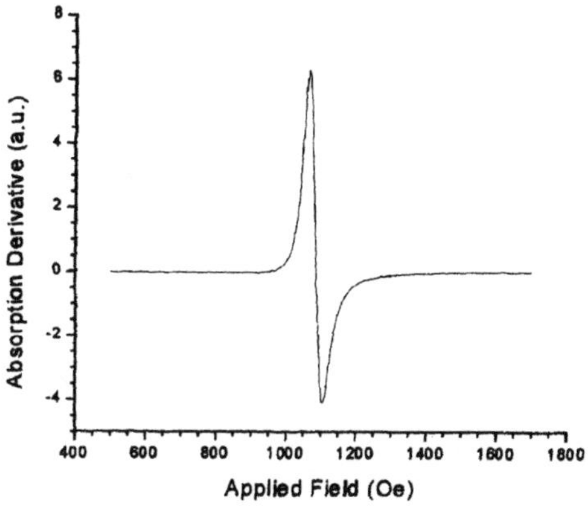

Figure 24. FMR spectrum of an unpatterned permalloy reference film of thickness t = 30 nm. After [17].

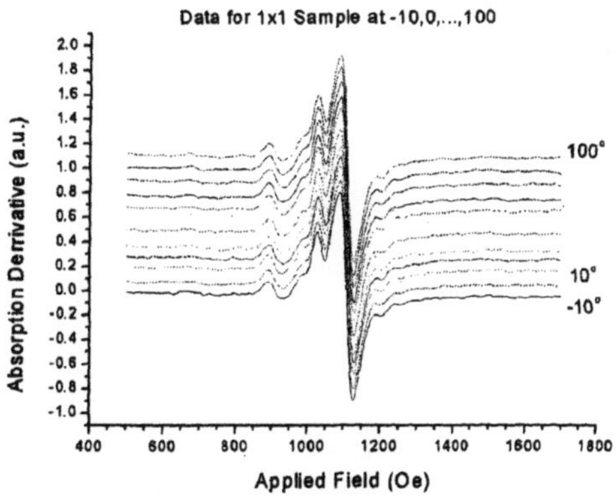

Figure 25. Figure 11d. FMR spectrum of the patterned film of Fig. 22 for various orientations of the applied magnetic field in the film plane. After [17].

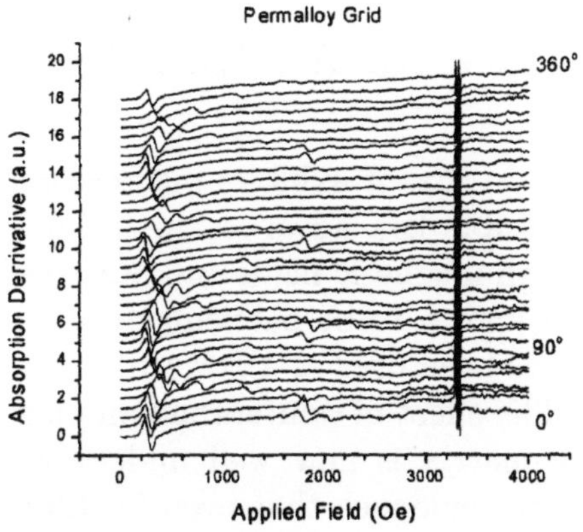

Figure 26. FMR spectra at various orientations of in-plane DC field for a permalloy film of 30 nm thickness patterned (EBL) with a square array of square antidots of 1.8 μm spacing and 0.8 μm width. After [17].

Figure 27. Schematic drawing (top-right) of antidot lattice topology (not to scale). After [17].

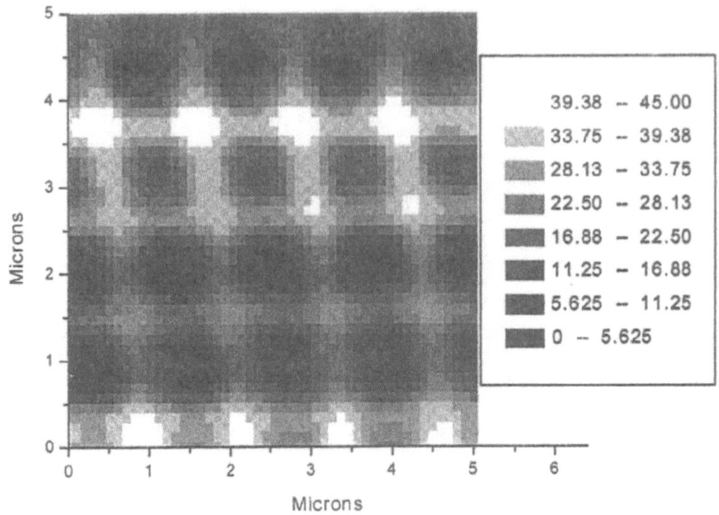

Figure 28. AFM topography of the film described in Fig. 26, showing thickness modulations (see gray scale in Å). After [17].

4. SUPERCONDUCTIVITY AS A PROBE OF MAGNETISM

The superconducting state is unique, in that it exhibits macroscopic quantum coherence that has made possible key advances in technology, including the production of very high magnetic fields and SQUIDs. Moreover, the superconducting state is known to be particularly sensitive to magnetic interactions [25], and has been used as a powerful probe of magnetism for many years [26–29]. The effect of dilute, random magnetic impurities in a superconducting matrix is now well understood [26, 27, 29]. However, there is significant evidence that even in dilute magnetic alloys the moments couple into small random clusters (either due to correlation effects between near-neighbor impurities resulting in glassy behavior, or due to clustering via phase segregation) [26, 28]. The exact mechanisms by which interacting magnetic clusters suppress superconductivity are not well understood [28].

Although the minimum feature size for a stable superconducting state is the zero-temperature coherence length ($\xi_o \approx 1 - 100$ nm), quantum confinement effects on flux quantization appear on the scale of the penetration depth ($\lambda \approx 100 - 1000$ nm), and this threshold can be enhanced even further ($\lambda_{eff} = \lambda^2/t$, where t = thickness) in thin film geometries. When superconductors are in intimate contact with a normal metal, "proximity effects" [25] can yield important information regarding the nature of both the interface and the electronic structure of the normal material. It is noteworthy that both ξ and λ diverge as T_c is approached from below [25], so that a superconducting film feature can serve as a probe of adjacent magnetic materials over a wide range of length scales. The full potential of studies of laterally confined superconductor/magnetic nanostructures has hardly been explored up to now. The effects of submicron MDL on a superconducting matrix film have just begun to be investigated by theorists (see Sec. 4.2) but there has been essentially no theoretical work exploring the effect of even smaller magnetic particles on superconductivity.

4.1 Flux Line Pinning in Patterned Superconducting Films

Early interest in patterned superconducting films stemmed from widespread use of artificial pinning centers to increase their maximum possible current capacity J_c, which is determined by the balance between Lorentz and pinning forces acting on flux lines (FL). Lorentz forces drive FL motion, which dissipates energy and destroys the perfect conductivity of the superconducting state. Isolated defects form local zones of reduced condensation energy that are efficient storage locations for FL (whose normal core diameter $\approx 2\xi$) within the superconducting condensate. These sites attract ("pin") FL and thereby oppose their motion and increase J_c.

The introduction of ordered lattices of antidots into superconducting thin films gives rise to extraordinary effects [30]. *Exact commensurability* between the pin array and the FL lattice (FLL) densities substantially enhances pinning and J_c at a precise matching field H_1 for which the equilibrium vortex density equals that of the pin array. At this special field, a "fluxoid" (consisting of *exactly one* flux quantum Φ_o) is generated by a supercurrent that circulates about each antidot. For a square ADL, $\mu_0 H_1 = \Phi_o/d^2$.

An antidot always attracts a FL when empty, but it has a finite capacity due to the gain in electromagnetic energy ($\propto H^2$) as the number n of flux quanta trapped by the antidot increases until it saturates at $n = n_s \approx \kappa D/4\lambda = D/4\xi(T)$, where the "saturation number" n_S indexes the highest possible matching field H_S [31, 32]. Therefore, any increase of the applied field in the range $H_1 < H < H_S$ will be accommodated by the creation of "multiquantum fluxoids" at the antidots. The precise spatial order of the pinning array (achieved by electron beam or laser interference lithography techniques) induces very sharp matching anomalies in M(H) and J_c (see Fig. 30). When H is increased just beyond H_S, single-Φ_o Abrikosov FL will be nucleated in the interstices between the antidots. These interstitial flux lines (IFL) are only weakly pinned, and their creation is accompanied by a sharp decrease in M(H) and J_c, and enhanced dissipation at $H = H_S$ [30–34].

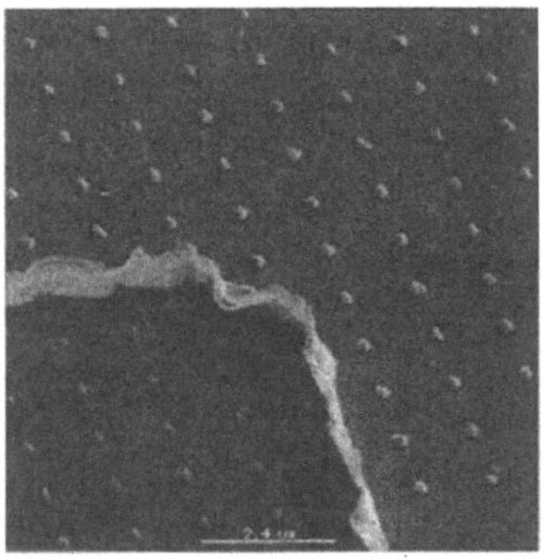

Figure 29. SEM picture of a square array of Ni dots embedded in a Nb film of thickness t = 95 nm and total width 80 μm. The Ni dots are cylindrical and of height 110 nm, with diameter D = 120 nm and spacing d = 2 μm. After [35].

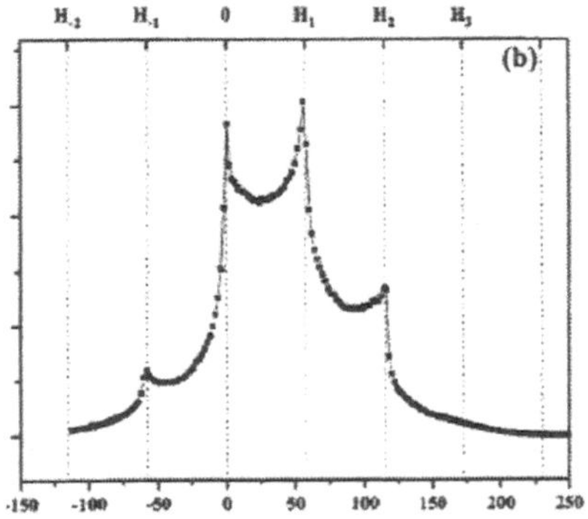

Figure 30. Critical current of a Nb film (T_c = 8.6 K) of thickness t = 95 nm with a triangular lattice of Ni dots of diameter D = 120 nm, as a function of applied perpendicular magnetic field at temperature T = 8.52 K. The upper legend indicates the calculated matching fields H_n = (57.5 Oe)n. After [36].

4.2 Magnetic Dot Lattices in Superconducting Matrix Films

Precision patterning of submicron MDL embedded in a matrix film offers highly controlled structures for systematic investigations of magnetic interactions within a variety of materials. The particular sensitivity of the properties of a superconducting film to MDL has been observed [35–43]. Recent electrical transport measurements (see Figs. 29, 30) of Nb films patterned with 3d transition element MDL [36, 41, 42] revealed enhancement and striking structure in the critical current density J_c(H) near $T_c \approx$ 8-9 K, but without any strong depression of the superconducting T_c! High sensitivity magnetic studies demonstrated that (see Fig. 31) the strong "matching" effects in **M(H)** and J_c(**H**) of Nb films

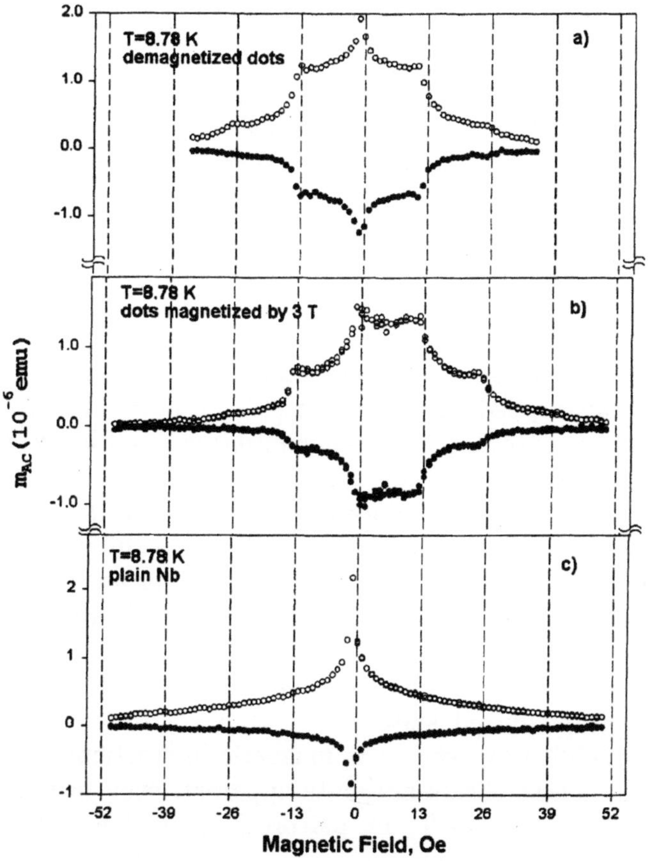

Figure 31. a) Real (filled circles) and imaginary (open circles) parts of the AC magnetic moment m_{AC} of a Nb film patterned with a demagnetized, square Ni MDL. Matching fields (vertical dashed lines) are symmetric with respect to zero applied DC magnetic field. b) Corresponding data for Ni dots magnetized by a perpendicular magnetic field H = 3 T. The second matching anomaly is enhanced in fields aligned with the remanent moment of the Ni dots, and is suppressed in the antialigned case. c) Corresponding data for an unpatterned Nb (larger area) reference film. After [43].

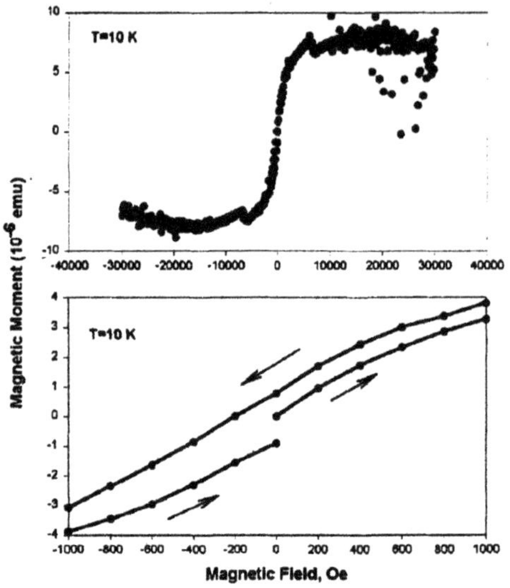

Figure 32. (Upper Panel) DC magnetic moment vs. applied magnetic field for a Nb film with a Ni MDL initially demagnetized at T = 10 K (above the Nb T_c = 8.8 K), after a linear substrate diamagnetic response is subtracted. The small cusp near the saturation field is similar to a domain reorientation effect observed in MOKE experiments on Fe films with a submicron ADL [13, 14]. (Lower Panel) Magnetic hysteresis near zero magnetic field. After [41].

with Ni MDL depend upon the orientation of the dot magnetization with respect to an applied magnetic field [43]. In particular, the *perpendicular component of remanent dot magnetization* (see Fig. 32) is responsible for observed asymmetries of the M(H) and J_c(H) data for Nb films patterned with lattices of Ni dots shown in Figs. 30 & 31 [36, 43]. The dot magnetization data shown in Fig. 32 also exhibit subtle cusps that are very similar to magnetic domain switching anomalies detected in the magneto-optical data of Vavassori et al. in Fe films patterned with *antidot* lattices [13, 14].

These initial results show that the stability and dynamics (e.g., FL pinning) of the mixed state of the superconducting matrix film are strongly modified by even weak magnetic interactions with *submicron-scale* dots. These effects are quite apparent, even in the absence of a significant reduction of T_c by pairbreaking, and would be *difficult to study by alternative means.*

Moreover, the behavior of Nb films patterned by MDL are very similar to a "magnetic pinning" effect investigated theoretically by Peeter's Antwerp Group [44]. The FL pinning by a magnetic dot is distinct from that of an empty AD, and is due to the additional gain in free energy attained when a magnetic fluxoid trapped at a MDL site possesses a moment parallel to the dot magnetization. Conversely, it is energetically unfavorable for a fluxoid pinned at a dot to have a polarization antiparallel to the dot magnetization. The net pinning potential is a result of the usual effect of an empty antidot (or normal state area displaced by the magnetic dot) in the superconductor [31, 32], plus the

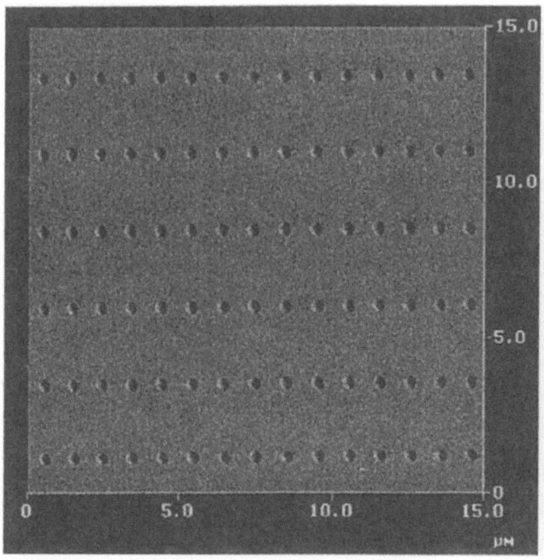

Figure 33. AFM image of a rectangular ADL (2.5/1 ratio) in a Nb film. The size scale is in μm. After [34].

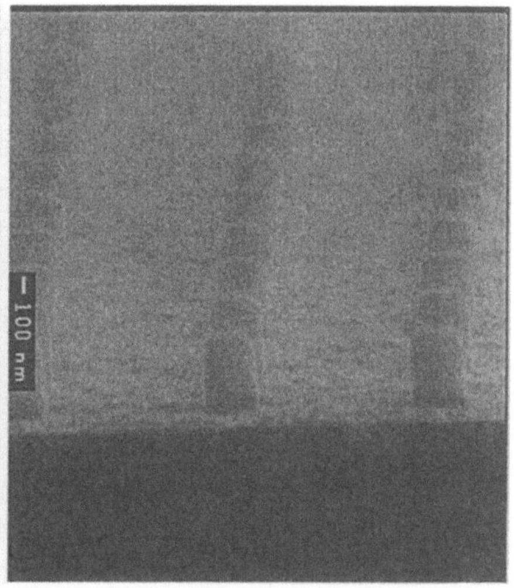

Figure 34. Figure 18. SEM image of rectangular array of photoresist pillars on Si/SiO_2 substrate. The sample is ready for the superconducting film deposition.

magnetic pinning of the fluxoid to the dot, and more subtle interactions between Abrikosov IFL and the fluxoid current distribution at IFL-dot separations $\approx \xi(\mathrm{T})$ [44, 45].

Much more needs to be done to build upon these initial results. Existing theories for the MDL have only addressed a single, isolated, "hard" (no shape anisotropy) magnetic dot in an infinite superconducting continuum, which clearly neglects the effects of interactions between magnetic dots. If $\lambda(T) > d$ (the MDL parameter), "collective pinning" effects due to fluxoid-fluxoid, fluxoid-IFL and IFL-IFL interactions are also expected to be relevant. A potentially strong proximity effect [25] between the magnetic dot material and the superconducting interstitial material may also play a role in the behavior of the system.

It is clear that films patterned with MDL present quite a large number of interesting scenarios for future investigation of the magnetic structure of small magnetic particles.

5. FABRICATION OF SMALL MAGNETIC FEATURES

Practical methods to fabricate patterned arrays of uniform magnetic nanostructures on size scales < 100 nm are essentially nonexistent at present if one demands the precision and control available in the ADL and MDL studies discussed above. New synthetic approaches must be investigated to develop methodologies to pattern novel nanoscale device structures on a highly reproducible basis. On the other hand, systematics gained from the study of highly controlled structures created via existing techniques is essential for understanding the more imperfect systems that are currently generated on size scales below 100 nm. We are currently using highly controlled, state-of-the-art methods for creating precise submicron film patterns, as described below.

5.1 Electron Beam Lithography

Systematic investigations of the mesoscale of magnetism can benefit from the controlled patterning of ordered lattices of antidots or magnetic dots embedded in high-quality thin films and multilayers. The simplest case of an ADL or MDL imbedded in a film can be created by a lift-off technique using electron beam lithography (EBL). The dot or antidot shape, diameter and separation can be systematically controlled down to dimensions $\approx$ 100 nm, and a typical separation between AD is d $\approx$ 1000 nm, and AD diameters are generally D $\approx$ 150-1000 nm. A Nb

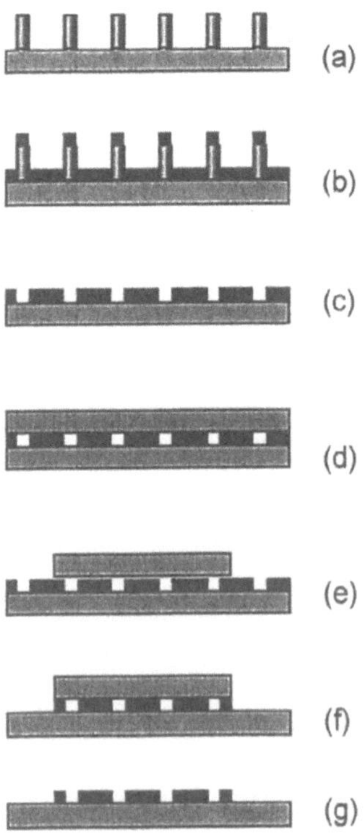

Figure 35. Steps are shown of a typical patterning of a thin film via electron beam lithography. (a) Resist pillars are put on Si/SiO_2 substrate via electron-beam writing (SEM image of pillars is shown in Fig. 34. (b) Deposition of thin superconducting film is done. (c) Lift-off of resist pillars produces an ADL in the film. (d) Film with ADL is covered with photoresist and prebaked. (e) Photoresist layer is patterned by UV lithography. (f) Using the patterned photoresist layer as an etch mask, bridges for transport measurements are etched in the patterned film. (g) Photoresist is stripped off.

Figure 36. "Stitching errors" between "writing fields" are visible as the broader spacing in a pattern of photoresist posts (500 nm spacing) patterned by the EBL technique. Typical write field sizes are of order 100 μm.

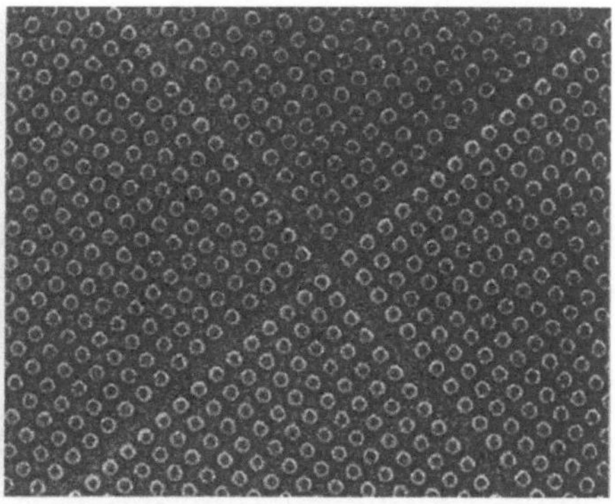

Figure 37. Close-up of the gap between write fields in Fig. 36.

film patterned with a rectangular ADL is shown in Fig. 33, and an SEM image of the lattice of electron beam resist pillars used to create the rectangular ADL is shown in Fig. 34. The steps involved in producing such a film are summarized in Fig. 35.

However, certain difficulties have prevented a wider use of EBL, including the moderate speed and high cost of equipment and software systems used to design and pattern the pinning arrays, and the size limitation on the "writing field", which is the maximum area over which small structures can be written without significant distortion. Magnetic measurements generally require film structures with critical dimensions in the range of 100-200 nm, but with overall areas of order ≈ 1 cm^2. Unfortunately, a limiting coefficient of approximately 1000 exists between the critical dimensions of the structures and the writing field, implying that in order to write 100 nm structures, a writing field of at most 100 x 100 μm^2 must be selected. Samples made on this scale are generally too small for standard magnetic susceptibility methods. This necessitates "multiple field" modes in which the sample stage of the electron beam writer is moved and "single-field blocks" are sequentially written side-by-side in order to create large-area samples. A major disadvantage of this method is the introduction of "stitching errors" at the boundaries between adjacent fields, as shown in Fig. 36 & 37. Unless carefully controlled, stitching errors lead to altered boundary conditions on current flow and field penetration in superconducting and magnetic patterned films. These imperfections make interpretation of data more difficult, and can lead to spurious effects. For example, "channel motion" of unpinned FL down the relatively wide margins between write fields can cause extra energy dissipation in the case of a superconducting film with a patterned ADL [30, 33, 34].

5.2 Laser Interference Lithography

Optical lithography is an older, but frequently used method for patterning thin films. However, it is limited in spatial resolution by the wavelength of light (of order 1 μm) used to expose photoresist. A state-of-art extension of optical lithography can be achieved by using short-

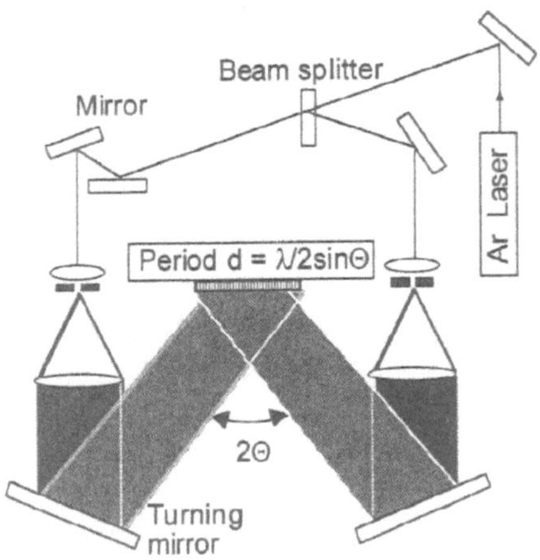

Figure 38. Laser Interference Lithography: A standing wave interference pattern of two laser beams produces alternating light and dark fringes with a spacing determined by their wavelength λ and the angle at which the beams intersect. By performing a second exposure of photoresist on a substrate, following a 90 degree rotation, a square pattern of dots of separation d is generated. This provides a quick, inexpensive method to produce periodic submicron structures without distortion over large areas (to 10 cm^2), and without a mask. After [20].

wavelength laser radiation and an interference technique (Fig. 38) that enables the fast production of large-area samples, and automatically eliminates stitching errors in the special case of periodic arrays of patterned features. We have used the interference method to successfully fabricate V and Nb thin films having ADL of 1 μm period and antidot diameters D = 0.10 - 0.4 μm [34, 46], and [Co4Å/Pt10Å]5 MDL with D = 300 nm [6], as illustrated in Fig. 40. We were able to reproducibly make structures as small as 100 nm after gentle oxygen ashing. Nevertheless, every patterning technique is subject to some disadvantages. Even an elegant and efficient method such as laser interference lithography suffers from unwanted modulations in surface topography due to interference patterns left in residual photoresist (Fig. 39).

Figure 39. SEM image of square lattice of photoresist posts on Si/SiO_2 substrate after developing. Note the 3-d interference "ripple" in the substrate plane and along post axes.

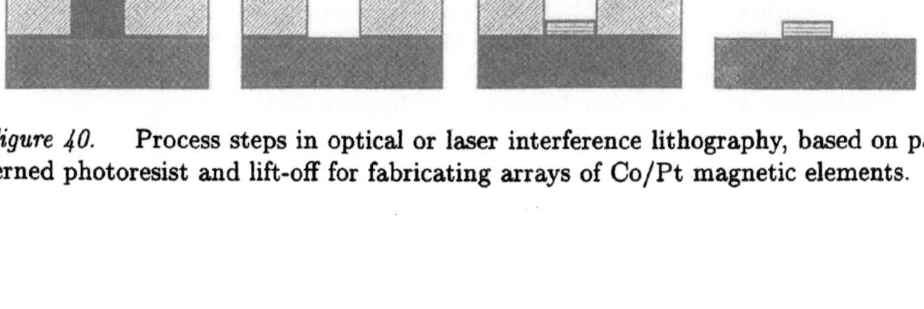

Figure 40. Process steps in optical or laser interference lithography, based on patterned photoresist and lift-off for fabricating arrays of Co/Pt magnetic elements.

Acknowledgments

Research at the University of Kentucky, University of Illinois-Chicago and Northwestern University was supported by U.S. Dept. of Energy Office of Science, Division of Materials Sciences Grant No. DE-FG02-97ER45653. Research at Northwestern was also supported by the U.S. Army Research Office.

The authors would like to acknowledge the many contributions to this research by their students and postdoctoral associates, including: J. Feller, S. Jung, D. J. Morgan, J. Sukkoo, A. N. Terentiev and D. B. Watkins. We also are grateful for the productive collaborations with colleagues and their groups at other institutions, including: S. D. Bader, G. W. Crabtree, M. Grimsditch, and U. Welp at Argonne National Laboratory; S. R. J. Brueck at University of New Mexico; R. D. Gomez at; P. J. Hesketh at University of Illinois-Chicago; and B. Terris of IBM Almaden Research Center.

The authors would like to thank Dr. Johannes Eisenmenger for assistance in the final preparation of the final manuscript.

References

[1] Mathieu, C., C. Hartmann, M. Bauer, O. Buettner, S. Riedling, B. Roos, S. O. Demokritov, and B. Hillebrands, "Anisotropic Magnetic Coupling of Permalloy Micron Dots Forming a Square Lattice", *Appl. Phys. Lett.* **70**, 2912 (1997).

[2] Kirk, K. J., J. N. Chapman and C. D. W. Wilkinson, "Switching Fields and Magnetostatic Interactions of Thin Film Magnetic Nano-Elements", *Appl. Phys. Lett.* **71**, 539 (1997).

[3] Dunin-Borkowski, R. E., M. R. McCartney, B. Kardynal, D. J. Smith, M. R. Scheinfein, "Switching Asymmetry in Closely Coupled Magnetic Nanostructure Arrays", *Appl. Phys. Lett.* **75**, 2641 (1999).

[4] Tsymbal, E. Y., "Theory of Magnetostatic Coupling in Thin-Film Rectangular Magnetic Elements", *Appl. Phys. Lett.*, **77**, 2740 (2000).

[5] Gomez, R. D., T. V. Luu, A. O. Pak, K. J. Kirk and J. N. Chapman, "Domain Configurations of Nanostructured Permalloy Elements", *J. Appl. Phys.* **85**, 6163 (1999).

[6] V. Metlushko, G. Crabtree, A. Berger, V. Vlasko-Vlasov, L. Rehn, M. Kirk, B. Ilic, B. Terris, S. Zhang, S. R. J. Brueck, Influence of Sub-Micron Patterning on Hysteresis Curves of Co/Pt Multilayers, The 8th Joint MMM-Intermag Conference Abstracts, San Antonio, TX, January 7-11, 2001

[7] Johnson, J. A., M. Grimsditch, V. Metlushko, P. Vavassori, B. Ilic, P. Neuzil, R. Kumar, "Magneto-Optic Kerr Effect Investigation of Cobalt and Permalloy Nanoscale Dot Arrays: Shape Effects on Magnetization Reversal", *Appl. Phys. Lett.* **77**, 44102 (2000).

[8] Koo, H., M. Dreyer, V. V. Metlushko and R. D. Gomez, "Systematic Investigation of the Magnetic Interactions of Small Permalloy Particles", submitted to *J. Appl. Phys.*

[9] Terris, B. D., L. Folks, D. Weller, J. E. E. Baglin, A. J. Kellock, H. Rothuizen and P. Vettiger, "Ion-Beam Patterning of Magnetic Films Using Stencil Masks", *Appl. Phys. Lett.* **75**, 403 (1999).

[10] Chappert, C., H. Bernas, J. Ferre, V. Kottler, J.-P. Jamet, Y. Chen, E. Cambril, T. Devolder, F. Rousseaux, V. Mathet, and H. Launois, "Planar Patterned Magnetic Media Obtained by Ion Irradiation, *Science* **280**, 1919 (1998).

[11] Thielen, M., S. Kirsch, H. Weinforth, A. Carl, and E. F. Wasserman, "Magnetization Reversal in Nanostructured Co/Pt Multilayer Dots and Films", *IEEE Trans. Mag.* **34**, 1009 (1998).

[12] Haast, M. A. M., J. R. Schuurhuis, L. Abelmann, J. C. Lodder, and Th. J. Popma, "Reversal Mechanism of Submicron Paterned CoNi/Pt Multilayers", *IEEE Trans. Mag.* **34**, 1006 (1998).

[13] Vavassori, P., V. Metlushko, M. Grimsditch, B. Ilic, P. Neuzil, R. Kumar, "Magneto-Optical Studies of Superlattice Dot Arrays", *Phys. Rev. B.* **61**, (2000) 5895.

[14] Vavassori, P., O. Donzelli, V. Metlushko, M. Grimsditch, B. Ilic, P. Neuzil and R. Kumar, "Magnetic Switching in Submicron-Scale Periodic Magnetic Arrays", *J.Appl. Phys.* **88**, 999 (2000).

[15] Koo, H., T. V. Luu, and R. D. Gomez and V. V. Metlushko, "Slow Magnetization Dynamics of Small Permalloy Islands", *J. Appl. Phys.* **87**, 5114 (2000).

[16] Torres, L., L. Lopez-Diaz and J. Iniguez, "Micromagnetic Tailoring of Periodic Antidot Permalloy Arrays for High Density Storage", *Appl. Phys. Lett.* **73**, 3766 (1998).

[17] Watkins, B., J. Feller, J. B. Ketterson, L. De Long, V. Metlushko, J. Sukkoo and V. Chandrasekhar, "Ferromagnetic Resonance of Permalloy Films with Patterned Array of Holes", *Bull. Am. Phys. Soc.* **46**, 182 (2001).

[18] Jung, S., B. Watkins, L. DeLong, J. B. Ketterson and V. Chandrasekhar, "Ferromagnetic Resonance in Periodic Particle Arrays", submitted to *Physical Review Letters* (2002).

[19] Martin, D. H., *Magnetism* (Cambridge, MA, M.I.T. Press, 1967); see Chapts. 1 & 7.

[20] Walker, L. R., "Spin Waves and Other Magnetic Modes", in: G. T. Rado and H. Suhl, *Magnetism* (Academic Press, New York, 1963), Vol. I, Chapt. 8, pp. 299-383.

[21] Kittel, C., *Introduction to Solid State Physics, 7th Edition* (JohnWiley and Sons, New York, 1996).

[22] Kittel, C., "Excitation of Spin Waves in a Ferromagnet by a Uniform rf Field", *Phys. Rev.* **110**, 1295 (1958).

[23] Voltairas, P. A., D. I. Fotiadis and C. V. Massalas, "Mode Mixing in Ferromagnetic Resonance in Magnetic Microspheres", *J. Appl. Phys.* **88**, 374 (2000).

[24] Arias R., and D. L. Mills, "Theory of Spin Excitations and the Microwave Response of Cylindrical Ferromagnetic Nanowires", *Phys. Rev. B* **63**, 134439 (2001).

[25] De Gennes, P. G., *Superconductivity of Metals and Alloys* (Benjamin, New York, 1966).

[26] Maple, M. B., "Paramagnetic Impurities in Superconductors", in: Rado, G. T. and Suhl, H., *Magnetism* (Academic Press, New York, 1973), Vol. V, Chapt. 10, pp. 289-325.

[27] Maple, M. B., "Superconductivity: A Probe of the Magnetic State of Local Moments in Metals", *Appl. Phys.* **9**, 179 (1976).

[28] Maple, M. B. and Fischer, Ø., Eds., *Superconductivity in Ternary Compounds II*, Springer-Verlag, Berlin, 1982); especially Chapt. 1; Maple, M. B., "Interplay between Superconductivity and Magnetism", *Physica B* **215**, 110 (1995).

[29] Maple, M. B., L. E. De Long and B.C. Sales, "Kondo Effect: Alloys and Compounds", in: K. A. Geschneidner, Jr. and L. Eyring, Eds., *Handbook on the Chemistry and Physics of Rare Earths* (North-Holland, Amsterdam, 1978), Vol. 1, Chapt. 11.

[30] Moshchalkov, V. V., M. Baert, V. V. Metlushko, E. Rosseel, M. J. Van Bael, K. Temst, Y. Bruynseraede and R. Jonckheere, "Quantum Design of Superconducting Critical Parameters: Fundamental Aspects", *J. de Physique IV Colloq.* **C3**, C3-335 (1996).

[31] Buzdin, A. I., "Multiple Quanta Vortices at Columnar Defects", *Phys. Rev. B* **47**, 11416 (1993).

[32] Mkrtchyan, G. S. and V. V. Shmidt, "Interaction between a Cavity and a Vortex in a Superconductor of the Second Kind", *Sov. Phys.: JETP* **34**, 195 (1972).

[33] Metlushko, V. V., L. E. De Long, M. Baert, E. Rosseel, M. J. van Bael, K. Temst, V. V. Moshchalkov and Y. Bruynseraede, "Supermatching Vortex Phases in Superconducting Thin Films with Antidot Lattices", *Europhys. Lett.* **41**, 333 (1998).

[34] Metlushko, V., U. Welp, G. Crabtree, R. Osgood, S. D. Bader, L. E. De Long, Zhao Zhang, S. R. J. Brueck, B. Ilic, K. Chung and P. J. Hesketh, "New Interstitial Flux Phases in Superconducting Niobium Film with Square Lattices of Artificial Pinning Centers", *Phys. Rev. B Rapid Commun.* **60**, R12585 (1999).

[35] Morgan, D. J., *Flux Pinning in Artificial Patterned Arrays*, Ph.D. Thesis, Northwestern University, 1997 (unpublished).

[36] Morgan, D. J. and J. B. Ketterson, "Asymmetric Flux Pinning in a Regular Array of Magnetic Dipoles", *Phys. Rev. Lett.* **80**, 3614 (1998).

[37] Geoffroy, O., D. Givord,and F. Ossart, "Magnetic and Transport Properties of Ferromagnetic Particulate Arrays Fabricated on Superconducting Thin Films", *J. Magn. Magn. Mater.* **121**, 223 (1993).

[38] Otani, Y., B. Pannetier, J. P. Nozieres, and D. Givord, "Magnetostatic Interactions between Magnetic Arrays and SuperconductingThin Films", *J. Magn. Magn. Mater.* **126**, 622 (1993).

[39] Van Bael, M. J., K. Temst, V. V. Moshchalkov and Y. Bruynseraede, "Magnetic Properties of Submicron Co Islands and their Use as Artificial Pinning Centers", *Phys. Rev. B* **59**, 14674 (1999).

[40] Otani, Y., Y. Nozaki, H. Miyajima, B. Pannetier, M. Ghidini, J. P. Nozieres, G. Fillion and P. Pugnat, "Magnetic Flux Penetration Process in Superconducting Nb Film Covered with Lithographic Array of Ferromagnetic Particles", *Physica C* **235-240**, 2945 (1994).

[41] Martin, J. I., M. Velez, J. Nogues and I. K. Schuller, "Flux Pinning in a Superconductor by an Array of Submicrometer Magnetic Dots", *Phys. Rev. Lett.* **79**, 1929 (1997).

[42] Martin, J. I., M. Velez, A. Hoffmann, I. K. Schuller and J. L. Vincent, "Artificially Induced Reconfiguration of the Vortex Lattice by Arrays of Magnetic Dots", *Phys. Rev. Lett.* **83**, 1022 (1999).

[43] Terentiev, A., D. B. Watkins, L. E. De Long, D. Morgan and J. Ketterson, "Observation of Magnetic Flux Pinning in a Thin Nb Film with a Square Lattice of Nickel Dots", *Physica C* **324**, 1 (1999).

[44] Marmorkos, I. K., A. Matulis and F. M. Peeters, "Vortex Structure around a Magnetic Dot in Planar Superconductors", *Phys. Rev. B* **53**, 2677 (1996).

[45] Priour, D. J. Jr., and H. A. Fertig, "Core Strings and Flux Spreading near Pinning Centers", submitted to *Physical Review Letters* (2002)

[46] Metlushko, V., U. Welp, G. W. Crabtree, Zhao Zhang, S. R. J. Brueck, B. Watkins, L. E. De Long, B. Ilic, K. Chung and P. J. Hesketh, "Non-Linear Flux Line Dynamics in Vanadium Films with Square Lattices of Submicron Holes", *Phys. Rev. B* **59**, 603 (1999).

Part 2

Phenomenological Studies of Correlation Effects

SPECTRAL DENSITY FUNCTIONAL APPROACH TO ELECTRONIC CORRELATIONS AND MAGNETISM IN CRYSTALS

A. I. Lichtenstein
NSRIM, University of Nijmegen
NL-6525 ED Nijmegen, The Netherlands
A.Lichtenstein@sci.kun.nl

M. I. Katsnelson
Institute of Metal Physics
620219 Ekaterinburg , Russia
Mikhail.Katsnelson@usu.ru

G. Kotliar
Serin Physics Laboratory, Rutgers University
Piscataway, New Jersey 08855, USA
kotliar@physics.rutgers.edu

Abstract A novel approach to electronic correlations and magnetism of crystals based on the realistic electronic structure calculations is reviewed. In its simplest form it is a combination of the "local density approximation" (LDA) and the dynamical mean field theory (DMFT) approaches. Using numerically exact QMC solution of effective DMFT multi-orbital quantum-impurity problem, a successful description of the electronic structure and finite temperature magnetism of transition metals has been obtained. We discuss a simplified perturbation LDA+DMFT scheme which combined the T-matrix and fluctuation-exchange approximation (TM-FLEX). We end with a discussion of cluster generalization of the non-local DMFT scheme and its applications to the magnetism and superconductivity of the high-T_c superconductors.

Electron Correlations and Materials Properties 2
Edited by Gonis *et al.*, Kluwer Academic/Plenum Publishers, 2002

1. Introduction

The theory of electronic structure and magnetism of solids historically was split into two distinct parts, namely, the model investigations of many-body effects and the calculations of the energy spectra and properties of specific compounds in the framework of density functional (DF) scheme [1, 2]. Recently, within the dynamical mean-field theory (DMFT, for a review see Ref. [3]) the correlation effects have been incorporated into realistic electronic structure calculations [4, 5, 6, 7, 8, 9]. This method has been successfully applied to a number of classical problems of solid state physics such as the finite-temperature magnetism of iron-group metals [10], $\alpha - \delta$ transition in plutonium [11], electronic structure of doped Mott insulators [9]. In contrast with standard DF theory, in this new approach known as "LDA+DMFT" [4] or "LDA++" [5] the total energy of the system (or, more accurately, thermodynamic potential Ω) is considered as a functional of the Green function instead of the denstity matrix [7, 8, 12, 13]. To stress this new feature more explicitely we will use the term "spectral density functional (SDF)". The analytical properties of the Green function garantee that the knowlegde of the spectral density is equivalent to the knowledge of the time-dependent Green function whereas the density matrix is just static value of the latter [14]. Here we will describe the basic ideas of SDF method, both in the framework of a standard DMFT and from a more general point of view, discuss possible cluster generalizations of the DMFT and consider the applications of SDF to the correlation effects and magnetism in transition metals.

2. Dynamical Mean Field Theory: an effective action perspective

A most economical approach to unifying the various dynamical mean field approximations in use, is provided by the effective action construction [15]. The idea is to select a set of variables which is relevant to the physics of the problem and to write down a functional of the relevant variables. The extremum of this functional yields the values of those variables in equilibrium and the value of this functional at stationarity gives the free energy of the system in equilibrium.

Density functional theory is the simplest example of this construction [16], here the total energy of the solid is expressed in terms of the density of the electrons. Another well tested example is the spin density functional theory in which the total energy is expressed in terms of the spin and charge densities. The construction of an explicit expression of the exact effective action functional is usually not available except for the case of very simple examples and the success of the method relies on the availability of good approximations to this functional. The LDA and the LSDA approaches have been extraordinarily successful for weakly correlated systems.

The dynamical mean field approach to model Hamiltonians of strongly correlated systems on a lattice can also be brought to such an effective action perspective. More important this construction can be easily generalized to incorporate the so called extended dynamical mean field approach (E-DMFT) [17], by constructing an effective action for both the local Greens function and the local density density and local spin spin autocorrelation function. The effective action approach also allowed the formulation of the E-DMFT approach in the case that spatial and spin or charge symmetries are broken. [8, 18].

The effective action approach, allows us a simple combination of the density functional theory and the DMFT. By Legendre transformation techniques one can construct a functional of the density and of the local spectral function of the heavy orbitals whose extremization would yield the exact density local spectral function of the heavy orbitals, and the total energy of the system. Again the exact form of this functional are not known in explicitly form, but useful approximations to it are available and can be used to study interesting problems.

This realistic DMFT, LDA+DMFT approach or LDA++ approach, was first implemented ignoring the coupling between the density and the local spectral function. This amounts to performing first a LDA calculation to derive a tight binding model Hamiltonian and then performing a DMFT calculation for the spectra of the DMFT Hamiltonian. This is very close in spirit to the philosphy of model Hamiltonian calculations. Recently a full implementation of the self consistent determination of the density and the local spectra was carried out by S. Savrasov [11], this is now closer in spirit to traditional first principles electronic structure calculations.

The effective action approach, can be generalized to clusters, if short range correlations need to be taken into account. In the context of model Hamiltonians, a functional of the restriction of the Green function to a given cluster is defined. The extremization of this functional give rise to cluster Dynamical mean field equations. There is no difficulty in constructing hybrid funcionals for the self consistent determination of both the density and the cluster Greens functions, namely a CDMFT+LDA method.

In the next section we motivate the approximate DMFT form of those functionals from a perspective of a reduced fermionic description.

3. Fermionic reduced description and dynamical mean-field theory

In the SDF approach two electron subsytems are introduced, one of them is described by the standard DF theory (usually this is the subsystem of *sp*-electrons) and dynamical interelectron correlations are taken into account for the another one (usually this is the subsystem of *d*- or *f*-electrons). Further simplifications can be connected with the local form of interelectron interactions (only Hubbard-type on-site electron correlations are considered), using some approximations like DMFT or its cluster generalizations [21, 22, 23], etc. From a general point of view all these approaches can be considered as specific cases of a "coarse graining" (reduced description) ideology [24] when all the variables describing the system can be separated into the "gross" and "slave" variables; the only assumption that a closed set of the equations of motion for the gross variables exist is sufficient to find an explicit form of these equations [25]. Here we demonstrate the coarse graining procedure for fermionic degrees of freedom for the system of interacting electrons in a crystal.

Let us start with the functional integral over the Grassman creation and annihilation electron fields c^+, c, where the "measure" is $\exp(-S)$,

$$S = -\int dx \int_0^\beta d\tau \left[c^+(x,\tau)\left(\frac{\partial}{\partial\tau} + \mu - H_0\right)c(x,\tau)\right] + S_{int}\left[c^+, c\right] \quad (1)$$

Here μ is the chemical potential, $\beta = 1/T$ is the inverse temperature, H_0 is a "one-particle" part of the Hamiltonian. In the SDF approach

$$H_0 = -\nabla^2/2 + V_{KS} \quad (2)$$

where V_{KS} is the Kohn-Sham self-consistent potential [1].

In spirit of a reduced description approach [24] we introduce "gross" variables $d = f \cdot c$ (more explicitly, $d_\mu = \sum_j f_{\mu j} c_j$ where j are site indices; e.g., in a standard DMFT $f_{\mu j} = \delta_{\mu 0}\delta_{j0}$). The effective action for the gross variables S_{eff} is defined by introducing a delta-functional $\delta(d - f \cdot c)\,\delta(d^+ - f^* \cdot c^+)$ [26] into the functional integral.

$$\exp\left(-S_{eff}\right) = \frac{1}{Z}\int DcDc^{+}\exp\left(-S\right) *$$

$$\int D\lambda D\lambda^{+}\exp\left[i\lambda\left(d^{+} - f^{*}\cdot c^{+}\right) + i\lambda^{+}\left(d - f\cdot c\right)\right] \qquad (3)$$

Unfortunately, this functional integral cannot be done exactly and approximations are needed. Here we consider only the simplest approximation to the derivation of the effective action, we replace $S_{int}\left[c^{+}, c\right] \rightarrow S_{int}^{CG}\left[d^{+}, d\right]$ but compensate for the omission of the interaction terms away from those in the gross variables, namely in the medium by adding a medium self energy in the standard DMFT

$$S \rightarrow \widetilde{S} = S + \sum_{i\geq 0}\int_{0}^{\beta}\int_{0}^{\beta} d\tau d\tau' c^{+}\left(\tau, i\right)\Sigma\left(\tau - \tau', i, i\right)c\left(\tau', i\right) \qquad (4)$$

In other words, the DMFT approach treats exactly the interaction terms *only in* those part of the interaction Lagrangian which can be written in terms of gross variables, while *the rest of the interaction terms are handled in a gaussian approximation* by replacing the interaction terms by a medium self energy.

Now the effective measure for the gross variables can be found as

$$\exp\left(-S_{eff}\right) = \frac{1}{Z}\int DcDc^{+}\exp\left(-\widetilde{S}_{CG} + d^{+}\Sigma d\right) *$$

$$\int D\lambda D\lambda^{+}\exp\left[i\lambda\left(d^{+} - f^{*}\cdot c^{+}\right) + i\lambda^{+}\left(d - f\cdot c\right)\right] \qquad (5)$$

where $Z = \int DcDc^{+}\exp\left(-\widetilde{S}_{CG}\right), \widetilde{S}_{CG} = S_{med} + S_{int}^{CG}\left[d^{+}, d\right].$

Passing to the Matsubara frequencies [14] one has

$$S_{med} = -T\sum_{\omega_{n}j} c^{+}\left(i\omega_{n}j\right)\left[i\omega_{n} + \mu - H_{0} - \Sigma\left(i\omega_{n}\right)\right]c\left(i\omega_{n}j\right) \qquad (6)$$

In the standard DMFT scheme the a correlations between $d(f)$-electrons on a central site $j = 0$ is treated exactly while the correlations in the medium is treated in the Gaussian approximation. It is sometimes stated in the literature that this assumption violates the translational invariance of the system. The effective action approach, is meant to construct a functional of the selected variables, so by construction it is not possible to discuss within this framework the issue of translation invariance. The "self-consistency condition" does the best possible job within a Gaussian approximation to restore the equivalence of the central site and the medium.

We show below that this reasoning in a cluster setting leads to the "cellular DMFT" approach [23]. In principle, the correlations between the gross and slave variables (e.g., the long-range part of the Coulomb interaction) also can be taken into account in a close analogy with the classical spin models [24]; the corresponding modification of the DMFT approach will be considered elsewhere. In accordance with the general scheme of the non-equilibriun statistical operator method [25], we add an "auxiliary field" conjugated to the Green function of the gross variables, namely, the term $d^{+}\bar{\Sigma}d$. These fields will restore the translational invariance for the fermionic Green function (see below, Eq.(11)).

Calculating the Gaussian integral over c, c^{+} one has

$$\exp\left(-S_{eff}\right) = \exp\left(-S_{int}^{CG}\left[d^{+}, d\right] + d^{+}\bar{\Sigma}d\right) * \\ \int D\lambda D\lambda^{+} \exp\left[i\lambda d^{+} + i\lambda^{+}d - \lambda^{+}f \cdot G_{med} \cdot f^{*}\lambda\right] \tag{7}$$

where the medium Green functions reads

$$G_{med} = \left[i\omega_n + \mu - H_0 - \Sigma\left(i\omega_n\right)\right]^{-1}. \tag{8}$$

At last, calculating the Gaussian integral over λ, λ^{+} we obtain

$$S_{eff}\left[d^{+}, d\right] = S_{int}^{CG}\left[d^{+}, d\right] - T\sum_{\omega_n} d^{+}\left(i\omega_n\right)\mathcal{G}_0^{-1}\left(i\omega_n\right)d\left(i\omega_n\right) \tag{9}$$

where

$$\mathcal{G}_0^{-1}\left(i\omega_n\right) = \left(f \cdot G_{med} \cdot f^{*}\right)^{-1} + \bar{\Sigma} \tag{10}$$

The self consistency condition that determinesΣ requires the equality of the local green function computed from the reduced description and from the medium agree

$$\langle Td(\tau)d^{+}(\tau')\rangle_S = \langle Td(\tau)d^{+}(\tau')\rangle_{S_{eff}} \tag{11}$$

Further, $\langle Tc(\tau)c^{+}(\tau')\rangle_S = G_{med}$ (because of the replacement $S_{int}\left[c^{+}, c\right] \to S_{int}^{CG}\left[d^{+}, d\right]$) and

$$\langle Td(\tau)d^{+}(\tau')\rangle_S = f \cdot G_{med} \cdot f^{*} \tag{12}$$

On the other hand, $\langle Td(\tau)d^{+}(\tau')\rangle_{S_{eff}} = \mathcal{G}$ satisfies the Dyson equation

$$\mathcal{G}^{-1} = \mathcal{G}_0^{-1} - \Sigma_{CG} = \left(f \cdot G_{med} \cdot f^{*}\right)^{-1} + \bar{\Sigma} - \Sigma_{CG} \tag{13}$$

and we have $\bar{\Sigma} = \Sigma_{CG}$ where Σ_{CG} is a "coarse-grained" self-energy. The most natural choise is $\Sigma_{CG} = f^* \cdot \Sigma \cdot f$.

To obtain the main equation of the DMFT [3] we have to choise f as a projection operator on the central site. In this case Eq.(13) will take the desired form

$$\mathcal{G}_0^{-1}(i\omega_n) = \left[\sum_{\mathbf{k}} \frac{1}{i\omega_n + \mu - t(\mathbf{k}) - \Sigma(i\omega_n)}\right]^{-1} + \Sigma(i\omega_n) \tag{14}$$

where the quasimomentum $\mathbf{k}$ runs the Brillouine zone, $t(\mathbf{k})$ is the Fourier transform of the Hamiltonian H_0 projected into the subspace of $d(f)$-electrons.

To consider posiible cluster generalizations one can choose let $f = PU^+P$ where P is the projection operator on the cluster and U is an unitary transformation of the variables in this cluster. Thus our result reads:

$$\mathcal{G}_0^{-1}(i\omega_n) = \left[\left(\frac{L_c}{L}\right)^d \sum_{\mathbf{k}_c} U^+ \frac{1}{i\omega_n + \mu - t(\mathbf{k}_c) - \Sigma} U\right]^{-1} + U\Sigma U^+ \tag{15}$$

where $\mathbf{k}_c$ runs the "new" supercell Brillouin zone, L and L_c are the sizes of the crystal and cluster, correspondingly. In terms of $\Sigma_{CG} = U\Sigma U^+$,

$$\mathcal{G}_0^{-1}(i\omega_n) = \left[\left(\frac{L_c}{L}\right)^d \sum_{\mathbf{k}_c} \frac{1}{i\omega_n + \mu - Ut(\mathbf{k}_c)U^+ - \Sigma_{CG}}\right]^{-1} + \Sigma_{CG} \tag{16}$$

As it was shown in Ref. [23] this equation for general basis set can be very useful in the optimisation of interaction problem within the cluster or cellular DMFT scheme. Another version of the cluster DMFT [22] will be described below.

4. Spectral density versus density functionals

In a standard DF theory the thermodynamic potential for non-correlated conduction-"c" electrons Ω^c is represented as a functional of the electron density $\rho(\mathbf{r})$ which is, generally speaking, a matrix in spin indices. Formally it can be represented as a thermodynamic potential of the Kohn-Sham quasiparticles [1], Ω_{sp}, minus the contribution of the so called "double counted" terms, Ω_{dc}:

$$\begin{aligned}
\Omega^c &= \Omega^c_{sp} - \Omega^c_{dc} \\
\Omega^c_{sp} &= -Tr\log[i\omega + \nabla^2/2 - V_{KS}] \qquad (17)\\
\Omega^c_{dc} &= \int V_{KS}(\mathbf{r})\rho(\mathbf{r})d\mathbf{r} - \int V_{ext}(\mathbf{r})\rho(\mathbf{r})d\mathbf{r} - \frac{1}{2}\int \frac{\rho(\mathbf{r})\rho(\mathbf{r}')}{|\mathbf{r}-\mathbf{r}'|}d\mathbf{r}d\mathbf{r}' - E_{xc}[\rho]
\end{aligned}$$

where $Tr = Tr_{\omega iL\sigma}, Tr_\omega$ is the sum over Matsubara frequencies $Tr_\omega ... = T\sum_\omega e^{i\omega 0^+} ...$, $\omega = \pi T(2n+1)$, $n = 0, \pm 1, ...$, T is the temperature, and $iL\sigma$ are site numbers (i), orbital quantum numbers ($L = l, m$) and spin projections σ, correspondingly, $V_{ext}(\mathbf{r})$ is the external potential, $E_{xc}[\rho]$ is the exchange-correlation energy and the Kohn-Sham effective potential is defined as

$$V_{KS}(\mathbf{r}) = V_{ext}(\mathbf{r}) + \int \frac{\rho(\mathbf{r}')}{|\mathbf{r}-\mathbf{r}'|} d\mathbf{r}' + \frac{\delta E_{xc}[\rho]}{\delta\rho(\mathbf{r})}. \tag{18}$$

In contrast with the standard density functional theory, the SDF approach deals with the real dynamical quasiparticles for correlated "d-electrons" defined via local Green functions rather than with Kohn-Sham "quasiparticles" which are, strictly speaking, only auxiliary states to calculate the total energy. Therefore, instead of the working with the thermodynamic potential Ω as a *density* functional we have to start from its general expression in terms of an exact Green function [27]

$$\begin{aligned} \Omega^d &= \Omega^d_{sp} - \Omega^d_{dc} \\ \Omega^d_{sp} &= -Tr\left\{\ln\left[\Sigma - G_0^{-1}\right]\right\} \\ \Omega^d_{dc} &= Tr\Sigma G - \Phi \end{aligned} \tag{19}$$

where G, G_0 and Σ are an exact Green function, its bare value and self-energy, Φ is the Luttinger generating functional (sum of the all connected skeleton diagrams without free legs), correspondingly. A complete SDF thermodynamic potential is equal to $\Omega = \Omega^c + \Omega^d$. We have to keep in mind also Dyson equation

$$G^{-1} = G_0^{-1} - \Sigma \tag{20}$$

and the variational identity

$$\Sigma = \frac{\delta\Phi}{\delta G}. \tag{21}$$

When neglecting the quasiparticle damping, Ω_{sp} will be nothing but the thermodynamic potential of "free" fermions but with exact quasiparticle energies. Formal analogy between Eqs.(17) and (19), (18) and (21) are obvious: the self-energy plays the role of the Kohn-Sham potential (without the external potential) and the Green function plays the role of the density matrix. As an example of this correspondence one can prove [7] in the framework of the SDF an useful identity known as the "local force theorem" basically in the same way as it has been done in the DF theory [28, 29].

For both parts of the thermodynamic potential, Ω^c and Ω^d, the local force theorem is based on extremum properties with respect to the variation of the density matrix and Green function, correspondingly. Further we will consider the contribution Ω^d (omitting the index d for brevity). In principle, fermionic reduced description scheme allows us to combined conduction "c" and correlated "d" part of the total SDF: $\Omega = \Omega^c + \Omega^d$ [13].

5. Effective exchange interactions

Let us discuss the problem of calculation of effective exchange interactions (J_{ij}) in correlated systems. In principle the J_{ij} parameters are not well defined for arbitrary magnetic systems, and the traditional way to study spin excitations related to the calculation of non-local freqency dependent spin-susceptibility [3, 31]. In this case the auxiliary space-time depended magnetic field is added to the the Hamiltonian: $\sigma\mathbf{h}(\mathbf{r},\tau)$ and the second derivative of free-energy with respect to magnetic field gives the interacting spin-susceptibility: $\chi^{-1} = \chi_0^{-1} - \Gamma$, where χ_0 is an empy-loop susceptibility and Γ is the vertex corrections [3, 31]. Here we consider a simple approximation of "rigid spin rotation" of spectral density for a small angle:

$$\delta\mathbf{e}_i = \delta\boldsymbol{\varphi}_i \times \mathbf{e}_i \tag{22}$$

here $\mathbf{e}_i$ is a general direction of constrained effective spin-dependent potential on site i and $\delta\boldsymbol{\varphi}_i$ is a rotation vector. In this case it is useful to write explicitly the spinor structure of the self-energy and Green functions:

$$\begin{aligned} \Sigma_i &= \Sigma_i^c + \mathbf{\Sigma}_i^s\sigma \\ G_{ij} &= G_{ij}^c + \mathbf{G}_{ij}^s\sigma \end{aligned} \tag{23}$$

where $\Sigma_i^{(c,s)} = \frac{1}{2}\left(\Sigma_i^{\uparrow} \pm \Sigma_i^{\downarrow}\right)$, $\mathbf{\Sigma}_i^s = \Sigma_i^s\mathbf{e}_i$, $\sigma = (\sigma_x, \sigma_y, \sigma_z)$ are Pauli matrices, $G_{ij}^c = \frac{1}{2}Tr_\sigma(G_{ij})$ and $\mathbf{G}_{ij}^s = \frac{1}{2}Tr_\sigma(G_{ij}\sigma)$. We suppose that the bare Green function G^0 does not depend on spin directions and all the spin-dependent terms including the Hartree-Fock terms are incorporated in the self energy. In the rigid spin approximation we assume that the unit vector $\mathbf{e}_i$ does not depend on the energy and orbital indices and represents the direction of the average local magnetic moment on the site i. Note, that thermodinamic potential Ω should be considered as a constrained SDF which depends on $\mathbf{e}_i$ as on external parameters (cf.

Ref. [32]). Then the variation of the thermodynamic potential with respect to small spin-rotation can be written as

$$\delta\Omega = \delta^*\Omega_{sp} + \delta_1\Omega_{sp} - \delta\Omega_{dc} \tag{24}$$

where δ^* is the variation without taking into account the change of the "self-consistent potential" (i.e. self energy) and δ_1 is the variation due to this change of Σ. Taking into account Eq. (21) it can be easily shown (cf. Ref. [27]) that

$$\delta_1\Omega_{sp} = \delta\Omega_{dc} = TrG\delta\Sigma \tag{25}$$

and hence

$$\delta\Omega = \delta^*\Omega_{sp} = -\delta^* Tr \ln\left[\Sigma - G_0^{-1}\right] \tag{26}$$

which is an analog of the "local force theorem" in the density functional theory [29].

In the case of rigid spin rotation the corresponding variation of the thermodynamic potential can be written as

$$\delta\Omega = \mathbf{V}_i \delta\varphi_i \tag{27}$$

where the torque $\mathbf{V}_i$ is equal to

$$\mathbf{V}_i = 2Tr_{\omega L}\left[\mathbf{\Sigma}_i^s \times \mathbf{G}_{ii}^s\right] \tag{28}$$

Basing on the expansion of this expression (28) in a sum of pairwise contributions one can obtain [7] useful formula for the effective magnetic interactions:

$$J_{ij} = -Tr_{\omega L}\left(\Sigma_i^s G_{ij}^{\uparrow} \Sigma_j^s G_{ji}^{\downarrow}\right) \tag{29}$$

and, correspondingly, for the stiffness tensor of a ferromagnet:

$$D_{\alpha\beta} = -\frac{2}{M} Tr_{\omega L} \sum_{\mathbf{k}} \left(\Sigma^s \frac{\partial G^{\uparrow}(\mathbf{k})}{\partial k_\alpha} \Sigma^s \frac{\partial G^{\downarrow}(\mathbf{k})}{\partial k_\beta}\right) \tag{30}$$

where M is the magnetic moment per unit cell. These results generalize the LSDA expressions of Ref. [29] to the case of correlated systems. Note that passing from Eq.(28) to Eq. (29) is not accurate, since the exchange parameters are connected with the second variations of the Ω-potential and use of the local force theorem can not be justified. Eq. (29) correspond to the "empty loop" approximation neglecting the vertex corrections. At the same time, for the stiffness the latter are absent and Eq.(30) appears to be exact provided that the self-energy and three-leg vertex are local (as in the DMFT) [12].

The fact that vertex corrections to the spin stiffness are absent within DMFT, is also suggested by the analogy between electric and spin transport developed in reference [19]. The spin wave stiffness can be obtained from the zero frequency limit of the spin conductivity. Within DMFT , the charge current current correlation function does not require vertex corrections [20] and can therefore be obained directly from the convolution of the one electron spectra. These arguments are independent of the spin strucure, and can therefore be used for the spin wave stiffness.

In order to elucidate the approximation behind the expression for the exchange parameters (Eq. (29)), we consider the energy of a spiral magnetic configuration with the rigid rotation of the spinor electron operators by the polar angles θ and φ:

$$c_{im} \to U(\theta_i, \varphi_i)\, c_{im}$$

where

$$U(\theta,\varphi) = \begin{pmatrix} \cos\theta/2 & \sin\theta/2\exp(-i\varphi) \\ -\sin\theta/2\exp(i\varphi) & \cos\theta/2 \end{pmatrix} \tag{31}$$

assuming that $\theta_i = \text{const}$ and $\varphi_i = \mathbf{qR_i}$ where $\mathbf{R_i}$ is the site lattice vector. Since we take into accounts only on-site correlation effects the interaction term in the Hamiltonian is invariant under that transformation and the change of the Hamiltonian is

$$\begin{aligned} \delta H &= \sum_{ij} Tr_{m\sigma}\left[t_{ij}c_i^+\left(U_i^+U_j - 1\right)c_j\right] = \delta_1 H + \delta_2 H \\ \delta_1 H &= \sin^2\tfrac{\theta}{2}\sum_k Tr_{m\sigma}\left[\left(t(\mathbf{k+q}) - t(\mathbf{k})\right)c_{\mathbf{k}}^+c_{\mathbf{k}}\right] \\ \delta_2 H &= \tfrac{1}{2}\sin\theta\sum_{ij} Tr_m\left[t_{ij}c_{i\downarrow}^+c_{j\uparrow}\right]\left(\exp(i\mathbf{qR}_i) - \exp\left(i\mathbf{qR}_j\right)\right) \end{aligned} \tag{32}$$

Consider further the case of small θ, we can calculate change of the total energy in the lowest order in θ corresponds to the first order in $\delta_1 H$ and the second order in $\delta_2 H$:

$$\begin{aligned} \delta E &= \frac{\theta^2}{4}\{\sum_k [t(\mathbf{k+q}) - t(\mathbf{k})]n_k - iTr_m\int\frac{d^4k}{(2\pi)^4}[t(\mathbf{k+q}) - t(\mathbf{k})] * \\ &\quad \gamma(k,q)\, G_\downarrow(k+q)\,[t(\mathbf{k+q}) - t(\mathbf{k})]G_\uparrow(k)\}, \end{aligned}$$

where $n_k = Tr_{m\sigma}\left\langle c_k^+ c_k\right\rangle$, q is four-vector with component $(\mathbf{q},0)$, and γ is the three-leg vertex. Our main approximation is neglecting of the vertex corrections ($\gamma = 1$). In this case the previous equation takes the following form:

$$\delta E = -\frac{\theta^2}{4}Tr_m\{i\int\frac{d^4k}{(2\pi)^4}[t(\mathbf{k+q}) - t(\mathbf{k})] * \tag{33}$$

$$G_{\downarrow}(\mathbf{k}+\mathbf{q})\left[G_{\downarrow}^{-1}(k+q)-G_{\uparrow}^{-1}(k)+t(\mathbf{k}+\mathbf{q})-t(\mathbf{k})\right]G_{\uparrow}(k)\}$$

Using the following consequence of the Dyson equation:

$$t(\mathbf{k}+\mathbf{q})-t(\mathbf{k})=G_{\uparrow}^{-1}(k)-G_{\downarrow}^{-1}(k+q)+\Sigma_{\uparrow}(E)-\Sigma_{\downarrow}(E) \tag{34}$$

one can rewrite the Eq. (33) in the form: $\delta E=\frac{\theta^2}{4}[J(0)-J(q)]$ with the exchange integrals corresponding to Eq. (29). We conclude that the expression for J_{ij} is accurate if the vertex corrections can be neglected. Note that in the limit of small **q** this can be justified rigorously, provided that the self-energy and three-leg scalar vertex are local. Therefore, the expression for the stiffness constant of the ferromagnet (Eq. (30)) appears to be exact in the framework of DMFT [12].

It should be stressed that the exchange integrals discussed here are just static characteristics connected with energy of inhomogeneous spin configurations. They determine the frequencies of spin excitation in itinerant electron systems only under adiabatic (rigid spin) approximation. For more general consideration one should calculate **q**- and ω- dependent spin susceptibility.

6. Electron correlations and finite-temperature magnetism in transition metals

Now we describe the applications of the SDF approach to a classical problem of finite-temperature magnetism of the iron-group transition metals. Despite a lot of attempts starting from seminal works by Heisenberg and Frenkel (for review of early theories see e.g. [30, 33, 34]) we have yet no complete quantitative theory describing their magnetic and spectral properties. The reason is that to describe the properties of Fe, Co, and Ni one has to solve the problem of taking into account of moderately strong electronic correlations where approaches developed both for weakly correlated systems such as normal-group metals and to higly correlated systems such as Mott insulators are not, generally speaking, reliable.There were a lot of attempts to introduce by some way correlation effects in band structure calculations of these metals [35, 36, 37, 38, 39], not referring to numerous purely model works. But the question of applicability of specific approximations such as the lowest order perturbation theory [36, 39], moment method [38] or three-body Faddeev equations [37] is not clear and one needs a reliable approach which would be checked carefully for model systems and demonstrate its applicability for moderately correlated systems. It has been demonstrated in Ref. [10] that the *ab initio* dynamical mean-field theory does give a very successful description of both correlation effects in the electron energy spectra and the finite-temperature magnetic properties of Fe and, especially, Ni. Here we present the corresponding results.

We start with the LDA Hamiltonian in the tight-binding orthogonal LMTO representation $H^{LDA}_{mm'}(\mathbf{k})$ [40], where m describes the orbital basis set containing $3d$-, $4s$- and $4p$- states, $\mathbf{k}$ runs over the Brillouin zone (BZ). The interactions are parameterized by a matrix of screened local Coulomb interactions $U_{mm'}$ and a matrix of exchange constants $J_{mm'}$, which are expressed in terms of two screened Hubbard parameters U and J, describing the average Coulomb repulsion and the interatomic ferromagnetic exchange respectively. We use the values $U = 2.3$ (3.0) eV for Fe (Ni) and the same value of the interatomic exchange, $J = 0.9$ eV for both Fe and Ni, a result of constrained LDA calculations [41, 4, 5]. These parameters, which are consistent with those of many studies result in a very good description of the physical properties of Fe and Ni.

As was discussed above, the DMFT maps the many-body system onto a multi-orbital quantum impurity, i.e. a set of local degrees of freedom in a bath described by the Weiss field function $\mathcal{G}$. The impurity action (here $n_{m\sigma} = c^{+}_{m\sigma}c_{m\sigma}$ and $\mathbf{c}(\tau) = [c_{m\sigma}(\tau)]$ is a vector of Grassman variables) is given by:

$$S_{eff} = -\int_0^\beta d\tau \int_0^\beta d\tau' Tr[\mathbf{c}^+(\tau)\mathcal{G}^{-1}(\tau,\tau')\mathbf{c}(\tau')] + \frac{1}{2}\sum_{m,m',\sigma}\int_0^\beta d\tau [U_{mm'}n^m_\sigma n^{m'}_{-\sigma} + (U_{mm'} - J_{mm'})n^m_\sigma n^{m'}_\sigma] \tag{35}$$

It describes the spin, orbital, energy and temperature dependent interactions of particular magnetic $3d$-atom with the rest of the crystal and is used to compute the local Greens function matrix:

$$\mathbf{G}_\sigma(\tau - \tau') = -\frac{1}{Z}\int D[\mathbf{c},\mathbf{c}^+]e^{-S_{eff}}\mathbf{c}(\tau)\mathbf{c}^+(\tau') \tag{36}$$

(Z is the partition function) and the impurity self energy $\mathcal{G}^{-1}_\sigma(\omega_n) - \mathbf{G}^{-1}_\sigma(\omega_n) = \mathbf{\Sigma}_\sigma(\omega_n)$.

The Weiss field function is required to obey the self consistency condition (14), which can be specified for a given case as

$$\mathbf{G}_\sigma(\omega_n) = \sum_{\mathbf{k}}[(i\omega_n + \mu)\mathbf{1} - \mathbf{H}^{LDA}(\mathbf{k}) - \mathbf{\Sigma}^{dc}_\sigma(\omega_n)]^{-1} \tag{37}$$

The local matrix $\mathbf{\Sigma}^{dc}_\sigma$ is the sum of two terms, the impurity self energy and a so-called "double counting " correction, E_{dc} which is meant to substract the average electron-electron interactions already included in the LDA Hamiltonian. For metallic systems we proposed the general form of dc-correction: $\mathbf{\Sigma}_{\sigma}^{dc}(\omega) = \mathbf{\Sigma}_{\sigma}(\omega) - \frac{1}{2}Tr_{\sigma}\mathbf{\Sigma}_{\sigma}(0)$. This is motivated by the fact that the static part of the correlation effects are already well described in the density functional theory. Only the d-part of the self-energy is presented in our calculations, therefore $\mathbf{\Sigma}_{\sigma}^{dc} = 0$ for s- and p- states as well as for non-diagonal $d - s,p$ contributions. In order to describe the finite temperature ferromagnetism of transition metals we use the *non* spin-polarized LDA Hamiltonian $\mathbf{H}^{LDA}(\mathbf{k})$ and accumulate *all* temperature-dependent spin-splittings in the self-energy matrix $\mathbf{\Sigma}_{\sigma}^{dc}(\omega_n)$.

We used the impurity QMC scheme (see Appendix) for the solution of the multiband DMFT equations [44]. The Hirsch discrete Hubbard-Stratonovich transformations [42, 43] introduces $(2M-1)M$ auxiliary Ising fields $S_{mm'}^{\sigma\sigma'}$, where M is the orbital degeneracy of the d-states and calculate $\mathbf{G}_{\sigma}(\tau)$ by an exact integration of the fermion degrees of freedom in the functional integral (Eq.(36)) [3]. In order to sample efficiently all the spin configurations in the multi-band QMC scheme, it is important to use "global" spin-flips: $[S_{mm'}^{\sigma\sigma'}] \rightarrow [-S_{mm'}^{-\sigma-\sigma'}]$ in addition to the local moves of the auxiliary fields. The number of QMC sweeps was of the order of 10^5. A parallel version of the DMFT program was used to sample the 45 Ising fields for $3d$-orbitals. We used 256 **k**-points in the irreducible part of the BZ for the k integration. 10 to 20 DMFT iterations were sufficient to achieve convergence far from the Curie point. Due to the cubic symmetry of the bcc-Fe and fcc-Ni lattices the local Green function is diagonal in the basis of real spherical harmonics. The spectral functions for real frequencies were obtained from the QMC data by applying the maximum entropy method [45].

Our results for the local spectral function for iron and nickel are shown in Figs.1 and 2, respectively. The SDF approach describes well all the qualitative features of the density of states (DOS), which is especially non-trivial for nickel. Our QMC results reproduce well the three main correlation effects on the one particle spectra below T_C [46, 47, 48]: the presence of a famous 6 eV satellite, the 30% narrowing of the occupied part of d-band and the 50% decrease of exchange splittings compared to the LDA results. Note that the satellite in Ni has substantially more spin-up contributions in agreement with photoemission spectra [48]. The exchange splitting of the d-band depends very weakly on temperature from T=0.6 T_C to T=0.9 T_C. Correlation effects in Fe are less pronounced than in Ni, due to its large spin-splitting and the characteristic bcc-structural dip in the density of states for spin-down states near Fermi level, which reduces the density of states for particle hole excitations.

Now we discuss the applications of the SDF approach to the description of finite-temperature magnetic properties of iron and nickel. While density functional theory, can in principle provide a rigorous description of the thermodynamic properties, at present there is no accurate practical implemention available. As a result the finite-temperature properties of magnetic materials are estimated following a simple suggestion [29], whereby constrained DFT at $T = 0$ is used to extract exchange constants for a *classical* Heisenberg model, which in turn is solved using approximation methods (e.g. RPA, mean field) from classical statistical mechanics of spin systems [29, 49, 50, 51]. The most recent implementation of this approach gives good values for the transition temperature of iron but not of nickel [52]. While these localized spin models give, by construction, at high temperatures a Curie-Weiss like magnetic susceptibility, as observed experimentally in Fe and Ni, they encountered difficulties in predicting the correct values of the Curie constants[53].

The uniform spin susceptibility in the paramagnetic state: $\chi_{q=0} = dM/dH$ was extracted from the QMC simulations by measuring the induced magnetic moment in a small external magnetic field. It includes the polarization of the impurity Weiss field by the external field [3]. The dynamical mean field results accounts for the Curie-Weiss law which is observed experimentally in Fe and Ni. As the temperature increases above T_C, the atomic character of the system is partially restored resulting in an atomic like susceptibility with an effective moment:

$$\chi_{q=0} = \frac{\mu_{eff}^2}{3(T - T_C)} \tag{38}$$

The temperature dependence of the ordered magnetic moment below the Curie temperature and the inverse of the uniform susceptibility above the Curie point are plotted in Fig. 3 together with the corresponding experimental data for iron and nickel[54]. The LDA+DMFT calculations describes the magnetization curve and the slope of the high-temperature Curie-Weiss susceptibility remarkably well. The calculated values of high-temperature magnetic moments extracted from the uniform spin susceptibility are $\mu_{eff} = 3.09\ (1.50)\mu_B$ for Fe (Ni), in good agreement with the experimental data $\mu_{eff} = 3.13\ (1.62)\mu_B$ for Fe (Ni)[54].

We have estimated the values of the Curie temperatures of Fe and Ni from the disappearance of spin polarization in the self-consistent solution of DMFT problem and from the Curie-Weiss law in Eq.(38). Our estimates $T_C = 1900\ (700)K$ are in reasonable agreement with experimental

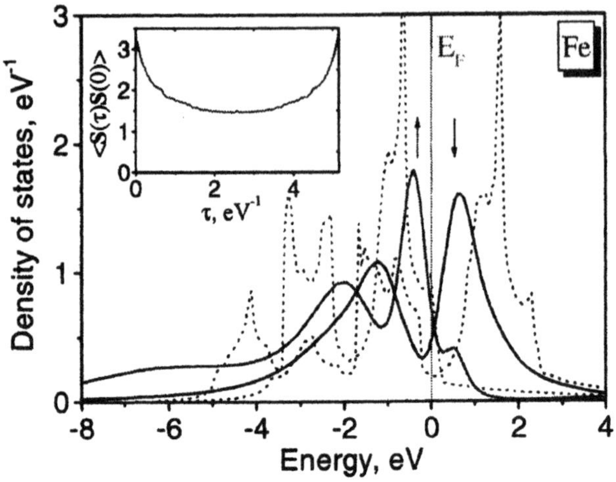

Figure 1. LDA+DMFT results for ferromagnetic iron ($T = 0.8\ T_C$). The partial densities of d-states (full lines) is compared with the corresponding LSDA results at zero temperature (dashed lines) for the spin-up (arrow-up) and spin-down (arrow-down) states. The insert shows the spin-spin autocorrelation function for T=1.2 T_C.

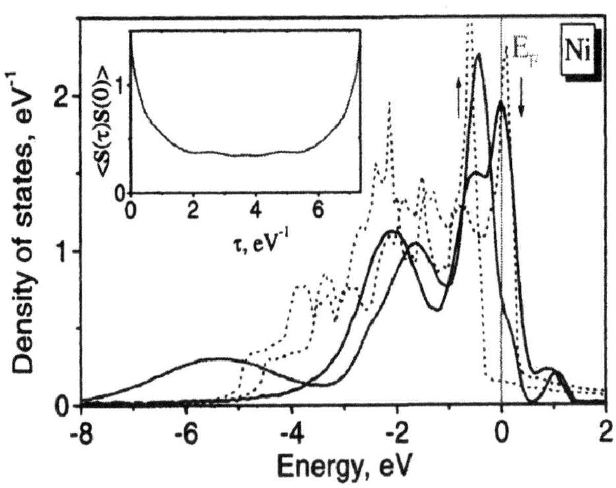

Figure 2. Same quantities as in Fig.1 for ferromagnetic nickel ($T = 0.9\ T_C$). The insert shows the spin-spin autocorrelation function for T=1.8 T_C.

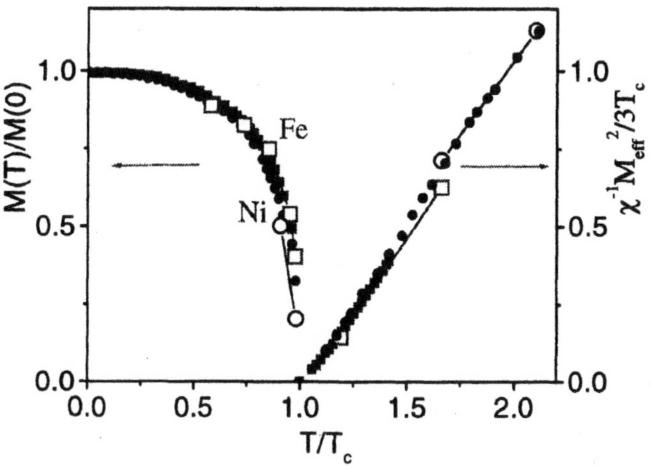

Figure 3. Temperature dependence of ordered moment and the inverse ferromagnetic susceptibility for Fe (open square) and Ni (open circle) compared with experimental results for Fe (square) and Ni (circle) (from Ref.[33]).

values of 1043 (631)K for Fe (Ni) respectively[54], considering the single site nature of the DMFT approach, which is not able to capture the reduction of T_C due to long wavelength spin waves. These effects are governed by the spin wave stiffness. Since the ratio of the spin wave stiffness (D) to T_C, T_C/a^2D is nearly a factor of 3 larger for Fe than for Ni[54] (a is the lattice spacing), we expect the DMFT T_C to be much higher than the observed Curie temperature in Fe than in Ni. Note that this is a consequence of the long-range oscilating character of exchange interactions in iron compare to short-range ferromagnetic exchange interactions in nickel [52]. Quantitative calculations demonstrating the sizeable reduction of T_C due to spin waves in Fe in the framework of a Heisenberg model were performed in Ref [52].

Within dynamical mean field theory one can also compute the local spin susceptibility defined by

$$\chi_{loc} = \frac{g_s^2}{3} \int_0^{\beta} d\tau \left\langle \mathbf{S}\left(\tau\right) \mathbf{S}(0) \right\rangle \tag{39}$$

where $g_s = 2$ is the gyromagnetic ratio and $\mathbf{S} = \frac{1}{2}\sum_{m,\sigma,\sigma'} c^{\dagger}_{m\sigma}\sigma_{\sigma\sigma'}c_{m\sigma'}$ is single-site spin operator and $\sigma = (\sigma_x, \sigma_y, \sigma_z)$ are Pauli matrices. It differs from the $q = 0$ susceptibility, by the absence of spin polarization in the Weiss field of the impurity model. Eq.(39) cannot be probed directly in experiments but it is easily computed in DMFT-QMC. Its behavior as function of temperature, gives a very intuitive picture of the degree of correlations in the system. In a weakly correlated system we expect Eq.(39) to be nearly temperature independent, while in a strongly correlated system we expect a leading Curie-Weiss behavior at high temperatures $\chi_{local} = \mu_{loc}^2/(3T + const)$ where μ_{loc} is an effective local magnetic moment. In the Heisenberg model with spin S, $\mu_{loc}^2 = S(S+1)g_s^2$ and for well-defined local magnetic moments (e.g., for rare earth magnets) this quantity should be temperature independent. For the itinerant electron magnets μ_{loc} is temperature-dependent, due to a variety of competing many body effects such as Kondo screening, the induction of local magnetic moment by temperature [34] and thermal fluctuations which disorders the moments [55]. All these effects are included in the DMFT calculations. The τ-dependence of the correlation function $\langle \mathbf{S}(\tau)\, \mathbf{S}(0) \rangle$ results in the temperature dependence of μ_{loc} and is displayed in the inserts on the Figs.1,2. Iron can be considered as a magnet with very well-defined local moments above T_C (the τ-dependence of the correlation function is relatively weak), whereas nickel is more itinerant electron magnet (stronger τ -dependence of the local spin-spin autocorrelation function).

The comparison of the values of the local and the $q = 0$ susceptibility gives a crude measure of the degree of short range order which is present above T_C. As expected, the moments extracted from the local susceptibility Eq.(39) are a bit smaller (2.8 μ_B for iron and 1.3 μ_B for nickel) than those extracted from the uniform mangetic susceptibility. This reflects the small degree of the short-range correlations which remain well above T_C [57]. The high-temperature LDA+DMFT clearly show the presence of a local-moment above T_C. This moment, is correlated with the presence of high energy features (of the order of the Coulomb energies) in the photomeission. This is also true below T_C, where the spin dependence of the spectra is more pronounced for the satellite rigion in nickel than for that the quasiparticle bands near the Fermi level (Fig. 2). This can explain the apparent discrepancies between different experimental determinations of the high-temperature magnetic splittings [56, 58, 59] as being the result of probing different energy regions. The resonant photoemission experiments [58] reflects the presence of local-moment polarization in the high-energy spectrum above Curie temperature in nickel, while the low-energy ARPES investigations [59] results in non-magnetic bands near the Fermi level. This is exactly the DMFT view on the electronic structure of transition metals above T_C. Fluctuating moments and atomic-like configurations are large at short times, which results in correlation effects in the high-energy spectra such as spin-multiplet splittings. The moment is reduced at longer time scales, corresponding to a more band-like, less correlated electronic structure near the Fermi level.

7. Approximate solution for the self-energy: TM-FLEX method

The QMC method described above is probably the most accurate way to solve the effective impurity problem in the DMFT. However, it is rather cumbersome and expensive computationally; besides that, it deals with the "truncated" two-indices interaction matrix (see Eq.(35)) instead of the complete four-indices one. Therefore a scheme has been proposed in Ref. [6] basing on a multiband spin-polarized generalization of the "fluctuating exchange" (FLEX) approximation by Bickers and Scalapino [60]. Original formulation of the FLEX approximation treats in the equal way both particle-hole (PH) and particle-particle (PP) channels. But their roles in magnetism are completely different. The interaction of electrons with spin fluctuations in PH channel leads to the most relevant correlation effects [34] whereas PP processes are important for the renormalizations of the effective interactions in spirit of T-matrix approach ("ladder aprroximation") by Galitskii [61] and Kanamori [62]. Therefore we used in Ref. [6] a "two-step" procedure when, at first, bare matrix vertex is replaced by T-matrix, and, secondly, PH channel processes with this effective interaction are taken into account explicitly. This approximation has high enough accuracy both for the Hubbard model and for real systems with moderate correlations $U < W/2$ where U is the Hubbard on-sire repulsion snergy and W is the bandwidth (see [6] and Refs therein). However, specific form of the approximation used in [6] can be improved further by taking into account the spin-dependence of the T-matrix. Here we present the formulation of this T-matrix-FLEX approximation.

Let us start, as in Refs. [5, 6], with the general many-body Hamiltonian for crystal in LDA+U scheme [63]:

$$
\begin{aligned}
H &= H_t + H_U \\
H_t &= \sum_{\lambda\lambda'\sigma} t_{\lambda\lambda'} c^{+}_{\lambda\sigma} c_{\lambda'\sigma} \\
H_U &= \frac{1}{2} \sum_{\{\lambda_i\}\sigma\sigma'} \langle \lambda_1 \lambda_2 | v | \lambda'_1 \lambda'_2 \rangle \, c^{+}_{\lambda_1\sigma} c^{+}_{\lambda_2\sigma'} c_{\lambda'_2\sigma'} c_{\lambda'_1\sigma},
\end{aligned}
\qquad (40)
$$

where $\lambda = im$ are the site number (i) and orbital (m) quantum numbers, $\sigma = \uparrow, \downarrow$ is the spin projection, c^+, c are the Fermi creation and annihilation operators, H_t is the effective single particle Hamiltonian from the LDA, corrected for the double-counting of average interactions among correlated electrons as it was described above, and the Coulomb matrix elements are defined in the standard way

$$\langle 12 \left| v \right| 34 \rangle = \int d\mathbf{r} d\mathbf{r}' \psi_1^*(\mathbf{r}) \psi_2^*(\mathbf{r}') v \left(\mathbf{r} - \mathbf{r}' \right) \psi_3(\mathbf{r}) \psi_4(\mathbf{r}'), \tag{41}$$

where we define for briefness $\lambda_1 \equiv 1$ etc. Following to [61] let us take into account the ladder (T-matrix) renormalization of the effective approximation:

$$\left\langle 13 \left| T^{\sigma\sigma'} \left(i\Omega \right) \right| 24 \right\rangle = \langle 13 \left| v \right| 24 \rangle - \frac{1}{\beta} \sum_{\omega} \sum_{5678} \langle 13 \left| v \right| 57 \rangle * \\ G^{\sigma}_{56} \left(i\omega \right) G^{\sigma'}_{78} \left(i\Omega - i\omega \right) \left\langle 68 \left| T^{\sigma\sigma'} \left(i\Omega \right) \right| 24 \right\rangle, \tag{42}$$

Further we rewrite the perturbation theory in terms of this effective interaction matrix.

At first, we take into account the "Hartree" and "Fock" diagrams with the replacement of the bare interaction by the T-matrix

$$\begin{aligned} \Sigma^{(TH)}_{12,\sigma} \left(i\omega \right) &= \frac{1}{\beta} \sum_{\Omega} \sum_{34\sigma'} \left\langle 13 \left| T^{\sigma\sigma'} \left(i\Omega \right) \right| 24 \right\rangle G^{\sigma'}_{43} \left(i\Omega - i\omega \right) \\ \Sigma^{(TF)}_{12,\sigma} \left(i\omega \right) &= -\frac{1}{\beta} \sum_{\Omega} \sum_{34} \langle 14 \left| T^{\sigma\sigma} \left(i\Omega \right) \right| 32 \rangle G^{\sigma}_{34} \left(i\Omega - i\omega \right) \end{aligned} \tag{43}$$

Note that $\Sigma^{(TH)} + \Sigma^{(TF)}$ contains exactly all the second-order contributions. Now we have to consider the contribution of particle-hole excitations to sigma. Similar to [6] we will replace in the corresponding diagrams the bare interaction by the static limit of T-matrix. However, we improve the approximation [6] by taking into account its spin dependence. When considering the particle-hole channel we replace in the Hamiltonian (40) $v \to T^{\sigma\sigma'}$ which is the solution of Eq.(42) at $\Omega = 0$. Eq. (43) is exact in the limit of low electron (or hole) density which is important for the criterion of magnetism e.g. in the case of nickel (with almost completely filled d- band).

Now we rewrite the effective Hamiltonian (40) with the replacement $\langle 12 \left| v \right| 34 \rangle$ by $\left\langle 12 \left| T^{\sigma\sigma'} (\Omega = 0) \right| 34 \right\rangle$ in H_U. To consider the correlation effects due to PH channel we have to separate density (d) and magnetic (m) channels as in [60]

$$d_{12} = \frac{1}{\sqrt{2}} \left(c^+_{1\uparrow} c_{2\uparrow} + c^+_{1\downarrow} c_{2\downarrow} \right)$$

$$\begin{aligned} m^0_{12} &= \frac{1}{\sqrt{2}} \left(c^+_{1\uparrow} c_{2\uparrow} - c^+_{1\downarrow} c_{2\downarrow} \right) \\ m^+_{12} &= c^+_{1\uparrow} c_{2\downarrow} \\ m^-_{12} &= c^+_{1\downarrow} c_{2\uparrow}, \end{aligned} \tag{44}$$

Then the interaction Hamiltonian can be rewritten in the following matrix form

$$H_U = \frac{1}{2} Tr \left(D^+ * V^{\parallel} * D + m^+ * V^{\perp}_m * m^- + m^- * V^{\perp}_m * m^+ \right) \tag{45}$$

where * means the matrix multiplication with respect to the pairs of orbital indices, e.g.

$$\left(V^{\perp}_m * m^+ \right)_{11'} = \sum_{34} \left(V^{\perp}_m \right)_{11',22'} m^+_{22'},$$

the supervector D defined as

$$D = \left(d, m^0 \right), D^+ = \begin{pmatrix} d^+ \\ m^+_0 \end{pmatrix},$$

and effective interactions have the following form:

$$\begin{aligned} \left(V^{\perp}_m \right)_{11',22'} &= - \left\langle 12 \left| T^{\uparrow\downarrow} \right| 2'1' \right\rangle, \\ V^{\parallel} &= \begin{pmatrix} V^{dd} & V^{dm} \\ V^{md} & V^{dd} \end{pmatrix}, \\ V^{dd}_{11',22'} &= \frac{1}{2} \sum_{\sigma\sigma'} \left\langle 12 \left| T^{\sigma\sigma'} \right| 1'2' \right\rangle - \frac{1}{2} \sum_{\sigma} \left\langle 12 \left| T^{\sigma\sigma} \right| 2'1' \right\rangle, \\ V^{mm}_{11',22'} &= \frac{1}{2} \sum_{\sigma\sigma'} \sigma\sigma' \left\langle 12 \left| T^{\sigma\sigma'} \right| 1'2' \right\rangle - \frac{1}{2} \sum_{\sigma} \left\langle 12 \left| T^{\sigma\sigma} \right| 2'1' \right\rangle, \\ V^{dm}_{11',22'} &= V^{md}_{22',11'} = \\ & \frac{1}{2} \left[\begin{array}{c} \left\langle 12 \left| T^{\uparrow\uparrow} \right| 1'2' \right\rangle - \left\langle 12 \left| T^{\downarrow\downarrow} \right| 1'2' \right\rangle - \left\langle 12 \left| T^{\uparrow\downarrow} \right| 1'2' \right\rangle + \\ \left\langle 12 \left| T^{\downarrow\uparrow} \right| 1'2' \right\rangle - \left\langle 12 \left| T^{\uparrow\uparrow} \right| 2'1' \right\rangle + \left\langle 12 \left| T^{\downarrow\downarrow} \right| 2'1' \right\rangle \end{array} \right] \end{aligned} \tag{46}$$

To calculate the PH contribution to the electron self-energy we first have to write the expressions for the generalized susceptibilities, both transverse $\chi^{\perp}$ and longitudinal $\chi^{\parallel}$. Corresponding expressions are the same as in [6] but with another definition of the interaction vertices. One has

$$\chi^{+-}(i\omega) = \left[1 + V_m^{\perp} * \Gamma^{\uparrow\downarrow}(i\omega)\right]^{-1} * \Gamma^{\uparrow\downarrow}(i\omega)\,, \tag{47}$$

where

$$\Gamma^{\sigma\sigma'}_{12,34}(\tau) = -G^{\sigma}_{23}(\tau)\, G^{\sigma'}_{41}(-\tau) \tag{48}$$

is an "empty loop" susceptibility and $\Gamma(i\omega)$ is its Fourier transform. The corresponding longitudinal susceptibility matrix has a more complicated form:

$$\chi^{\parallel}(i\omega) = \left[1 + V^{\parallel} * \chi_0^{\parallel}(i\omega)\right]^{-1} * \chi_0^{\parallel}(i\omega), \tag{49}$$

and the matrix of bare longitudinal susceptibility:

$$\chi_0^{\parallel} = \frac{1}{2}\begin{pmatrix} \Gamma^{\uparrow\uparrow} + \Gamma^{\downarrow\downarrow} & \Gamma^{\uparrow\uparrow} - \Gamma^{\downarrow\downarrow} \\ \Gamma^{\uparrow\uparrow} - \Gamma^{\downarrow\downarrow} & \Gamma^{\uparrow\uparrow} + \Gamma^{\downarrow\downarrow} \end{pmatrix}, \tag{50}$$

in the *dd*-, dm^0-, m^0d-, and m^0m^0- channels ($d, m^0 = 1, 2$ in the super-matrix indices). An important feature of these equations is the coupling of longitudinal magnetic fluctuations and of density fluctuations. It is absent in one-band Hubbard model due to the absense of the interaction of electrons with parallel spins. For this case Eqs. (47,49) coinsides with the well-known result [64].

Now we can write the particle-hole contribution to the self-energy. According to [6] one has

$$\Sigma^{(ph)}_{12,\sigma}(\tau) = \sum_{34,\sigma'} W^{\sigma\sigma'}_{13,42}(\tau)\, G^{\sigma'}_{34}(\tau)\,, \tag{51}$$

with p-h fluctuation potential matrix:

$$W^{\sigma\sigma'}(i\omega) = \begin{bmatrix} W^{\uparrow\uparrow}(i\omega) & W^{\perp}(i\omega) \\ W^{\perp}(i\omega) & W^{\downarrow\downarrow}(i\omega) \end{bmatrix}, \tag{52}$$

where the spin-dependent effective potentials are defined as

$$W^{\uparrow\uparrow} = \frac{1}{2} V^{\parallel} * \left[\chi^{\parallel} - \chi_0^{\parallel}\right] * V^{\parallel}$$

$$W^{\downarrow\downarrow} = \frac{1}{2} V^{\parallel} * \left[\tilde{\chi}^{\parallel} - \tilde{\chi}_0^{\parallel}\right] * V^{\parallel}$$

$$W^{\uparrow\downarrow} = V_m^{\perp} * \left[\chi^{+-} - \chi_0^{+-}\right] * V_m^{\perp}$$

$$W^{\downarrow\uparrow} = V_m^{\perp} * \left[\chi^{-+} - \chi_0^{-+}\right] * V_m^{\perp}.$$

where $\tilde{\chi}^{||}, \tilde{\chi}_0^{||}$ differ from $\chi^{||}, \chi_0^{||}$ by the replacement of $\Gamma^{\uparrow\uparrow} \Leftrightarrow \Gamma^{\downarrow\downarrow}$ in Eq.(50). We have substracted the seconf-order contributions since they have already taken into account in Eq.(43).

Our complete expression for the self energy is

$$\Sigma = \Sigma^{(TH)} + \Sigma^{(TF)} + \Sigma^{(ph)} \tag{53}$$

This expression takes into account accurately spin-polaron effects because of the interaction with magnetic fluctuations [6], energy dependence of T-matrix which is important to describe the satellite effects in Ni [35], contains exact second-order terms in v and is rigorous (because of the first term) for almost filled or almost empty bands. In spirit of the DMFT approach we have to use $\mathcal{G}_0$ instead of G in all the expressions when calculating self- energy on a separated central site. It should be noted that this TM-FLEX scheme is not conserved (or "Φ-derivable") therefore one need to inforce the Luttinger theorem by introducing the μ_0 for the bath Green function as in iterative preturbation theory[65]

We have started from the spin-polarized LSDA band structure of ferromagnetic nickel within the TB-LMTO method [40] in the minimal s, p, d basis set and used numerical orthogonalization to find the H_t part of our starting Hamiltonian. We take into account Coulomb interactions only between d-states. Semiempirical analysis of the appropriate interaction value gives $U \simeq 2 - 4$ eV. The difficulties with choosing the correct value of U are connected with complicated screening problems, definitions of orthogonal orbitals in the crystal, and contributions of the intersite interactions. In the quasiatomic (spherical) approximation the full U-matrix for the $d-$shell is determined by the three parameters U, J and δJ or equivalently by effective Slater integrals F^0, F^2 and F^4 [63]. For example, $U = F^0$, $J = (F^2 + F^4)/14$ and we use the simplest way of estimating δJ or F^4 keeping the ratio F^2/F^4 equal to its atomic value 0.625 [63]. Note that the value of intra-atomic (Hund) exchange interaction J is not sensitive to the screening and approximately equals to 0.9 eV in different estimations [63].

The resulting DOS for Ni (Fig. 4) shows that spin-polarized TM-FLEX calculations approximatelly reproduce the satellite structure and reduction of the band width in satisfactory agreement with exact QMC-result(Fig. 2)

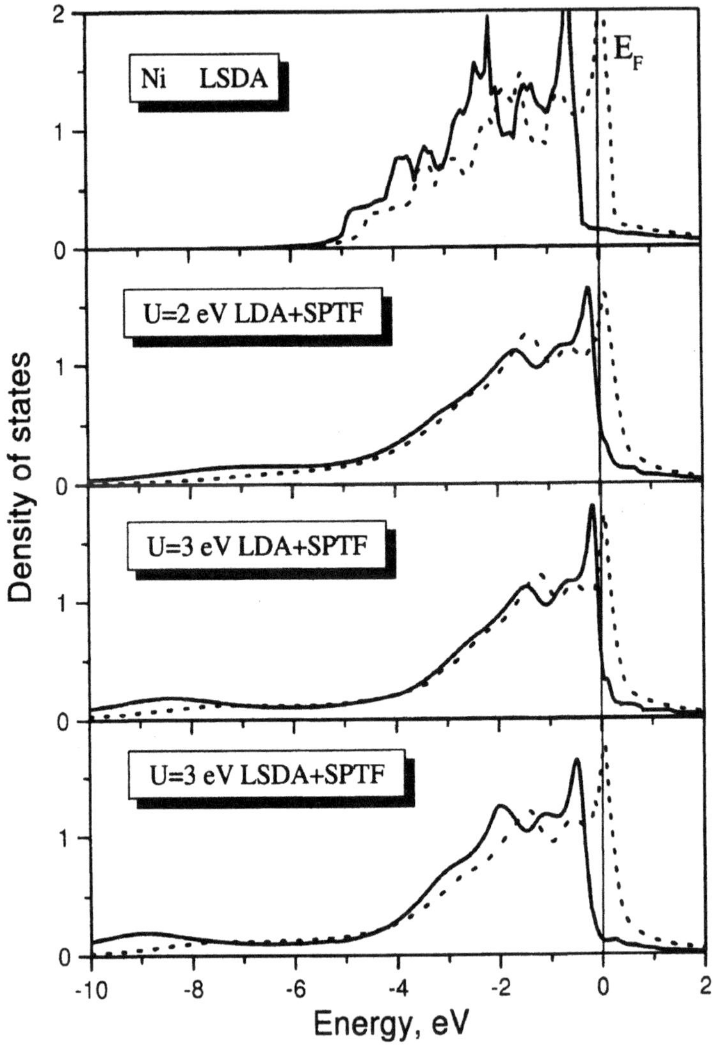

Figure 4. Spin-up (full lines) and spin-down (dased lines) density of d-states for ferromagnetic nickel in the LSDA and the LDA+SPTF (LSDA+SPTF) calculations for different average Coulomb interaction U with $J = 1$ eV and temperature T=200 K.

8. Cluster DMFT appoach: antiferromagnetism and superconductivity

As it was stressed above the SDF approach does not necessarily connected with the standard DMFT scheme; one can use more general choise of fermion coarse-grained variables, e.g., considering the case of the effective cluster instead of the effective impurity. It is especially important for the problems where intersite correlations are involved from the beginning such as d-wave superconducting pairing [21, 22] or charge ordering [66]. Here we consider, following Ref. [22], one of the cluster generalizations of the DMFT, and its application to the problem of magnetism and high-temperature superconductivity (HTSC) of copper-oxide compounds.

Microscopic theory of high-temperature superconducting cuprates is still far from the final understanding [67, 68, 69]. One of the most important recent experimental achievements was the discovery of the pseudogap (PG) phenomenon above superconducting transition temperatures [70] and existence of a sharp 41-meV resonance below T_c related with some collective antiferromagnetic excitations [71]. Thus, an interplay of an antiferromagnetism (AFM) and d-wave superconductivity (d-SC) in cuprate could be a natural way of discussing different HTSC phenomena. This require a quantitative electronic structure theory including two different type of the order parameters: AFM and d-SC. Within such approach one can in principle analyze the phase diagram of HTSC compounds and resolve the long-standing problem of competition between antiferromagnetism and d-wave superconductivity in cuprates [72, 73].

A standard theoretical tool for cuprates electronic structure consists of the two-dimensional Hubbard model [67]. We start with the extended-hopping Hubbard model on the square lattice:

$$H = \sum_{ij} t_{ij} c^{+}_{i\sigma} c_{j\sigma} + \sum_{i} U_i n_{i\uparrow} n_{i\downarrow}$$

where t_{ij} is an effective hopping and U_i local Coulomb interactions. We have chosen the nearest-neighbor hopping $t = 0.25$ eV and the next nearest hopping $t'/t = -0.15$ for the model of $La_{2-x}Sr_xCuO_4$[74]. The total band width is W=2 eV and all Coulomb parameters set to be U=1.2 eV ($U/W = 0.6$). Let us introduce the "super-site" as an 2×2 square plaquet. The numeration of the atoms in the super-site is shown in the Fig.5. It is useful to introduce the superspinor $C^{+}_{i} = \{c^{+}_{i\alpha}\}$ where $\alpha = 0, 1, 2, 3$ (the spin indices are not shown). Taking into account the spin degrees of freedom, this is the 8-component superspinor creation operator. Then the crystal Green function for the Hubbard model can be rewritten as

$$G(\mathbf{k},i\omega) = [i\omega + \mu - h(\mathbf{k}) - \mathbf{\Sigma}(i\omega)]^{-1} \tag{54}$$

where $h(\mathbf{k},i\omega)$ is the effective hopping supermatrix with the self-energy corrections. For simplicity we will write all the formulas in the nearest-neighbor approximations:

$$h(\mathbf{k}) = \begin{pmatrix} 0 & tK_x^+ & 0 & tK_y^+ \\ tK_x^- & 0 & tK_y^+ & 0 \\ 0 & tK_x^- & 0 & tK_x^- \\ tK_y^- & 0 & tK_x^+ & 0 \end{pmatrix} \tag{55}$$

where $K_{x(y)}^{\pm} = 1 + \exp\left(\pm ik_{x(y)}a\right)$, a is the lattice constant, and each element is 2×2 matrix in the spin space. Within the cluster-DMFT approach we introduce intra-atomic self-energy Σ_0 and inter-atomic self-energies Σ_x, Σ_y and the both functions are of the intra-site nature in the sense of our super-site:

$$\Sigma(i\omega) = \begin{pmatrix} \Sigma_0 & \Sigma_x & 0 & \Sigma_y \\ \Sigma_x^* & \Sigma_0 & \Sigma_y & 0 \\ 0 & \Sigma_y^* & \Sigma_0 & \Sigma_x^* \\ \Sigma_y^* & 0 & \Sigma_x & \Sigma_0 \end{pmatrix} \tag{56}$$

For the small 2×2 cluster it is usefull to introduce the translationally invariant (**k**-dependent) self-energy and rewrite $h(\mathbf{k}) + \mathbf{\Sigma}(i\omega) -> h(\mathbf{k},i\omega)$ where

$$h(\mathbf{k},i\omega) = \begin{pmatrix} \Sigma_0 & t_x K_x^+ & 0 & t_y K_y^+ \\ t_x^* K_x^- & \Sigma_0 & t_y K_y^+ & 0 \\ 0 & t_y^* K_x^- & \Sigma_0 & t_x^* K_x^- \\ t_y^* K_y^- & 0 & t_x K_x^+ & \Sigma_0 \end{pmatrix}. \tag{57}$$

The effective Hamiltonian is defined through the renormalized energy dependent hoppings: $t_x = t + \Sigma_x$, $t_y = t + \Sigma_y$. The functions $\Sigma_0(i\omega)$, $\Sigma_x(i\omega)$, $\Sigma_y(i\omega)$ are found self-consistently within the cluster DMFT scheme and for the d-wave superconduction state $\Sigma_x \neq \Sigma_y$. It is straightforward to generalize this scheme for a next-nearest neighbor hopping as well as the long-range Green function and the self-energy. In this case we can renormalized also the second-nearest hopping: $t_{xy} = t' + \Sigma_{xy}$ for the 2×2 cluster, where Σ_{xy} (or Σ^{02}) is the non-local self-energy in xy direction.

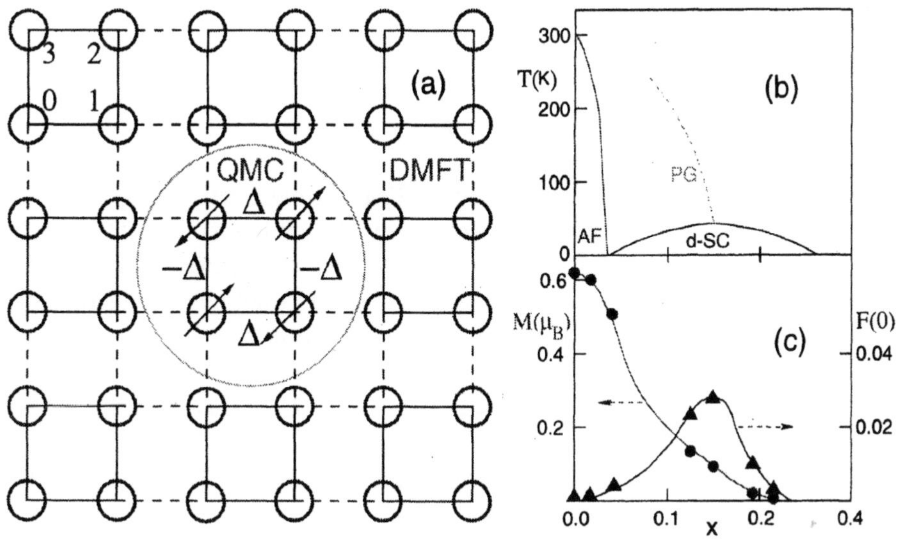

Figure 5. (a) Schematic representation of antiferromagnetic d-wave 2x2 periodically repeated clust er; (b) generic phase diagram of HTSC materials; (c) The calculated values of two order parameters: local magnetic moment M and d-SC equal time Green function $F^{01}(\tau = 0) \equiv F(0)$ for different hole doping (x) at the inverse temperature $\beta = 60$ eV$^{-1}(T = 190K)$.

In the cluster version of the DMFT scheme one can write the matrix equation for the bath Green function matrix $\mathcal{G}$ which describe an effective interactions with the rest of crystal:

$$\mathcal{G}^{-1}(i\omega) = G^{-1}(i\omega) + \Sigma(i\omega),$$

where the local cluster Green function matrix is equal to $G_{\alpha\beta}(i\omega) = \sum_{\mathbf{k}} G_{\alpha\beta}(\mathbf{k},i\omega)$, and summation is run over the Brillouin zone of the square lattice. If instead of Eq.(57) we use Eq.(55), it would corresponds to a free cluster DMFT scheme or the simplest case of the so-called cellular-DMFT. Note that in the we use translationally invariant self-energy obtained from the cluster DMFT scheme (Eq.(57)) or the simplest version of so-called cellular DMFT approach [23]. We believe that for a given choise of the small (2×2) cluster, the renormalization of the hopping parameter by Σ_x, Σ_y terms is physically essential. The present "matrix" form of a cluster DMFT with the self-energy which is not periodic inside the cluster allow us to study a multicomponent ordered state. Unfortunately, in contrast with the so-called Dynamical

Cluster Approximation (DCA) [21] or cellular-DMFT approach [23] (see above, Section 1), we were unable to prove casuality of this approach for arbitrary band structure and interaction parameters. However, the casuality of the Green function for the *realistic* choise of the parameters has been checked numerically.

In this case we have the standard DMFT problem with four "orbital" states per super-site. It has been solved by the multi-orbital QMC technique described above (Section 3). We used the generalized Nambu technique [75] to analyze the coexistence of the magnetic ordering and superconductivity. Let us introduce the superspinor

$$\Psi_i^+(\tau) \equiv (\psi_{1i}^+, \psi_{2i}^+, \psi_{3i}^+, \psi_{4i}^+) = \left(c_{i\uparrow}^+, c_{i\downarrow}^+, c_{i\uparrow}, c_{i\downarrow}\right)$$

and the anomalous averages describing the (collinear) antiferromagnetism $\left\langle c_{i\uparrow}^+ c_{j\downarrow}\right\rangle$ and the superconductivity $\Delta_{ij} = \langle c_{i\downarrow} c_{j\uparrow}\rangle$.

The generalization of the Hirsch-Fye QMC-algorithm [43] for superconducting problem [76] have been used. In the 4-spinor case a discrete Hubbard-Stratonovich transformation has the following form:

$$\begin{aligned} &\exp[-\Delta\tau U_i n_{i\uparrow} n_{i\downarrow} + \frac{\Delta\tau U_i}{2}(n_{i\uparrow} + n_{i\downarrow})] = \\ &\frac{1}{2}\sum_{\sigma=\pm 1} \exp[\lambda_i \sigma(\psi_{1i}^+\psi_{1i} - \psi_{2i}^+\psi_{2i} - \psi_{3i}^+\psi_{3i} + \psi_{4i}^+\psi_{4i})] \end{aligned} \quad (58)$$

where $\lambda_i = \frac{1}{2}\text{arccosh}[\exp(\frac{1}{2}\Delta\tau U_i)]$.

Since we take into account only the singlet pairing, there are following nonzero elements of the d-SC energy gap parameters: $\Delta = \Delta_{01} = -\Delta_{12} = \Delta_{23} = -\Delta_{30}$. One can chose Δ_{ij} to be real and therefore symmetric: $\Delta_{ij} = \Delta_{ji}$. Separating normal and anomalous parts of the Green function we have

$$G\left(\mathbf{k},\tau,\tau'\right) = \begin{pmatrix} G\left(\mathbf{k},\tau,\tau'\right) & F\left(\mathbf{k},\tau,\tau'\right) \\ F^+\left(\mathbf{k},\tau,\tau'\right) & -G\left(-\mathbf{k},\tau',\tau\right) \end{pmatrix} \quad (59)$$

where $G\left(\mathbf{k},\tau,\tau'\right) = -\left\langle T_\tau C_{\mathbf{k}}\left(\tau\right) C_{\mathbf{k}}^+\left(\tau'\right)\right\rangle$ and $F\left(\mathbf{k},\tau,\tau'\right) = -\left\langle T_\tau C_{\mathbf{k}}\left(\tau\right) C_{-\mathbf{k}}\left(\tau'\right)\right\rangle$ are the matrices in spin and "orbital" space. It is convenient to expand the anomalous Green function in Pauli matrices $F = \left(F^0 + \mathbf{F}\sigma\right) i\sigma^y$ and use the symmetry properties [77]:

$$\begin{aligned} F^0\left(\mathbf{k},\tau,\tau'\right) &= F^0\left(-\mathbf{k},\tau',\tau\right) \\ \mathbf{F}\left(\mathbf{k},\tau,\tau'\right) &= -\mathbf{F}\left(-\mathbf{k},\tau',\tau\right). \end{aligned} \quad (60)$$

then a 4×4 spinor formalism is reduced to 2×2 one in the collinear antiferromagnetic case with the d-wave superconductivity with the following spin-matrix form of the local Green function for the super-site:

$$G(\tau,\tau') = \begin{pmatrix} G_{\uparrow}(\tau,\tau') & F(\tau,\tau') \\ F(\tau,\tau') & -G_{\downarrow}(\tau',\tau), \end{pmatrix} \quad (61)$$

and the QMC formalism for the antiferromagnetic superconducting state is equivalent to the previous non-magnetic one [76]. Using the discretization of $[0,\beta]$ interval with L-time slices: $\Delta\tau = \beta/L$ ($\beta = 1/T$ is an inverse temperature) the G_{σ}- and F- Green functions become the matrices of $2NL$ dimension, where N is the number of atoms in the cluster. After Fourier transform to the Matsubara frequencies the Green function matrix has the following form:

$$G(i\omega) = \begin{pmatrix} G_{\uparrow}(i\omega) & F(i\omega) \\ F(i\omega) & -G_{\downarrow}^{*}(i\omega) \end{pmatrix} \quad (62)$$

In superconducting states the self-energy defined as [3]:

$$\mathcal{G}^{-1}(i\omega) - G^{-1}(i\omega) = \begin{pmatrix} \Sigma_{\uparrow}(i\omega) & S(i\omega) \\ S(i\omega) & -\Sigma_{\downarrow}^{*}(i\omega) \end{pmatrix}, \quad (63)$$

and the inverse crystal Green function matrix is equal to:

$$G^{-1}(\mathbf{k},i\omega) = \begin{pmatrix} i\omega + \mu - h(\mathbf{k},i\omega) & s(\mathbf{k},i\omega) \\ s(\mathbf{k},i\omega) & i\omega - \mu + h^{*}(\mathbf{k},i\omega) \end{pmatrix} \quad (64)$$

where $s(\mathbf{k},i\omega)$ translationally invariant anomalous part of the self-energy $S(i\omega)$ similar to Eq.(55).

The two-component order parameters state which include the Neel antiferromagnetism and d-wave superconductivity (Fig.5a) lowered the symmetry of the effective cluster-DMFT problem. A self-consistent DMFT cluster problem with AFM and d-SC general order parameters have been solved within the QMC scheme for 8x8 matrix Green function with L=64 time slices. The resulting two order parameters for $\beta = 60$ eV^{-1} (T=190 K) and $t' = 0$ presented in the Fig.5c together with the generic HTSC phase diagram (Fig.5b) as function of the hole doping. In this case the ordered magnetic moment is directly related with imaginary-time Green function $G_{\sigma}(\tau)$: $M = G_{\uparrow}^{00}(0) - G_{\downarrow}^{00}(0)$ and for the d-SC order parameter we chose a positive value of superconducting imaginary time Green function $F^{01}(0)$. It is important, that we find no serious sign-problem for all QMC calculations with various doping level, probably due to "stabilized" antiferromagnetic dynamical

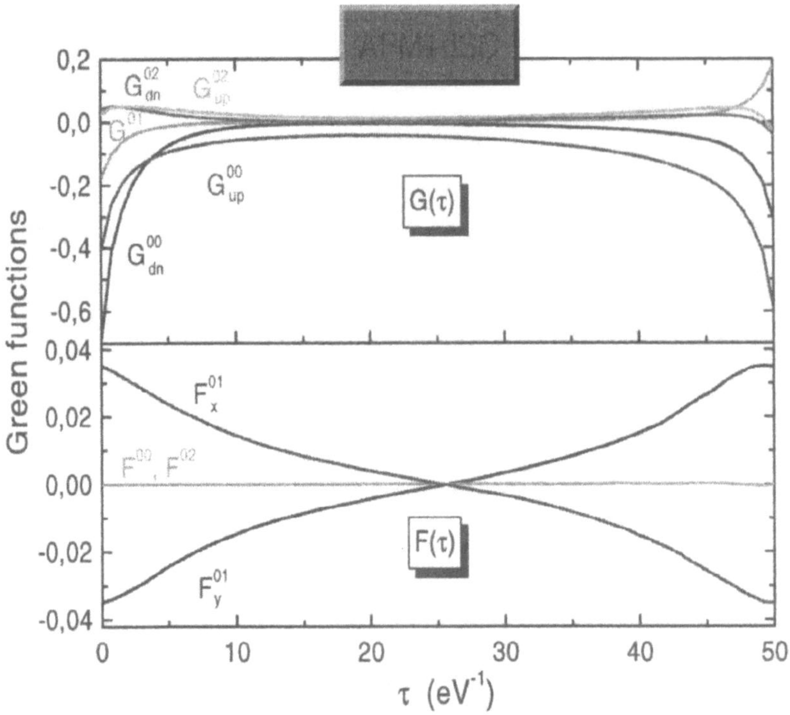

Figure 6. Imaginary time normal (G_σ) and superconducting (F) Green functions for the 2×2 cluster DMFT solution with second-nearest neighbor hopping and inverse temperature $\beta = 50$ eV^{-1} (T=230 K).

mean fields acting on the atoms in our 2×2 cluster. Note that the AFM cluster-DMFT solution exist for a much higher doping concentration then experimental AFM ordered state and describe a dynamical mean-field version of AFM-spin fluctuations related to pseudogap phenomena (the PG-region on the Fig.5b). The maximum of d-SC order parameter corresponds to the doping level of about 15% in agreement with the generic HTSC phase diagram. The d-SC order parameter is zero close to the undoped region (x=0), due to presence of a large AFM-gap. When the AFM-gap is closed ($x \sim 5\%$) the d-SC states developed but for $x > 20\%$ d-gap decrease again since AFM spin-fluctuations around (π, π) point disappear [68]. The precise characteristic of the phase diagram including the interactions between AFM and d-SC order parameters demands an extensive cluster-DMFT calculations for the different temperatures and doping.

We would like to note that the existing of d-SC cluster-DMFT solution for such high temperature does not necessary means that the superconducting transition temperature is larger then 190K in our model. A crude estimation shows that d-SC solution disappear at T=300 K for x=0.15 and AFM solutions for x=0 become unstable at the temperature just above 1000 K. This could be the sign of a "local" AFM solution and a local d-wave solution, like local moments in magnetic systems [6]. Due to a multiscale nature of the problem under considerations: essentially different energies connected with local moment formation, long-range magnetic order, local d-wave pairs within the 2×2 plaquet, and finally coherent superconductivity, it is difficult to distinguish a real long-range ordering from slow dynamical fluctuations in our QMC simulations. We plane to separate these energy scales analytically and estimate superconductiong transition temperature in a future publication.

The role of next-nearest hopping is to lowering of the van-Hove singularity [74] which increase the density of state at the Fermi level for the hole-doped case and favored the d-SC solution for a moderate correlation strength. There is also change in the spin-fluctuation spectrum related with the broadening of AFM-peak near (π, π) point due to formation of so-called extended van-Hove singularities with increasing of the t'. We show one of the AFM-dSC solution on the Fig.6 with the next nearest-neighbor hopping for the 10% doping level and β = 50 eV^{-1}. Resulting local magnetic moment is M=0.28 μ_B and d-SC order parameter $F(0)$=0.036. One can see that the superconduction order parameter is really of the $d_{x^2-y^2}$ symmetry since diagonal elements (F^{00}) as well as the next nearest-neighbors elements (F^{02}) are all equal to zero and only the nearest neighbor superconducting Green functions (F^{01}) are non-zero and change the sign for F_x and F_y components. The normal local Green function (G^{00}) (plotted for the spin-up atom in the Fig.6) as well as (G^{02}) are spin-split, while the nearest-neighbor Green function (G^{01}) has no spin-splitting due to AFM spin symmetry (see Fig.5). The absence of magnetic polarization in the non-diagonal G-function along the x(y)-directions suppress the magnetic pair-breaking and makes the AFM-dSC coexistence possible.

9. Summary

Spectral density functional (SDF) approach allows to study the correlation effects in solids basing on the realistic electronic structure calculations. Among all possible applications we have chosen the magnetism of transition metals and high-temperature superconductivity. From these two examples one can see already all the main advantages of the new approach in comparison with the standard density functional theory. First, we can describe now the spectral density transfer phenomena (e.g., the formation of 6 eV satellite in Ni), the quasiparticle damping and other effects connected with the frequency dependence of the self-energy; they are absent completely not only in th DF approach but also in the Hartree-Fock, LDA+U, or self-interaction corrections approximation (see, e.g., [63]). Second, we can describe adequately the contribution of the Bose degrees of freedom (e.g., spin fluctuations) to the electronic structure and thermodynamic properties. In the DF-based calculations the temperature is really taken into account only via the thermal expansion and the Fermi distribution function [78]. It was the main reason of the failure of the standard band theory for the description of finite-temperature effects in magnetic metals. We show that the SDF gives a satisfactory solution of this problem.

The most of real applications of the SDF approach are connected with the single-site dynamical mean field theory. At the same time, for a number of problems this can be not sufficient and some generalizations to take into account the non-local effects are necessarym for example, cluster ones. The first attempts of such generalizations are already leading to some interesting results in the theory of high-T_c superconductivity [21, 22] but a lot of additional work is required in the area of cluster dynamical mean field theories to reach the level of understanding that was reached in single site DMFT.

Acknowledgments

The work was supported by the Netherlands Organization for Scientific Research (NWO project 047-008-16) and partially supported by Russian Basic Research Foundation (grant 00-15-96544). GK is supported by the NSF under grantNSF DMR 0096462 , by the division of Basic Sciences of the Department of Energy under grant DE-FG02-99ER45761, and by the ONR under grant N000149910653.

Appendix: Multi-orbital QMC scheme

The multi-orbital DMFT problem and general cluster DMFT scheme can be reduced to the general impurity action:

$$S = -\int_0^\beta d\tau \int_0^\beta d\tau' \sum_{i,j} c_i^+(\tau)\mathcal{G}_{ij}(\tau-\tau')c_j(\tau') + \frac{1}{2}\int_0^\beta d\tau \sum_{i,j} n_i(\tau)U_{ij}n_j(\tau)$$

where $i = \{m, \sigma\}$ - orbital (site) and spin. Without spin-orbital coupling we have: $\mathcal{G}_{ij} = \mathcal{G}^{\sigma}_{m,m'}\delta_{\sigma\sigma'}$.

The auxiliary fields Green-function QMC use the discrete Hubbard-Stratanovich transformation inroduced by Hirsch[42]

$$\exp\{\Delta\tau U_{ij}[n_i n_j - \frac{1}{2}(n_i + n_j)]\} = \frac{1}{2}\sum_{S_{ij}=\pm 1} \exp\{\lambda_{ij} S_{ij}(n_i - n_j)\}$$

where $S_{ij}(\tau)$ are the auxiliary Ising fields for each pair of orbitals and time slice with the strength:

$$\lambda_{ij} = \text{arccosh}[\exp(\frac{\Delta\tau}{2}U_{ij})]$$

Using Hirsch transofrmation one can integrated out fermionic fields in the path integral[3] and resulting partition function and Green function matrix have the following form:

$$\begin{aligned} Z &= \frac{1}{2^{N_f L}} \sum_{S_{ij}(\tau)} \det[\widehat{G}^{-1}(S_{ij})] \\ \widehat{G} &= \frac{1}{Z}\frac{1}{2^{N_f L}} \sum_{S_{ij}(\tau)} \widehat{G}(S_{ij}) \det[\widehat{G}^{-1}(S_{ij})] \end{aligned}$$

where N_f is the number of Ising fields, L is the number of time slices, and $\widehat{G}(S_{ij})$ is the Green function in the auxiliary Ising fileds:

$$\begin{aligned} G_{ij}^{-1}(S) &= \mathcal{G}_{ij}^{-1} + \Delta_i \delta_{ij} \delta_{\tau\tau'} \\ \Delta_i &= (e^{V_i} - 1) \\ V_i(\tau) &= \sum_{j(\neq i)} \lambda_{ij} S_{ij}(\tau) \sigma_{ij} \end{aligned}$$

here we introduce the generalized Pauli matrix:

$$\sigma_{ij} = \{ \begin{array}{c} +1, i < j \\ -1, i > j \end{array}$$

For efficient calculations of the Green function in arbitrary configuration of Ising fields $G_{ij}(S)$ we use the following Dyson equation [43]:

$$G' = [1 + (1 - G)(e^{V' - V} - 1)]^{-1} G$$

The QMC important sampling scheme allowed us to integrated over the Ising fields with the abs($\det[\widehat{G}^{-1}(S_{ij})]$) as a stochastic weight[43, 3]. For a single spin-flip S_{ij}, the determinant ratio calculated as following:

$$\begin{aligned} \det[\widehat{G}]/\det[\widehat{G}'] &= R_i R_j - R_{ij} \\ R_i &= 1 + [1 - G_{ii}(\tau,\tau)]\Delta_i(\tau) \\ R_j &= 1 + [1 - G_{jj}(\tau,\tau)]\Delta_j(\tau) \\ R_{ij} &= G_{ij}(\tau,\tau)\Delta_j(\tau) G_{ji}(\tau,\tau)\Delta_i(\tau) \end{aligned}$$

and the Green function matrix updated in the standard maner[43, 3]:

$$\begin{aligned} G'_{i_1 j_2}(\tau_1,\tau_2) &= G_{i_1 j_2}(\tau_1,\tau_2) + [G_{i_1 i}(\tau_1,\tau) - \delta_{i_1 i}\delta_{\tau_1,\tau}]\Delta_i(\tau)/R_i(\tau) G_{i j_2}(\tau,\tau_2) \\ G^{new}_{i_1 j_2}(\tau_1,\tau_2) &= G'_{i_1 j_2}(\tau_1,\tau_2) + [G'_{i_1 j}(\tau_1,\tau) - \delta_{i_1 j}\delta_{\tau_1,\tau}]\Delta_j(\tau)/R_j(\tau) G'_{j j_2}(\tau,\tau_2) \end{aligned}$$

References

[1] P. Hohenberg and W. Kohn, Phys. Rev.**136**, B864 (1964); W. Kohn and L. J. Sham, **140**, A1133 (1965).

[2] R. O. Jones and O. Gunnarsson, Rev. Mod. Phys. **61**, 689 (1989).

[3] A. Georges, G. Kotliar, W. Krauth, and M. Rozenberg, Rev. Mod. Phys.**68**, 13 (1996).

[4] V. I. Anisimov, A. I. Poteryaev, M. A. Korotin, A. O. Anokhin, and G. Kotliar, J. Phys.: Condens. Matter **9**, 7359 (1997).

[5] A. I. Lichtenstein and M. I. Katsnelson, Bull. Am. Phys. Soc. **42**, 573 (1997); Phys. Rev. B **57**, 6884 (1998).

[6] M. I. Katsnelson and A. I. Lichtenstein, J. Phys.: Condens. Matter **11**, 1037 (1999).

[7] M. I. Katsnelson and A. I. Lichtenstein, Phys. Rev. B **61**, 8906 (2000).

[8] R. Chitra and G. Kotliar, Phys. Rev. B **62**, 12715 (2000).

[9] K. Held, L. A. Nekrasov, N. Blumer, V. I. Anisimov, and D. Vollhardt, Int. J. Modern Phys. B **15**, 2611 (2001).

[10] A. I. Lichtenstein, M. I. Katsnelson, and G. Kotliar, Phys. Rev. Lett. **87**, 067205 (2001).

[11] S. Y. Savrasov, G. Kotliar, and E. Abrahams, Nature **410**, 793 (2001).

[12] A. I. Lichtenstein and M. I. Katsnelson, in: *Band Ferromagnetism. Ground State and Finite-Temperature Phenomena* (Lecture Notes in Physics, Springer, Berlin, 2001), ed. by K. Baberschke, M. Donath, and W. Nolting, p. 75.

[13] S. Y. Savrasov and G. Kotliar, cond-mat/0106308.

[14] A. A. Abrikosov, L. P. Gorkov, and I. E. Dzyaloshinski, *Methods of Quantum Field Theory in Statistical Physics* (Dover, New York, 1975).

[15] R. Fukuda, T. Kotani, and S. Yokojima, Prog. Theor. Phys. **92**, 833 (1994); R. Fukuda, M. Komachiya, S. Yokojima, Y. Suzuki, K. Okumura, and T. Inagaki, Prog. Theor. Phys. Suppl. **121**, 1 (1996).

[16] M. Valiev and G. Fernando, Phys. Lett. A **227**, 265 (1997).

[17] H. Kajueter, Rutgers University Ph.D. Thesis (1996); H. Kajueter and G. Kotliar, Rutgers University preprint (1996); Q. Si and J.L. Smith, Phys. Rev. Lett **77**, 3391 (1997); R. Chitra and G. Kotliar, Phys. Rev. Lett. **84**, 3500 (2000).

[18] S. Pankov and G. Kotliar, cond-mat/0112083.

[19] P. Chandra, P Coleman, and A. Larkin, J. Phys.: Condens. Matter **2**, 7933 (1990).

[20] A. Khurrana, Phys. Rev. Lett., to be published.

[21] M. H. Hettler, S. Mukherjee, M. Jarrell, and H. R. Krishnamurthy, Phys. Rev. B **61**, 12739 (2000); T. Maier, M. Jarrell, T. Pruschke, and J. Keller, Phys. Rev. Lett. **85**, 1524 (2000).

[22] A. I. Lichtenstein and M. I. Katsnelson, Phys. Rev. B **62**, R9283 (2000).

[23] G. Kotliar, S. Y. Savrasov, G. Palsson, and G. Biroli, Phys. Rev. Lett. **87**, 186401 (2001)

[24] V. V. Dobrovitski, M. I. Katsnelson, and B. N. Harmon, J. Magn. Magn. Mater. **221**, L235 (2000); cond-mat/0111324.

[25] D. N. Zubarev, *Nonequilibrium Statistical Thermodynamics* (Consultant Bureau, New York, 1974).

[26] L. D. Faddeev and A. A. Slavnov, *Gauge Fields: Introduction to Quantum Theory* (Benjamin, Reading Mass., 1980).

[27] J. M. Luttinger and J. C. Ward, Phys. Rev.**118**, 1417 (1960); see also G. M. Carneiro and C. J. Pethick, Phys. Rev. **B11**, 1106 (1975).

[28] A. R. Mackintosh and O. K. Andersen, in: *Electron at the Fermi Surface*, ed.M. Springford (Univ. Press, Cambridge, 1980), p.145.

[29] A. I. Liechtenstein, M. I. Katsnelson, V. P. Antropov, and V. A. Gubanov, J. Magn. Magn. Mater. **67**, 65 (1987).

[30] C. Herring *Magnetism*, vol. 4 (Academic Press, New York, 1966), ed. by G. T. Rado and H. Suhl.

[31] F. Aryasetiawan and K. Karlsson, Phys. Rev. B **60**, 7419 (1999).

[32] G. M. Stocks, B. Ujfalussy, X. Wang, Y. Wang, D. M. C. Nicholson, W. A. Shelton, A. Canning, and B. L. Gyorffy, Philos. Mag. B **78**, 665 (1998).

[33] S. V. Vonsovsky, *Magnetism* (Wiley, New York, 1974).

[34] T. Moriya, *Spin Fluctuations in Itinerant Electron Magnetism* (Springer, Berlin, 1985).

[35] A. Liebsch, Phys. Rev. B **23**, 5203 (1981).

[36] G. Treglia, F. Ducastelle, and D. Spanjaard, J. Phys. (Paris) **43**, 341 (1982).

[37] F. Manghi, V. Bellini, and C. Arcangelli, Phys. Rev. B **56**, 7149 (1997); F. Manghi, V. Bellini, J. Osterwalder, T. J. Kreutz, P. Aebi, and C. Arcangeli Phys. Rev. B **59**, R10409 (1999).

[38] W. Nolting, S. Rex, and S. Mathi Jaya, J. Phys. : Condens. Matter **9**, 1301 (1987).

[39] M. M. Steiner, R. C. Albers, and L. J. Sham, Phys. Rev. B **45**, 13272 (1992).

[40] O. K. Andersen and O. Jepsen, Phys. Rev. Lett. **53**, 2571 (1984).

[41] T. Bandyopadhyay and D. D. Sarma, Phys. Rev. B **39**, 3517 (1989).

[42] J. E. Hirsch, Phys. Rev. B **28**, 4059 (1983).

[43] J. E. Hirsch and R. M. Fye, Phys. Rev. Lett. **25**, 2521 (1986).

[44] K. Takegahara, J. Phys. Soc. Japan **62**, 1736 (1992); M. J. Rozenberg, Phys. Rev. B **55**, R4855, (1997).

[45] M. Jarrell and J. E. Gubernatis, Physics Reports**269**, 133 (1996).

[46] M. Iwan, F. J. Himpsel, and D. E. Eastman, Phys. Rev. Lett. **43**, 1829 (1979).

[47] W. Eberhardt and E. W. Plummer, Phys. Rev. B **21**, 3245 (1980).

[48] K. N. Altmann, D. Y. Petrovykh, G. J. Mankey, N. Shannon, N. Gilman, M. Hochstrasser, R. F. Willis, and F. J. Himpsel, Phys. Rev. B **61**, 15661 (2000).

[49] N. M. Rosengaard and B. Johansson, Phys. Rev. B **55**, 14975 (1997).

[50] S. V. Halilov, H. Eschrig, A. Y. Perlov, and P. M. Oppeneer, Phys. Rev. B **58**, 293 (1998).

[51] V. P. Antropov, M. I. Katsnelson, B. N. Harmon, M. van Schilfgaarde, and D. Kusnezov, Phys. Rev. B **54**, 1019 (1996).

[52] M. Pajda, J. Kudrnovsky, I. Turek, V. Drchal, and P. Bruno, Phys. Rev. B **64**, 174402 (2001).

[53] J. B. Staunton and B. L. Gyorffy, Phys. Rev. Lett. **69**, 371 (1992).

[54] *Ferromagnetic materials*, vol. 1, ed. by E. P. Wolfarth (North-Holland, Amsterdam, 1986).

[55] V. Yu. Irkhin and M. I. Katsnelson, Physics - Uspekhi **37**, 659 (1994).

[56] E. Kisker, K. Schröder, M. Campagna, and W. Gudat , Phys. Rev. Lett. **52**, 2285 (1984); A. Kakizaki, J. Fujii, K. Shimada, A. Kamata, K. Ono, K.H. Park, T. Kinoshita, T. Ishii, and H. Fukutani, Phys. Rev. Lett. **72**, 2781 (1994).

[55] V. Yu. Irkhin and M. I. Katsnelson, Physics - Uspekhi **37**, 659 (1994).
[56] E. Kisker, K. Schröder, M. Campagna, and W. Gudat , Phys. Rev. Lett. **52**, 2285 (1984); A. Kakizaki, J. Fujii, K. Shimada, A. Kamata, K. Ono, K.H. Park, T. Kinoshita, T. Ishii, and H. Fukutani, Phys. Rev. Lett. **72**, 2781 (1994).
[57] H. A. Mook and J. W. Lynn, J. Appl. Phys. **57**, 3006 (1985).
[58] B. Sinkovic, L. H. Tjeng, N. B. Brookes, J. B. Goedkoop, R. Hesper, E. Pellegrin, F. M. F. de Groot, S. Altieri, S. L. Hulbert, E. Shekel, and G. A. Sawatzky , Phys. Rev. Lett. **79**, 3510 (1997).
[59] T. J. Kreutz, T. Greber, P. Aebi, and J. Osterwalder, Phys. Rev. B **58**, 1300 (1998).
[60] N. E. Bickers and D. J. Scalapino, Ann. Phys. (N.Y.) **193**, 206 (1989).
[61] V. M. Galitskii, ZhETF**34**, 151, 1011 (1958)
[62] J. Kanamori, Prog. Theor. Phys. **30**, 275 (1963).
[63] V. I. Anisimov, F. Aryasetiawan, and A. I. Lichtenstein, J. Phys.: Condens. Matter **9**, 767 (1997).
[64] T. Izuyama, D. Kim, and R. Kubo, J. Phys. Soc. Japan **18**, 1025 (1963).
[65] H. Kajueter and G. Kotliar , Phys. Rev. Lett. **77**, 131 (1996).
[66] V. V. Mazurenko, A. I. Lichtenstein, M. I. Katsnelson, I. Dasgupta, T. Saha-Dasgupta, and V. I. Anisimov, cond-mat/0107200.
[67] P.W. Anderson, *The Theory of Superconductivity in the High-T_c Cuprate Superconductors* (Univ. Press, Princeton, 1997).
[68] D. J. Scalapino, Physics Reports **251**, 1 (1994), J. Low Temp. Phys. **117**, 179 (1999).
[69] J. Schmalian, D. Pines, and B. Stojkovic, Phys. Rev. Lett. **80**, 3839 (1998).
[70] A.G. Loeser, Science **273**, 325 (1996); H. Ding, Nature **382**, 51 (1996).
[71] G. Aeppli, T. E. Mason, S. M. Hayden, H. A. Mook, and J. Kulda, Science **278**, 1432 (1997); H. A. Mook, P. C. Dai, S. M. Hayden, G. Aeppli, T. G. Perring, and F. Dogan, Nature **395**, 580 (1998).
[72] E. Demler and S.C. Zhang, Nature **396**, 733 (1998).
[73] E. Dagotto, Rev. Mod. Phys. **66**, 763 (1994).
[74] O.K.Andersen, A. I. Liechtenstein, O. Jepsen, and F. Paulsen, J. Phys. Chem. Solids **56**, 1537 (1995).
[75] J. R. Schrieffer, *Theory of Superconductivity* (Benjamin, New York, 1964); S. V. Vonsovsky, Yu. A. Izyumov, and E. Z. Kurmaev, *Superconductivity of Transition metals, Their Alloys and Compounds* (Springer, Berlin, 1982).
[76] A. Georges, G. Kotliar, and W. Krauth, Z. Phys. B **92**, 313 (1993).
[77] G. E. Volovik and L. P. Gor'kov, ZhETF **88**, 1412 (1985).
[78] T. Jarlborg, Rep. Prog. Phys. **60**, 1305 (1997).

ORIGINS OF HEAVY QUASIPARTICLES

Peter Fulde

Max Planck Institute for the Physics of Complex Systems

Nöthnitzer Str. 38, 01187 Dresden, Germany

fulde@mpipks-dresden.mpg.de

Abstract Two new microscopic models for the appearance of heavy quasiparticles have been recently suggested. They concern UPt_3 and LiV_2O_4 and are based on a partial localization of $5f$ electrons and on geometrical frustrations, respectively.

1. Introduction

A hallmark of strongly correlated metallic electron systems is the appearance of low-energy scales. In an ordinary metal the characteristic scale is provided by the Fermi energy ϵ_F. This quantity is typically of the order of 5 eV and the associated temperature is approximately 50000 K. In distinction, systems with strongly correlated electrons are found have temperature scales of the order of a few hundred K or less. One consequence is that the systems have low-temperature thermodynamics properties resembling those of free electrons with strongly renormalized, i.e., large effective masses. Therefore these metals are often called heavy-fermion systems and it is of interest to know which microscopic processes are responsible for their existence. Originally the Kondo effect was thought to be the only source of heavy-quasiparticle formation [1]. However in the meantime several different processes have been identified as candidates for heavy quasiparticle behavior. A possibly incomplete list is the following:

Kondo effect:	$CeAl_3$, $CeCu_2Si_2$		[1]
Zeeman effect:	$Nd_{2-x}Ce_xCuO_4$;	$x \leq 0.2$	[2, 3]
Charge ordering:	Yb_4As_3		[4, 5, 6]
Quantum critical point:	YMn_2		[7]
Partial electron localization:	UPt_3		[8, 9]
Geometrical frustration:	LiV_2O_4		[10, 11]

The last two scenarios have been proposed only recently and we will limit ourselves in the following to their discussion.

2. UPt_3 - a system with partially localized $5f$ electrons

Consider UPt_3. This compound is one of the most studied heavy-fermion systems [12, 13]. The γ coefficient of the low-temperature specific heat $C=\gamma T$ is $\gamma=420$ mJ $(\text{mole}\cdot K^2)^{-1}$ and the Pauli susceptibility is similarly enhanced as compared with the one of ordinary metals like sodium. Heavy quasiparticles have been observed in de Haas-van Alphen (dH-vA) experiments [14, 15] but the observed masses are by a factor of 20 larger than the calculated band masses based on the local-density approximation (LDA) to density functional theory [16]. The dH-vA frequencies are rather well reproduced within that approximation. This failure of LDA in predicting the observed masses indicates that the correlations are strong and raises the question for their adequate description.

We have recently suggested that in UPt_3 one is dealing with localized as well as delocalized $5f$ electrons [9]. This proposal was motivated by accurate quantum chemical calculations on $U(C_8H_8)_2$. Thereby both types of $5f$ electrons were found to exist, giving rise to low-lying excitations. They correspond to intraatomic rearrangements of the localized $5f$ electrons [17]. The mass enhancement of the quasiparticles results then from virtual excitations of the internal degrees of freedom of the localized $5f$ electrons by the delocalized ones, which are part of the conduction electron system. Note the similarity of this source of the mass enhancement of conduction electrons to the one in Pr metal [18].

We consider two $5f$ electrons as being localized and put them into $j_z = \pm\frac{1}{2}$ and $\pm\frac{5}{2}$ orbitals. This is consistent with the ground state of UPd_2Al_3 [19, 20], a different but closely related U compound. Electrons with $j_z = \pm\frac{3}{2}$ are treated as being delocalized. Performing a LDA band-structure calculation, thereby keeping the localized $5f$ electrons as fixed yields the measured dH-vA frequencies, a satisfying feature of the proposed theory. This agreement is not obtained when $5f$ electrons with $j_z = \pm\frac{5}{2}$

or $\pm\frac{1}{2}$ are delocalized instead. We want to point out that this result is obtained without any adjustments of the bands. The band mass is found to be twice as large as in the conventional LDA calculation where all $5f$ electrons are treated as itinerant. Thus a discrepancy to the observed quasiparticle mass by a factor of 10 remains to be explained. The Fermi surface is formed by two bands which are doubly degenerate and have partially $5f$ $(j, j_z) = (\frac{5}{2}, \pm\frac{3}{2})$ character. One band gives rise to open orbits for a magnetic field along the c-axis in agreement with magnetoresistance measurements [21]. The other one gives rise to a cylindrical surface with six arms in the a-b plane. Inside that surface is a Γ-centered hole surface in form of an ellipsoid. In addition there are ellipsoids centered at H points.

The missing factor of 10 in the quasiparticle mass is explained by the all important excitation of internal degrees of freedom of the localized $5f$ electrons by the delocalized ones. It is given by

$$\frac{m^*}{m} = 1 + 4a^2 N(0)\ 2\frac{|\ M\ |^2}{\delta} \quad . \tag{1.1}$$

Here a denotes the weight of the delocalized $5f$ orbital in the conduction-electron states near the Fermi energy, and N(0) is the density of states per spin direction of the latter. The transition-matrix element between the ground state of the localized $5f$ electrons and the low-energy excited state is denoted by M and δ is that excitation energy. In order to determine the multiplet structure of the localized f^2 states we diagonalize a 6 x 6 Coulomb matrix based on the six two-particle states which can be build from $|j = \frac{5}{2}; j_z = \pm\frac{5}{2}\rangle$ and $|j = \frac{5}{2}; J_z = \pm\frac{1}{2}\rangle$ states. The matrix elements are determined according to Condon-Shortley [22]. It is found that the ground state is doubly degenerate with total angular momentum $J = 4$ and $J_z = \pm 3$. The degeneracy is lifted by the crystal field giving rise to two singlets

$$|\ \Gamma_{3,4}\rangle = \frac{1}{\sqrt{2}}(|\ J = 4; J_z = 3\rangle \pm |\ J = 4; J_z = -3\rangle) \quad . \tag{1.2}$$

We assume that the center of gravity of the broadened energy splitting is of order δ=20 meV. This is the same order of magnitude as the splitting which has been assumed for UPd_2Al_3 [19] or UBe_{13}. More recently for UPd_2Al_3 a smaller splitting of 7 meV has been proposed based on a fit of the susceptibility and other physical properties [20]. For URu_2Si_2 a splitting of the lowest CEF states of 6.5 meV was suggested. Note that in all these cases no crystalline field splitting has been directly observed by inelastic neutron scattering. Instead the above values were derived indirectly. In the present theory the next-higher states are 0.2 eV above the two singlets and can be neglected. This way we find for the transition matrix element

$$M = \langle f'; \frac{5}{2}, \frac{3}{2} \mid \otimes \langle \Gamma_4 \mid U_{\text{coul}} \mid \Gamma_3 \rangle \otimes \mid f'; \frac{5}{2}, \frac{3}{2} \rangle = 0.19 \text{eV} \quad . \qquad (1.3)$$

For more details we refer to [9]. With N(0) $\simeq$ 15.5 states/(eV cell), $|M|^2 = 36 \cdot 10^{-3}$ eV^2, $4a^2$=0.13 and δ=20 meV we find that m^*/m_b=8.3 which is close to the missing enhancement factor. Note that all parameters appearing in (1.1) have been *calculated* except δ which we have estimated as explained above. When a smaller value is chosen instead the quasiparticle mass increases according to (1.1). We believe that the above theory supplies a satisfactory explanation of the physical origin of the heavy quasiparticle mass in UPt_3.

3. LiV_2O_4 - a frustrated magnetic system

Next we consider LiV_2O_4. This material shows many of the low-temperature thermodynamic properties of heavy-fermion systems. The specific heat has a γ coefficient of $\gamma \simeq$0.4 J (mole $\cdot$ $K^2)^{-1}$ and a broad maximum around 16 K [10]. When integrated up to 60 K the associated entropy is of order R $ln2$ where R is the gas constant. So when heated up to that temperature almost one excitation per V ion takes place, a result inconceivable for a partially filled conduction band with a width of a few eV. The spin susceptibility is similarly enhanced again with a broad shallow maximum at 16 K [23]. No magnetic ordering has been found down to 20 mK. The low-temperature resistivity is of the form $\rho(T)=\rho_0+AT^2$ with a large A coefficient. The system represents the rare case of heavy-fermion quasiparticles not involving f electrons but 3d electrons instead. LDA band-structure calculations show that the conduction band is of 3d t_{2g} character. The calculated band mass is by a factor of 25 smaller than the one corresponding to the large γ coefficient [24] . Again this points towards strong electronic correlations.

The model we have suggested for explaining the large γ coefficient is based on the cubic spinel structure of LiV_2O_4. Within that structure the V ions form a pyroclore lattice, i.e., one consisting of corner-sharing tetrahedra. We assume strong on-site Coulomb repulsions so that the V ions have 50 % d^1 and 50 % d^2 configurations. In the case of a d^2 configuration a S=1 state forms due to a strong Hund's rule coupling. The on-site repulsions suppress configurations d^0, d^3 etc. but obviously cannot explain the large effective masses. Therefore we assume that also the interactions of electrons on neighboring sites are strong. This implies that for an average d-count per V ion of 1.5, d^1-d^1 as well as d^2-

d^2 configurations repel each other when situated on neighboring sites. The repulsions are minimized when on each tetrahedron two sites are in a d^1 (spin $\frac{1}{2}$) and two sites are in a d^2 (spin 1) configuration. We call this the tetrahedron rule which goes back to Verwey [25] and Anderson [26]. One checks easily that all configurations $|\phi_n\rangle$ which satisfy the tetrahedron rule consist of chains and rings of S=$\frac{1}{2}$ and S=$\frac{1}{2}$ sites (see Fig 1).

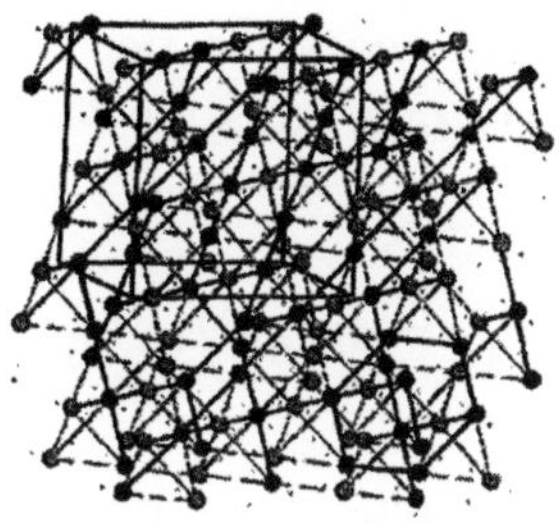

Figure 1.1. Example of a configuration $|\phi_n\rangle$ obeying the tetrahedron rule. Solid dots: spin $\frac{1}{2}$ sites; empty dots: spin 1 sites. They form chains and rings each. Notice a ring consisting of six spin $\frac{1}{2}$ sites in the upper right corner.

Within that manifold the ground state is a superposition of all those states, i.e.,

$$| \psi_0\rangle = \sum_n \alpha_n \mid \phi_n\rangle \quad . \tag{1.4}$$

A ring consists of at least 6 sites of one type, i.e., S=$\frac{1}{2}$ or S=1 sites and possible ring sizes increase in units of two. Let us pick out an arbitrary configuration $|\phi_\nu\rangle$ like the one shown in Fig 1. Assume that the spin $\frac{1}{2}$ chains (or rings) would not couple to the spin 1 chains (rings). Then the corresponding excitations of that particular $|\phi_\nu\rangle$ would lead to a linear specific heat because we are dealing with independent spin $\frac{1}{2}$ Heisenberg chains separated by independent spin 1 chains. The γ value is closely related to the coupling constant between neighboring spins $\frac{1}{2}$. Next we consider the case that there is also a coupling between a spin $\frac{1}{2}$ and a neighboring spin 1 as well as between two neighboring spins 1. With the help of a LDA + U calculation we can determine (or better estimate) the different coupling constants [10]. We find a value of J_1=24 meV for the interaction of two neighboring spins 1 and a value of J_3=3 meV for the antiferromagnetic coupling of two neighboring spins $\frac{1}{2}$. The coupling of a spin 1 with a spin $\frac{1}{2}$ is intermediate. Because of the geometrical frustration inherent in the pyroclore structure and the fact that $J_3 \simeq 8J_1$ the spin $\frac{1}{2}$ chains cannot couple with each other via spin 1 chains. This would require spin flips within the spin 1 chains which are difficult to achieve because of the large value of J_1 and the fact that a spin 1 chain has a gap in the excitation spectrum (Haldane gap). Consequently we are dealing again with uncoupled spin $\frac{1}{2}$ chains (rings) and a linear specific heat $C=\gamma T$ results from the spin degrees of freedom. The γ coefficient is given by [21]

$$\gamma = \frac{2}{3}\frac{k_B R}{J_3} \tag{1.5}$$

taking into account that only one V ion per formula unit is contributing to a spin $\frac{1}{2}$ chain. Similarly one finds for the spin susceptibility of the spin $\frac{1}{2}$ chains

$$\chi_{sp} = 4\mu_{\text{eff}}^2 R/(\pi^2 J_3) \tag{1.6}$$

where μ_{eff} is the effective magnetic moment. In order to explain the observed size of γ a value of J_3/k_B=13.3 K is required. Up to a factor of 2.6 this agrees with the calculated value of J_3=3 mV or J_3/k_B=35 K. Note that the observed Sommerfeld-Wilson ratio of 1.7 is relatively close to the value of 2 following from (1.5) and (1.6). One should keep also in mind that electron-phonon interactions contribute to γ in the form of $\gamma = \gamma_0(1 + \lambda_{e-p})$ where γ_0 is the value without that interaction. But they also change the Sommerfeld-Wilson ratio accordingly since λ_{e-p} does not contribute to χ_{sp}. In order to compare the measured Sommerfeld-Wilson ratio with the theoretical one, it must be kept in mind that the measured spin susceptibility contains also contributions from the S=1 chains. The theory of the specific heat of Heisenberg spin $\frac{1}{2}$ chains predicts a maximum in $\gamma(T)=C(T)/T$ which is not observed. But a corresponding maximum in χ_{sp} is observed around $T\simeq16$ K which would imply a value of $J_3 \approx 27$ K. Although the agreement between theory and experiment is not perfect the model seems to contain the right ingredients in view of the relatively small discrepancies. The latter are much bigger when one starts out from a conventional band picture. As explained before, the discrepancy between calculated and required quasiparticle mass is a factor of 25. The above considerations have used an arbitrary but fixed configuration $|\phi_\nu\rangle$ for the calculations. But it must be kept in mind that LiV_2O_4 is a metal. When an electron hops from a S=1 to a S=$\frac{1}{2}$ site the two sites change their configurations. With a minimum of three such changes one can go over from one configuration $|\phi_n\rangle$ to another one $|\phi_m\rangle$ contained in $|\psi_0\rangle$. This is seen by considering a ring of six alternating spin $\frac{1}{2}$ and spin 1 sites. Those hops will predominantly take place between site with nearly the same spin direction because of Hund's rule coupling. They favor a ferromagnetic spin alignment. Thereby it is unimportant whether we deal with direct hopping processes or indirect ones involving oxygen ions. We note that spin-spin interactions mediated by oxygen ions via superexchange are antiferromagnetic. These dynamical effect, including those which result in occasional violations of the tetrahedron rule have been left out in the model described afore.

Acknowledgments

I would like to thank Profs. G. Zwicknagl, A. N. Yaresko and A. Zvyagin for many fruitful discussions.

References

[1] Hewson, A. C. (1993). *The Kondo Problem to Heavy Fermions.* Cambridge University Press, Cambridge and references therein.

[2] Brugger, T., T. Schreiner, G. Roth, P. Adelmann, and G. Czjzek (1993). Phys. Rev. Lett. **71**, 2481.

[3] Fulde, P., V. Zevin, and G. Zwicknagl (1993). Z. Phys. B **92**, 133.

[4] Ochia, A., T. Suzuki, and T. Kasuya (1990). J. Phys. Soc. Jpn. **59**, 4129.

[5] Fulde, P., P. Thalmeier, and B. Schmidt (1995). Europhys. Lett. **35**, 109.

[6] Kohgi, M., K. Iwasa, J.-M. Mignot, A. Ochiai, and T. Suzuki (1997). Phys. Rev. B **56**, R 11388.

[7] Pinettes, C., and C. Lacroix (1994). J. Phys. Cond. Mat. **6**, 10093.

[8] Stewart, G. R., Z. Fisk, J. O. Willis, and J. L. Smith (1984). Phys. Rev. Lett. **52**, 679.

[9] Zwicknagl, G., A. N. Yaresko, and P. Fulde (to appear). Phys. Rev. B, Rapid Comm.

[10] Kondo, S., D. C. Johnston, C. A. Swenson, F. Borsa, A. V. Mahajan, L. L. Miller, T. Gu, A. I. Goldman, M. B. Maple, D. A. Gajewski, E. J. Treemann, N. R. Dilley, R. P. Dickey, J. Merrin, K. Kojima, G. M. Luke, Y. J. Kemura, O. Chaissem, and J. D. Jorgensen (1997). Phys. Rev. Lett. **78**, 3729.

[11] Fulde, P., A. N. Yaresko, A. Zvyagin, and Y. Grin (2001). Europhys. Lett. **54**, 779.

[12] Stewart, G. (1984). Rev. Mod. Phys. **56**, 755.

[13] Grewe, N., and F. Steglich (1991). Handbook Phys. and Chem. Rare Earths, eds. K. A. Gschneidner, Jr., and L. Eyring, Vol. **14**, 343. (Elsevier, Amsterdam).

[14] Taillefer, L., R. Newbury, G. G. Lonzarich, Z. Fisk, and J. L. Smith (1987). J. Magn. Magnet. Mat. **63+64**, 372.

[15] Kimura, N., T. Komatsubara, D. Aoki, Y. Onuki, Y. Haga, E. Yamamoto, H. Aoki, and H. Harima (1998). J. Phys. Soc. Jpn. **67**, 2185.

[16] Norman, M. R., R. C. Albers, A. M. Boring, and N. E. Christensen (1988). Solid State Comm. **68**, 245.

[17] Liu, W., M. Dolg, and P. Fulde (1997). J. Chem. Phys. **107**, 3584.

[18] White, R., and P. Fulde (1981). Phys. Rev. Lett. **47**, 1540.

[19] Krimmel, A., A. Loidl, R. Eccleston, C. Geibel, and F. Steglich (1996). J. Phys. Cond. Mat. **8**, 1677.

[20] Sato, N. K., N. Aso, K. Miyake, R. Shiina, P. Thalmeier, G. Varelogiannis, C. Geibel, F. Steglich, P. Fulde, and T. Komatsubara (2001). Nature **410**, 340.

[21] Lonzarich, G. (unpublished)

[22] Condon, E. U., and G. H. Shortley (1957). *The Theory of Atomic Spectra*, Cambridge University Press, Cambridge.

[23] Kondo, S., D. C. Johnston, and L. L. Miller (1999). Phys. Rev. B **59**, 2609.

[24] Eyert, V., K. H. Hock, S. Horn, A. Loidl, and P. S. Riseborough (1999). Europhys. Lett. **46**, 762.

[25] Verwey, E. J. W., and P. W. Haaymann (1941). Physica **8**, 979.

[26] Anderson, P. W. (1956). Phys. Rev. **102**, 1008.

[27] Bonner, J. C., and M. E. Fisher (1964). Phys. Rev. A **135**, 640.

ORBITAL PROPERTIES OF VANADIUM IONS IN MAGNETICALLY ORDERED V_2O_3

S. W. Lovesey, K. S. Knight and D. S. Sivia*

1. INTRODUCTION

Vanadium sesquioxide (V_2O_3) displays a number of electronic, magnetic and structural properties that are challenging to interpret and explain[1–7]. At room temperature V_2O_3 has the corundum structure with space group $R\bar{3}c$, and it is metallic and paramagnetic. On reducing the temperature the corundum structure distorts to a monoclinic structure with space group *I2/a*. The structural transition in the temperature range 150 – 160K is strongly first-order and ferroelastic. Accompanying the transition are a change from metallic to insulating behaviour and the onset of antiferromagnetic order[3]. The metal-insulator transition is viewed as a classic Mott transition, in which spin and orbital degrees of freedom are independent.

Recently, results from resonant x-ray Bragg diffraction experiments[8–10] have added to the wealth of knowledge about V_2O_3. We put forward a detailed theoretical explanation of some key aspects of the observed x-ray diffraction pattern[11] in which intensity is enhanced by the resonance at the K-edge of a vanadium ion. The success of our model in explaining the resonant x-ray diffraction data permits us to draw specific conclusions about the orbital magnetization present in V_2O_3.

The resonance enhanced intensity is related to properties of the valence shell that accommodates the photo-ejected core electron in the intermediate state of the elastic scattering process. Absorption at the K-edge and an electric-dipole event (E1) gives access to valence states with atomic p-like character, and an electric quadrupole event (E2) at the same edge gives access to d-like states (1s $\rightarrow$ 3d). We analyse in terms of a pure E2 event diffraction data collected with the primary energy tuned to a feature in the pre-edge (5.464 keV and λ = 2.27Å) of the vanadium K-shell. The associated Bragg intensity is therefore attributed directly to properties of the 3d-shell of a vanadium ion. As a function of the primary energy, the observed intensity is adequately described by a

*ISIS Facility, Rutherford Appleton Laboratory, Oxfordshire, OX11 0QX, UK.

Electron Correlations and Materials Properties 2
Edited by Gonis *et al.*, Kluwer Academic/Plenum Publishers, 2002

single oscillator. In this instance, we can confidently ascribe the intensity to orbital degrees of freedom and spin degrees of freedom are not explicitly contributing.

The data of interest has been gathered for Bragg reflections that are forbidden by the space group, and often called charge-forbidden reflections. In general, such reflections are due to spatial anisotropy in the charge distribution or a long-range magnetic order that is not indexed on the chemical structure. When long-range magnetic order is absent, the diffraction at charge-forbidden reflections is usually referred to as Templeton-Templeton scattering[12, 13]. Intensities measured for different settings of the crystal rotated about the Bragg wavevector can provide direct information on the nature of the anisotropy and spatial symmetry. Should the Bragg wavevector coincide with an axis of rotation symmetry the azimuthal-angle (ψ) meshes with the rotational periodicity, and the Bragg intensity displays a periodicity in ψ equal to twice the degree of the axis of rotation[14, 15]. Such a configuration can be achieved in diffraction from the corundum structure[11, 14] and the observed azimuthal-angle scan is six-fold periodic. Broadly speaking, complexity in azimuthal-angle scans observed for magnetically ordered V_2O_3[8–10] is due to the low spatial symmetry of the monoclinic structure adopted by the crystal. While the symmetry of the site occupied by a resonant vanadium ion carries an imprint of a three-fold axis of rotation, monoclinic Bragg wavevectors generally do not coincide with the trigonal axis of the corundum structure and it is this mismatch, between the wavevectors and the axis of rotation, that is largely responsible for the seemingly complex pattern of the azimuthal-angle scans.

There are two main ingredients in the calculated structure factors. First, knowledge of the chemical and magnetic structures. Information on the crystal structure, which is based on observations reported by Dernier and Marezio[2], is gathered in section 2 together with the configuration of magnetic moments[3]. The second ingredient is a model of resonance-enhanced Bragg diffraction. We employ an atomic model which should be adequate for a one-component intensity profile like the ones observed with V_2O_3 at the vanadium K-shell pre-edge[8, 9]. After adopting a single oscillator to describe the intensity profile the amplitude attached to the oscillator can be calculated without further approximation.[16] The atomic model for the event $1s \rightarrow 3d$, which is of interest to us here, shows that the amplitude is constructed from spherical tensors that describe the orbital properties of the 3d-shell. After putting together the two ingredients it only remains to describe the geometry of the experiment, namely, the orientation of the crystal (a function of the azimuthal angle) in the frame of reference used for the polarization and wavevector of the primary and diffracted beams of x-rays. The calculation of structure factors for unrotated ($\sigma'\sigma$) and rotated ($\pi'\sigma$) channels of scattering is outlined in section 3 together with examples of our interpretation of x-ray diffraction data. Thereafter, in section 4, we characterize the orbital properties of a vanadium ion by exploiting our success with the interpretation. Our conclusions are found in section 5.

2. CHEMICAL AND MAGNETIC STRUCTURES

At room temperature vanadium sesquioxide has a trigonal (corundum) structure with space group 167 ($R\bar{3}c$). Lowering the temperature of the material induces distortions which include tilting of the trigonal (hexagonal-c) axis and reduction of the point-group symmetry of sites occupied by vanadium ions from $3(C_3)$ to that of no symmetry.

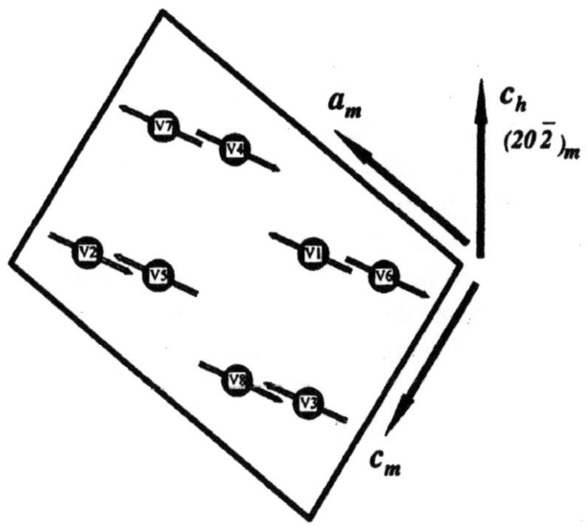

Figure 1. Positions of the eight vanadium ions in the monoclinic unit cell together with the configuration of their magnetic moments which lie in the $\mathbf{a}_m - \mathbf{c}_m$ plane of the diagram. The monoclinic Bragg wavevector $(20\bar{2})_m$ is parallel to the trigonal axis $\mathbf{c}_h$, and $\mathbf{b}_m$ is normal to the plane of the diagram and parallel to $\mathbf{a}_h$.

The space group of the low temperature monoclinic structure is number 15($I2/a$) in which vanadium ions occupy sites 8 (f). This is a body-centred cell and Bragg wavevectors $\tau_m(hkl)$ for charge reflections have the necessary condition $h + k + l$ an even integer. (Miller indices h, k and l are integers.) The antiferromagnetic configuration of vanadium magnetic moments, displayed in Fig. 1, consists of sheets of moments with ferromagnetic alignment within $(010)_m$ layers, or hexagonal (110) layers, and moment reversal between adjacent layers. The moments are orientated along some easy-axis in these layers, and we take ϕ as the canting angle with respect to the trigonal axis.

The trigonal basis vectors are $\mathbf{a}_h = a(1, 0, 0)$, $\mathbf{b}_h = a(-½, ½\sqrt{3}, 0)$ and $\mathbf{c}_h = c(0, 0, 1)$ and the volume of the unit cell = $a^2c\sqrt{3}/2$. Following Dernier and Marezio[2] in the use of an I - centred cell, from these vectors we generate monoclinic basis vectors $\mathbf{a}_m = (0, 2a/\sqrt{3}, c/3)$, $\mathbf{b}_m = \mathbf{a}_h$ and $\mathbf{c}_m = (0, a/\sqrt{3}, -c/3)$, and the volume of the cell = $a^2c/\sqrt{3}$. The corresponding Bragg wavevector $\tau_m(hkl) \equiv (hkl)_m$ is,

$$\boldsymbol{\tau}_m(hkl) = (k, (h+l)/\sqrt{3}, (h-2l)a/c)/a. \qquad (1)$$

We note that $(l0\bar{l})_m$ is parallel to $\mathbf{c}_h$ and $(2lkl)_m$ is normal to $\mathbf{c}_h$.

Referring to Fig. 1, the position coordinates of vanadium ions labelled (1) and (5) are (x, y, z) and $(-x, -y, -z)$, respectively, with $x = 0.3439$, $y = 0.0012$ and $z = 0.2993$[2]. The positions of the pair (2) and (6) are related by a body-centre translation to the pair (1) and (5). The position coordinates of (3) and (7) are $(½ - x, y, -z)$ and $(½ + x, -y, z)$, respectively, and the pairs (4) and (8), and (3) and (7) are related by the body-centre translation. The body-centre translation $(½,½,½)_m = (1, \sqrt{3}, 0)a/2$ and $(½,½,½)_m \cdot \tau_m(hkl) = ½(h + k + l)$. In writing out spatial phase factors in the structure factor one finds two angles $\nu = 2\pi (x, y, z)_m \cdot \tau_m(hkl) = 2\pi(xh + yk + zl)$ and $\varepsilon = 4\pi yk \sim 0$.

3. THE STRUCTURE FACTOR FOR BRAGG DIFFRACTION BY MAGNETICALLY ORDERED V_2O_3

As a function of the energy of the primary x-rays, the observed intensity of diffraction from V_2O_3 which is enhanced by electric-quadrupole (E2) resonance has a distinct shape[8, 9] which is adequately described by a single oscillator. In this treatment, the integrated intensity is calculated without approximation[16], and for a pure E2 absorption event 1s → 3d the intensity is proportional to the orbital moments of the 3d valence shell. The orbital moments are represented by atomic tensors $\langle \mathbf{T}^{(K)} \rangle$ of rank K = 0, 1, 2, 3 and 4. In the structure factor, which is a scalar quantity, atomic tensors appear in a scalar product with a tensor, denoted by $\mathbf{H}^{(K)}$, that describes the conditions of the primary and diffracted beams of x-rays. It is convenient to consider the quantity,

$$\Psi_Q^{(K)} = \sum_{\mathbf{d}} \langle T_Q^{(K)}(\mathbf{d}) \rangle \exp(i\boldsymbol{\tau}\cdot\mathbf{d}), \tag{2}$$

where $-K \le Q \le K$, $\mathbf{d}$ defines the position of a resonant vanadium ion in the unit cell and τ is the Bragg wavevector for the reflection in question. With this notation the structure factor is[16, 17],

$$F(\tau) = \sum_K (2K+1)^{1/2}\mathbf{H}^{(K)}.\boldsymbol{\Psi}^{(K)} = \sum_{K,Q} (2K+1)^{1/2}(-1)^Q H_{-Q}^{(K)}\Psi_Q^{(K)} . \tag{3}$$

The components $H_Q^{(K)}$ for unrotated ($\sigma'\sigma$) and rotated ($\pi'\sigma$) radiation are tabulated[17]. In Equ. (3) the tensor $\mathbf{H}^{(K)}$ and $\Psi^{(K)}$ are evaluated in the coordinate system that defines the experimental geometry. Referring to Fig. 2, σ-polarization is perpendicular to the plane of scattering and parallel to the z-axis, and $\hat{\mathbf{q}} - \hat{\mathbf{q}}' = -2\sin\theta\,\hat{\mathbf{x}}$ and $\hat{\mathbf{q}} + \hat{\mathbf{q}}' = 2\cos\theta\,\hat{\mathbf{y}}$ where θ is the Bragg angle.

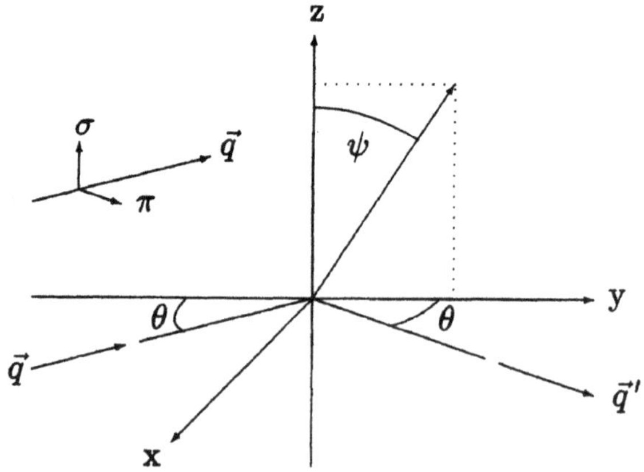

Figure 2. The orthogonal states of polarization in the primary beam of x-rays with σ-polarization normal to the plane of scattering defined by the primary (**q**) and diffracted (**q**′) wavevectors. X-rays are deflected through an angle 2θ and $\tau_m = \mathbf{q} - \mathbf{q}'$. Orthogonal axes ($x$, y, z) are related to the geometry of the experiment; the x-axis is parallel to $-\tau_m$ and the z-axis is normal to the plane of scattering and parallel to σ-polarization. An azimuthal-angle scan is rotation of the crystal by ψ about τ_m, and the origin ψ = 0 is defined by a specified plane normal to the plane of scattering.

We find it convenient initially to consider $\Psi_Q^{(K)}$ with respect to a second set of orthogonal axes (x,y,z) that include $\mathbf{a}_h = (a,0,0) = \mathbf{b}_m$ and $\mathbf{c}_h = (0,0,c)$. Thereafter we apply to $\Psi_Q^{(K)}$ rotations that describe the orientation of the crystal in the coordinates of the experimental geometry that apply to $\mathbf{H}^{(K)}$.

Following the discussion in section 2, the chemical environments of ions in the pairs (1) & (2), (5) & (6), (7) & (8) and (3) & (4) are identical. Moreover, the axes of quantization of ions in a pair are oppositely aligned and, in consequence, their atomic tensors differ by a phase factor $(-1)^K$. Odd-rank tensors are purely magnetic and $\langle \mathbf{T}^{(1)} \rangle = \langle \mathbf{L} \rangle / \sqrt{30}$ and $\langle \mathbf{T}^{(3)} \rangle = \langle \Lambda \rangle / 3\sqrt{70}$ where $\langle \mathbf{L} \rangle$ and $\langle \Lambda \rangle$ are, respectively, the orbital magnetic moment and the orbital octupole moment of the 3d-valence shell of a resonant vanadium ion. Vanadium ions at sites (1) & (5) are related by inversion. The atomic tensor is unchanged by inversion and thus $< T_Q^{(K)}(1) > = < T_Q^{(K)}(5) >$. A similar relation holds for (3) & (7).

Ions at sites (1) & (7) are related by an a-glide that includes reflection in the $\mathbf{a}_m$ - $\mathbf{c}_m$ plane, which is normal to $\mathbf{a}_h = \mathbf{b}_m$. Reflection in the $\mathbf{a}_m$ - $\mathbf{c}_m$ plane amounts to the change $x \to -x$, and with it $< T_Q^{(K)} > \to (-1)^K < T_{-Q}^{(K)} >$. For $K = 1$ $(L_x, L_y, L_z) \to (L_x, -L_y, -L_z)$, so that reflection in the $\mathbf{a}_m$ - $\mathbf{c}_m$ plane is accompanied by a change in the orientation of the orbital magnetization. The configuration of moments determined by Moon[3] is preserved if the polarity of the local field is now reversed, leading to $(L_x, -L_y, -L_z) \to (-L_x, L_y, L_z)$, for according to Moon $\langle L_x \rangle = 0$ and the moments are confined to the $\mathbf{a}_m$ - $\mathbf{c}_m$ plane. Reversing the polarity of the local field, which is sometimes described as time-reversal, introduces in the atomic tensor a phase factor $(-1)^K$. The appropriate relation between tensors at sites (1) and (7) is seen to be the union of reflection in the $\mathbf{a}_m$ - $\mathbf{c}_m$ plane and time-reversal, and our discussion leads to the result $\langle T_Q^{(K)}(7) \rangle = \langle T_{-Q}^{(K)}(1) \rangle$; for simplicity of notation, we hereafter write $< T_Q^{(K)} > = < T_Q^{(K)}(1) >$.

In the monoclinic structure, sites occupied by the vanadium ions possess no spatial symmetry and one can choose any set of axes for the associated atomic tensors. When proceeding from the corundum structure to the monoclinic one the average vanadium-oxygen distance remains essentially constant. This aspect of the structural transition, and others mentioned in section 2, indicate that in the monoclinic structure the potential field experienced by a vanadium ion is principally referred to the trigonal axis, $\mathbf{c}_h$, and the field is almost subject to the requirement of a three-fold axis of rotation. An atomic tensor invariant with respect to a three-fold rotation is zero unless the projection index Q has the values 0 or $\pm$ 3. Additions to the potential field, which reduce the symmetry from C_3 to one of no spatial symmetry, lift the restriction on the allowed values of Q.

3.1 Templeton-Templeton scattering

For the model of V_2O_3 we have described, in the previous section, $\Psi_Q^{(K)}$ can be different from zero for even values of $K+(h+k+l)$, so there is a selection rule on scattering that links the sum of Miller indices and the rank of atomic tensors. In this subsection we interpret data indexed by even $(h+k+l)$ and tensors in the corresponding structure factor have an even rank. The appropriate expression for $\Psi_Q^{(K)}$ is,

$$\Psi_Q^{(K)} = 4\cos(\nu)\{<T_Q^{(K)}> + (-1)^h <T_{-Q}^{(K)}>\}. \tag{4}$$

Note that for h odd the contributing tensors have rank $K = 2$ and $K = 4$ and there is no contribution in $\Psi_Q^{(K)}$ with $K = 0$ or $Q = 0$. Diffraction indexed by even $(h+k+l)$ and odd h is Templeton-Templeton scattering.

An example of such scattering has been observed at $(10\bar{1})_m$. Because this Bragg wavevector is parallel to our chosen principal axis $\mathbf{c}_h$ an azimuthal-angle scan, which is rotation of the crystal around the Bragg wavevector, incorporates what amounts to another selection rule that limits harmonics of the azimuthal angle ψ to values set by the projection index Q of atomic tensors. In consequence, the structure factor that describes an azimuthal-angle scan at $(10\bar{1})_m$ is a linear combination of $<T_Q^{(K)}>\cos(Q\psi)$.

A confrontation between predictions of the theory and the data requires the input of several parameters. Some can be derived from geometrical considerations, such as Euler angles for orientating the crystal in the geometry of the experiment and the Bragg angle, and are believed to be estimated fairly well; others are experimental in nature, particularly the zero-setting for the azimuthal angle, but are thought to be calibrated reasonably well; some stem from material properties, and little is known about their values a priori. These varying degrees of uncertainty are easily taken into account within the Bayesian framework of data analysis[18] and, with suitable commonly-used simplifying assumptions, lead to a somewhat generalized form of least-squares: in addition to a quadratic cost function for a mismatch with the measurements, there is also a penalty for deviating from those parameter-values within the theory which are thought to be known fairly well. The best fit to data (e.g. Fig. 3) utilizes a robust simplex algorithm[19] for carrying out the (local) optimization.

Our interpretation of the azimuthal-angle scan at $(10\bar{1})_m$ is illustrated in Fig. 3. The data are collected in the rotated $(\pi'\sigma)$ channel of scattering. The material properties that enter our interpretation are $<T_Q^{(2)}>$ with $Q = 1$ and 2 and $<T_Q^{(4)}>$ with $Q = 1, 2, 3$ and 4. We find that the coeffficients of $\cos(\psi)$ and $\cos(3\psi)$, by far, are the largest, and in creating Fig.3 $<T_4^{(4)}>$ is set equal to zero. One finds the coefficients of $\cos(3\psi)$ and $\cos(2\psi)$ differ by more than an order of magnitude, and the coefficients of $\cos(3\psi)$ and $\cos(\psi)$ are in the ratio ~ 0.2.

3.2 Magnetic scattering

Here we are concerned with space-group forbidden reflections indexed by odd $(h+k+l)$. The corresponding atomic tensors are odd with respect to their rank and they vanish with loss of long-range magnetic order. The structure factor is constructed from $\Psi_Q^{(K)}$ defined in Equ. (4). (The diffraction pattern observed when neutrons are scattered

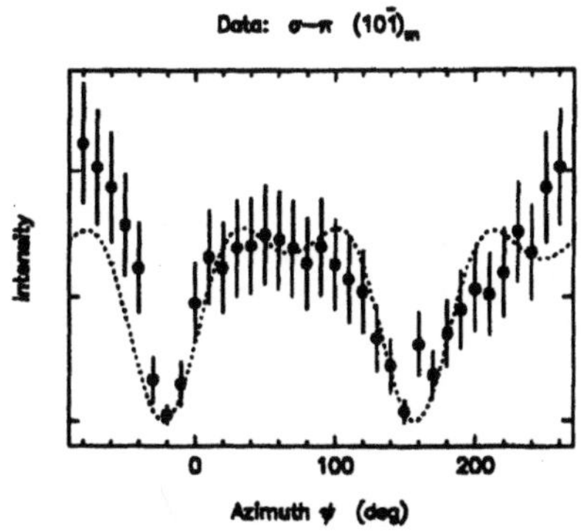

Figure 3. Data gathered in the rotated polarization channel at the (10 $\bar{1}$)$_m$ reflection from V_2O_3 with 2.8% Cr doping, held at a temperature of 100K. The continuous curve is generated from, Equ. (4) with a Bragg angle $\theta = 14.1°$ ($\lambda = 2.27$Å). The origin of the azimuthal-angle scan is specified by a plane in reciprocal – space normal to the plane of scattering. The plane in question is defined by (10 $\bar{1}$)$_m$ and (211)$_m$ and on our scale for ψ the origin is at – 60°.

by magnetically ordered V_2O_3 is indexed by odd $(h+k+l)$ and even h.) Figure 4 contains examples of our interpretations of resonant x-ray Bragg diffraction.

We have successfully compared calculated intensities with data gathered in azimuthal-angle scans at space-group forbidden reflections $(2\bar{2}1)_m$, $(111)_m$ and $(30\bar{2})_m$. As a function of the primary energy of the x-rays the signal observed at 5.464 keV is fully consistent with a single-oscillator model. From this premiss, the x-ray experiments in question reveal exclusively the orbital magnetization of V_2O_3, and quantities not directly obtainable with other experimental techniques.

Calculated structure factors conform to established chemical and magnetic structural properties. At space-group forbidden reflections, the structure factors contain the following additional features.

(a) Reflections are purely magnetic, being associated with atomic moments of the vanadium 3d-valence shell that are odd with respect to the reversal of time. In general, such moments vanish with the loss of long-range magnetic order. Our prediction for V_2O_3 is borne out in experiments which show that intensities decrease with increasing temperature, and vanish at the Néel temperature.

(b) A selection rule on the order of the electric absorption event and Miller indices $h\ k\ l$. With $h + k + l$ odd (a space-group forbidden reflection) reflections with h odd do not contain a contribution from an E1 (electric dipole) event, and the finding is in accord with observations. In the reported calculation, the selection rule stems from the actual configuration of magnetic moments and the fact that they are contained in the $\mathbf{a}_m$ - $\mathbf{c}_m$ plane. The same fact is utilized in arriving at the general form of the structure factor, based on the established chemical structure augmented by the operation of time reversal.

(c) With h odd diffraction is by anisotropic components of the orbital octupole moment of a vanadium ion. Such diffraction might be regarded as the magnetic equivalent of Templeton and Templeton scattering from a charge distribution.

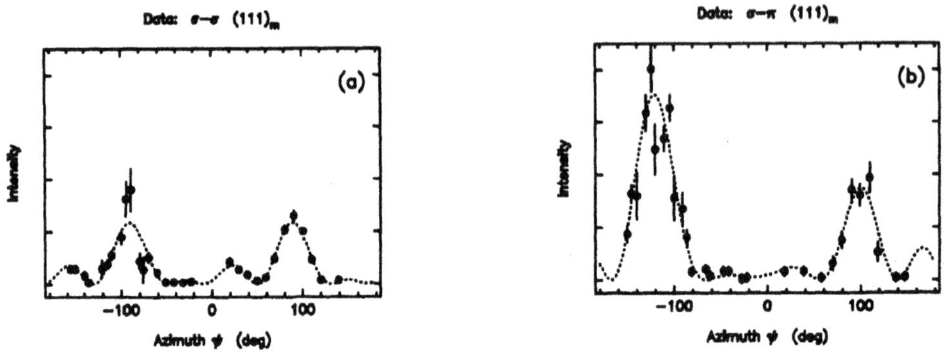

Figure 4. Data on Cr-doped V_2O_3 held at temperature = 0.55 T_N = 100K, and τ_m = $(111)_m$ with $\theta = 20.8^0$. From data in the unrotated channel we find $r = 0.33 \pm 0.04$ and $t = -0.35 \pm 0.15$, and from data in the rotated channel we find $r = 0.36 \pm 0.02$ and $t = -1.00 \pm 0.13$.

(d) Intensities with h even can be used to determine the canting angle, ϕ, of the magnetic easy-axis. Our analysis of data gathered at the pre-edge (5.464 keV) and $(2\bar{2}1)_m$ gives $\phi \sim 70°$ and the value is consistent with the analysis of intensities observed in magnetic neutron diffraction[3]. This outcome at the pre-edge, and related success in explaining other features, confirm that it is due mainly to an E2 resonance event.

(e) As a function of the azimuthal-angle the calculated intensity in the channel with unrotated polarization ($\sigma'\sigma$) is two-fold periodic. This prediction and additional detail in calculated ($\sigma'\sigma$) azimuthal-angle scans is found in intensities collected at Bragg reflections $(2\bar{2}1)_m$, $(111)_m$ and $(30\bar{2})_m$.

(f) In general, the dependence on azimuthal-angle of intensities in the rotated channel ($\pi'\sigma$) show no systematic features. With h odd the ($\pi'\sigma$) azimuthal-angle scans have a quite complicated pattern which essentially reflects the low symmetry (monoclinic) of the crystal below the Néel temperature.

4. A CHARACTERIZATION OF THE ORBITAL PROPERTIES OF A V ION

We construct a model wavefunction to characterize the orbital properties of a vanadium ion deduced from our interpretation of data collected on magnetically ordered V_2O_3 by resonant x-ray diffraction. The model is based on the ^3F term of V^{3+} $(3d^2)$, and orbital components $|M\rangle$ with $-3 \leq M \leq 3$. Should the V ion occupy a site that contains an axis of three-fold rotation symmetry (C_3) components in the wavefunction must differ by $M = \pm 3$, and $\langle T^{(K)}_Q \rangle$ can be different from zero for $Q = 0$ and ± 3. In the absence of spatial symmetry no selection rules operate on M and all Q are allowed. On using a wavefunction,

$$|G\rangle = a|-2\rangle + b|-1\rangle + c|0\rangle + d|1\rangle, \quad (5)$$

non-zero values of b and c represent the influence of reducing the V ion site symmetry from C_3 to that of no spatial symmetry. Our interpretation of the data can be characterized by $|G>$ on taking a, b and d purely real and $c = c' + ic''$, and the normalized wavefunction contains four independent coefficients.

In addition to normalization of $|G>$, coefficients in the wavefunction are constrained by requiring the orbital magnetic moment to lie in the plane $\mathbf{a}_m - \mathbf{c}_m$ and enclose an angle ϕ with $\mathbf{c}_h$. The orbital magnetic moment $< L_0 >_m$ satisfies,

$$< L_0 >_m \cos\phi = d^2 - b^2 - 2a^2, \tag{6}$$

together with,

$$ab\sqrt{5} + c'(b+d)\sqrt{6} = 0. \tag{7}$$

The material quantities that enter our interpretation of data collected at magnetic reflections include ratios of components of the octupole moment, namely,

$$r = \sqrt{3/5} < \Lambda_{+1} >' / < \Lambda_{+3} >' = b/2d, \tag{8}$$

$$t = \sqrt{6} < \Lambda_{+2} >'' / < \Lambda_{+3} >' = - c''\sqrt{3}/d \tag{9}$$

$$\text{and,} < \Lambda_0 >_m = 3\,(a^2 + b^2 - d^2) \cos\phi\,(3 - 5\cos^2\phi)/2. \tag{10}$$

Other quantities of interest are components of the quadrupole moment, e.g.,

$$< Q_0 > = 3\{3(b^2 + d^2) + 4\,|c|^2\}/10. \tag{11}$$

There remains the question as to how well the experimental data constrains the three unknown coefficients in $|G>$. If the answer is fairly tightly, then we can make useful predictions about material quantities, like $<Q_0>$, which do not appear in structure factors for space-group forbidden reflections.

The coefficients are related to quantities inferred from x-ray and neutron diffraction data. These quantities are ϕ, r, t, $< L_0 >_m$ and the ratio $< \Lambda_0 >_m / 3 < L_0 >_m$. We have two, or more, estimates of ϕ, r, t, $< L_0 >_m$ and the ratio $< \Lambda_0 >_m / 3 < L_0 >_m$, and wish to know what restrictions these impose on the allowed values of a, b and d, say. Before proceeding to that, however, let us first consider the intermediate step of combining the different estimates of the same quantity into a single entity. For example, what does $r = 0.81 \pm 0.11$, 1.24 ± 0.24, 0.33 ± 0.04 and 0.36 ± 0.02 tell us about the possible value of r? Well, assuming independence between the four relevant data sets, the usual elementary analysis[18], leading to a weighted mean combination, yields the summary $r = 0.37 \pm 0.02$. While this result is understandable given the large spread in the size of the error bars on the four estimates of r, it does not sit well with our common sense intuition. A more reasonable representation of our state-of-knowledge about the value of r is obtained by making the more pessimistic assumption that the four quoted error-bars only

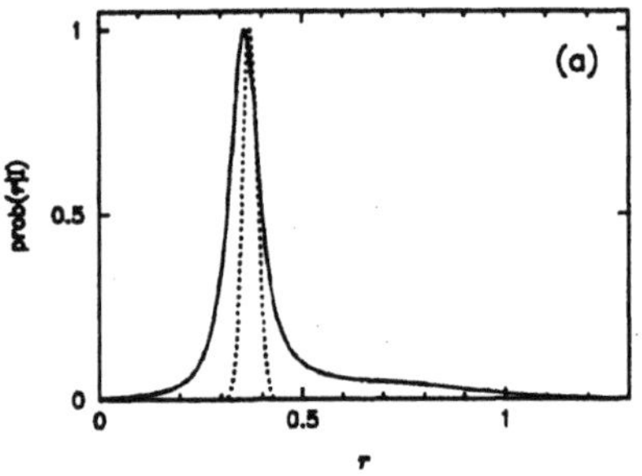

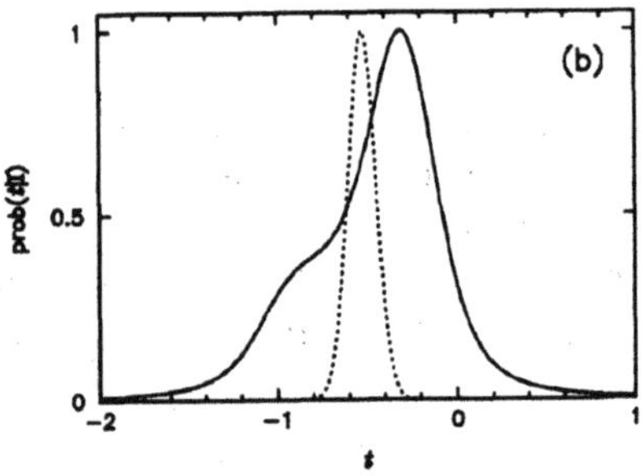

Figure 5. The solid lines show the p.d.f.'s representing our resultant knowledge of r and t given by several different pieces of information; being based on somewhat pessimistic assumptions, they are less constraining, and more realistic, than the equivalent conventional analysis indicated by the dotted lines. It should be noted that the p.d.f.'s are not normalized, but have been scaled to have a maximum of unity.

define a lower bound on the related uncertainties. Following this line of thought in a more formal[20] leads to the resultant probability density function (p.d.f.) for r shown by the solid curve in Fig. 5(a); for comparison, the implicit Gaussian p.d.f. pertaining to the conventional analysis mentioned earlier is shown with a dotted line. Although a value of r around 0.37 is still most probable, we are now not prepared to rule out the possibility that it may be of order unity quite so strongly.

A similar analysis of three values of t inferred from x-ray diffraction data (-0.35 ± 0.15, -1.00 ± 0.13 and -0.19 ± 0.13) leads to the resultant p.d.f. for t shown in Fig. 5(b). The pairs of "measurements" for ϕ ($66.1° \pm 2.1°$ and $75.7° \pm 1.8°$), $< L_0 >_m$ (-0.5 and -0.8) and $< \Lambda_0 >_m/3 < L_0 >_m$ (0.00 ± 0.05 and -0.06 ± 0.01) were also combined in our "conservative" manner but, in these cases, the resulting p.d.f.'s were approximated by Gaussians defined by $\phi = 71° \pm 5°$, $< L_0 >_m = -0.65 \pm 0.20$ and $< \Lambda_0 >_m /3 < L_0 >_m = -0.060 \pm 0.025$.

Our state of knowledge about the vandium wavefunction is then encapsulated in three-dimensional p.d.f. for the parameters a, b and d conditional on our "noisy constraints" on r, t, ϕ, $< L_0 >_m$ and $< \Lambda_0 >_m / 3 < L_0 >_m$, and all the implicit assumption and approximation, jointly labelled as I: prob(a, b, $d \mid$ I). This can be computed numerically on a three-dimensional grid spanning the space $-1 \leq a \leq 1$, $-1 \leq b \leq 1$ and $-1 \leq d \leq 1$, as follows:

(i) for each point, calculate r, t, ϕ, and so on, corresponding to the associated values of a, b and d;
(ii) read off the probabilities for those values of r, t, ϕ, etc., from Fig 5 and related graphs not shown;
(iii) assign the product of these probabilities to the grid-point being considered.

The result will be a 3-dimension grid for the likelihood of a, b and d which, with the assumption that we know nothing about the values of these parameters *a priori*, is proportional to the p.d.f., prob(a, b, $d \mid$ I), that we seek.

Carrying out the analysis above, we found that there are four regions of high probability in our 3-dimension hypothesis-space. They are all symmetry-related in that the constraints of the data leave an ambiguity with regard to the signs of a, b and d.

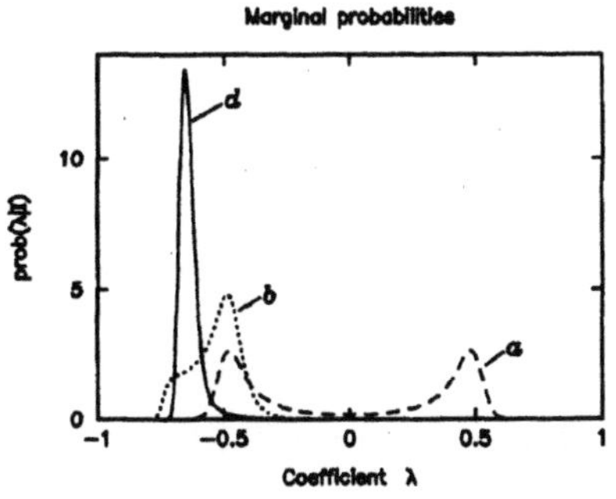

Figure 6. The marginal p.d.f.'s for a, b, and d, representing the constraints imposed on these expansion coefficients by our limited knowledge of r, t, ϕ , $< L_0 >_m$ and $< \Lambda_0 >_m / 3 < L_0 >_m$.

Since the absolute phase of the wavefunction is arbitrary, we choose to impose the condition $d \leq 0$. Due to a correlation between b and d, principally through the relationship $r = b/2d$, this restriction is sufficient to imply that $b \leq 0$, but the sign of a remains undetermined. The marginal probabilities for a, b and d are shown in Fig.6.

Given prob(a, b, d | I) we can make inferences about any quantity x, say, which is related to a, b and d, so that $x = \mathrm{f}(a, b, d)$, by using the rules of probability theory:

$$\begin{aligned} \mathrm{prob}(x \mid \mathrm{I}) &= \sum_{a,b,d} \mathrm{prob}(x,a,b,d \mid \mathrm{I}) \\ &= \sum_{a,b,d} \mathrm{prob}(x \mid a,b,d,\mathrm{I})\, \mathrm{prob}(a,b,d \mid \mathrm{I}) \\ &= \sum_{a,b,d} \delta(x - \mathrm{f}(a,b,d))\, \mathrm{prob}(a,b,d \mid \mathrm{I})\,, \end{aligned} \tag{13}$$

where δ is a Dirac delta-function. An example of such a calculation is illustrated in Fig.7, which shows the marginal probabilities for the real and imaginary part of the wavefunction expansion coefficient $c = c' + ic''$. While marginal p.d.f.'s are very useful for conveying our state-of-knowledge about one parameter irrespective of the others, they

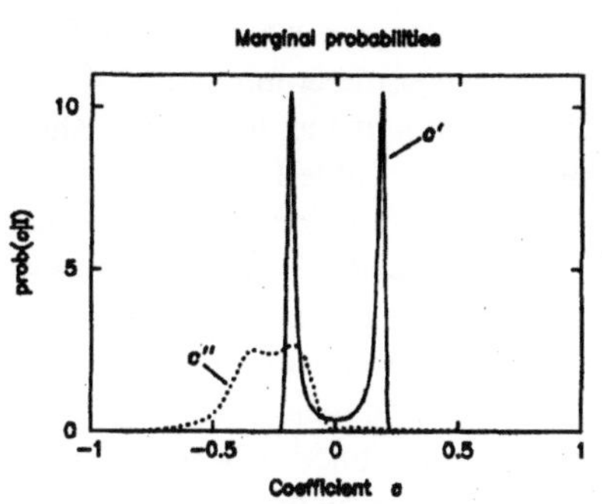

Figure 7. The marginal p.d.f.'s for the real and imaginary parts of the expansion coefficient c, inferred from our probabilistic analysis of the model wavefunction.

Table I. The table contains some values of orbital moments of the vanadium ion at site (1) in Fig.(1), with respect to orthogonal axes that include as components $\mathbf{a}_h = (a,0,0)$ and $\mathbf{c}_h = (0,0,c)$. Moments are proportional to an atomic tensor, namely, <**Q**> = $3\sqrt{35/2}$ <**T** $^{(2)}$>, <**Λ**> = $3\sqrt{70}$ <**T** $^{(3)}$> and <**Y**> = $18\sqrt{70}$ <**T** $^{(4)}$>. Note that $<T_Q^{(K)}>^* = (-1)^Q <T_{-Q}^{(K)}>$. Values of the moments are derived from estimates of the coefficients in the model wavefunction given in Equ. (5).

$$<Q_0> = 3\{3(b^2+d^2) + 4\,|c|^2\}/10 = 0.78 \pm 0.08$$

$$<Q_{+1}>' = 3\{c'(d-b)\sqrt{2} - ab\sqrt{15}\}/10 = -0.25 \pm 0.06$$

$$<Q_{+2}>'' = 3ac''/\sqrt{5} = 0.13 \pm 0.11$$

$$<\Lambda_0> = 3(d^2 - b^2 - a^2) = -0.20 \pm 0.14$$

$$<\Lambda_{+1}>' = -3c'(b+d) = 0.56 \pm 0.14$$

$$<\Lambda_{+2}>'' = -3ac'' = -0.29 \pm 0.24$$

$$<\Lambda_{+3}> = 3\sqrt{2}ad = 1.09 \pm 0.35$$

$$<\mathrm{Y}_0> = 6\{-7a^2 + b^2 + 6\,|c|^2 + d^2\} = 1.12 \pm 4.75$$

$$<\mathrm{Y}_{+1}>' = 6\sqrt{2}\{-4ab + c'(d-b)\sqrt{(15/2)}\} = -7.46 \pm 1.91$$

$$<\mathrm{Y}_{+2}>'' = 6\sqrt{3}\,ac'' = 1.02 \pm 0.83$$

$$<\mathrm{Y}_{+3}> = 6\sqrt{14}\,ad = 5.77 \pm 1.82$$

$$<\mathrm{Y}_{+4}> = 0.$$

do, of course, only give a partial picture of the situation. For example, they do not tell us the signs of a and c' are anticorrelated, so that $c' > 0$ if $a < 0$ and vice versa. To avoid this particular source of ambiguity, the remaining results in this paper assume that $a \leq 0$; the alternative choice, of $a \geq 0$, lead to equivalent symmetry-related conclusions.

Rather than displaying prob($x\,|\,$I), we often try to summarize its salient features by quoting a "best" estimate x_0 and a measure of its uncertainty σ (i.e. $x = x_0 \pm \sigma$); the most common practice is to use the mean, $x_0 = <x> = \int x \text{ prob } (x\,|\,\mathrm{I})dx$, and the variance, $\sigma^2 = <(x - x_0)^2>$. We have done this for some components of the quadrupole <**Q**>, octupole <**Λ**> and hexadecapole <**Y**> moments of the 3d valence shell of the V ion at site (1) in Fig.1. Our findings are summarized in Table I.

Looking at entries in Table I, it is evident that we have a good knowledge of all moments with the projections $Q = +1$ and $+3$, yet some moments with even Q are known very poorly, e.g. $<\mathrm{Y}_0>$. The estimates in Table I serve as tests to be met by an *ab initio* calculation of orbital properties of a vanadium ion in V_2O_3. Our inferred model wavefunction $|\mathrm{G}>$ is illustrated graphically in Fig.8. We display isosurfaces of $<|<\mathbf{r}\,|\,\mathrm{G}>|^2>$, or the expected wavefunction when averaged over prob(a, b, $d\,|\,$I). In addition, we provide two colour contour sections through the 3-dimension orbital density and its uncertainty equivalent.

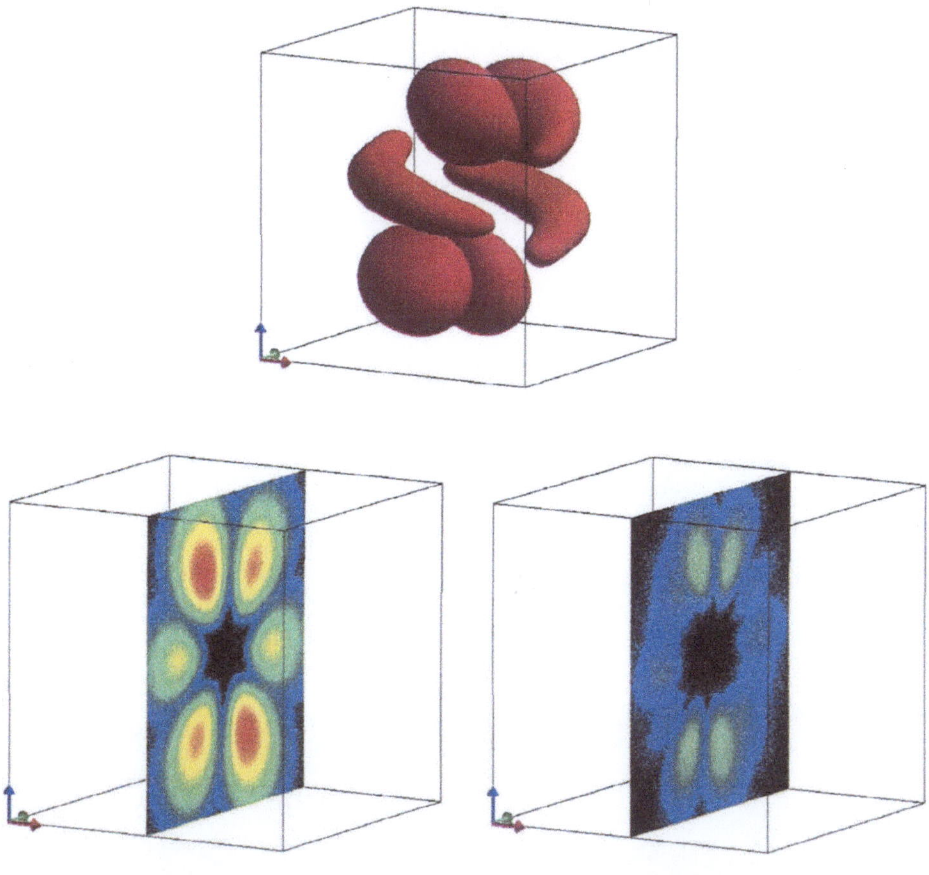

Figure 8. (a) An isosurface of the expected electron-density inferred from our probabilistic analysis of the model wavefunction. (b) An equivalent colour-contour slice through the 3-dimensional density, and (c) the corresponding 1- σ uncertainties displayed on the same scale. The red, green and blue arrows indicate the spatial *x*, *y*, and *z* axes of the crystal, respectively.

With regard to the nature of the ordering of orbital moments in the unit cell relevant information is contained in the relation between tensors for sites (1) and (7). (As before, we will use othogonal axes (x, y z) with $x(z)$ parallel to $\mathbf{a}_h(\mathbf{c}_h)$.) Applied to the quadrupole moment one finds that, for the two sites in question $< Q_{yz} >$ and $< Q_{xx} - Q_{yy} >$ have one sign, whereas $< Q_{+1}>' \propto - < Q_{xz} >$ and $< Q_{+2} >'' \propto < Q_{xy} >$ are opposite in sign at the two sites. All diagonal atomic tensors $< T_0^{(K)} >$ are of one sign.

5. CONCLUSIONS

We have added a successful interpretation of Templeton-Templeton x-ray scattering to our earlier success with magnetic scattering by magnetically ordered V_2O_3. Expressed in terms of trigonal axes, that include $\mathbf{a}_h$ and $\mathbf{c}_h$, a vanadium ion in magnetically ordered V_2O_3 bears an imprint of the three-fold rotation symmetry associated with the high-symmetry (corundum) structure adopted above the Néel temperature. In addition, the resonant x-ray Bragg diffraction data we have interpreted is consistent with an orbital magnetic moment confined to the plane $\mathbf{a}_m - \mathbf{c}_m$, which is normal to $\mathbf{a}_h = \mathbf{b}_m$, and inclined with respect to $\mathbf{c}_h$ by an angle $\phi \sim 70°$. However, resonant x-ray diffraction experiments are not suitable for the determination of the magnitude of the orbital moment. (The moment can be measured by diffraction of x-rays with an energy well above the resonance region or neutron diffraction.)

The successful outcome of our interpretation has been exploited to characterize orbital properties of a vanadium ion. To this end, we have used a model wavefunction created from the V ion ground state determined by Hund's rules. Coefficients in the model are determined by fitting to material properties that enter our interpretation of experimental date. One use of our inferred model wavefunction is to estimate orbital moments. The moments characterize the 3d valence shell and the ordering of orbitals in the magnetic unit cell.

Acknowledgements

We are grateful to Dr. Luigi Paolasini for guidance in the interpretation of the experiments he and his colleagues have performed on V_2O_3, and access to unpublished data.

REFERENCES

1. D. B. McWhan, T. M. Rice, and J. P. Remeika, Phys. Rev. Lett. **23**, 1334 (1969).
2. P. D. Dernier and M. Marezio, Phys. Rev. **B2**, 3771 (1970).
3. R. M. Moon, Phys. Rev. Lett. **25**, 527 (1970).
4. D. B. McWhan et al., Phys. Rev. **B7**, 1920 (1973).
5. N. F. Mott, Metal-Insulator Transitions (Taylor and Francis, London, 1990)
6. T. M. Rice, Spectroscopy of Mott Insulators and Correlated Metals, edited by A. Fujimori and Y. Tokura (Springer, Berlin, 1995).
7. F. Mila et al., Phys. Rev. Lett. **85**, 1714 (2000).
8. L. Paolasini et al., Phys. Rev. Lett. **82**, 4719 (1999).
9. L. Paolasini et al., submitted to J. Elect. Spect. & Related Phenomena (2000).
10. L. Paolasini, private communication.
11. S. W. Lovesey and K. S. Knight, J. Phys.: Condens. Matter **12**, L367 (2000).
12. D. H. Templeton and L. K. Templeton, Phys. Rev. **B49**, 14850 (1994).

13. D. H. Templeton, Acta Cryst. **A54**, 158 (1998).
14. K. D. Finkelstein, Q. Shen and S. Shastri, Phys. Rev. Lett. **69**, 1612 (1992).
15. Y. Tanaka et al., J. Phys.: Condens. Matter **11**, L505 (1999).
16. S. W. Lovesey, J. Phys.: Condens. Matter **10**, 2505 (1998).
17. S. W. Lovesey, K. S. Knight and E. Balcar, to appear in Phys. Rev. **B** (2001).
18. D. S. Sivia, Data Analysis – A Bayesian Tutorial (Oxford University Press, Oxford, 1996).
19. J. A. Nelder and R. Mead, Computer J. **7**, 308 (1965).
20. D.S. Sivia, Dealing with duff data, MaxEnt **96**, eds. M. Sears, V. Nedeljkovic, N.E. Pendock and, S. Sibisi, University of Witwatersand, Johannesburg, 1997.

THEORY OF MANGANITES

Adriana Moreo
NHMFL and Physics Department
Florida State University, Tallahassee, FL, U.S.A.
adriana@magnet.fsu.edu

Elbio Dagotto
NHMFL and Physics Department
Florida State University, Tallahassee, FL, U.S.A.
dagotto@magnet.fsu.edu

Abstract The colossal magneto resistance observed in manganites has recently attracted considerable attention. Important advances in the understanding of the properties of these materials have been achieved by applying newly developed numerical techniques to the study of model Hamiltonians proposed decades ago. The numerical discovery of electronic and structural phase separation, confirmed by experimental results, has been of particular relevance. The implicances of this phenomenon to the present understanding of the physics of manganites is discussed. Estimations of the resistivity in the mixed phase regime allows us to conclude that the intrinsic inhomogeneities of the Mn-oxides are crucial to understand their curious magneto transport behavior. The general aspects of the influence of quenched disorder on the competition between ordered states separated by a first-order transition are investigated and a new scale T^* for cluster formation is predicted.

Keywords: Manganites, phase separation

Introduction

Materials that present the phenomenon of "colossal" magnetoresistance are currently under much experimental investigation due to their potential technological applications. Typical compounds that have this phenomenon are ferromagnetic (FM) metallic oxides of the form $R_{1-x}X_xMnO_3$ (where $R = La, Pr, Nd$; $X = Sr, Ca, Ba, Pb$) [1, 2]. As an example, a decrease in resistivity of several orders of magnitude has been reported in thin films of $Nd_{0.7}Sr_{0.3}MnO_3$ at magnetic fields of 8 Teslas [3]. The relative changes in resistance for the manganites can be as large as $\Delta R/R \sim 100,000\%$, while in magnetic superlattices Co/Cu/Co the enhancement is about 100%. This result suggests that manganites indeed have technological potential since large changes in resistance can be obtained at fixed temperature upon the application of magnetic fields, opening an alternative route for next generation magnetic storage devices. However, since the development of La-manganite sensors is still at a very early stage, a more fundamental approach to the study of manganites is appropriate and, thus, theoretical guidance is needed. The existence of correlation effects in the fairly dramatic magnetic, transport, and magneto-transport properties of doped La-manganites reinforces this notion.

The early theoretical studies of models for manganites concentrated their efforts on the existence of ferromagnetism. The so-called "Double Exchange" (DE) model [4, 5] explained how carriers improve their kinetic energy by forcing the localized spins to become ferromagnetically ordered (this phenomenon is quite reminiscent of the Nagaoka phase discussed in models for cuprates). However, in spite of this successful explanation of the existence of ferromagnetism at low temperature several features of the experimental phase diagram of manganites remain unclear, and they are likely beyond the DE model. Actually, the phase diagram of $La_{1-x}Ca_xMnO_3$ is very rich with not only ferromagnetic phases, but also regions with charge-ordering and antiferromagnetic correlations at $x > 0.5$ [6], and a poorly understood "normal" state above the critical temperature for ferromagnetism, T_c^{FM}, which has insulating characteristics at $x \sim 0.33$. Finding an insulator above T_c^{FM} is a surprising result since it would have been more natural to have a standard metallic phase in that regime which could smoothly become a ferromagnetic metal as the temperature is reduced.

The rich phase diagram of the manganites described above, plus the several experimental indications of strong correlations in the system, deserves a systematic theoretical study using state-of-the-art computational techniques. These methods are unbiased and can provide useful information on models for manganites in a regime of couplings that cannot be handled perturbatively or exactly. However, the large number of degrees of freedom and associated couplings of a full Hamiltonian model for manganites makes this approach quite cumbersome. In principle the two e_g active orbitals per Mn-ion must be included, in addition to the t_{2g} electrons. Also phonons should be incorporated to fully describe these materials. However, as a first step towards a theoretical understanding of the behavior of models for manganites, it is convenient to work only in the electronic sector (i.e. leaving aside phonons) and with just one orbital per site (i.e. keeping only one e_g orbital of the two available, which in practice amounts to assuming a static Jahn-Teller (JT) distortion). Another simplification is obtained by assuming the t_{2g} degrees of freedom to be localized i.e. no mobility is given to these electrons. They basically provide a spin background in which the e_g electrons move, with a Hund term that couples all active electrons per Mn-ion. Under these assumptions a rich phase diagram is obtained.[7] The addition of phonons allows for the interesting interplay of spin, charge and orbital degrees of freedom,[8] and consideration of the effects of disorder becomes crucial to understand many of the experimental results.[9]

The organization of the article is as follows: in Section 1 the theoretical models that will be applied to the study of the manganites are introduced, the results are presented in Section 2 and Section 3 is devoted to the summary.

1. The Models

The ferromagnetic Kondo Hamiltonian [4, 10] is defined as

$$H = -t \sum_{\langle \mathbf{ij} \rangle \sigma} (c^{\dagger}_{\mathbf{i}\sigma} c_{\mathbf{j}\sigma} + h.c.) - J_H \sum_{\mathbf{i}\alpha\beta} c^{\dagger}_{\mathbf{i}\alpha} \sigma_{\alpha\beta} c_{\mathbf{i}\beta} \cdot \mathbf{S_i} + H_{AF}, \qquad (1)$$

where $c_{\mathbf{i}\sigma}$ are destruction operators for electrons at site $\mathbf{i}$ with spin σ, and $\mathbf{S_i}$ is the total spin of the t_{2g} electrons, assumed localized. The first term is the electron transfer between nearest-neighbor Mn-ions, $J_H > 0$ is the ferromagnetic Hund coupling, the number of sites is L, and the rest of the notation is standard. The electronic density of e_g electrons, denoted by $\langle n \rangle$, is adjusted using a chemical potential μ. In most of the results shown below the spin $\mathbf{S_i}$ will be considered classical (with $|\mathbf{S_i}| = 1$). The appropriateness of this simplification has been verified in one dimension (1D) where both classical and quantum mechanical t_{2g} spins have been studied. [7] Phenomenologically $J_H \gg t$. Finally, a small coupling between the t_{2g}-spins is needed to account for the AF character of the real materials even when all La is replaced by Ca. This classical Heisenberg term is $H_{AF} = J_{AF} \sum_{\langle \mathbf{ij} \rangle} \mathbf{S_i} \cdot \mathbf{S_j}$, where J_{AF} is around 0.05 [11], a value compatible with experiments [12].

The Hamiltonian used to consider orbital degrees of freedom has three contributions $H_{KJT} = H_K + H_{JT} + H_{AF}$. The first term is

$$H_K = - \sum_{\langle \mathbf{ij} \rangle \sigma ab} t_{ab} (c^{\dagger}_{\mathbf{i}a\sigma} c_{\mathbf{j}b\sigma} + h.c.) - J_H \sum_{\mathbf{i}a\alpha\beta} \mathbf{S_i} \cdot c^{\dagger}_{\mathbf{i}a\alpha} \sigma_{\alpha\beta} c_{\mathbf{i}a\beta}, \qquad (2)$$

where $\langle \mathbf{ij} \rangle$ denotes nearest-neighbor lattice sites, $a, b = 1, 2$ are the two e_g-orbitals, $d_{x^2-y^2}$ and $d_{z^2-r^2/3}$, $\mathbf{S_i}$ and $J_H > 0$ have been described above and the rest of the notation is standard. None of the results described below depends crucially on the set $\{t_{ab}\}$ selected [13].

The coupling with JT-phonons is through [14]

$$H_{JT} = \lambda \sum_{\mathbf{i}ab\sigma} c^{\dagger}_{\mathbf{i}a\sigma} Q^{ab}_{\mathbf{i}} c_{\mathbf{i}b\sigma} + \frac{1}{2} \sum_{\mathbf{i}} (Q^{(2)}{}^2_{\mathbf{i}} + Q^{(3)}{}^2_{\mathbf{i}}), \tag{3}$$

where $Q^{11}_{\mathbf{i}} = -Q^{22}_{\mathbf{i}} = Q^{(3)}_{\mathbf{i}}$, and $Q^{12}_{\mathbf{i}} = Q^{21}_{\mathbf{i}} = Q^{(2)}_{\mathbf{i}}$. These phonons are assumed to be classical. λ is the electron-phonon coupling. This approximation has been used and discussed in previous literature [14], where it was concluded that at temperatures of the order of the critical ones (room temperature), or a sizable fraction of them, the use of classical phonons captures the important physics of the model [15]. The reader is referred to Ref. [14] for further details. Certainly at very low-temperatures the quantum character of the phonons is important, but this is not the range of temperatures that concerns us. Note that here $T = 1/10$ is about 200-300K [7]. In this study we will set $t = t_{aa} = 1$.

It is not easy to estimate the values of the parameters in the models described above from comparisons with experiments, particularly because the theoretical calculations are notoriously difficult. With this caveat here some estimations of the parameters are presented. [16] $J_H \approx$ 2eV; [17] $t \approx 0.2-0.5$ eV; [18] $\lambda \approx 1.6$; [19] and $J_{AF} \approx 0.1t \approx 200$K. [12]

To study these Hamiltonians a Monte Carlo (MC) technique will be used. The trace over the e_g-electrons is carried out exactly using library subroutines for a fixed background of t_{2g}-spins and phonons. This background is selected based on a Metropolis MC procedure [7]. The CPU time of the technique grows rapidly with the number of sites L, but the method has the important advantage that it does not have sign-problems at any T [7]. Finally, to analyze orbital correlations the pseudospin operator $\mathbf{T_i} = \frac{1}{2}\sum_{\sigma ab} c^{\dagger}_{\mathbf{i}a\sigma}\sigma_{ab}c_{\mathbf{i}b\sigma}$ will be used, while for spin correlations the operator is standard. The Fourier-transform of the pseudospin correlations is defined as $T(\mathbf{q}) = \frac{1}{L}\sum_{\mathbf{l},\mathbf{m}} e^{i\mathbf{q}\cdot(\mathbf{l}-\mathbf{m})}\langle \mathbf{T_m} \cdot \mathbf{T_l}\rangle$, with a similar definition for the spin structure factor $S(\mathbf{q})$.

2. Results

2.1 Phase Diagrams

The phase diagram for the one orbital model (Eq.(1)) is shown in Fig.1.[7] In both 1D and 2D and at low temperatures clear indications of (i) ferromagnetism, (ii) incommensurate (IC) correlations, and (iii) phase separation (PS) were identified. For D=1 and 2, finite size effects were small, although the PS-IC boundary in 2D was difficult to obtain accurately. PS was obtained by calculating the density $\langle n \rangle$ vs μ. If

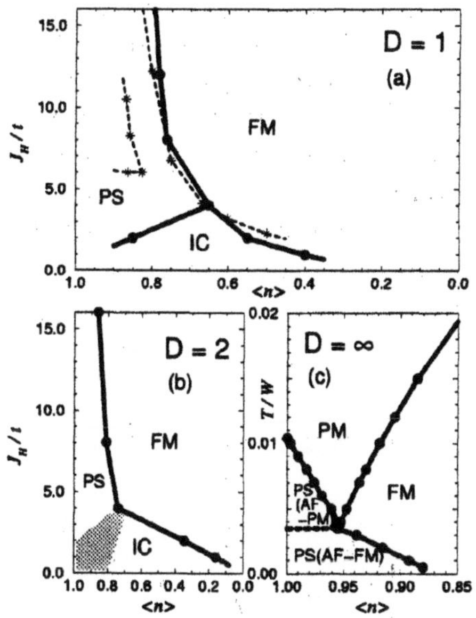

Figure 1. Phase diagram of the FM Kondo model reported here. [7] FM, IC, PM and PS denote regimes with FM correlations, incommensurate correlations, paramagnetic correlations, and with phase separation between undoped AF and hole-rich FM regions, respectively. (a) was obtained with MC simulations at $T = t/75$ using chains with L=20, 30 and 40 sites. Full circles correspond to classical spins. The stars and dashed lines are DMRG results obtained with t_{2g} spins S=3/2 on chains with up to L=16 sites keeping 48 states, and using $(2/3)J_H$ as coupling in the Hund term; (b) are MC results for classical spins at $T = t/50$ using 4×4, 6×6 and 8×8 clusters. In the shaded region a PS-IC crossover was observed but the actual boundary position is difficult to find; (c) corresponds to results in the D=∞ limit at a fixed coupling $J_H/W = 4.0$ and as a function of T and $\langle n \rangle$ (W is defined in the text). The "PS (AF-PM)" region denotes PS between undoped AF and hole-rich PM regions.

a discontinuity in $\langle n \rangle$ exists, then the system phase separates between the two densities at the extremes of the discontinuity. This occurs because the intermediate densities cannot be stabilized for any value of the chemical potential. For more details see ref. [7, 20]. Results are also available in small 3D clusters and qualitatively they agree with those in Figs.1a-b. In 1D we also obtained results with quantum t_{2g} spins S=3/2. In this case the PS regime was studied calculating the compressibility (κ) using DMRG (Density Matrix Renormalization Group), and the FM regime monitoring the ground state spin quantum number with a Lanczos procedure. A good agreement with results in the classical limit was obtained. PS for S=3/2 (and S=1/2) is between hole undoped AF and hole-rich non-fully-saturated FM regions. In Fig.1c results in D=∞ are shown at $J_H/W = 4$, where W is the half-width of the e_g density of states $D(\epsilon) = (2/\pi W)\sqrt{1-(\epsilon/W)^2}$. In agreement with the predictions for D=1 and 2, at low temperatures PS between undoped AF and hole-rich FM regions was observed. For $\langle n \rangle < 0.88$ and $T/W \ll 1$ the ground state becomes FM. The quantitative similarities between the results obtained for D=1,2 and ∞ led us to believe that these conclusions are also valid for 3D manganite systems.

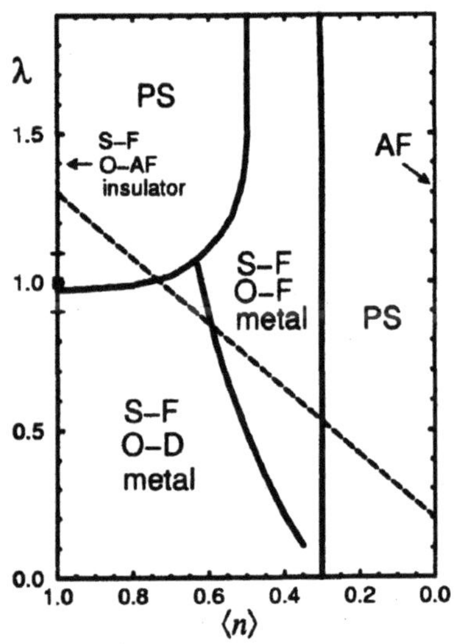

Figure 2. Phase diagram of the two-orbital model for manganites in one-dimension and T $\sim$ 0 including Jahn-Teller phonons, obtained with Monte Carlo techniques. [8] S-F labels a spin-ferromagnetic configuration. O-F, O-AF, and O-D denote a state where the orbital degrees of freedom are ordered uniformly, staggered or they are disordered, respectively; PS indicates a phase separated state, and AF is an antiferromagnetic state. The Hund-coupling is $J_H = 8$, the Heisenberg coupling between localized classical spins $J_{AF} = 0.05$, both in units of the hopping among the same orbitals. The meaning of the dashed line is explained in the text.

The consideration of the two e_g orbitals (Eq.(2)) and the addition of JT phonons (Eq.(3)) enriches considerably the phase diagram. Results at low temperature for a one-dimensional system at large Hund coupling are presented in Fig.2.[8, 20] The phase diagram is rich and includes a variety of phases such as metallic and insulating regimes with orbital order.

The latter can be uniform, with the same combination of orbitals at every site, or staggered, with two combinations alternating between the even- and odd-sites of the lattice at $\langle n \rangle = 1$. This "orbital order" is crucial in manganites and it has been found experimentally. Of special importance for the discussion here are the regions of unstable densities. Phase separation appears at small e_g-densities between an electron-undoped AF-state and a metallic uniform-orbital-ordered FM-state. The latter phase itself coexists at larger densities and intermediate values of λ with an insulating ($\langle n \rangle$=1) staggered-orbital-ordered FM-state, in an orbital-induced PS process.[8] The overall results are qualitatively similar to those obtained with other model parameters, and in studies of 2D and 3D systems.

If phenomenologically one assumes that λ/t decreases with hole doping, the dashed-line in Fig.1 runs through the proper series of experimentally observed phases [21], namely, an insulating staggered orbital-ordered state at x=0, a charge-segregated regime at small x, a metallic orbital-disordered FM-phase at a higher density, and finally the x $\sim$ 0.5 orbitally ordered FM-state compatible with A-type AF-order in dimensions D $<$ 3 [22]. If it were possible to complete the phase diagram of $La_{1-x}Sr_xMnO_3$ by synthesizing x>0.6 samples, the calculations predict a new mixed-phase, involving the A-type AF-metal with x $<$ 1 and a G-type AF-insulator with x = 1, where large MR effects could potentially occur.

2.2 Electronic Phase Separation

In the previous subsection we have seen that overall, PS tendencies are strong both in the one- and two-orbital models, and over a wide range of couplings. Similar tendencies have been recently observed including large on-site Hubbard interactions [23], which is reasonable because at intermediate and large electron-phonon coupling a negligible probability of on-site double-occupancy was found.[8]

To study this phenomenon in the grand-canonical ensemble, where the Monte Carlo simulations are performed, it is convenient to analyze $\langle n \rangle$ versus μ. If $\langle n \rangle(\mu)$ is discontinuous then there are densities that can not be estabilized, regardless of the value of μ. The results shown in Fig.3 obtained at $J_H/t = 12$ clearly show that indeed phase separation occurs in the FM Kondo model. The discontinuity is between a density corresponding to the antiferromagnetic regime $\langle n \rangle = 1.0$ and ~ 0.77 where ferromagnetic correlations start developing.[7] In the case of the canonical ensemble these results can be rephrased as follows: if the system is initially setup with a density in the forbidden band, it will spontaneously

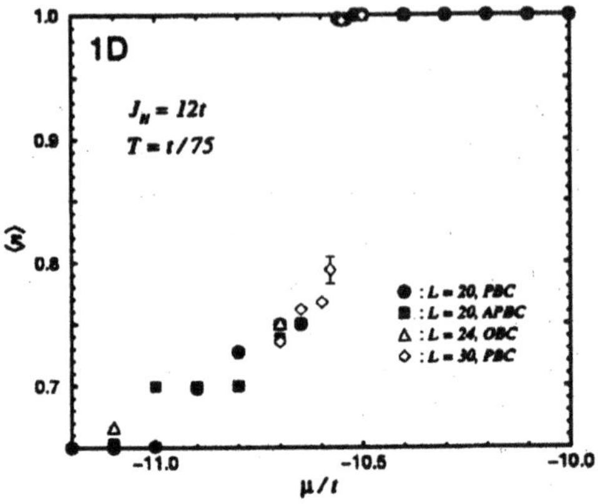

Figure 3. Electronic density $\langle n \rangle$ vs the chemical potential μ obtained with the Monte Carlo technique applied to the one dimensional FM Kondo model with classical spins. Coupling, temperature, lattice sizes, and boundary conditions are indicated. The discontinuity suggests that some densities are unstable, signaling the presence of phase separation. [7]

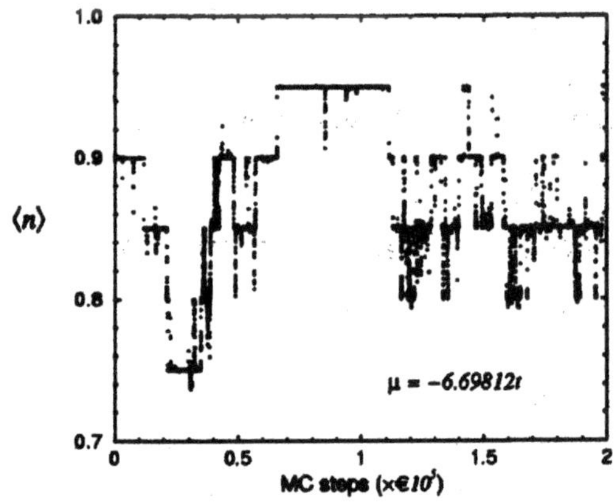

Figure 4. Monte Carlo time evolution of the density $\langle n \rangle$ at the particular value of μ where the discontinuity takes place working at $J_H/t = 8$, $T = t/75$, using $L = 20$ sites, and PBC.[7] Frequent tunneling events are observed showing the competition between two states as in a first order phase transition.

separate into two regions having (i) antiferromagnetic (AF) correlations and no holes, and (ii) FM correlations and most of the holes.

The existence of phase separation can also be deduced from the actual Monte Carlo runs since they require large amounts of CPU time for convergence in the vicinity of the critical chemical potential μ_c. The reason is that in this regime there are two states in strong competition.

Qualitatively this effect can be visualized analyzing $\langle n \rangle$ as a function of Monte Carlo time. Fig.4 shows such a time evolution when the chemical potential is fine tuned to its critical value $\mu_c \sim -6.69812t$ at $J_H/t = 8$. Wild fluctuations in $\langle n \rangle$ are observed with frequent tunneling events covering a large range of densities. A change in μ as small as 0.001 or even smaller reduces drastically the frequency of the tunneling events, and makes the results more stable although certainly strong fluctuations remain in a finite window near μ_c.

What consequences will PS have for experiments? The macroscopic separation of two phases with different densities, and thus different charges, should actually be prevented by long-range Coulombic interactions, which were not incorporated into the one- and two-orbital models discussed thus far. Even including screening and polarization effects, a complete separation leads to a huge energy penalization. This result immediately suggests that the two large regions involved in the process will break into smaller pieces to spread the charge more uniformly. These pieces are hereafter referred to as polarons if they consist of just one-carrier in a local environment that has been distorted by its presence. This distortion can involve nearby spins (magnetic polaron), nearby ions (lattice polaron), or both, in which case this object will be simply referred to as a "polaron." However, the terms "clusters" or "droplets" are reserved for extended versions of the polarons, characteristics of a PS regime, containing several carriers inside a common large magnetic distortion or lattice distortion or both. The present discussion suggests that in the regime of unstable densities the inclusion of extended Coulomb interactions will lead to a stable state, with clusters of one phase embedded in the other (see also [24]). It is expected that the competition between the attractive DE tendencies among carriers and the Coulomb forces will determine the size and shape of the resulting clusters. Either sizable droplets or polarons may arise as the most likely configuration. [25] The stable state resulting from the inclusion of extended Coulomb interactions on an otherwise PS unstable regime will be referred to as a charge-inhomogeneous (CI) state. However, the ideas presented here will still be described as the "PS scenario," with the understanding that only microscopic phase segregation, i.e., nanometer-size clusters, is the resulting net effect of the DE-Coulomb competition. Related ideas have been previously discussed in the context of the cuprates [27], with attractive interactions generated by antiferromagnetism or phonons. An exception to the existence of only purely microscopic effects occurs if the competing phases have approximately the same density, as observed experimentally at x = 0.5 (discussed below in 2.3.). In this case, large-scale

PS can be expected. Note also that the CI-state is certainly different from the metastable states that arise in a standard first-order transition.

2.3 Structural Phase Separation Due to Disorder

Recently, important new experimental information about the microscopic properties of manganites has been reported. Electronic diffraction and transport techniques were applied to $La_{5/8-y}Pr_yCa_{3/8}MnO_3$, material that changes from a ferromagnetic (FM) metal to a charge-ordered (CO) insulating state at y$\sim$0.35[28]. At low temperatures in this regime the unexpected coexistence of *giant* clusters of FM and CO phases was observed[28]. Similar results were also reported using scanning tunneling spectroscopy applied to $La_{0.7}Ca_{0.3}MnO_3$[29]. The clusters found in both these experiments were as large as 0.1 μm=1000Å$\sim$250 a, with $a\sim$4Å the Mn-Mn distance. These experimental results are *not* caused by chemical inhomogeneities, and the results are believed to correspond to equilibrium properties[28, 29]. The metal-insulator FM-CO transition occurs percolatively, varying either temperature or magnetic fields. These results rule out the picture of homogeneously distributed small polarons to describe manganites in the CMR region.

The discovery of huge coexisting FM-CO clusters in manganite single-crystals is puzzling. The only theoretical framework with similar characteristics is the phase separation (PS) scenario involving phases with different electronic densities[20]. The electronic PS ideas could be successful in describing manganites at, e.g., small hole density, where nanometer size inhomogeneities have been reported[20], and at high densities x$\sim$1 [30]. However, the micrometer clusters at intermediate densities[28, 29] appear to require an alternative explanation since the energy cost of charged μm-size domains is too large to keep the structure stable.

The random location and shape of the clusters observed experimentally[28, 29] also differs from the regularly-spaced charge arrangement observed in PS regimes when long-range interactions are included[26]. A novel framework involving *equal-density* large clusters is needed to rationalize the results of Refs.[28, 29]. Searching for an explanation of the coexistence of large clusters in manganese-oxides, we have combined (i) strong coupling interactions, necessary to produce the ordered phases, and (ii) quenched disorder. The latter is caused by the random chemical replacement of ions, such as La and Pr, with different sizes. This replacement affects the hopping of e_g-electrons and the exchange J_{AF} between the t_{2g}-spins due to the buckling of the Mn-O-Mn bonds near Pr.[31] Considering hopping and exchange couplings fluctuating about

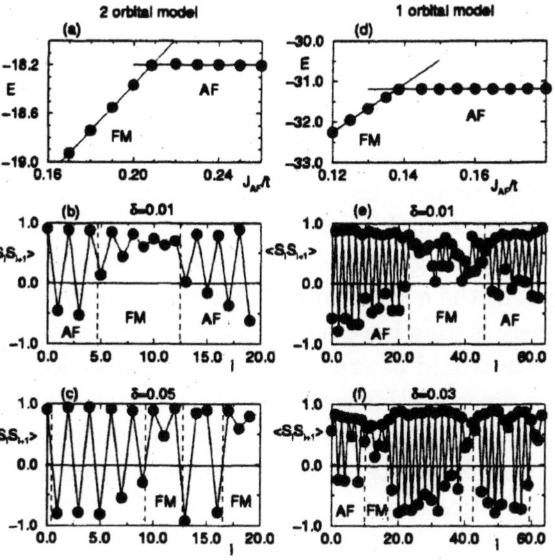

Figure 5. a-c) are MC results for the two-orbital model with $\langle n\rangle$=0.5, T=1/100, $J_H=\infty$, λ=1.2, t=1, PBC, and L=20. (a) is the energy per site vs J_{AF}/t for the non-disordered model; (b) MC averaged nearest-neighbor t_{2g}-spins correlations vs position along the chain (denoted by i) for one set of random t^{α}_{ab} and J_{AF} couplings (J_{AF}/t at every site is between 0.21-δ and 0.21+δ, with δ=0.01). FM and AF regions are shown; (c) Same as (b) but with δ=0.05; (d-f): results for the one-orbital model with $\langle n\rangle$=0.5, T=1/70, $J_H=\infty$, t=1, OBC, and L=64. (d) is energy per site vs J_{AF} for the non-disordered model, showing the FM-AF states level crossing at $J_{AF}\sim$0.14; (e) are the MC averaged nearest-neighbor t_{2g}-spin correlations vs position for one distribution of random hoppings and t_{2g} exchanges, such that J_{AF}/t is between 0.14-δ and 0.14+δ, with δ=0.01; (f) Same as (e) but with δ=0.03.

non-disordered values of interest, we have observed the appearance of coexisting giant clusters of equal-density FM and AF phases in realistic manganite models, as described below.

To present our main results first consider the two-orbital model (Eq.(2-3)), described extensively in previous work [8]. The focus of our studies will be on *first-order* transitions, which in the two-orbital model occur in several locations in parameter space in any dimension according to recent studies.[32, 20, 8] Let us start with 1D systems. Among the possible 1D first-order transitions, efforts are here focussed on the FM-AF transition at fixed x=0.5 and large λ, as a function of J_{AF}/t (similar results were found at other densities). The AF phase studied has a four-spin unit cell. The state is insulating, with vanishing Drude weight. Nearest-neighbors correlations among the t_{2g}-spins are used to distinguish among the FM and AF phases. In Fig.5a the energy per site (E) vs J_{AF}/t for the non-disordered model is shown. The $dE/d(J_{AF}/t)$ discontinuity indicates the first-order character of the transition at $J_{AF}/t|_c \approx 0.21$. Disorder is introduced in t_{ab}^{α} and J_{AF} such that J_{AF}/t becomes effectively random in the interval $J_{AF}/t|_c-\delta$ to $J_{AF}/t|_c+\delta$. Results for one fixed set of couplings are shown at δ=0.01 in Fig.5b (other sets lead to similar results). The MC averaged correlations in Fig.5b already show one of the our main results, namely the remarkable formation of coexisting large FM and AF clusters in the ground state, of order 10a each for δ=0.01 (a is the lattice spacing). This occurs even though J_{AF}/t at each link rapidly changes at the a scale since different sites are *uncorrelated* in the disorder. Naively it may have been expected that at every link either the FM or AF phases would be stable depending on the local J_{AF}/t, as it occurs for strong disorder. However, at weak disorder this would produce a large interface energy and the order parameter cannot follow the rapid J_{AF}/t oscillations from site to site. As a consequence, structures much larger than the lattice spacing emerge, with a size regulated by δ (compare the results in Fig.5c at δ=0.05 with Fig.5b). The effect occurs only near first-order transitions, i.e. the same weak disorder in other regions is not important[33]. Qualitatively similar results were found to appear also in other first-order transitions of the two-orbital model, such as for the FM-CO(CE-state) level crossing[32] using 4×4 and 4×4×2 clusters. Note also that here and in all the discussion below it has been checked that the MC results correspond to *equilibrium* properties: a variety of different starting configurations leads to the same results.

The well-known one-orbital model (Eq.(1))[20] presents the same behavior as the two-orbital one, and larger lattices can be studied. This model also has a first-order transition at x=0.5 as J_{AF}/t varies[11]. It involves equal-density metallic FM and insulating AF states, the latter with a similar spin structure as the AF state of Figs.5a-c. To investigate disorder effects, J_{AF} at every site is randomly selected in the interval $[J_{AF}^c-\delta, J_{AF}^c+\delta]$[34], where $J_{AF}^c \sim 0.14$ is the critical first-order transition coupling at $J_H=\infty$, t=1, and T=1/70 in the non-disordered limit (Fig.5d). In Fig.5e, results of a MC simulation corresponding to a representative set of random couplings (t, J_{AF}) centered at 0.14 are shown. As in the two-orbital case, FM and AF clusters, this time as large as 20a, are obtained. If the range of possible (t, J_{AF}) increases, the cluster size decreases (Fig.5f)[35].

2.4 Resistivity

Estimations of the resistivity ρ_{dc} are notoriously difficult in the manganites and thus, the behavior of ρ_{dc} remains unexplained, although it is central to their physics.

Our goal in this subsection is to present a rationalization of the ρ_{dc} vs T curves of Mn-oxides based on the currently prevailing phase-separated/percolative framework for these compounds.

The analysis necessarily involves phenomenological considerations, since percolation cannot be addressed on sufficiently large lattices using accurate microscopic models. However, the μ-meter clusters in experiments [28, 29, 36] strongly suggest that a coarse-grain approach should be sufficient. In addition, results of microscopic calculations presented below are consistent with those of the macroscopic approach. The main concept introduced here is summarized in Figs.6a-b.[37] The manganite state in the CMR regime is assumed to be percolative, with metallic filaments across the sample (Fig.6a). Percolation indeed occurs in models [38], and in many experiments.[28, 29, 36] The insulating and metallic (percolative) regions are assumed to have resistances $R_I(T)$ and $R_M^{per}(T)$, respectively, as sketched in Fig.6b. R_M^{per} is large at T=0 due to the complex shape of the conducting paths, and grows with T as in any metal, eventually diverging when the percolative path melts with increasing T. Note that at room temperature $R_I < R_M^{per}$ and, thus, most of the conduction in this regime occurs *through the insulator*. On the other hand, R_I is so large at low-T that current can only flow through the percolative paths. This suggests a simple two parallel resistances description (Fig.6b), where a peak in the effective resistance at intermediate T is natural.

To substantiate this idea, first consider results obtained using a *random-resistor-network* that mimics the prominent FM-CO mixtures [28, 29, 36] in Mn-oxides in the CMR regime. Two-dimensional (2D) and three-dimensional (3D) square and cubic clusters are used, with link resistivities randomly selected as metallic (ρ_M) or insulating (ρ_I), with a fixed metallic fraction p (which in, e.g., $(La_{5/8-y}Pr_y)Ca_{3/8}MnO_3$ (LPCMO) it is proportional to the amount of La [28, 29, 36]). The lattice spacing of this effective network is comparable to the FM or CO domain size, much larger than the Mn-Mn distance. The actual values of ρ_M and ρ_I vs T were directly taken from LPCMO data [28, 29, 36] (y=0.00 and 0.42, respectively), and for simplicity they are used in both 3D and 2D clusters. Other materials were tried and the analysis below does not depend qualitatively on the reference compounds.

The Kirchoff equations for the network were solved iteratively on large clusters using well-known techniques,[39] and the net resistivity was found. Typical results are shown in Fig.6c. Only the limiting cases p=0 (all insulator) and 1 (all metal) are taken from experiments. As expected, a percolative regime exists between p=0.4 and 0.5, where ρ_{dc}(T=0) is as large as in LPCMO and other materials.[40] ρ_{dc} has insulating behavior at room-T, even for p as high as 0.65, while at low-T

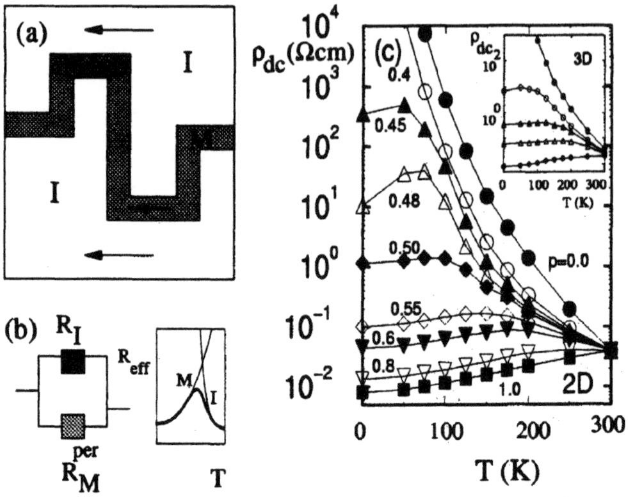

Figure 6. (a) Mixed-phase state near percolation.[37] Arrows indicate conduction through insulating or metallic regions depending on T (see text). (b) Two-resistances model for Mn-oxides. Effective resistance R_{eff} vs T (schematic) arising from the parallel connection of metallic (percolative) R_M^{per} and insulating R_I resistances. (c) Net resistivity ρ_{dc} of a 100×100 cluster vs T, at the indicated metallic fractions p. Inset: Results for a 20^3 cluster with (from top) p=0.0, 0.25, 0.30, 0.40 and 0.50. In both cases, averages over 40 resistance configurations were made (errors the size of the points). The p=1 and 0 limits are from Ref.[28, 29, 36]. Results on 200×200 clusters (not shown) indicate that size effects are negligible.

a (bad) metallic behavior is observed. A broad peak appears at intermediate T's and p's. Similar results exist in 3D (inset of Fig.6c).

It is remarkable that Fig.6c is already in good qualitative agreement with some Mn-oxide experiments, such as for $La_{0.96-y}Nd_yK_{0.04}MnO_3$, [41] or even non-manganite materials, such as $CaFe_{1-x}Co_xO_3$ where an AF-FM competition occurs.[42] However, many manganites present a more pronounced ρ_{dc} peak at intermediate T's. To reproduce this feature, it is necessary to introduce a percolative process not only as p (or x) varies, but also as T changes. This is reasonable since the metallic component triggered by ferromagnetism is sensitive to T, and the FM clusters shrink in size as T increases. This proposal was tested qualitatively using two models: (i) the Random Field Ising model (RFIM), that describes the disorder-induced PS,[38] with spin up and down crudely representing the competing metal and insulator, and (ii) the well-known one-orbital model[20, 7] (with parameters t, J_H, and J' representing the e_g-hopping, Hund coupling, and Heisenberg exchange among t_{2g}-spins, respectively). The latter is supplemented by a term $\sum_i \phi_i n_i$, with ϕ_i

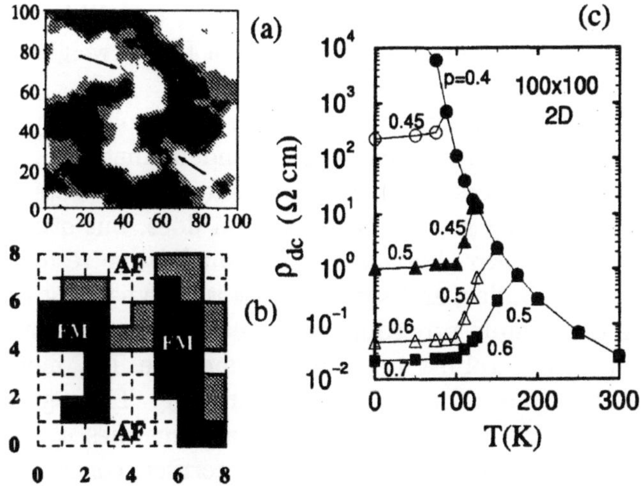

Figure 7. (a) MC results for the Random Field Ising Model at T=1.6 (J=1), and a fixed set of random fields taken from [-1.0,1.0]. Shown is a 100×100 subset of a 500×500 cluster with periodic boundary conditions (PBC). Black, white, and grey denote regions with Ising spin mean-value $\langle\sigma\rangle$ in the intervals (0.66,1.0), (-1.0,-0.66), and (-0.66,0.66), respectively. The arrows indicate regions where percolation occurs as T decreases. (b) MC results for the one-orbital model on a PBC 8×8 cluster, J_H=8, J'=0.0, t=1, density $\langle n\rangle$=0.875, $\rangle$=0.875, and a (fixed) realization of disorder (W=0.5) with percolative characteristics at T$\sim$0. Black and white denote FM and AF regions, respectively. Grey domains are FM at T$\sim$0, but the nearest-neighbor (NN) spin correlations become much smaller (PM) at T=0.05. (c) Net ρ_{dc} of a 100×100 cluster as in Fig.1c, but with p changing with T (representative values indicated). Averages over 40 resistance configurations are shown. Reproduced from Ref. [37].

randomly taken from [-W,W] and n_i the on-site density atsite i. This disorder generates coexisting clusters near first-order FM-AF transitions, as explained in Ref.[38].

In Fig.7a, it is shown a portion of a typical Monte Carlo (MC) simulation of the RFIM at T=1.6 on a 500×500 cluster, with fixed random fields taken from [-1.0,1.0] (J=1 is the FM Ising coupling). These parameters are the same as in Ref.[38]. Three main domains were found: spins-up (black), spins-down (white), and regions with a small spin expectation value (grey). The generation with T of these paramagnetic (PM) areas in the surface of the up and down domains weakens the percolative tendencies of the RFIM: for example, in Fig.7a domains connected at T=0, become disconnected at finite-T. Similar behavior occurs in the microscopic one-orbital case, as shown in Fig.7b for a disordered configuration with percolative features.[43] Increasing T from $\sim$0.0 to

0.05t decouples the two FM regions. FM, AF, and PM regimes dominate at finite T, as in the RFIM. The T=0.05t Drude weight (not shown) is much smaller than at T$\sim$0.0.

To incorporate the indications of T-induced percolation (Figs.7a-b) in the phenomenological approach, a T-dependent metallic fraction p(T) is needed. p must decrease as T grows, should vary rapidly near the Curie temperature T_C as the magnetization does, but otherwise its T-dependence is unknown. Fortunately, the qualitative results using several functions are similar, and a typical case is shown in Fig.7c. The ρ_{dc}'s obtained by this simple procedure now clearly resemble those found in experiments, with a robust peak at intermediate T's, moving toward lower T's as the system becomesmore insulating. This agreement with experiments is unlikely to be accidental, and justifies *a posteriori* our assumptions. Note that if our approach is correct, consistency requires that in Mn-oxides above T_C there should exist (disconnected) FM clusters on an insulating matrix, since p does not drop abruptly to 0 at T_C. A new temperature scale T* is predicted, with those FM clusters existing in the range $T_C \leq T \leq T^*$. The density of states in this regime likely has a ***pseudogap***, based on previous investigations of mixed-phase states [20, 7, 38].

2.5 Effect of Intrinsic Inhomogeneities on Transport Properties

In this subsection, the general aspects of the competition between two ordered states in the presence of quenched disorder is investigated. These states are assumed sufficiently "different" that their low-T transition in the clean limit is abrupt, with *first-order* characteristics. From these considerations, predictions emerge for doped transition metal oxides (TMOs), where intrinsic disorder is caused by chemical substitution (e.g. $La^{3+} \rightarrow Sr^{2+}$ creates randomly located hole-trapping centers, while their different ionic sizes produce a distribution of electronic hopping amplitudes). In the case of Mn-oxides, a rationalization of the CMR effect is provided here, with observable predictions including a new scale T* for cluster formation. The calculations are mainly carried out using a two dimensional (2D) toy model of Ising spins, but data in three dimensions (3D) have also been gathered. The simplicity of the results suggest that our conclusions should be valid for a variety of models with competing orders. The actual Hamiltonian employed here is defined on a square/cubic lattice (spacing a=1) and contains antiferromagnetic (AF) couplings J_1, J_2, and J_4connecting spins at distances 1, $\sqrt{2}$, and $\sqrt{5}$, respectively. More than one coupling is certainly needed to generate two competing T=0 states, and J_1 and J_2 are the natural ones. However, the clean-limit first-order transition between those states was found to be more robust if a small $J_4 \sim 0.2 J_1$ coupling is added. The resulting competing states O_1 and O_2 are an AF state for low J_2/J_1, and a "collinear" AF state with rows (or columns) of spins up and down for large J_2/J_1 [44]. As explained later, the reader should consider these states as the analogs of the metallic and insulating phases in Mn-oxides, with J_2 acting as $\langle r_A \rangle$.

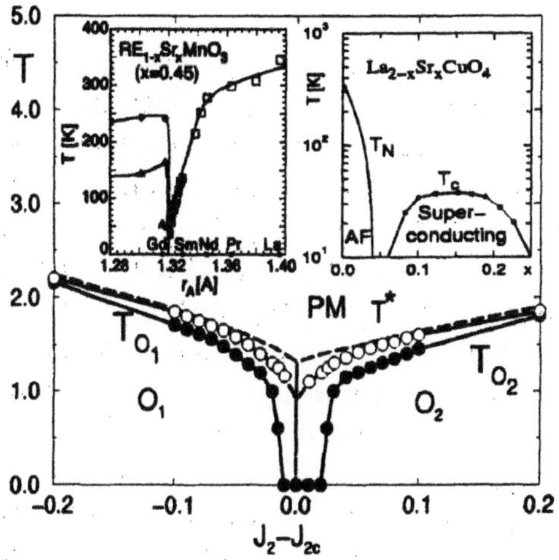

Figure 8. Phase diagram of the 2D J_1-J_2-J_4 toy model used to analyze the competition between ordered states.[45] Results on 32^3 lattices (not shown) lead to a qualitatively similar phase diagram. J_1=1 is the scale, and J_4 is fixed to 0.2. Disorder is incorporated such that $J_2(ij)$ at a link joining sites i and j is uniformly distributed between J_2-W/2 and J_2+W/2. Filled (open) circles corresponds to W=1.5 (0.75). Dashed lines are the result without disorder W=0, and T* denotes the clean-limit transition. A Metropolis algorithm was used, on up to 512^2 lattices, calculating (i) the largest ordered cluster size and (ii) the order parameters for AF and collinear phases with a spin structure factor maximized at momenta (π, π) and $(\pi, 0)$-$(0, \pi)$, respectively. Ten or more realizations of disorder were used, sufficient for large systems due to self-averaging. Insets are the phase diagrams of Mn-oxides in the FM-CO competition region Ref.[46] and of the single-layer Cu-oxide in standard notation Ref. [47].

The toy model phase diagram, without disorder, is shown in Fig.8, [45] and it has the expected shape: the ordering temperatures decrease and meet at the clean-limit critical coupling $J_{2c}=0.7J_1$, and the low-T transition was found to be clearly first-order. The most important result in Fig.8 is the influence of disorder on the clean-limit diagram. The first-order transitions become continuous with sufficiently large disorder.

Critical temperatures far from J_{2c} are not affected much by the disorder strengths considered. However, a *drastic* reduction is observed near J_{2c}. In fact, the Monte Carlo (MC) results suggest that the obtained phase diagram is similar to the insets in Fig.8 for Mn- and Cu-oxides. With increasing disorder strength W, either a first-order line separating the competing phases still survives at J_{2c} (open circles), as in manganites, or a disordered region of finite J_2 width opens at T=0 (filled circles), as in single-layer cuprates. Note that the ordering temperatures exactly meet at T=0 for only one fine-tuned value of W . However, the overall shape of the phase diagram resembling QCP behavior exists in a robust range of W.

Intuitively, it is clear that sufficiently strong quenched disorder will smooth first-order transitions [48]. At J_{2c}, this should occur at infinitesimal W in 2D [49], while a finite disorder strength is needed in 3D. Then, at J_{2c} and with finite temperature, a paramagnetic state must be generated with growing disorder. Note also that the toy model studied here is classical, and the understanding of QCP-like behavior in the insets of Fig.8 does not appear to need a quantum formalism.

Figure 8 is the result of a systematic computational effort. Keeping J_2 constant and varying T, three regimes are observed: **(1)** A high-T regime, where the system is disordered after MC time averaging. **(2)** A novel intermediate region $T_{O_1}<T<T^*$ with preformed clusters, but with uncorrelated order parameters giving a globally *paramagnetic* state. The clusters appear mostly static over long times, but tunneling to other cluster configurations was observed. **(3)** A low-T regime where the clusters from (2) grow in size, although the disorder is *uncorrelated* from link to link, and percolate upon cooling. Note the remarkable result that clusters with different signs for the order parameter are separated by thin regions of the competing phase, providing an interesting mechanism for stabilizing domain walls. Note also that the results of Fig.6 suggest that systems in which disorder creates QCP-like behavior must have a crossover scale T^*, remnant of the clean-limit ordering transition, where coexisting clusters starts forming upon cooling.

Let us adapt the results of Fig.8, believed to be qualitatively general, to Mn-oxides. A natural "translation" relates order O_1 with FM, with order parameters pointing in different directions for different clusters, while O_2 corresponds to CO. The state that appears bellow the dashed line and above the open circles in Fig.8 is characterized by small and disconnected clusters of all the competeing phases. Translating these states into Mn-oxide language [45] leads to Fig.9a, our proposed state for manganites in the CMR regime. Here FM clusters are preformed, but their moment orientation is uncorrelated due to the shielding effect of insulating regions, leading to a globally vanishing magnetization [51]. Note also that the "depth" of the QCP-like feature is not universal, as its value depends on the disorder strength [52].

To confirm the relevance of the state in Fig.9a to the CMR phase of the manganites, a resistor network calculation was set up. Translating to manganite language, as in the previous paragraph, the MC generated configurations were mapped into a resistance grid (see caption of Fig.9b). Up (down) spins define regions of low (high) conductivity, and the disordered regions are insulating. The Kirchoff equations were solved exactly, leading to results in Fig.9b.

In agreement with intuition, the non-percolated state has a *large resistance* for both spins up and down. On the other hand, the percolated low-T or disordered high-T states have far better conductance. Note that the resistance peak intensity increases as the ordering temperature is reduced by varying J_2, the analog of $\langle r_A \rangle$, closer to the QCP-like regime [53].

The rotation of the large moments of the preformed FM clusters (Fig.9a) should occur with small magnetic fields. These effects are mimicked in the toy model using a staggered external field H_s which favors O_1 clusters with order parameter $M_s>0$ to the detriment of O_1 clusters with $-M_s$ or O_2 regions. Results Fig.9c, the main results of this paper, are similar to those found in Mn-oxides, and a huge MR ratio $[R(0) - R(H_s)]/R(H_s)$ of $\sim 4\times 10^3$ % was obtained at $H_s=10^{-2}$.

The general prediction of a new scale T^* for uncorrelated cluster formation, below which disorder induces QCP-like behavior (Fig.8), has implications for manganites. In fact, in Mn-oxides there are already experimental indications of a new temperature scale [54]. Note that Fig.8 suggests a T^* that, unlike the ordering temperatures, is not very dependent on couplings. This new scale should be ubiquitous in low-bandwidth manganites [55].

The phase diagram in Fig.8 is representative of more realistic models. Fig.9d contains MC results for the one-orbital model [56], in which the coupling J_{AF} between localized spins is varied to induce a metal-insulator transition [16]. Without disorder, the $T\sim 0$ transition is known to be first-order between FM and AF$(0,\pi)$ or AF$(\pi,0)$ states, the latter having rows or columns of spins up and down [16]. The "characteristic" ordering temperatures T^*_C, at which spin correlations become robust upon cooling, are shown vs. J_{AF}. With disorder, a reduction in the spin correlations suggests the depletion of critical temperatures in the transition region, as in Fig.8.

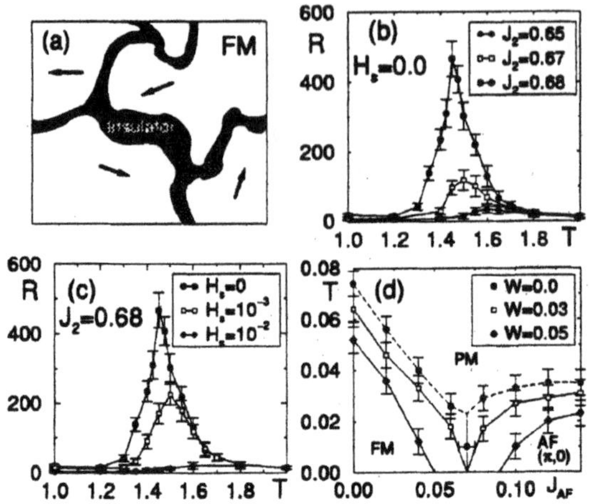

Figure 9. (a) Proposed state for Mn-oxides in the CMR regime. (b) Resistance of the toy model after the equivalence to manganite states is used (see text), using a 256^2 lattice, W=1.5, at the couplings indicated. The calculation is carried out by transforming a spin configuration into a resistor network, with nodes centered at plaquettes (# nodes = 1/4 # sites) and resistors between them (see details in Ref. [45]). The algorithm used to obtain the total conductance is exact [50]. (c) Resistance (arbitrary units) vs T, at external fields H_s indicated, using a 256^2 lattice, J_2=0.68, and W=1.5. (d) T_C^* results for the one-orbital manganite model using 8^2 and 16^2 clusters, density x=0.5, infinite Hund coupling, and hopping t=1. With disorder, the J_{AF} couplings are randomly distributed between J_{AF}-W and J_{AF}+W (W indicated). In practice, T_C^* was defined when the spin correlations at distance $\sqrt{2}$ dropped below 40% of the maximum value. Results reproduced from Ref.[45].

3. Summary

The novel regime of electronic and structural phase separation is particularly interesting, and possible consequences of its existence in manganites can be envisioned. Experimentally, phase separation can be detected using neutron diffraction techniques if the two coexisting phases have different lattice parameters as it occurs in $La_2CuO_{4+\delta}$, a Cu-oxide with hole-rich and hole-undoped regions [57]. NMR and NQR spectra, as well as magnetic susceptibility measurements, can also be used to detect phase separation [58, 59] since a splitting of the signal appears when there are two different environments for the ions. In fact, inhomogeneities in lanthanum manganites were reported by J. W. Lynn et al. using neutron scattering already. [60] Recently De Teresa et al. [61] using a combination of volume thermal expansion, magnetic susceptibility and small-angle neutron scattering measurements found indications of magnetic heterogeneity in these compounds. Goodenough and Zhou describe these results as a new form of phase segregation [62]. Then, a large amount of experimental information reviewed more extensively in Ref.[20] suggests that the tendency to form large clusters of ferromagnetically aligned spins in an antiferromagnetic background is a real possibility for the manganites.

In the 2-orbital Kondo model including Jahn-Teller phonons the phase diagram includes regions of electronic phase separation both at large and small e_g-density. One of them corresponds to a novel regime where PS is driven by the *orbital*, rather than spin, degrees of freedom. In the limit of small $\langle n \rangle$, the PS observed could be transformed by long-range Coulomb interactions into a charge-ordered state. Note that the PS observed in electron-doped Sr_2MnO_4 [59] and other compounds [57] is also compatible with our results [8].

Tendencies toward charge inhomogeneous states exist in real manganese oxides all around the FM-phase in the temperature-density phase diagram (see Fig.5 in Ref.[20]). The computer simulations have shown that the region with PS tendencies substantially influences the stable FM-phase by increasing its compressibility, an aspect that can be tested experimentally. This provides a rationalization for the experimental observation of a large MR effect at the boundaries of the FM-phase. The presence of short-range charge correlations is certainly a crucial feature of the PS scenario.

The giant clusters in manganites [28, 29] are conjectured to be caused by quenched disorder in the couplings (t^{α}_{ab} and J_{AF}), induced by chemical substitution, and which affect transitions that otherwise would be of first-order without disorder. The mixed-phase state presented in Figs.5a-f involves clusters with equal electronic density, complementing the electronic PS scenario which involves regions with different densities[20]. Although non-disordered models remain crucial to determining the competing tendencies in manganites, disordering effects appear necessary to reproduce the subtle percolative nature of the metal-insulator transition and the conspicuous presence of μm domains in these compounds.[28, 29] The present observations are very general, and the formation of coexisting giant clusters when two states are in competition through first-order transitions should be a phenomenon frequently present in transition-metal-oxides and related compounds.[63]

The manganite resistivity was studied in the framework of phase separation using a semi-phenomenological approach. At large (room) T, conduction predominantly through the insulating regions occurs leading to $d\rho/dT<0$, while at low T the metallic filaments near percolation carry the current. The large MR induced by magnetic fields is caused by a small change in the fraction of metallic clusters and a small increase in the insulating region conductivity, effects which severely affect transport in a percolative regime.[65] Other materials, such as EuB_6,[64] with transport properties similar to those of manganites are likely in a similar mixed-phase/percolated state.

If the phenomenological approach described here is correct, it is unavoidable to conclude that above T_C there should still be (disconnected) FM clusters on an insulating background in Mn-oxides. Their existence defines a new temperature scale, T^* for analogy with the cuprates, such that the FM clusters exist in the range $T_C \leq T \leq T^*$. The density of states of such mixed-phase regime likely has pseudogap characteristics. Also the present analysis leads to the conclusion that the metallic state reached from the insulator upon the application of magnetic fields is not homogeneous but must still have a substantial fraction of insulating clusters. This is at least partially supported by the large T=0 resistivity of such state. These predictions should be tested with experimental techniques.

Disorder acting on a system with a low-T first-order transition separating ordered phases creates quantum-critical-like features with coexisting clusters of both phases, a T^* scale at which local clusters start forming, and large effects upon small external perturbations. When these considerations are applied to Mn-oxides, large magnetoresistance is obtained. When applied to cuprates,[45] a tentative explanation of the large proximity effect is provided. "Colossal" effects in TMO's could be more common than previously anticipated.

The field of manganites is already mature and much progress has been achieved in recent years in the understanding of their properties. However, much work still remains to be done. The above mentioned percolative approach must be confirmed experimentally, as well as the existence of a T^* new temperature scale. Finally, the notorious changes in resistivity upon the application of modest fields must also be understood. Progress in this direction has been recently presented by Mayr et al. using, once again, a percolated state.[37] Intrinsic inhomogeneities are without doubt a key feature of manganites and theories that do not take them into account cannot reproduce properly the physics of these compounds.

Acknowledgments

The authors are supported by NSF under grant DMR-9814350. Additional support is provided by the National High Magnetic Field Laboratory and MARTECH.

References

[1] S. Jin, et al., Science **264**, 413 (1994), and references therein.

[2] Y. Tokura et al., J. Appl. Phys. **79** (8), 5289 (1996).

[3] G. C. Xiong et al., Appl. Phys. Lett. **66**, 1427 (1995), and references therein.

[4] C. Zener, Phys. Rev. **82**, 403 (1951); P. W. Anderson and H. Hasegawa, Phys. Rev. **100**, 675 (1955).

[5] P. G. de Gennes, Phys. Rev. **118**, 141 (1960).

[6] P. E. Schiffer, A. P. Ramirez, W. Bao, and S-W. Cheong, Phys. Rev. Lett. **75**, 3336 (1995); A. P. Ramirez et al., Phys. Rev. Lett. **76**, 3188 (1996); C. H. Chen and S-W. Cheong, Phys. Rev. Lett. **76**, 4042 (1996). See also H. Y. Hwang, S-W. Cheong, P.G. Radaelli, M. Marezio, and B. Batlogg, Phys. Rev. Lett. **75**, 914 (1995); P. G. Radaelli, D.E. Cox, M. Marezio, and S-W. Cheong, Phys. Rev. **B 55**, 3015 (1997).

[7] S. Yunoki, J. Hu, A. Malvezzi, A. Moreo, N. Furukawa, and E. Dagotto, Phys. Rev. Lett.**80**, 845 (1998); E. Dagotto, S. Yunoki, A. Malvezzi, A. Moreo, J. Hu, S. Capponi, D. Poilblanc and N. Furukawa, Phys. Rev. **B 58**, 6414 (1998).

[8] S. Yunoki, A. Moreo and E. Dagotto, Phys. Rev. Lett.**81**, 5612 (1998).

[9] A. Moreo, M. Mayr, A. Feiguin, S. Yunoki and E. Dagotto, Phys. Rev. Lett. **84**, 5568 (2000).

[10] N. Furukawa, J. Phys. Soc. Jpn.**63**, 3214 (1994); **64**, 2754 (1995).

[11] S. Yunoki and A. Moreo, Phys. Rev. **B58**, 6403 (1998).

[12] T. G. Perring et al., Phys. Rev. Lett.**78**, 3197 (1997).

[13] Most of the work in one-dimension (1D) has been performed using $t_{11} = t_{22} = 2t_{12} = 2t_{21}$ (set T_1), but results have also been obtained with $t_{11} = t_{22}$ and $t_{12} = t_{21} = 0$ (T_2), as well as with the hopping that takes into account the proper orbital overlap, namely $t_{11} = 3t_{22} = \sqrt{3}t_{12} = \sqrt{3}t_{21}$ (T_3) (see S. Ishihara et al., Phys. Rev. **B56**, 686 (1997)). In two-dimensions (2D), the set T_1 in both directions was used, but also the combination of T_3 in the y-direction and $t_{11} = 3t_{22} = -\sqrt{3}t_{12} = -\sqrt{3}t_{21}$ (T_4) in the x-direction. Finally, in three-dimensions (3D) T_4 was used in the x-direction, T_3 in the y-direction, and $t_{11} = t_{12} = t_{21} = 0$, $t_{22} = 4/3$ (T_5) in the z-direction.

[14] A. J. Millis, P. B. Littlewood, and B. I. Shraiman, Phys. Rev. Lett.**74**, 5144 (1995); H. Röder, J. Zang and A. R. Bishop, Phys. Rev. Lett.**76**, 1356 (1996); A. J. Millis, B. Shraiman and R. Mueller, Phys. Rev. Lett.**77**, 175 (1996).

[15] Note that estimations of T_c^{FM} using quantum and classical phonons lead to very similar results [14].

[16] E. Dagotto, T. Hotta and A. Moreo, Physics Reports, **344**, 1 (2001).

[17] Y. Okimoto, T. Katsufuji, T. Ishikawa, A. Urushibara, T. Arima, and Y. Tokura, Phys. Rev. Lett.**75**, 109 (1995) .

[18] T. Arima, Y. Tokura, and J.B. Torrance, Phys. Rev. **B48**, 17006 (1993).

[19] D.S. Dessau, and Z.-X. Shen, contribution to *Colossal Magnetoresistance Oxides* edited by Y. Tokura (Gordon & Breach, Monographs in Condensed Matter Science, London, 1999).

[20] A. Moreo et al., Science **283**, 2034 (1999).

[21] T. Akimoto et al., Phys. Rev. **B 57**, R5594 (1998).

[22] Similar results were reported within a mean-field approximation in R. Maezono, S. Ishihara, and N. Nagaosa, Phys. Rev. **B 57**, R13993 (1998).

[23] Y. Endoh, K. Hirota, Y. Murakami, T. Fukuda, H. Kimura, H.Nojiri, K. Kaneko, S. Ishihara, S. Okamoto, and S. Maekawa, preprint; K. Kubo, D. M. Edwards, A. C. M. Green, T. Momoi, and H. Sakamoto, preprint, cond-mat/9811286; Y. Motome, H. Nakano and M. Imada, preprint, cond-mat/9811221; Y. Motome and M. Imada, in preparation; N. Nagaosa, private communication.

[24] E. L. Nagaev, phys. stat. sol. (b),186 (1994) 9; and references therein.

[25] Some numerical results already support this picture. [26]

[26] A. L. Malvezzi *et al.*, Phys. Rev.**B59**, 7033 (1999).

[27] K. A. Müller, and G. Benedek, eds., proceed. of the conf. *Phase Separation in Cuprate Superconductors*, World Scientific (1993); V. J. Emery and S. A. Kivelson, Physica **C209**, 597 (1993); C. Castellani, C. DiCastro, and M. Grilli, Phys. Rev. Lett.**75**, 4650 (1995); J. M. Tranquada et al., Nature, 375 (1995) 561; E. Dagotto et al., Phys. Rev.**B49**, 3548 (1994); and references therein. See also L. P. Gor'kov and A. Sokol, JETP Lett.**46**, 42 (1987).

[28] M. Uehara *et al.*, Nature **399**, 560 (1999).

[29] M. Fäth *et al.*, Science **285**, 1540 (1999).

[30] J. J. Neumeier and J. L. Cohn, preprint.

[31] S.-W. Cheong, and H. Y. Hwang, in *Colossal Magnetoresistance Oxides*, ed. Y. Tokura, Monogr. in Condensed Matter Sci., Gordon & Breach, London, 1999.

[32] S. Yunoki *et al.*, Phys. Rev. Lett.**84**, 3714 (2000).

[33] At this early stage in the calculations it is difficult to quantify how close to the first-order transition one must be in real manganites to observe the effect.

[34] Simulations with other distributions of random numbers lead to similar results.

[35] Open boundary conditions (OBC) were used in Figs.5d-f, and periodic boundary conditions (PBC) in Figs.5a-c, to show that large cluster formation occurs in both cases.

[36] M. Ibarra, and J. De Teresa, JMMM **177-181**, 846 (1998); D. Louca and T. Egami, Phys. Rev. B**59**, 6193 (1999).

[37] M. Mayr, A. Moreo, J. Verges, J. Arispe, A. Feiguin and E. Dagotto. Phys. Rev. Lett.**86**, 135 (2001).

[38] A. Moreo *et al.*, Phys. Rev. Lett. **84**, 5568 (2000).

[39] S. Kirkpatrick, Rev. Mod. Phys. **45**, 574 (1973).

[40] Since ρ_I(T∼0) of y=0.42 LPCMO is finite, the net ρ_{dc}(T=0) does not diverge below the naively expected critical point p_c=0.50 of a simple 2D network with $\rho_I=\infty$.

[41] R. Mathieu *et al.*, cond-mat/0007154.

[42] S. Kawasaki *et al.*, J. Phys. Soc. Jpn. **67**, 1529 (1998).

[43] "Percolative" configurations are rare in small clusters.

[44] D. P. Landau, Phys. Rev. B**21**, 1285 (1980).

[45] J. Burgy et al., cond-mat/0107300.

[46] Y. Tokura *et al.*, talk APS March meeting, Seattle, 2001.

[47] J. B. Torrance *et al.*, Phys. Rev. B**40**, 8872 (1989).

[48] For previous work see Y. Imry and M. Wortis, Phys.Rev. B**19**, 3580 (1979); J. Cardy, cond-mat/9806355; H. G. Ballesteros *et al.*, Phys. Rev. B**61**, 3215 (2000).

[49] Y. Imry and S. Ma, Phys. Rev. Lett. **35**, 1399 (1975).

[50] D. Frank and C. Lobb, Phys. Rev. B**37**, 302 (1988).

[51] Typical cluster sizes are not universal, but depend on J_2 (analog of $\langle r_A \rangle$), and W.

[52] While the Sr-based Mn-oxide x=0.45 has a clear T_C depletion (Fig.8), results for other Mn-oxides may not be as prominent.

[53] Consistent with the experimentally well-known resistivity increase with decreasing T_C in Mn-oxides [2, 16, 28, 29, 36].

[54] Studies of LCMO at x=0.30 have reported anomalies at T$\sim$400K (De Teresa *et al.*, Ref. [28, 29, 36]). Competing inhomogeneities in bilayers exist at similar temperatures [D. N. Argyriou *et al.*, Phys. Rev. B**60**, 6200 (1999)]. Even in the context of EuO, a T^* well above the T_C has been identified [C. S. Snow *et al.*, cond-mat/0011527].

[55] Mixed-phase regimes have a *pseudogap* in the density of states [A. Moreo *et al.*, Phys. Rev. Lett. **83**, 2773 (1999)].

[56] J. L. Alonso *et al.*, Nucl. Phys. B **596**, 587 (2001).

[57] P. G. Radaelli, J. D. Jorgensen, R. Kleb, B. A. Hunter, F. C. Chou, and D. C. Johnston, Phys. Rev.**B49**, 6239 (1994) and references therein.

[58] P. C. Hammel, A. P. Reyes, S-W. Cheong, Z. Fisk, and J. E. Schirber, Phys. Rev. Lett.**71**, 440 (1993), and references therein.

[59] It is interesting to observe that experimental studies of $Sr_{2-y}La_yMnO_4$ have reported evidence for electronic phase separation and charge ordering at small y. See W. Bao, et al., preprint. See also J. Q. Li et al., Phys. Rev. Lett.**79**, 297 (1997) for phase segregation in $La_{1-x}Sr_xFeO_3$.

[60] J. W. Lynn et al., Phys. Rev. Lett.**76**, 4046 (1996).

[61] J. M. De Teresa et al., Nature **386**, 256 (1997).

[62] J. B. Goodenough and J.-S. Zhou, Nature **386**, 229 (1997).

[63] Our results should apply to any manganite with an AF-FM competition, and also to non-manganite compounds with similar phenomenology as those described in G. Cao *et al.*, Phys. Rev.**B56**, 5387 (1997) and Ref.[64]. Our results are also related to the relaxor FM picture of Cr-doped $Nd_{1/2}Ca_{1/2}MnO_3$ (T. Kimura *et al.*, Phys. Rev. Lett. **83**, 3940 (1999)).

[64] S. Yoon *et al.*, Phys. Rev.**B58**, 2795 (1998).

[65] In our approach large ρ_{dc} changes can be obtained whenever two ordered states are in competition, even if both are insulating, if their ρ_{dc}'s are sufficiently different.

THE KONDO-LATTICE MODEL FOR CERIUM COMPOUNDS

B. Coqblin[1], M. A. Gusmão[2], J. R. Iglesias[2], C. Lacroix[3] , S. G. Magalhães[4], A. Ruppenthal[2], A. A. Schmidt[5] and Alba Theumann[2]

[1]*Laboratoire de Physique des Solides, Université Paris-Sud, 91405-Orsay, France*
[2]*Instituto de Fisica, U.F.R.G.S., 91501-970 Porto Alegre, RS, Brazil*
[3]*Laboratoire L. Néel, CNRS, BP 166, 38042- Grenoble Cedex 09, France*
[4]*Departamento de Fisica, U.F.S.M., 97119-900 Santa Maria, Brazil*
[5]*Departamento de Matematica, U.F.S.M., 97119-900 Santa Maria, Brazil*

Abstract Many cerium alloys and compounds are characterized by *heavy-fermion* behaviour due to the Kondo effect. In the case of a lattice, there is a strong competition between the Kondo effect and magnetic ordering, which has been described by the Doniach diagram. We describe here the Kondo-lattice model within a mean-field treatment of the Hamiltonian, including both the s-f exchange intra-site interaction and the nearest neighbour inter-site f-f interaction in the three following cases: first, we treat the non-magnetic case and discuss the effects of conduction band filling on the occurrence of the Kondo effect; second, we describe the antiferromagnetic to Kondo transition; and third, the spin-glass to Kondo transition, both as a function of the intra-site interaction for a finite inter-site interaction. Comparison with experimental data in cerium systems is presented.

1. INTRODUCTION

The physics of strongly correlated electron systems is presently very active, and many books [1,2], review papers [3–6], or conference proceedings [7–9] are available on that subject. Most of such systems contain anomalous rare earths or actinides. In the series of rare-earth metals [10,11], most of them have a normal behaviour corresponding to a $4f^n$ configuration with an integer number n of 4f electrons, and a resulting valence of 3. But three of them, cerium, europium and ytterbium metals, can have at normal pressure a valence which can be different from 3. Indeed, cerium metal has a well known anomalous phase diagram showing two f.c.c. phases, α and γ, with different lattice parameters and magnetic properties [12]. There is a considerable number of anomalous alloys or compounds with these three rare earths, but also with three other, namely praseodymium, samarium and thulium, as observed for example in $PrSn_3$ [13], SmS [14], or TmSe [15]. In the series of actinides, uranium, neptunium and plutonium metals are close to be magnetic, with a high 5f density of states at the Fermi level [16], while many uranium compounds have also an anomalous behaviour and can be, for example, magnetic and superconducting at low temperatures [17].

Cerium is certainly the most striking case, with a very peculiar phase diagram for the metal [12], and more than one hundred anomalous compounds [18]. We define, as usual, the valence as the average number of conduction electrons, here equal to $4-n_f$, where n_f is the average number of 4f-electrons. Thus, roughly speaking, cerium systems are divided in two groups: the "intermediate-valence" ones, where cerium has clearly a valence different from 3, and the "Kondo" systems, where the valence is close to 3 and Kondo effect is present due to the proximity of the 4f and Fermi levels. However, the separation between these two regimes is not always very sharp. Moreover, some experimental features, such as the phase diagram of cerium metal, have been theoretically accounted for by both intermediate valence or Kondo models [19–22].

The Kondo effect is usually described as the screening of the localized spins by the conduction electrons via an antiferromagnetic coupling between them [23]. The properties of cerium Kondo compounds have already been extensively reviewed. At "high temperatures", i.e., at temperatures above the Kondo temperature T_K, perturbation theory holds and the properties of cerium Kondo alloys and compounds are well accounted for by the resonant scattering mechanism, including crystal field effects [24, 25]. On the contrary, the $T = 0$ solution of the single Kondo impurity is a nonmagnetic singlet and at "low temperatures" ($T << T_K$), a "heavy-fermion" behaviour has been observed in many cerium compounds [26, 27].

Then, in the case of a lattice, there is a strong competition between the Kondo effect and magnetic ordering, arising from the RKKY (Ruderman-Kittel-Kasuya-Yosida) interaction between rare-earth atoms at different lattice sites. This situation has been first described by the Doniach diagram [28], which describes the variation of the Néel temperature and of the Kondo temperature with increasing antiferromagnetic intra-site exchange interaction $|J_K|$ between localized and conduction-electron spins. It is shown that the actual Néel temperature obtained for the lattice first increases with $|J_K|$ and then goes through a maximum, before vanishing at a quantum critical point. Such a behaviour has been observed with increasing pressure (which enhances $|J_K|$) in many cerium compounds, such as $CeAl_2$ [29], $CePd_2Al_3$ [30] and $CeRh_2Si_2$ [31], or with the relative concentration of cerium in the case of alloys such as $Ce(Ru_xRh_{1-x})_2Si_2$ [32]. Above the "Quantum Critical Point" (QCP) corresponding to the magnetic–non magnetic instability (or to the disappearance of the Néel temperature), several different behaviours exist, including the standard Fermi Liquid and different Non-Fermi Liquid (NFL) behaviours. Several questions are still unsolved: first, the precise nature of the different NFL phases, second, the variation with pressure of the Kondo temperature in the non magnetic regime, and finally, the striking observation of a possibly superconducting phase in many cerium compounds close to the QCP [33]. Many ytterbium compounds [34, 35] have been recently studied at high pressures. In this case, pressure has an opposite role to that observed in cerium compounds: some compounds, such as $YbCu_2Si_2$ [35], are non-magnetic at normal pressure and become magnetic with an increasing Néel temperature while the Kondo temperature is decreasing rapidly with increasing pressure. A magnetic–non magnetic transition has also been observed as a function of the concentration x or pressure in europium compounds, such as $Eu(Pd_{1-x}Au_x)_2Si_2$ [36] or $EuNi_2(Ge_{1-x}Si_x)_2$ [37], but this transition is accompanied by a valence change for europium.

Thus, the Kondo temperature can be dramatically affected by the presence of magnetism in the case of the Kondo lattice, and we will discuss this feature in the present paper. In order to describe such an effect, we have introduced a Hamiltonian including both the s-f exchange intra-site interaction and a nearest-neighbour inter-site f-f interaction, and we have studied the Kondo-lattice model within a mean-field treatment of this Hamiltonian. Here, we will present the results for the Kondo-lattice model in the nonmagnetic heavy-fermion case. We will study both the effect of the inter-site f-f interaction and of a number of conduction electrons smaller than $n_f = 1$. In fact, considering the Kondo effect as the screening of the localized spins by the conduction electrons, in the case of the Kondo lattice an "exhaustion" problem [38] arises if the number of conduction electrons is not sufficient to screen all the lattice spins.

On the other hand, it is well known that there is a transition between a Kondo regime and a magnetically ordered phase. Then, we have considered the possibility of a classical antiferromagnetic order and studied the transition between this order and the Kondo state, which can be relevant to many cerium Kondo compounds. On the other hand, some disordered cerium ternary alloys, such as $Ce(Ni_{1-x}Cu_x)$, show "spin-glass" behaviour and can present a transition between this peculiar order and the Kondo state. Thus, we present here a Kondo-lattice model with both an intra-site Kondo interaction and an inter-site Ising interaction, with a random distribution of the interactions in order to describe the spin-glass behaviour. The resulting spin glass–Kondo transition can account for some specific transitions encountered in disordered cerium alloys.

2. THE "REVISITED" DONIACH DIAGRAM AND CONDUCTION BAND FILLING EFFECTS

The competition between the Kondo effect on each atom and the RKKY interaction is well described by the well-known "Doniach diagram", shown in Fig. 1. Let us call T_{KO} the Kondo temperature for a single impurity and T_{NO} the Néel (or Curie, in a few cases) temperature in the absence of Kondo effect. The usual exchange Hamiltonian between localized and conduction-electron spins is given by

$$H = -J_K \, \mathbf{s}_c \cdot \mathbf{S}_f \, . \tag{1}$$

T_{KO} is proportional to $\exp(-1/\rho|J_K|)$ and T_{NO} to $\rho|J_K|^2$, where ρ is here the density of states for the conduction band at the Fermi energy. Thus, for small $\rho|J_K|$ values, T_{NO} is larger than T_{KO} and the system tends to order magnetically, with often a reduction of the magnetic moment due to the Kondo effect; on the contrary, for large $\rho|J_K|$, T_{KO} is larger than T_{NO} and the system tends to become non magnetic. The

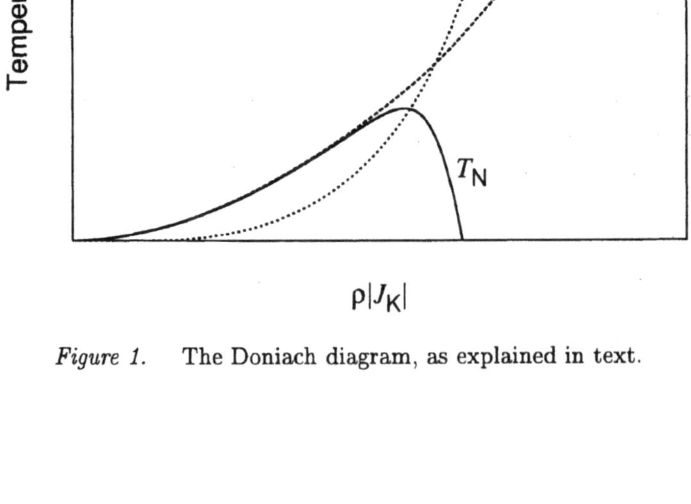

Figure 1. The Doniach diagram, as explained in text.

actual ordering temperature T_N, therefore, increases initially with increasing $\rho|J_K|$, then passes through a maximum and tends to zero at a critical value $\rho|J_K|_c$ corresponding to the so-called "quantum critical point" (QCP) of the Doniach diagram. Such a behaviour of T_N has been experimentally observed with increasing pressure in several Kondo compounds, such as $CeAl_2$ [29], $CePd_2Al_3$ [30], $CeRh_2Si_2$ [31], CeAg [39], CePtGa [40], $CeIn_3$ [41], $CePd_2Si_2$ [42], $CeAu_2Si_2$ [43] and $CeCu_2$ [43], or with the relative concentration of cerium in the case of alloys such as $Ce(Ru_xRh_{1-x})_2Si_2$ [32]. A maximum is observed in CeAg or $CePd_2Al_3$, but in all these compounds, the Néel temperature decreases and tends to zero at high pressures. Thus, we can conclude that the variation of the Néel temperature predicted by the Doniach diagram is well observed experimentally in many cerium compounds. We can also say that the Néel temperature is starting from zero at a given pressure and increasing rapidly with pressure in $YbCu_2Si_2$ [35] or related ytterbium compounds, which can be considered as another check of the Doniach diagram.

On the other hand, the one-impurity model predicts an exponential increase of the Kondo temperature with $\rho|J_K|$ and consequently with increasing pressure in cerium compounds and decreasing pressure in ytterbium compounds. This behaviour is well followed in many compounds, but deviations seem to occur in some cerium compounds. Firstly, as already explained [44], the electronic specific heat constant γ is increasing with pressure and passing through a maximum in $CeRh_2Si_2$ [31] at 14 kbar, which is clearly larger than the 9 kbar pressure of the QCP. It results that the Kondo temperature deduced from $1/\gamma$ is roughly constant above the QCP and does not increase exponentially with pressure. The resistivity of $CeRu_2Ge_2$ has also been measured at high pressures [45, 46], revealing a behaviour as AT^2 at very low temperatures above the QCP (estimated to be 67 kbar [45] or 87 kbar [46]). The deduced Kondo temperature, which is given by $A^{-1/2}$, is increasing very slowly with pressure, or even saturating at high pressures, and does not show any exponential behaviour above the QCP. Thus, the actual Kondo temperature observed in a lattice can be different and much smaller than the Kondo temperature derived for the single-impurity case.

Another interesting experimental fact concerns the observation of short-range magnetic correlations in cerium compounds which do not present any long-range magnetic order. Extensive neutron diffraction experiments have been performed in $CeCu_6$ [47], $CeInCu_2$ [48], $CeRu_2Si_2$ [32,47,49] or $Ce_{1-x}La_x\,Ru_2Si_2$ [32,49–51] single crystals at low temperatures. The last system is non magnetic and shows short-range magnetic correlations for $x = 0$, while an antiferromagnetic order appears at roughly $x = 0.08$. The neutron scattering spectra have been analyzed as the sum of a **q**-independent, single-site, quasi-elastic contribution associated with the single-site Kondo effect and an inter-site **q**-dependent contribution associated with short-range magnetic correlations between different cerium sites. Thus, short-range magnetic correlations between neighbouring cerium atoms develop in many cerium compounds at low temperatures. The temperature $T_{\rm cor}$ at which these correlations appear is clearly larger than the Kondo temperature T_K: $T_{\rm cor}\sim$ 60–70 K and $T_K\sim$ 20 K in $CeRu_2Si_2$ [47,49], and $T_{\rm cor}\sim$ 10 K and $T_K\sim$ 5 K in $CeCu_6$ [47].

Very recently, another example of such an anomalous behaviour was provided by the resistivity, magnetic susceptibility and specific-heat measurements of Umeo, Takabatake et al. [77] on the compound $Ce_2Rh_3Ge_5$. This compound orders antiferromagnetically below T_N roughly equal to 5 K at normal pressure and, with increasing pressure, the Néel temperature T_N is suppressed and disappears at the critical pressure P_c=0.45 GPa. The low-temperature electronic specific-heat constant γ, which is equal to 0.22 J/K^{2}molCe at normal pressure, increases continuously with pressure, even above P_c. The analysis developed in Ref. [77] yields a very weak increase of T_K below and above P_c, and contradicts the exponential increase derived from the Doniach diagram. Moreover, short-range magnetic correlations manifest themselves in broad maxima of the magnetic susceptibility and specific heat observed above P_c. Thus, the experiments recently performed on the compound $Ce_2Rh_3Ge_5$ have given a good support for our theoretical model [44], which we will describe below. However, a non-Fermi liquid behaviour is observed in both the resistivity and magnetic susceptibility of $Ce_2Rh_3Ge_5$ above the critical pressure, and still exists above $3P_c$. This fact is in contrast with the recovery of Fermi-liquid behaviour for a pressure larger than $2P_c$ in many cerium compounds such as Ce_7Ni_3 [78]. Thus, this new example shows clearly that the situation of cerium compounds just above the QCP is still a matter of discussion.

The Doniach diagram [28,52] was initially obtained using the mean-field approximation for the so-called "Kondo necklace", which is a one-dimensional equivalent of the Kondo-lattice model. In this approach, at $T = 0$, there is a finite critical value J_c of the exchange integral corresponding to a transition from the antiferromagnetic order to a Kondo-singlet state. A similar result was obtained by a real-space renormalization-group treatment of the Kondo necklace [53], but the various approaches yield different results: either a finite J_c value or a zero one [54]. Other treatments of the one-dimensional Kondo-lattice model by real-space renormalization [55] or by exact diagonalization [56] yield $J_c = 0$, indicating that the intrasite exchange interaction appears to be insufficient to produce magnetic ordering. Although this problem is controversial, we think that it is necessary to add an inter-site exchange interaction to the classical intra-site exchange interaction in order to have a full description of the interplay between Kondo effect and magnetic ordering.

The resulting Hamiltonian of the system is

$$H = E_0 \sum_{i\sigma} n^f_{i\sigma} + \sum_{\mathbf{k}\sigma} \epsilon_{\mathbf{k}} n^c_{\mathbf{k}\sigma} - J_K \sum_i \mathbf{s}^c_i \cdot \mathbf{S}^f_i - J_H \sum_{\langle ij \rangle} \mathbf{S}^f_i \cdot \mathbf{S}^f_j \,, \qquad (2)$$

where $\epsilon_{\mathbf{k}}$ is the energy of the conduction band, E_o the energy of the f-electrons, J_K the negative Kondo coupling between the spin of a localized f-electron, $\mathbf{S}^f_i$, and the spin of a conduction electron, $\mathbf{s}^c_i$, at the same site, and J_H the interaction between nearest neighboring f-spins. Here, we start from a zero f bandwidth, a conduction band of width $2D$ and a constant density of states. We also take a negative intrasite Kondo exchange and an antiferromagnetic intersite exchange interaction [44,57].

For spin-1/2 localized moments, i.e., a single f-electron per site, we can treat the Hamiltonian within an extended mean-field approximation, in which we keep only the following correlators [44, 57]: $\langle c^*_{i\sigma} f_{i\sigma} \rangle$, which describes the intra-site Kondo correlation, and $\langle f^*_{i\sigma} f_{i+\delta,\sigma} \rangle$, which describes the inter-site correlation between two neighbouring atoms. After performing this mean-field approximation, one deals with a one-electron Hamiltonian representing two hybridized bands: the f-band, which now exhibits a finite bandwidth proportional to the short-range f-f correlation $\Gamma = \langle f^*_{i,\sigma} f_{i+\delta,\sigma} \rangle$, and the conduction band. The magnitude of the hybridization gap is proportional to $\lambda^2 = \langle c^*_{i\sigma} f_{i\sigma} \rangle^2$, which is also a measure of the Kondo effect as the Kondo correlation function is proportional to λ^2.

We have first treated the half-filled case with a number of conduction electrons $n = 1$, and then we consider the general case of $n < 1$. The number of f-electrons n_f is always taken equal to 1. The half-filled case has been described in Ref. [44] and reviewed in Ref. [3]. In our mean-field approximation, T_K and T_{cor} are defined as the temperatures at which λ and Γ tend respectively to zero. We have shown that the Kondo temperature T_K for the lattice can be much smaller than the Kondo temperature T_{K0} for the single impurity, as observed in some cerium compounds. Moreover, in our previous work [44], we have also shown that the temperature T_{cor} for the occurrence of short-range magnetic correlations lies above the Kondo temperature T_K. Thus, these two theoretical results have provided a good explanation for the previously described anomalous behaviour observed in cerium compounds.

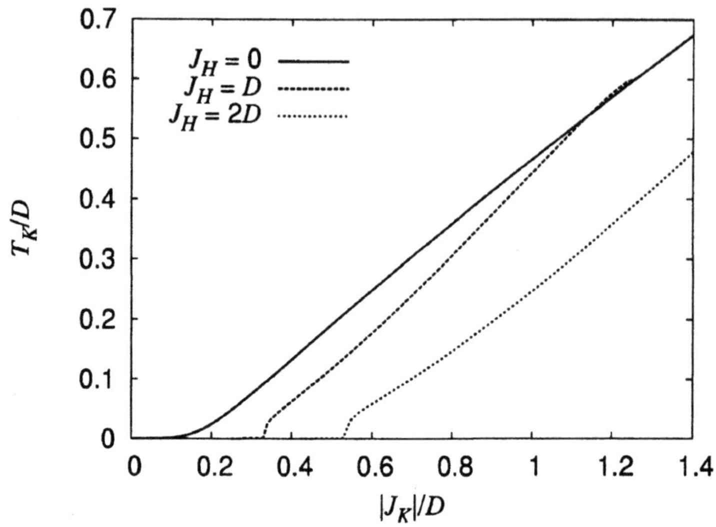

Figure 2. Variation of the Kondo temperature T_K with $|J_K|$ for $|J_H| = 0$, D and $2D$.

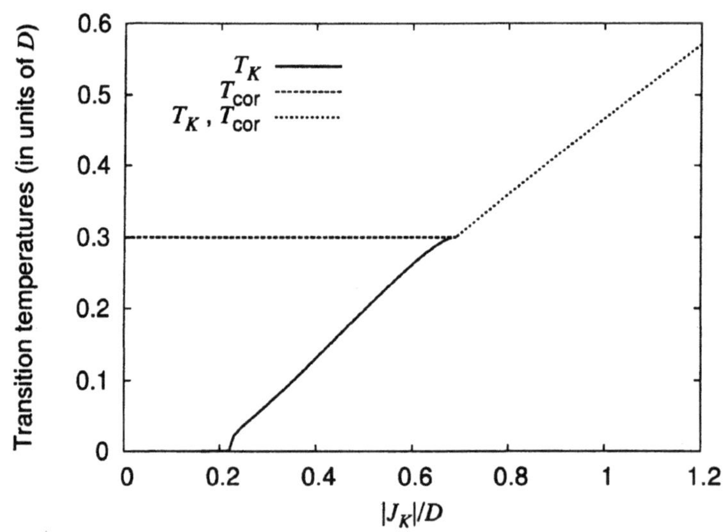

Figure 3. Plot of the Kondo temperature T_K and the correlation temperature $T_{\rm cor}$ versus $|J_K|/D$ for $J_H/D = -0.5$.

Then, we have considered the case of a number of conduction electrons n smaller than 1 and we will present here results only for this case $n < 1$. Figure 2 gives the curves of the actual Kondo temperature T_K versus $|J_K|$ for several J_H values. We see clearly that T_K decreases rapidly with increasing $|J_H|$. Figure 3 gives the two temperatures, T_{cor} and T_K, and we see clearly that short-range magnetic correlations occur above the Kondo temperature, in good agreement with experiment. In Ref. [44], we have used another approximation taking J_H proportional to J_K^2, but the physical results are still similar.

When $n < 1$ in the Kondo lattice, there appears an "exhaustion" problem, which means that there are not enough conduction electrons to screen the localized spins [38]. The stability of the Kondo effect when n decreases has been studied numerically at finite temperatures for some particular values of J_H, and the temperature dependence of the magnetic susceptibility has been computed for different values of n and J_H, showing clearly that both the reduction of n and the enhancement of J_H tend to suppress the Kondo effect [57, 58]. The influence of conduction-band filling on the Kondo effect has been studied again in Ref. [59], and we present both numerical results in Figure 4 and analytical expressions for two different specific limits: the case of small J_H and that of small λ. If $n < 1$ the Fermi energy crosses the resulting lower-energy hybridized band, which can have a maximum or not as a function of $\epsilon_{\mathbf{k}}$, depending on the relative values of the different parameters.

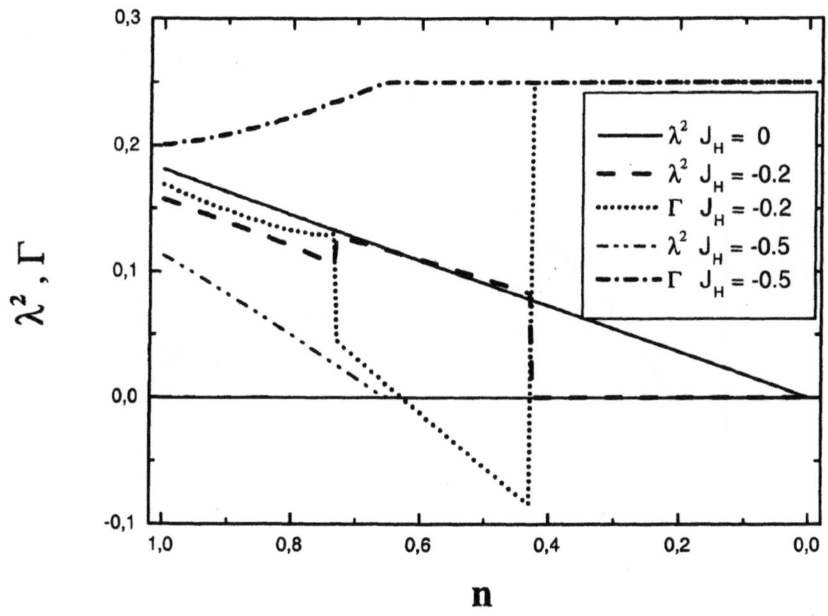

Figure 4. Plot of λ^2 and Γ versus the band filling n for $J_K = -1$ and different values of J_H.

For small values of J_H, we obtain a set of self-consistent equations for λ and Γ:

$$\lambda^2 = \frac{n_c D^2(1-B)^2}{J_K^2} \frac{u}{(1-u)^2}, \tag{3}$$

$$\Gamma = \frac{n\Delta}{4D(1-B)} - \frac{1-n}{4} + \frac{J_K \lambda^2}{D(1-B)}, \tag{4}$$

D being the half width of conduction band, $B = zJ_H\Gamma/D$ (where z is the number of nearest neighbors of a site), $u = \exp\left[2D(1-B)/J_K\right]$ and $\Delta = \sqrt{D^2(1-B)^2 + 4J_K^2\lambda^2/n}$.

These equations have been solved numerically: for small values of $|J_H|$, we found a discontinuous transition of λ as a function of n, from $\lambda \neq 0$ to $\lambda = 0$. This case corresponds in Fig. 4 to the curve for $J_K = -1$ and $J_H = -0.20$. For larger values of $|J_H|$, the transition from the Kondo to the non-Kondo region is continuous, as shown in Fig. 4 ($J_H = -0.5$). In the latter case we are able to expand the energy in powers of λ, and

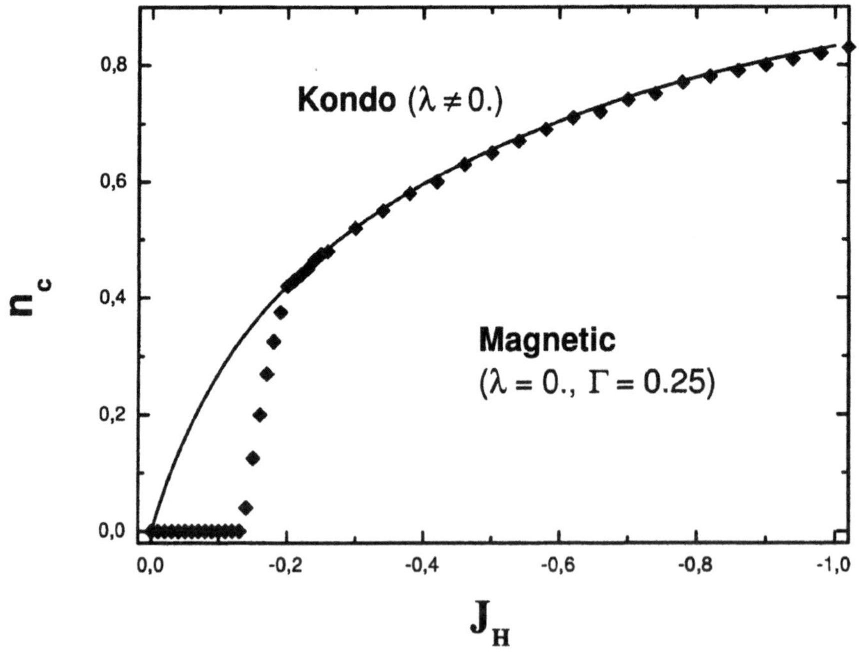

Figure 5. Phase diagram showing the Kondo and magnetic phases for varying critical band filling n_c and J_H.

minimizing it we obtain $\lambda^2 = p/2a$, with

$$p = 2J_K + \frac{J_K^2}{D(1-B)} \ln \frac{(1-B)^2 - (1-n)^2}{-B(1-n)^2} \tag{5}$$

and

$$a = -\frac{J_K^2}{4D(1-B)} \left[\frac{1}{B(1-n)^2} + \frac{1}{(1-B)^2 - (1-n)^2} \right] . \tag{6}$$

λ vanishes at a critical band filling given by

$$n_c = 1 - (1-B)\sqrt{\frac{u}{u-B}} \tag{7}$$

Figure 5 shows the phase diagram for $J_K = -1$. For $|J_H| < 0.13$, the Kondo phase is stable for all values of the band filling, while for $|J_H| > 0.2$ the Kondo phase is stable only for $n > n_c$ given by Eq. (7). Figure 5 shows that Eq. (7) is very close to the numerical result. Finally, for intermediate values $0.13 < |J_H| < 0.20$, the transition from the Kondo to the normal phase is discontinuous and has been obtained numerically.

Thus, our mean-field calculation shows clearly that small n and large $|J_H|$ tend to decrease the Kondo temperature and finally to suppress the Kondo effect, yielding a "magnetic" phase with $\lambda = 0$ and large short-range magnetic correlations. In fact, in this region a long range magnetic order should certainly be stabilized, but this was not taken into account in this approach. In contrast, both λ and Γ are different from zero in the Kondo phase.

3. THE ANTIFERROMAGNETIC–KONDO TRANSITION

In the preceeding section, we have studied the "non-magnetic" region of the Doniach diagram. In this section, we will study the problem of the transition between the magnetically ordered state (generally an antiferromagnetic one) which exists for small values of $|J_K|$ and the non-magnetic Kondo state for large values of $|J_K|$ [60]. We describe here the competition between antiferromagnetic ordering and Kondo effect, within the classical model for the Kondo lattice, including a Heisenberg-like antiferromagnetic exchange between nearest-neighbouring localized spins. We use again the Hamiltonian given by Eq. (2). In the previous section, we decoupled both Kondo and Heisenberg interactions by introducing two mean-field parameters associated with the local Kondo effect and with short-range magnetic correlations. Here, since we are interested in a magnetically ordered state, we use a different mean-field decoupling, keeping the previously used Kondo mean-field parameter λ, but introducing as a mean-field parameter the staggered magnetization M, instead of the previously used parameter Γ associated with short-range magnetic correlations.

We assign the localized spins to f electrons, subject to the restriction $\langle n_i^f \rangle = 1$. In the antiferromagnetic (AF) case, dividing the system in two sublattices, A and B, the mean-field Hamiltonian takes on the form

$$H_{MF} = H_A + H_B + H_{AB} \tag{8}$$

where

$$H_A = \sum_{i\sigma} \left[(E_o + \sigma M z J_H) n_{i\sigma}^f - \mu n_{i\sigma}^c + 2J_K \lambda (c_{i\sigma}^* f_{i\sigma} + f_{i\sigma}^* c_{i\sigma}) \right] , \tag{9}$$

$$H_B = \sum_{i\sigma} \left[(E_o - \sigma M z J_H) n_{i\sigma}^f - \mu n_{i\sigma}^c + 2J_K \lambda (c_{i\sigma}^* f_{i\sigma} + f_{i\sigma}^* c_{i\sigma}) \right] , \tag{10}$$

$$H_{AB} = -t \sum_{\langle ij \rangle} (c_{i\sigma}^* c_{j\sigma} + \text{H.c.}) . \tag{11}$$

The mean-field parameters satisfy the self-consistency conditions

$$\lambda = \frac{1}{2} \langle c_{i\sigma}^* f_{i\sigma} + f_{i\sigma}^* c_{i\sigma} \rangle , \quad M = \langle n_{i\uparrow}^f - n_{i\downarrow}^f \rangle / 2 . \tag{12}$$

The reference energy E_0 and the chemical potential μ have been introduced as Lagrange multipliers for the constraints $\langle n_i^f \rangle = 1$ and $\langle n_i^c \rangle = n$, where n is the filling of the conduction band.

The above Hamiltonian is numerically diagonalized and we obtain self-consistent solutions for λ and M, with different values of the physical parameters. As usual in the mean-field approximation, the Kondo phase is characterized by a non-zero value of the parameter λ and the antiferromagnetic (AF) order is related to a non-zero magnetization. The Kondo temperature T_K and the Néel temperature T_N are defined here as the temperatures at which the corresponding order parameters (λ or M) go to zero. Our results, shown in Figure 6, are presented in a plot similar to the Doniach diagram, where we plot the transition temperatures versus $|J_K|$ for a large J_H = - 0.5, in units of D, which is equal to half the conduction-band width. We choose the conduction electron density $n = 0.9$, in order to remain near half-filling but avoiding the Kondo-insulator situation which occurs for $n = 1$. For small values of $|J_K|$, we obtain antiferromagnetic order at low temperatures, with a constant Néel temperature since it depends only on J_H. On the contrary, for large values of $|J_K|$, we obtain a Kondo regime, with T_K rapidly increasing.

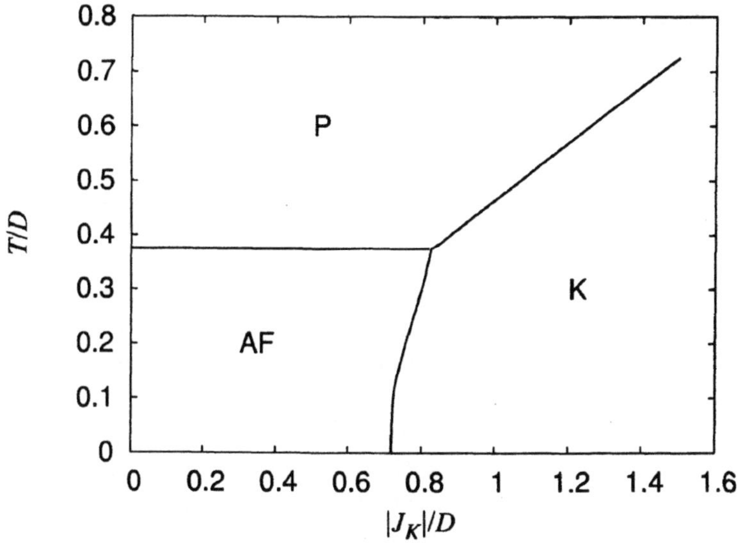

Figure 6. Phase diagram in T/D versus $|J_K|/D$ for $J_H = -0.5D$ and $n = 0.9$, showing the antiferromagnetic (AF), Kondo (K) and normal paramagnetic (P) regions.

Thus, we have obtained here a phase diagram showing a transition with increasing $|J_K|$ from antiferromagnetic ordering to a Kondo state, as in the Doniach diagram. This result is in agreement with the experimental situation of many cerium compounds which change from magnetic order to heavy-fermion behavior under pressure. However, we cannot account for two important points of the Doniach diagram: first, we never have a stable "mixed" phase where both λ and M have non-zero values –the occurrence of a first-order transition here is an artifact of the mean-field approximation, and one expects that this deficiency could probably be avoided by taking fluctuation corrections into account; second, the shape of the obtained theoretical phase diagram is different from the Doniach diagram, because there is no critical point and moreover the Néel temperature is constant, due to the absence of mixed phase. Improvements of this approach to include fluctuations around the mean-field solution might lead to a better description of the nature of the AF-Kondo transition.

4. SPIN-GLASS PHASE IN KONDO LATTICE SYSTEMS

In the Doniach diagram, above the QCP, a very strong heavy-fermion character is observed experimentally, with several possible behaviours, i.e., the classical Fermi-liquid behaviour with eventually a reduced Kondo temperature, or different Non-Fermi-Liquid (NFL) behaviours in the case of cerium or ytterbium compounds.

However, in the case of disordered cerium alloys, the disorder can yield a Spin Glass (SG) phase in addition to the NFL behaviour at low temperatures around the QCP. The magnetic phase diagram of $CeNi_{1-x}Cu_x$ has been extensively studied [61, 62]. CeCu is antiferromagnetic below 3.5 K and CeNi is a nonmagnetic compound with an intermediate valence. The low-temperature antiferromagnetic phase changes, around $x = 0.8$, to a ferromagnetic one which finally disappears around $x = 0.2$. At higher temperatures, a spin-glass state is deduced from all measured bulk properties, such as the ac susceptibility. For example, for $x = 0.6$, the SG state exists between 2 K and the Curie temperature $T_c = 1.1$ K. At $x = 0.2$, there exists below 6 K only a SG state, which transforms to a Kondo-lattice system before reaching an intermediate valence for CeNi. Thus, there exits a SG-Kondo transition at a small x value in the $CeNi_{1-x}Cu_x$ system.

Another disordered cerium system, the $CeCoGe_{3-x}Si_x$ alloys, has been studied for different x values, by different experimental techniques [63]. $CeCoGe_3$ is antiferromagnetic below 21 K, $CeCoSi_3$ is an intermediate valence compound, and the QCP of these alloys lies around $x = 1.5$. The muon spin relaxation experiments for $x = 1.1$ and 1.2 show that some cerium moments are frustrated and freeze like in a spin glass, while the other Ce moments form a disordered antiferromagnetic state. Thus, near the QCP, a spin-glass state can exist in these alloys, in addition to the NFL behavior. The alloys $Ce_2Au_{1-x}Co_xSi_3$ have been recently studied and present an unusual magnetic phase diagram [64]. Low-temperature heat capacity, ac susceptibility and magnetoresistance show that there are successively, with increasing x at low temperatures, a spin-glass phase up to roughly $x = 0.45$, then an antiferromagnetic phase up to roughly $x = 0.9$, and finally a non-magnetic Kondo lattice. Thus, the phase diagram is different from the previous cases, since there is no direct SG-Kondo transition, but SG-AF-Kondo transitions. The magnetic properties of Ce-Ni-Ge compounds are also presently extensively studied, and Ce_2NiGe_3 has been recently found to be a new Kondo-lattice compound showing spin glass behaviour [65].

The SG-Kondo transition has been recently studied theoretically [66]. We use a Hamiltonian that couples the localized spins of a Kondo lattice with an additional long-range random interaction, like in the Sherrington-Kirkpatrick spin-glass model [67]. A similar Hamiltonian has been considered to describe the NFL behaviour and the QCP at $T = 0$, but different approximations have been used there [68]. Another calculation using the same Hamiltonian and the Popov-Fedotov representation has also been proposed, and a second-order SG transition with a transition temperature depleted by the Kondo effect was obtained [69].

We present here a brief summary of the model and results of Ref. [66], and detailed calculations can be found there. In the present model, the localized spins of the Kondo lattice are immersed in a bath of conduction electrons. The Kondo effect is studied in a quantum static approximation which is basically equivalent to the previously used mean-field approximation [44]. When the classical exchange coupling of the localized moments to the conduction-band electrons is added to the fermionic Ising quantum spin glass, the problem is no longer soluble, and the static ansatz, which corresponds to an approximation similar to mean-field theory, is considered. Moreover, we use functional-integral techniques where the spin operators are represented by bilinear combinations of fermionic (anticommuting) Grassmann fields.

The model Hamiltonian is

$$H = \sum_{\mathbf{k},\sigma} \epsilon_{\mathbf{k}} n_{\mathbf{k}\sigma} + E_0 \sum_{i,\sigma} n^f_{i\sigma} - J_K \sum_i \left(S^+_{fi} S^-_{ci} + S^-_{fi} S^+_{ci} \right) - \sum_{i,j} J_{ij} S^z_{fi} S^z_{fj} \,. \tag{13}$$

We take here $J_K < 0$ as in the previous sections. The coupling J_{ij} in the preceeding Hamiltonian is an independent random variable with Gaussian distribution of zero mean and variance $(4J)^2/N$. Functional integration techniques have proved to be a suitable approach to describe phase transitions in disordered quantum many-particle systems. The static approximation consists here in neglecting time fluctuations of the order parameter. Combined with the neglecting of space fluctuations, it leads to the usual mean-field approximation. We can also remark that the limiting case $J_k = 0$ in the Hamiltonian (13) yields a pure quantum Ising SG, for which the static approximation is exact [70], while the other limit, $J = 0$, reproduces the result of Sec. 2 without any intersite interaction. Thus, the use of the static approximation has an interpolation character and provides reliable results to describe critical behaviour at finite temperatures that do not present any QCP, as it is the case in Ce(Ni,Cu) alloys.

Detailed calculations can be found in Ref. [66]. We just recall that the partition function in the Lagrangian formulation is expressed as

$$Z = \int D(\varphi^* \varphi)(\psi^* \psi) e^A \,, \tag{14}$$

where the action A is given by

$$A = \sum_{i,\omega} \int_0^\beta d\tau \left\{ \left(\frac{\partial}{\partial \tau} \varphi_{i\sigma}^* \right) \varphi_{i\sigma} + \left(\frac{\partial}{\partial \tau} \psi_{i\sigma}^* \right) \psi_{i\sigma} \right\} - \int_0^\beta d\tau \, H(\tau) \,. \tag{15}$$

In both expressions, $\phi_{i\sigma}(\tau)$ and $\psi_{i\sigma}(\tau)$ are anticommuting, complex Grassmann variables associated respectively to the conduction and localized electron fields, with τ an imaginary time and β the inverse absolute temperature. We use here a mean-field approximation and the two order parameters are the usual Kondo one, $\lambda = \langle c_{i\sigma}^* f_{i\sigma} \rangle$, and the spin-glass one, $q = \overline{\langle S_i \rangle^2}$. Using a static, replica-symmetric *ansatz*, we derive a mean-field expression for the free energy and the saddle-point equations for the order parameters.

The phase diagram in the T vs $|J_K|$ plane, for fixed value of J, is presented in Figure 7 and shows three different phases. At high temperatures, the "normal" phase is paramagnetic with both λ and q vanishing.

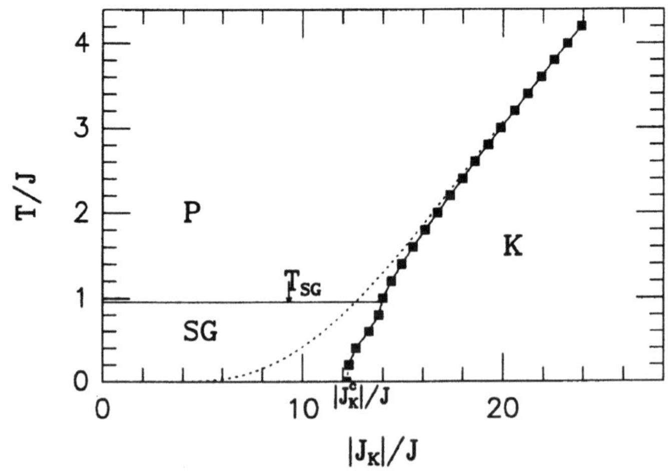

Figure 7. Phase diagram in T/J versus $|J_K|/J$ for $J = 0.05D$, showing the spin glass (SG), Kondo (K) and normal paramagnetic (P) regions. The dotted line represents the Kondo temperature for $J = 0$.

When the temperature is lowered, for not too large values of the ratio $|J_K|/J$, a second-order transition line is found, at $T = T_{SG}$, to a spin-glass phase with $q > 0$ and $\lambda = 0$. Finally, for large values of the ratio $|J_K|/J$, we have a second-order transition line to the "Kondo phase" with a non-zero λ value and $q = 0$. The transition line from the spin-glass phase to the Kondo phase, for temperatures smaller than T_{SG}, is a first-order one and it ends at $|J_K^c|$ at $T = 0$. Comparison between the theoretical Figure 7 and the magnetic phase diagram of $CeNi_{1-x}Cu_x$ is made, as usual, by relating an increase of $|J_K|$ with decreasing x, corresponding to a decrease of the distance between the 4f localized level and the Fermi level. Thus, the results shown in Figure 7 can explain the transition from a spin-glass state to a Kondo state observed above the Curie temperature for small x values in $CeNi_{1-x}Cu_x$ alloys.

5. CONCLUSIONS

We have presented here three different features of the Kondo lattice, namely the nonmagnetic region of the Doniach diagram, the antiferromagnetic to Kondo transition, and the spin glass to Kondo transition occurring in disordered alloys. We have studied the different problems as a function of the intrasite Kondo exchange coupling, of the intersite exchange coupling, and of the number of conduction electrons relative to the number of f-electrons, which remains fixed to 1. We use here a mean-field approximation with two order parameters in each case. In all cases, we have a mean-field Kondo correlator, but in addition to it the other order parameters could be either the short-range magnetic correlation, the staggered magnetization, or the spin-glass order parameter.

In the first case, it is well established that the Kondo temperature for the lattice is clearly different, and generally much smaller than the single impurity Kondo temperature. The existence of such two different Kondo temperatures is now well established by various works [71–76] using different theoretical approaches in the Kondo lattice and also in the Periodic Anderson Model. Another interesting feature concerns the derivation of the correlation temperature below which short-range magnetic correlations appear, in good agreement with neutron sctattering experiments in cerium compounds. It is also very clearly established that both the increase of the intersite coupling and the decrease of the number of conduction electrons tend to decrease the Kondo effect in such compounds.

The two other sections deal with transitions occurring from a Kondo state to a magnetic state when the intrasite coupling is decreased, and we found a sharp first-order transition in the two cases. However, agreement with experiment is poor for the AF–Kondo transition, because our predicted transition is a sharp one, without any occurrence of a QCP as in the Doniach diagram. Clearly, it is necessary to improve the theoretical approach to have a better description of the experimental situation. On the other hand, the SG–Kondo transition is also predicted to be a sharp one, which seems to agree with the situation of Ce(Ni,Cu) alloys. However, the present model cannot account for more complex experimental situations where, for example, there is successively SG, AF, and Kondo phases.

In conclusion, we can say that the mean-field treatment of the Hamiltonian describing both intrasite and intersite couplings gives fairly good results in comparison with experiment. Improvement of the model can be looked for in two directions: by considering more than two mean-field correlators to describe, for example, SG-AF-Kondo phase diagram, and taking fluctuations around the mean-field solution into account, therefore going beyond the present solutions.

References

[1] A. C. Hewson, *The Kondo Problem to Heavy Fermions*, Cambridge University Press (1993).

[2] P. Fulde, *Electron correlations in molecules and solids*, Solid-State Sciences 100, Springer (1995).

[3] B. Coqblin, B. H. Bernhard, J. R. Iglesias, C. Lacroix and K. Le Hur, in *Electron Correlations and Materials Properties*, ed. by A. Gonis, N. Kioussis and M. Ciftan, Kluwer Academic/Plenum Publishers, pp. 225-250 (1999) and references therein.

[4] N. Grewe and F. Steglich, in *Handbook on the Physics and Chemistry of Rare-earths*, ed. by K. A. Gschneidner Jr. and L. Eyring, Elsevier Science B. V., vol. **14**, 343 (1991).

[5] J. G. Sereni, in *Handbook on the Physics and Chemistry of Rare-earths*, ed. by K. A. Gschneidner Jr. and L. Eyring, Elsevier Science B. V., vol. **15**, 1 (1991).

[6] Y. Onuki and A. Hasegawa, in *Handbook on the Physics and Chemistry of Rare-earths*, ed. by K. A. Gschneidner Jr. and L. Eyring, Elsevier Science B. V., vol. **20**, 1 (1995).

[7] Proceedings of the SCES'96 Conference, Zürich, North Holland Elsevier, Physica B, **230-232** (1997).

[8] Proceedings of the SCES'98 Conference, Paris, North Holland Elsevier, Physica B, **259-261** (1999).

[9] Proceedings of the SCES'99 Conference, Nagano, North Holland Elsevier, Physica B, **281-282** (2000).

[10] B. Coqblin, *The electronic structure of Rare-Earth Metals and Alloys : the magnetic Heavy Rare-Earths*, Academic Press, New-York (1977).

[11] B. Coqblin, in *Magnetism of Metals and Alloys*, ed. by M. Cyrot, North Holland Publishing Company, Chap. 3, pp. 295-377 (1982).

[12] A. Jayaraman, Phys. Rev. **137**, A179 (1965).

[13] P. Lethuillier and P. Haen, Phys. Rev. Lett. **35**, 1391 (1975).

[14] M. B. Maple and D. Wohlleben, Phys. Rev. Lett. **27**, 511 (1971).

[15] P. Haen, F. Lapierre, J. M. Mignot, R. Tournier and F. Holtzberg, Phys. Rev. Lett. **43**, 304 (1979).

[16] R. Jullien, E. Galleani d'Agliano and B. Coqblin, Phys. Rev. B **6**, 2139 (1972).

[17] C. Geibel et al., Z. Phys. B **83**, 305 (1991) and **84**, 1 (1991).

[18] B. Coqblin, J. Arispe, J. R. Iglesias, C. Lacroix and K. Le Hur, J. Phys. Soc. Japan Suppl. B 65, **64** (1996).

[19] B. Coqblin and A. Blandin, Adv. in Phys. **17**, 281 (1968).

[20] R. Ramirez and L. M. Falicov, Phys. Rev. B **3**, 2425 (1971).

[21] J. W. Allen and R.M. Martin, Phys. Rev. Lett. **49**, 1106 (1982).

[22] M. Lavagna, C. Lacroix and M. Cyrot, J. Phys. F **13**, 1007 (1983).

[23] J. Kondo, Theoret. Phys. (Kyoto) **32**, 37 (1964).

[24] B. Coqblin and J. R. Schrieffer, Phys. Rev. **185**, 847 (1969).

[25] B. Cornut and B. Coqblin, Phys. Rev B **5**, 4541 (1972).

[26] K. G. Wilson, Rev. Mod. Phys. **47**, 773 (1975).

[27] P. Nozieres, J. Low Temp. Phys. **17**, 31 (1974).

[28] S. Doniach, Proceedings of the "Int. Conf. on Valence Instabilities and Related Narrow-band Phenomena", ed. By R. D. Parks, Plenum Press, 168 (1976).

[29] B. Barbara, H. Bartholin, D. Florence, M.F. Rossignol and E. Walker, Physica B **86-88**, 177 (1977).

[30] F. Nolting, A. Eichler, S. A. M. Mentink and J.A. Mydosh, Physica B **199-200**, 614 (1994).

[31] T. Graf, J. D. Thompson, M. F. Hundley, R. Movshovich, Z. Fisk, D. Mandrus, R. A. Fischer and N. E. Phillips, Phys. Rev. Lett. **78**, 3769 (1997).

[32] J. Flouquet, S. Kambe, L. P. Regnault, P. Haen, J.P. Brison, F. Lapierre and P. Lejay, Physica B **215**, 77 (1995).

[33] F. Steglich, J. Magn. Magn. Mater. **226-230**, 1 (2001).

[34] E. Bauer, Adv. in Phys. **40**, 417 (1991).

[35] K. Alami-Yadri, H. Wilhelm and D. Jaccard, Physica B **259-261**, 157 (1999).

[36] C. U. Segre, M. Croft, J. A. Hodges, V. Murgai, L. C. Gupta and R. D. Parks, Phys. Rev. Lett. **49**, 1947 (1982).

[37] H. Wada, M. F. Hundley, R. Movshovich and J. D. Thompson, Phys. Rev. **59**, 1141 (1999).

[38] P. Nozieres, Eur. Phys. J. **B 6**, 447 (1998).

[39] A. Eiling and J. S. Schilling, Phys. Rev. Lett. **46**, 364 (1981).

[40] Y. Uwatoko, T. Ishii, G. Oomi and S. Malik, Physica B **206-207**, 199 (1995).

[41] J. Flouquet, P. Haen, P. Lejay, P. Morin, D. Jaccard, J. Schweizer, C. Vettier, R. A. Fisher and N.E. Phillips, J.M.M.M. **90-91**, 377 (1990).

[42] P. Link and D. Jaccard, Physica B **223-224**, 303 (1996).

[43] P. Link and D. Jaccard, Physica B **230-232**, 31 (1997) ; D. Jaccard, P. Link, E. Vargoz and K. Alami-Yadri, Physica B **230-232**, 297 (1997).

[44] J. R. Iglesias, C. Lacroix and B. Coqblin, Phys. Rev.**B 56**, 11820 (1997).

[45] S. Sullow, M.C. Aronson, B.D. Rainford and P. Haen, Phys. Rev. Lett. **82**, 2963 (1999)

[46] H. Wilhelm, K. Alami-Yadri, B. Revaz and D. Jaccard, Phys. Rev. B **59**, 3651 (1999).

[47] J. Rossat-Mignod, L. P. Regnault, J. L. Jacoud, C. Vettier, P. Lejay, J. Flouquet, E. Walker, D. Jaccard and A. Amato, J.M.M.M. **76-77**, 376 (1988).

[48] J. Pierre, P. Haen, C. Vettier and S. Pujol, Physica B **163**, 463 (1990).

[49] L. P. Regnault, W. A. C. Erkelens, J. Rossat-Mignod, P. Lejay, J. Flouquet, Phys. Rev. B **38**, 4481 (1988).

[50] J. M. Mignot, J. L. Jacoud, L. P. Regnault, J. Rossat-Mignod, P. Haen, P. Lejay, P. Boutrouille, B. Hennion and D. Petitgrand, Physica B **163**, 611 (1990).

[51] L. P. Regnault, J. L. Jacoud, J. M. Mignot, J. Rossat-Mignod, C. Vettier, P. Lejay and J. Flouquet, Physica B **163**, 606 (1990).

[52] S. Doniach, Physica B **91**, 231 (1977).

[53] R. Jullien, P. Pfeuty, A. K. Bhattacharjee and B. Coqblin, J. Appl. Phys. **50**, 7555 (1979).

[54] H. Tsunetsugu, M. Sigrist and K. Ueda, Rev. Modern Phys. **69**, 809 (1997).

[55] R. Jullien and P. Pfeuty, J. Phys. F **11**, 353 (1981).

[56] H. Tsunetsugu, Y. Hatsugai, K. Ueda and M. Sigrist, Phys. Rev. B **46**, 3175 (1992).

[57] A. R. Ruppenthal, J. R. Iglesias and M. A. Gusmão, Phys. Rev. B **60**, 7321 (1999).

[58] B. Coqblin, M. A. Gusmão, J. R. Iglesias, C. Lacroix, A. Ruppenthal and Acirete S. Da R. Simoes, Physica B **281-282**, 50 (2000).

[59] C. Lacroix, J. R. Iglesias and B. Coqblin, presented at SCES'01, Ann Arbor, USA (August 2001), to be published in Physica B (2002).

[60] B. Coqblin, M. A. Gusmão, J. R. Iglesias, A. Ruppenthal and C. Lacroix, J. Magn. Magn. Mater. **226-230**, 115 (2001).

[61] J. Garcia Soldevilla, J. C. Gomez Sal, J. A. Blanco, J. I. Espeso and J. Rodriguez Fernandez, Phys. Rev. B **61**, 6821 (2000).

[62] J. C. Gomez Sal, J. Garcia Soldevilla, J. A. Blanco, J. I. Espeso, J. Rodriguez Fernandez, F. Luis, F. Bartolomé and J. Bartolomé, Phys. Rev. B **56**, 11741 (1997).

[63] D. Eom, M. Ishikawa, J. Kitagawa and N. Takeda, J. Phys. Soc. Japan, **67**, 2495 (1998).

[64] S. Majumdar, E. V. Sampathkumaran, St. Berger, M. Della Mea, H. Michor, E. Bauer, M. Brando, J. Hemberger and A. Loidl, to be published.

[65] D. Huo, J. Sakurai, T. Kuwai, Y. Isikawa and Q. Lu, to be published.

[66] Alba Theumann, B. Coqblin, S.G. Magalhães and A. A. Schmidt, Phys. Rev. B **63**, 054409 (2001).

[67] D. Sherrington and S. Kirkpatrick, Phys. Rev. Lett. **35**, 1792 (1975).

[68] A. M. Sengupta and A. Georges, Phys. Rev. B **52**, 10295 (1995).

[69] M. N. Kiselev and R. Oppermann, Pis'ma Zh. Eksp. Teor. Fiz. **71**, 359 (2000).

[70] Alba Theumann and Mercedes Vieira Gusmão, Phys. Lett. A **105**, 311 (1984).

[71] M. A. Continentino, G. M. Japiassu and A. Troper, Phys. Rev. B **39**, 9734 (1989).

[72] A. N. Tahvildar-Zadeh, M. Jarrell and J. K. Freericks, Phys. Rev. B **55**, R 3332 (1997).

[73] A. J. Arko, J. J. Joyce, J. Sarrao, Z. Fisk, J. L. Smith, J. D. Thompson, M. Hundley, A. Menovsky, A. Tahvildar-Zadeh and M. A. Jarrell, in "Electron Correlations and Materials Properties", ed. by A. Gonis, N. Kioussis and M. Ciftan, *Kluwer Academic/ Plenum Publishers*, pp. 33-58 (1999), *and references therein.*

[74] S. Burdin, A. Georges and D. R. Grempel, Phys. Rev. Lett. **85**, 1048 (2000).

[75] B. H. Bernhard, C. Lacroix, J. R. Iglesias and B. Coqblin, Phys. Rev. B **61**, 441 (2000).

[76] A. H. Castro-Neto and B. A. Jones, Phys. Rev. B **62**, 14975 (2000).

[77] K. Umeo, T. Takabatake, T. Suzuki, S. Hane, H. Mitamura and T. Goto, Phys. Rev. B **64**, 144412 (2001).

[78] K. Umeo, T. Takabatake, H. Ohmoto, T. Pietrus, H. v. Löhneysen, K. Koyama, S. Hane and T. Goto, Phys. Rev. B **58**, 12095 (1998).

A GEOMETRIC APPROACH TO CORRELATED SYSTEMS

Jamal Berakdar[1]

1. INTRODUCTION

Over the past few decades there has been an impressive and a steady progress in computational material science [1, 2, 3]. This development is fueled by the ever growing computational resources and by the demand for yet more precise information on technologically relevant material properties, such as the optical, transport and magnetic characteristics. A microscopic description of these properties entails the knowledge of the quantum spectrum of the system under study. Thus for real materials one has to deal with the notoriously difficult many-body problem in a computationally acceptable manner. For this purpose remarkably successful and efficient conceptual schemes have been developed where the multi-particle system is mapped onto a one body problem for a particle moving in an effective (non local) field created by all the other constituents of the system [1, 2, 4]. Usually, this effective field is further simplified according to certain recipes such as the local density approximation within the density functional theory [5]. These computationally manageable concepts have rendered possible the routine and accurate calculation of a wealth of static material properties, such as the ground state energies. On the other hand, however, it has been observed that the ground-state of certain compounds, e.g. transition metal oxides, is not described adequately within a single particle picture [2]. In addition, for the theoretical description of the excitation spectrum [6] and for the treatment of dynamical processes, such as many particle reactive scattering, methods have to be envisaged that go beyond the effective single particle model.

For example, correlated many-body states are a prerequisite for the theoretical formulation of recent measurements of the double and triple electronic

[1]Max-Planck-Institut für Mikrostrukturphysik, Weinberg 2, 06120 Halle, Germany.

excitation of localized or delocalized electronic compounds by one ultraviolet photon [7, 8, 9]. Recent technological advances have made it even possible to explore in full details the many-body continuum spectrum of four and more interacting particles [9, 10, 11, 12, 13, 14, 16] where numerical calculations are absent. Thus, it is desirable to develop, for interacting systems, a genuine many-body theory (in a sense specified below) that is conceptually sound and allows for systematic approximations that can be implemented numerically. This is particularly important in view of the current trend in miniaturization techniques that ultimately aim at the fabrication of atomic-size systems whose features are controlled primarily by the quantal behaviour of a finite number of correlated particles.

In this theoretical work a strategy is presented that enables us to derive approximate expressions for the spectrum of many-body, interacting finite systems by utilizing the knowledge of the properties of other reference systems with a less number of interactions. The treatment is not perturbative, i.e. it does not rely on pre-assumptions regarding the strength of a certain interaction or characteristic parameter in the system. Furthermore, the method provides a systematic and mathematically sound scheme for approximations that are computationally tractable and allows, when desirable, to incorporate well established single particle and perturbative approaches.

2. FORMAL DEVELOPMENT

The fundamental quantity from which we derives the microscopic properties of N body quantum systems is the Green operator $G^{(N)}$ which is the resolvent of the total Hamiltonian. It is defined by the Lippmann-Schwinger integral equation $G^{(N)} = G_0 + G_0 U^{(N)} G^{(N)}$ where G_0 is the Green operator of a reference N particle system which is usually chosen as a noninteracting system. U stands for total interaction and is given formally by $U^{(N)} = G^{-1} - G_0^{-1}$. Equivalently one may determine the dynamical behaviour of a system by means of the transition operator $T^{(N)}$ which satisfies the integral equation $T^{(N)} = U^{(N)} + U^{(N)} G_0 T^{(N)}$. These integral equations for $G^{(N)}$ and $T^{(N)}$ provide a natural framework for perturbative treatments, as they can be expanded in terms of interaction $U^{(N)}$ or some parts of it which are deemed small. However, for $N \geq 3$ the application of the above Lippmann-Schwinger equations (and those for the state vectors) is hampered by mainly two difficulties: 1.) as shown in Refs.[17, 18] the Lippmann-Schwinger equations for the state vectors do not have a unique solution, and 2.) as shown by Faddeev [19, 20] the kernel of these integral equations $K = G_0 U^{(N)}$ is not a square integrable operator for $N \geq 3$, i.e. the norm $\|K\| = [\text{Tr}(KK^\dagger)]^{1/2}$ is not square integrable. The kernel K is also not compact. The origin of these obstacles lies in the presence of the so-called disconnected diagrams that correspond to processes in which one of the N particles is a spectator, i.e. not correlated with the other $N-1$ particles. Therefore, for three-particle systems Faddeev and others [19, 22, 23] suggested an alternative set of equations for $G^{(N)}$ and $T^{(N)}$ that are free from disconnected diagrams (the behaviour of these equations for infinite range potentials, such as Coulomb potentials, is still the subject of current research [21]).

The scope of this study is three fold: **(a)** We aim at deriving a recursive procedure that expresses the Hamiltonian of N interacting body systems in terms of Hamiltonians of systems with a reduced number of interactions.
(b) This reduction scheme should not be a perturbative one, and the diagrams should be arranged in a way that disconnected terms are excluded. The procedure should be linear in N, i.e. it should relate $G^{(N)}$ and $T^{(N)}$ to $G^{(N-M)}$ and $T^{(N-M)}$ (where $M = 1, 2, \cdots, N-2$) in a linear manner in N.
(c) The method should provide a systematic and mathematically sound scheme for approximations that could be computationally realized (here we evaluate the continuum spectrum of four interacting Coulomb particles and two particle excitations at surfaces).

To this end we consider a nonrelativistic system consisting of N interacting particles. Total potential is assumed to have the form $U^{(N)} = \sum_{j>i=1}^{N} v_{ij}$, where v_{ij} stands for a two-particle interaction without any further specification of its analytical form.

Based on geometrical considerations (cf. Fig.1) one can express the potential $U^{(N)}$ exactly in terms of recurrence relations:

$$U^{(N)} = \frac{1}{N-2}\sum_{j=1}^{N} u_j^{(N-1)}, \tag{1}$$

$$u_j^{(N-1)} = \frac{1}{N-3}\sum_{k=1}^{N-1} u_{jk}^{(N-2)}, \; j \neq k. \tag{2}$$

In these equations $u_j^{(N-1)}$ is the total potential for a system of $N-1$ interacting particles, i.e. $u_j^{(N-1)}$ is obtained from $U^{(N)}$ by detaching particle j from the N particle system. In terms of the pair interactions v_{mn} the following relation applies $u_j^{(N-1)} = \sum_{m>n=1}^{N} v_{mn}$, $m \neq j \neq n$. Since the kinetic energy operator is an additive single particle operator the expansions (1,2) are reflected in similar recursion relations for the total Hamiltonian. Fig. 1 illustrates how, according to Eqs.(1,2), the potential of a system of six interacting particles is expressed in terms of potentials of five correlated particles. The latter potentials can be further expressed in terms of four body potentials (cf. Eq.(2)). This procedure is repeated until a potential with a desired number of interactions is reached.

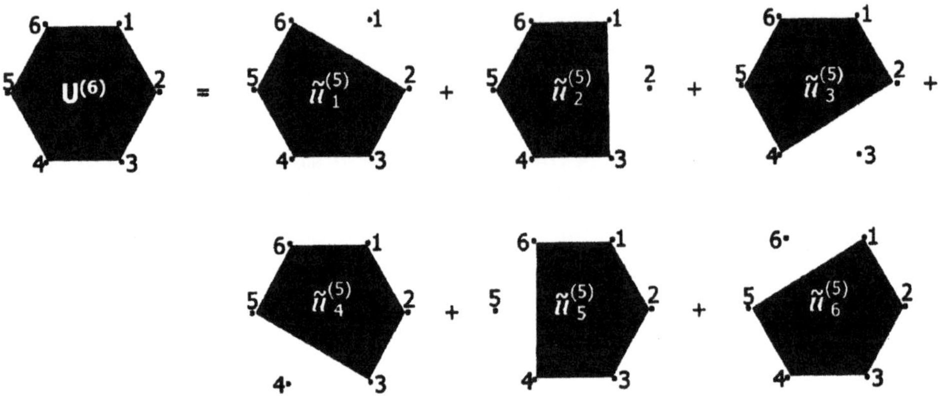

Figure 1: A pictorial interpretation of the total potential expansion (1) for six interacting particles enumerated and marked by the full dots at the corners of the hexagon. The hexagon indicates the full potential $U^{(6)}$ of the six correlated particles. Each pentagon symbolizes the full five body potential $\tilde{u}_j^{(5)} = \left(u_j^{(5)}\right)/4$ of those five particles that are at the corners of the pentagon. The particle being not at a corner of a pentagon is free (disconnected).

From Fig. 1 it is clear that this "minimal geometric reduction" scheme (Eqs.(1,2)) treats all interactions on equal footing and provides maximal flexibility to reduce systematically the N body potential (Hamiltonian) to sums of $N - M$ potentials (Hamiltonians) with $M = 1, 2, \cdots, N-2$.

The geometric expansion (Eqs.(1,2)), as visualized in Fig.1, has a wide ranging consequences in that the transition and the Green operators can be expanded along the same lines!.

This can be seen as follows: According to the decomposition (1), the integral equation for the transition operator can be written as

$$T^{(N)} = \sum_{j=1}^{N} T_j^{(N-1)} \tag{3}$$

$$T_j^{(N-1)} = \tilde{u}_j^{(N-1)} + T^{(N)} G_0 \tilde{u}_j^{(N-1)}, \; j = 1, \cdots, N. \tag{4}$$

Here we introduced the scaled potentials $\tilde{u}_j^{(N-1)} = \left(u_j^{(N-1)}\right)/(N-2)$. The physical meaning of the operators (4) is illustrated in Fig. 2 for the system depicted in Fig. 1.

The transition operator $t_j^{(N-1)}$ for a system of $N-1$ particles interacting via the scaled potential $\tilde{u}_j^{(N-1)}$ is

$$t_j^{(N-1)} = \tilde{u}_j^{(N-1)} + \tilde{u}_j^{(N-1)} G_0 t_j^{(N-1)}.$$

With this relation Eq.(4) can be reformulated to yield

$$\begin{aligned} T_j^{(N-1)} &= t_j^{(N-1)} + t_j^{(N-1)} G_0 T^{(N)} - t_j^{(N-1)} G_0 \left(\tilde{u}_j^{(N-1)} + \tilde{u}_j^{(N-1)} G_0 T^{(N)} \right) \\ &= t_j^{(N-1)} + t_j^{(N-1)} G_0 \left(T^{(N)} - T_j^{(N-1)} \right) = t_j^{(N-1)} + t_j^{(N-1)} G_0 \sum_{k \neq j}^{N} T_k^{(N-1)}. \end{aligned} \tag{5}$$

Eq.(5) can be expressed compactly in the matrix form

$$\begin{pmatrix} T_1^{(N-1)} \\ T_2^{(N-1)} \\ \vdots \\ T_{N-1}^{(N-1)} \\ T_N^{(N-1)} \end{pmatrix} = \begin{pmatrix} t_1^{(N-1)} \\ t_2^{(N-1)} \\ \vdots \\ t_{N-1}^{(N-1)} \\ t_N^{(N-1)} \end{pmatrix} + [\mathbf{K}^{(N-1)}] \begin{pmatrix} T_1^{(N-1)} \\ T_2^{(N-1)} \\ \vdots \\ T_{N-1}^{(N-1)} \\ T_N^{(N-1)} \end{pmatrix}. \tag{6}$$

The kernel $[\mathbf{K}^{(N-1)}]$ is a matrix operator and is given by

$$[\mathbf{K}^{(N-1)}] = \begin{pmatrix} 0 & t_1^{(N-1)} & t_1^{(N-1)} & \cdots\cdots & t_1^{(N-1)} \\ t_2^{(N-1)} & 0 & t_2^{(N-1)} & \cdots\cdots & t_2^{(N-1)} \\ \cdots & \cdots & \cdots & \cdots & \cdots \\ t_{N-1}^{(N-1)} & \cdots\cdots & t_{N-1}^{(N-1)} & 0 & t_{N-1}^{(N-1)} \\ t_N^{(N-1)} & \cdots\cdots & t_N^{(N-1)} & t_N^{(N-1)} & 0 \end{pmatrix} G_0. \tag{7}$$

From Eq.(2) we deduce that $t_j^{(N-1)}$ is expressible in terms of transition operators of the $N-2$ interacting subsystems as $t_j^{(N-1)} = \sum_{k \neq j}^{N-1} T_k^{(N-2)}$. The operators $T_k^{(N-2)}$ are deduced from Eq.(6) with N being replaced by $N-1$.

From the relation $G^{(N)} = G_0 + G_0 T^{(N)} G_0$ we conclude that the Green operator of the interacting N particle system has the form

$$G^{(N)} = G_0 + \sum_{j=1}^{N} G_j^{(N-1)}.$$

The reduced operators $G_j^{(N-1)}$ are related to the Green operators $g_j^{(N-1)}$ of systems in which only $N-1$ particles are correlated by virtue of $\tilde{u}_j^{(N-1)}$. This relation is readily deduced from the above derivation for transition operator and reads

$$\begin{pmatrix} G_1^{(N-1)} \\ G_2^{(N-1)} \\ \vdots \\ G_{N-1}^{(N-1)} \\ G_N^{(N-1)} \end{pmatrix} = \begin{pmatrix} g_1^{(N-1)} - G_0 \\ g_2^{(N-1)} - G_0 \\ \vdots \\ g_{N-1}^{(N-1)} - G_0 \\ g_N^{(N-1)} - G_0 \end{pmatrix} + [\tilde{\mathbf{K}}^{(N-1)}] \begin{pmatrix} G_1^{(N-1)} \\ G_2^{(N-1)} \\ \vdots \\ G_{N-1}^{(N-1)} \\ G_N^{(N-1)} \end{pmatrix}, \qquad (8)$$

where the kernel is now redefined as $[\tilde{\mathbf{K}}^{(N-1)}] = G_0[\tilde{\mathbf{K}}^{(N-1)}]G_0^{-1}$.

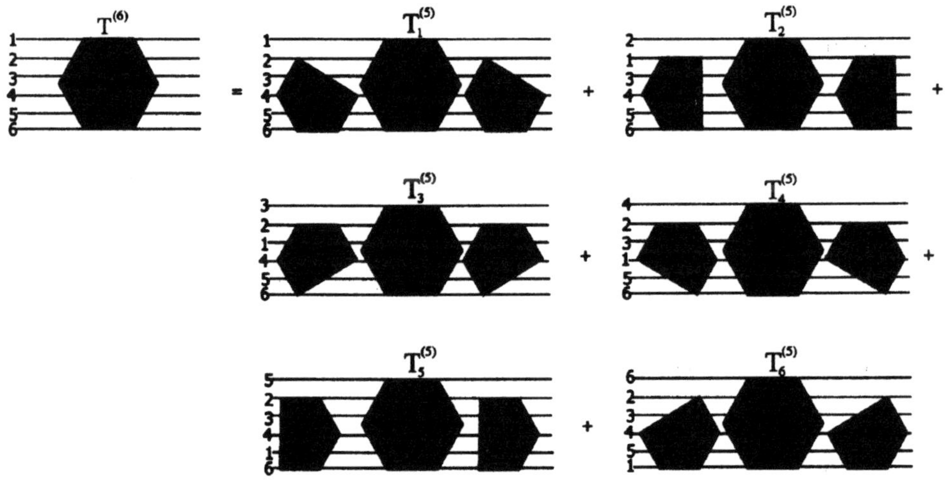

Figure 2: A diagrammatic representation of Eq.(4) for a system of six correlated particles (cf. Fig. 1). The hexagons and the pentagons (with a specific orientation) label the same potentials as explained in Fig. 1. Each of the pictures stands for a transition operator of the six body system (the particles are labeled by straight lines). E. g., the diagram $T_1^{(5)}$ means that the five particles $2, 3, 4, 5$ and 6 interact first, propagate and then all 6 particles interact with each other. The system then propagates and finally the five particles $2, 3, 4, 5$ and 6 interact again.

From Eqs.(6,8) we conclude that if the Green operator of the interacting $N-1$ body system is known the Green operator of the N particles can then be deduced by solving a set of N linear, coupled integral equations (namely Eqs.(6,8)). According to the above equations, if only the solution of the $N-M$ problem is known where $M = 1, 2, \cdots, N-2$ we have to perform a hierarchy of calculations starting by obtaining the solution for the $N - M + 1$ problem and repeating the procedure to reach the solution of the N body problem.

It is straightforward to show that any iteration of the kernels of Eqs. (6,8) is free of disconnected terms since the disconnected terms occurs only in the off-diagonal elements of $[\mathbf{K}^{N-M}]$ and $[\tilde{\mathbf{K}}^{N-M}]$. For $N = 3$ the present scheme reduces to the well established Faddeev equations.

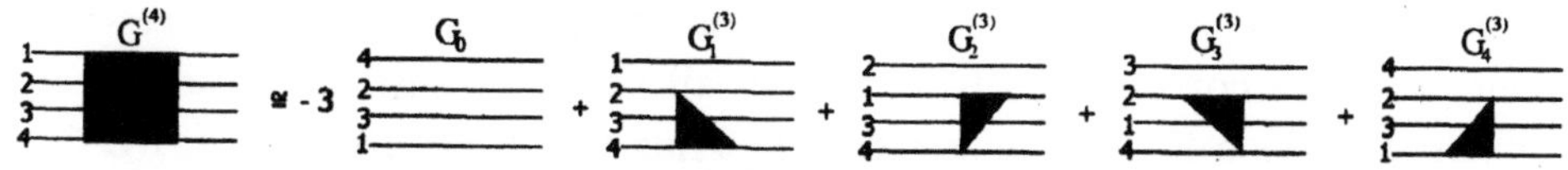

Figure 3: A diagrammatic representation of the approximation (9). The square represent the four-body total potential whereas the triangles are the scaled, total three-body potentials of those three particles whose lines cross the respective triangle.

3. EXCITATION SPECTRUM OF FOUR CHARGED PARTICLES

To demonstrate the applicability of the present approach we consider the four charged-particle problem. This is of a particular interest, since a substantial amount of knowledge on the three particle problem has been accumulated over the years whereas theoretical studies on the four body problem are still scare. On the other hand, in recent years an impressive progress has been made on the experimental side [9, 10, 11, 12, 13, 15] which renders possible a detailed insight into the four body continuum spectrum.

Using the method outlined above of the four body Green operator can be expressed in terms three body Green operators for which approximate solutions may be employed. For $N = 4$ the first iteration of Eq.(8) yields

$$G^{(4)} = \sum_{j=1}^{4} g_j^{(3)} - 3G_0, \qquad (9)$$

where $g_j^{(3)}$ is the Green operator of an interacting three body system (particle j is noninteracting). For the latter Green function we utilize the method developed in Ref.[24] (it is important to note in this context that the *three-*

body Green function defined here is called in quantum field theory the *two-particle* Green function which is defined through the Bethe-Salpeter equation). According to the theory in Ref.[24] the Hamiltonian of a general three body system reduces to a sum of three commuting Hamiltonians $h_k^{(2)}$ in which only two particles are interacting (particle k is free). The Green operators $g_j^{(3)}$ in Eq.(9) can therefore be written as

$$g_j^{(3)} \approx G_0^{-2} \prod_{k=1}^{(3)} g_k^{(2)}, k \neq j \in \{1,2,3,4\},$$

where $g_k^{(2)}$ is the resolvent of $h_k^{(2)}$ (this approximation for the three-body Green functions corresponds to summing all the diagrams in the ladder approximation which is done numerically). Thus we obtain from Eq.(9) $G^{(4)} = \left[\sum_{j=1}^{4} G_0^{-2} \prod_k^3 g_k^{(2)}\right] - 3G_0,\ j \neq k$.

Now we employ this four-body Green function for the description of three excited, interacting electrons (or two electrons and a positron) that moves in the Coulomb field of a residual positively charged ion. Such a state is achieved following the electron and positron impact double ionization.

Figs. 3(a,b) show the results for the electron and the positron impact double ionization of the ground state of atomic helium along with the experimental data [10, 15] and a full numerical evaluation of the first order perturbation series (first Born term) within a convergent close coupling (CCC) method [25]. The first Born approximation (FBA) corresponds to one term in Eq.(9) where the projectile motion is decoupled from the rest of the system. As a result of the perturbative treatment the calculations within the FBA are insensitive to the projectile charge state.

As seen from Fig. 4, the present model is capable of describing adequately the experimental cross sections in shape and magnitude. The origin of the peaks observed in the spectrum has been discussed in Refs. [15, 26]. The present model and the FBA theory yield similar results since the experiments are performed in the validity range of the perturbation theory. However, the differences between the positron and the electron impact case indicate that the Born limit is not yet fully reached.

5. THE TWO-PARTICLE EXCITATION AT SURFACES

Now let us turn to the application of the theory to extended systems. In particular the method will be applied to evaluate the simultaneous two-particle excitation at surfaces upon the absorption of a single VUV photon. Such experiments have been done recently [8] on Ni(001) and Cu(001). In the experiment one resolves the wave vectors $\mathbf{k}_1$ and $\mathbf{k}_2$ (i.e. the energies E_1, E_2

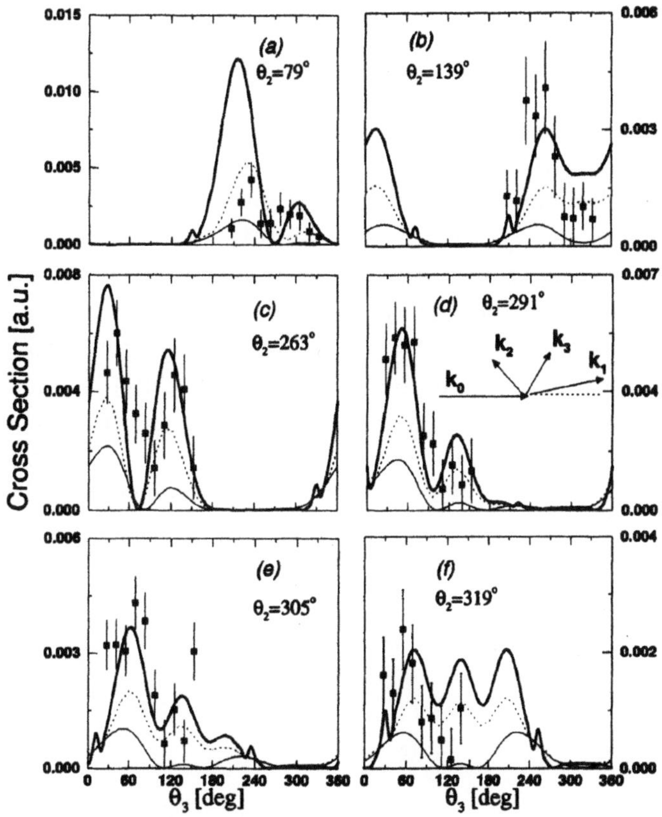

Figure 4: The fully resolved double ionization cross section of He($^1S^e$) following electron (solid lines) or positron (dotted line) impact. The scattering geometry is shown by the inset in (d). $\mathbf{k}_0$ and $\mathbf{k}_1$ are the initial and final state momenta of the projectile while $\mathbf{k}_2$ and $\mathbf{k}_3$ refer to the momenta of the two ejected electrons. The incident energy is $5.6\,keV$ and $k_2^2/2 = k_3^2/2 = 10\,eV$. All angles are measured with respect to $\hat{\mathbf{k}}_0$. The projectile is scattered through an angle of 0.45^o. The emission angle θ_2 of one of he electron is fixed at the value indicated on the figures while the cross section is scanned as function of the emission angle θ_3 of the second electron. The thick solid (dotted) line is the result of the present model for electron (positron impact) whereas the light solid curve is the outcome of the CCC method within the first Born approximation [25]. The data (full square [15]) are on absolute scale.

and emission angles $\hat{\mathbf{k}}_1, \hat{\mathbf{k}}_2$) of the two photoelectrons emitted simultaneously from the sample upon the absorption of a single VUV photon with energy $\hbar\omega$. The probability for such a reaction is given by [28, 27] ($E = E_1 + E_2$)

$$W(\hat{\mathbf{k}}_1, E_1, \hat{\mathbf{k}}_2, E_2, \hbar\omega) \propto \langle \mathbf{k}_1, \mathbf{k}_2 | \Pi_{12}^- \, \Delta \, \Im \Pi_{12}(E - \hbar\omega) \Delta^\dagger \, \Pi_{12}^- | \mathbf{k}_2, \mathbf{k}_1 \rangle. \quad (10)$$

Here $\Pi_{12}^{\pm}$ is the particle-particle (p-p) propagator [29] appropriate for the description of the two interacting photoelectrons in the presence of the surface.

The method presented in this paper can be utilized upon the following assumptions: 1. We treat the two electrons as independent ones in which case the single particle Green function for each electron can be deduced by conventional methods, such as density functional theory (we employ the layer Korringa Kohn Rostoker technique). 2. The electrostatic interaction between the two electrons is assumed local and is renormalized according to the Thomas-Fermi theory of screening. With this information we are now in a position to write down the p-p propagator as sum of three quasi-particle propagators. More details concerning the range of validity of the approximations will be given elsewhere. In Fig.5 the two-electron energy correlation function is shown along with the experimental results on Cu(001). As evident, the emission probability of one of the photoelectrons with certain energy depends strongly of the properties of the second photoelectrons which endorses the importance of a realistic treatment of the coupling between the electrons.

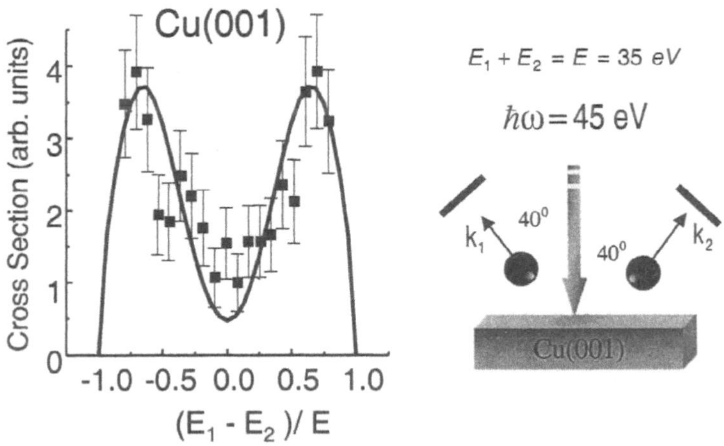

Figure 5: The one-photon two-electron emission from Cu(001). The photon with energy $\hbar\omega = 45$ eV is linear polarized and its wave vector is normal to sample. The two photoelectrons are detected co-planar with the photon wave vector and under emission angles of 40° to the left and to the right of the incident photon direction (cf. inset). The total energy of the two escaping electrons is $E = 35$ eV. The emission cross section is scanned as function of sharing of the total energy E between the two photoelectrons

Acknowledgment

I would like to thank N. Fominykh, O. Kidun, J. Henk and P. Bruno for valuable discussions.

References

[1] A. Gonis *Theoretical materials science : tracing the electronic origins of materials behavior*, (Warrendale, Pa : Materials Research Society, 2000)

[2] I. Turek, V. Drchal, J. Kudrnovský, M. Šob, and P. Weinberger, *Electronic Structure of Disordered Alloys, Surfaces, and Interfaces*, (Kluwer Academic Pub., Boston, London, Dordrecht, 1997)

[3] K. Ohno, K. Esfarjani, and Y. Kawazoe, *Computational Material Science*, (Springer, Berlin, 1999)

[4] P. Fulde, *Electron Correlation in Molecules and Solids*, Springer Series in Solid-State Sciences, Vol. 100, (Springer Verlag, Berlin, Heidelberg, New York, 1991)

[5] R. M. Dreizler and E. K. U. Gross *Density Functional Theory* (Springer Verlag, Berlin, Heidelberg, New York, 1990)

[6] L. Hedin, J. Phys. C **11** R489 R528 (1999)

[7] O. Schwartzkopf, B. Krässig, J. Elmiger, and V. Schmidt Phys. Rev. Lett **70**, 3008 (1993)

[8] R. Herrmann, S. N. Samarin, H. Schwabe, and J. Kirschner, Phys. Rev. Lett.**81**, 2148 (1998)

[9] R. Wehlitz et al Phys. Rev. Lett. **81**, 1813 (1998)

[10] I. Taouil, A. Lahmam-Bennani, A. Duguet, L. Avaldi Phys. Rev.Lett. **81**, 4600 (1998)

[11] A. Dorn et al Phys. Rev. Lett. **82**, 2496 (1999)

[12] M. Unverzagt et al Phys. Rev. Lett. **76**, 1043 (1996)

[13] B. El-Marji, J. P. Doering, J. H. Moore, M. A. Coplan Phys. Rev. Lett. **83**, 1574 (1999)

[14] R. Moshammer et al Phys. Rev. Lett. **84**, 447 (2000)

[15] A. Lahmam-Bennani et al Phys. Rev. A **59**, 3548 (1999)

[16] A. Dorn et al Phys. Rev. Lett. **86**, 3755 (1999)

[17] B. A. Lippmann, Phys. Rev. **102**, 264 (1956)

[18] L.L. Foldy and W. Tobocman, Phys. Rev. **105**, 1099 (1957)

[19] L.D. Faddeev, Soviet Phys. JETP **12**, 1014 (1961)

[20] L.D. Faddeev *Mathematical Aspects of the Three-Body Problem* (Davey, New York, 1965)

[21] A. M. Mukhamedzhanov, E. O. Alt, G. W. Avakov, Phys. Rev. C **61**, 064006 (2000)

[22] O.A. Yakubovsky, Sov. J. Nucl. Phys. **5**, 937 (1967)

[23] S.P. Merkuriev and L.D. Faddeev *Quantum Scattering Theory for Systems of few Particles* (Nauka, Moscow, 1985)

[24] J.Berakdar, Phys. Lett. A **277**, 35 (2000)

[25] A. Kheifets et al J.Phys.B **32**, 5047 (1999)

[26] J. Berakdar, Phys. Rev. Lett. **85** , 4036 (2000)

[27] N. Fominykh, J. Henk, J. Berakdar, P. Bruno, H. Gollisch, R. Feder , Solid State Commun. **113**, 665 (2000)

[28] J. Berakdar, Phys. Rev. B. **58**, 9808 (1998)

[29] E. K. U. Gross, E. Runge, *Vielteilchentheorie* (Teubner, Stuttgart, 1986)

ON SINE-GORDON VORTICES IN HIGH-TEMPERATURE SUPERCONDUCTORS

Sine-Gordon Vortices in Superconductors

O. Hudak[1]

1 INTRODUCTION

Sine-Gordon vortices in superconductors play important role [1]. Severals years ago a magnetic-field-induced superconductive state in heavy fermion systems [2] and [3] was studied. It was shown using the symmetry analysis that there may exists in the free energy expansion a four order coupling between the complex superconducting order parameter, which is in general one to three dimensional, and the magnetization. Depending on the strength of this coupling a superconductive state induced by the magnetic field [2] and [3] may be induced increasing the magnetic field. Experimentally in that time it was reported that in $CePb_3$ there exists a ferromagnetic phase and simultaneously there is an evidence for presence of superconductivity. The Kondo-lattice superconductivity was studied in this connection, [3] and [4]. Superconducting states of different symmetry s, p, d were found to exist in the Kondo-lattice systems. A phase transition from the normal phase to the superconducting phase in $CePb_3$ type materials occurs in our phenomenological model through creation of a vortex lattice, in which vortices are now superconducting regions induced by an external magnetic field. Increasing the magnetic field the density of vortices and their shape become larger and the superconducting phase increases its volume in the material.

In high-temperature superconductors, which are layered materials, the interlayer distance is larger than the in-layer lattice constant. The interlayer distance is comparable with the in-layer penetration length.

Our results in [2] - [4] lead to study of formation of a vortex in the plane state and to study of interaction of such vortices in neighbouring layers. In the presence of an in-layer vortex in-plane and out-of-plane magnetic field structures

[1] e-mail: hudako@mail.pvt.sk , Department of Theoretical Physics, Faculty of Mathematics and Physics, Comenius University, Mlynska dolina F2, SK- 845 01 Bratislava, and Department of Physics, Faculty of Materials and Technologies, Slovak Technical University, Paulinska 16, SK- 974 15 Trnava, Slovak Republic

exist. Note that the vortex state may exist in zero magnetic field [5]. These magnetic fields influence neighbouring planes and thus we study vortices and their in-plane and out-of-plane interactions due to to the inductance coupling. Note that in zero external magnetic field a selfconzistent process may exists. Creation of a vortex in one plane leads to magnetic fields which influence the neighbouring planes, in these neighbouring planes these fields may induce creation of vortices the magnetic fields of which influence the vortex in the original plane: its existence and properties. It follows then that in layered materials one may expect that formation of vortices may stabilize the superconducting state of different from clasical symmetry in zero external magnetic field. Note that in nonzero external magnetic fields the selfconsistent process may be modified.

It is the aim of this paper to describe the mentioned mechanism and to formulate a general problem: behaviour of such vortices in zero and non-zero magnetic field. The vortices of the in- and inter- layer type should be compared as concerning their theoretical properties in zero and non-zero fields with existing experiments concerning study of vortices in high-temperature superconductors [6], [7], [8] and [9]. It was found that the anisotropy of the interplane-vortices in these materials is intrinsic. This enables us to compare experimental results concerning properties of the vortices in the layered materials with those which we found theoretically. This is the reason that here we give also a review of superconducting states with vortices: the superconducting states in which single vortex exists, the superconducting states in which lines and lattices of vortices exist, the superconducting states in which pairs of vortex-antivortex type are present and form a kind of liquid /Kosterlitz-Thouless-like states for a weak interlayer coupling /for which at larger distances the energy is linearly dependent on the distance vortex-antivortex, for smaller distances there exists a minimum distance in which the vortex-antivortex pair exist. The pair however does not anihilate! Further we consider the superconducting states in which solitons and strings of vortices, which represent some kind of structures induced by internal mechanisms /twinning of the crystals, etc./ or by surface, may exist.

Intrinsic Josephson effect in layered systems was described in [10]. A summary of the Josephson effect with respect to the high-temperature superconductors is given in [11]. The time dependent equations for the gauge invariant phase differences of the layered order parameters for superconductive states and for a general vector potential of the electromagnetic field were discussed in [1]. The interlayer coupling within the mentioned model equations is then studied with its influence on the vortices in layers. These vortices in neighbouring layers are coupled. We use in this study a small parameter expansion method, and an exact method of solution of the corresponding equations to compare different macroscopic superconducting states, their free energy and their properties.

It is also natural to study in general superconducting states of different type in which homogeneous, soliton-like and vortex-lattice-like and vortex-liquid-like states exist, and which are macroscopically observed in non-zero magnetic field, in a gauge-invariant form [10] for superconductors with its order parameter and a given vector potential. The corresponding Lagrange-Euler equations are nonlinear coupled sine-Gordon equations in the approximation of the constant amplitude. Many authors neglect the nonlinearity describing the vortex configurations in nonzero /and in zero/ magnetic field due to difficulties of the mathematical origin and mainly due to expectation that the nonlinearity is negligible around the core of the vortex. This would be a natural expectation for the case in which a characteristic length L of the system, or the intervortex distance L_v in the case of a vortex-lattice are smaller than the characteristic length λ of the Josephson system. The corresponding free energy then would depend logarithmically on L, the characteristic length of the sample. However the nonlinearities in order parameter interactions are physically essential as it was shown by the author [5] in 1982. They lead in the case of a Josephson junction to anisotropic vortices the free energy of which depends linearly on L. Such a vortex has its topological charge which is conserved, in fact the vortex has characteristics of the well known Dirac magnetic monopole, see [5].

The Lagrange-Euler equations for the superconducting state and vector potential in the constant amplitude approximation for the scalar (s - type) superconducting order parameter reduce in the model studied here in this paper to equations which are connected with the two-dimensional sine-Gordon equation; p and d symmetric states are not studied here. The two-dimensional sine-Gordon equations solutions of which describe ground states occur also in other physical systems: in quantum antiferromagnets, in classical two-dimensional XY model with in-plane magnetic field [12] - [13], in models describing steps in crystal growth [14] and [15], and in models describing defects in incommensurate systems [16] and [17]. There are many interesting mathematical problems which are related to vortex-like configurations in two-dimensional nonlinear equations: in our paper from 1982 we introduced a new method for solving two-dimensional sine-Gordon equation [5] transforming this equation into a set of two second order ordinary nonlinear non-coupled equations, then we have found vortex-like solution and studied its topological and other properties. Multi - (Resonant-Soliton) solutions and vortex-like solutions to two- and three - dimensional sine-Gordon equations were studied in [18]. Interpretation of the structure of a vortex in terms of the solitons was studied in [19]. The multi-vortex solutions were studied in [20]. Exact solutions as rational-exponential solutions were studied in [21]. Vortices and solution of the boundary value problem by the inverse scattering transform were for this type of equation studied in [22]. The singular solutions of the elliptic sine-Gordon equation were studied by numerical methods in [23]. Numerical studies of dynamical isoperimeter pattern was studied in [24]. Thus in our paper a review of vortex configurations in two-dimensional sine-Gordon systems in connection with high-temperature superconductors is given.

Further in our paper new results are described as concerning the influence of neighbouring planes on the vortex state in a given plane and new exact solutions of the coupled Lagrange-Euler sine-Gordon equations are found, their properties and free energy are compared. We have found that the three-dimensional superconducting state with a vortex localized in the same point in every plane is a stable state in zero external magnetic field for the layered materials. The same holds for a lattice of vortices case.

1.1 SUPERCONDUCTING PLANES COUPLED BY JOSEPHSON EFFECT

Layers in layered superconductors of the high-temperature type are coupled due to intrinsic Josephson effect. The Lawrence - Doniach functional for the order parameter for such a system has the form /it is the free energy functional/ [1]

$$F(\Psi_n(\mathbf{r}), \mathbf{A}(\mathbf{R})) = \frac{\mathbf{H_c^2 s}}{4\pi} \sum_{\mathbf{n}} \int \mathbf{dr}[\zeta_{\mathbf{ab}}^{\mathbf{2}} \mid (-\mathbf{i}\nabla + \frac{2\pi}{\mathbf{\Phi_0}} \mathbf{A_n})\mathbf{\Psi_n}) \mid^{\mathbf{2}} - \tag{1}$$

$$- \mid \Psi_n) \mid^2 + \frac{1}{2} \mid \Psi_n) \mid^4 +$$

$$+r(\mid \Psi_n) \mid^2 + \mid \Psi_{n+1}) \mid^2 - \Psi_n \Psi_{n+1}^* \exp(-i\chi_{n,n+1}) - \Psi_n^* \Psi_{n+1} \exp(+i\chi_{n,n+1}))$$

$$+ \int d\mathbf{R} \frac{B^2}{8\pi}$$

Here the order parameter in the n-th layer is $\Psi_n(\mathbf{r}) = \mid \Psi_n(\mathbf{r}) \mid \exp(i\Phi_n](\mathbf{r}))$, here z=ns, and s is an interlayer distance, $\mathbf{R} = (\mathbf{r}, z)$, $\nabla = \frac{\partial}{\partial \mathbf{r}}$, H_c is the bulk critical field, ζ_{ab} is the coherence length in the ab plane which is perpendicular to the c-direction, $\Phi_0 = \frac{\pi \hbar c}{|e|}$, $A_n(\mathbf{r}) = (\mathbf{A_{nx}}, \mathbf{A_{ny}})$ ... an average value of the vector potential over the distance $[(n - \frac{1}{2})s, (n + \frac{1}{2})s$ along the c-axis, $\mathbf{B} =$ rot $(\mathbf{A})$, and $\chi_{n,n+1} = \frac{2\pi}{\Phi_0} \int_{ns}^{(n+1)s} dz A_z$.

If we denote r, where $r << 1$, the Josephson coupling constant between neighbouring layers order parameters, the coherence length in the c-direction is given by

$$\zeta_c^2 = \frac{s^2 r}{2} \tag{2}$$

The in-plane penetration depth is given by

$$\lambda_{ab}^2 = \frac{\Phi_0^2}{8\pi^2 H_c^2 \zeta_{ab}^2} \tag{3}$$

The anisotropy ratio is given by

$$\frac{\lambda_{ab}}{\lambda_c} = \frac{\zeta_{ab}}{\zeta_c} \tag{4}$$

The interlayer coupling is weak, $\zeta_c << s$. Let $\phi_n(\mathbf{r})$ be a change of the order parameter phase around a vortex. The constant amplitude approximation $\mid \Psi \mid^2 = 1$ is considered first, then we consider the constant amplitude approximation in which the density of superconducting pairs is temperature dependent.

1.2 LAWRENCE-DONIACH FUNCTIONAL FOR THE ORDER PARAMETER

The constant amplitude approximation $\mid \Psi \mid^2 = 1$ leads to the Lagrange - Euler equations minimizing the free energy

$$F(\Phi_n(\mathbf{r}), \mathbf{A}(\mathbf{R})) = \mathbf{E_0} \sum_{\mathbf{n}} \int \mathbf{dr}[\frac{\mathbf{1}}{\mathbf{2}} \mid (\nabla\phi_{\mathbf{n}} + \frac{2\pi}{\mathbf{\Phi_0}} \mathbf{A_n}) \mid^{\mathbf{2}} \tag{5}$$

$$- \frac{1}{\lambda_J^2}(1 - \cos(\phi_{n,n+1})) + \int d\mathbf{R} \frac{B^2}{8\pi}$$

where $E_0 = \frac{\Phi_0^2 s}{16\pi^3 \lambda_{ab}^2}$. The gauge invariant difference of phases between neighbouring layers is

$$\phi_{n,n+1}(\mathbf{r}) = \mathbf{\Phi_n} - \mathbf{\Phi_{n+1}} - \frac{2\pi}{\Phi_0}\int_{ns}^{(n+1)s} dz A_z(\mathbf{r},z) \tag{6}$$

The interlayer Josephson current is

$$J_{n,n+1}(\mathbf{r}) = J_0 \sin(\phi_{n,n+1}(\mathbf{r})) \tag{7}$$

where $J_0 = \frac{\Phi_0 cs}{8\pi^2\lambda_J^2\lambda_{ab}^2}$.
The equation for the vector potential $\mathbf{A}_n$ has the form

$$(\mathbf{rotrotA})_z = \frac{4\pi}{c} j_z = \tag{8}$$

$$\frac{\Phi_0 s}{2\pi\lambda_{ab}^2\gamma^2}\sum_n f_{n,n+1}(z) sin\phi_{n,n+1}$$

where $f_{n,n+1}(z)$ vanishes everywhere except for $ns < z < (n+1)s$, where it is unity.

1.3 LAGRANGE-EULER EQUATIONS FOR THE ORDER PARAMETER

The Lagrange-Euler equations for the phase difference $\phi_{m,m+1}$ are the sine-Gordon coupled equations

$$-\sum_m L_{n,m}\nabla^2\phi_{m,m+1} + \frac{1}{\lambda_J^2}\sin(\phi_{n,n+1}) = 0 \tag{9}$$

The interlayer inductance $L_{n,m}$ has the form

$$L_{n,m} = \int_0^{2\pi}\frac{dq}{2\pi}\frac{\cos(n-m)q}{2(1-cos(q)) + \frac{s^2}{\lambda_{ab}^2}} = \frac{\lambda_{ab}}{s}(1-\frac{s}{\lambda_{ab}})^{|n-m|} \tag{10}$$

The boundary conditions for the Lagrange - Euler equations (9) are

$$(\nabla_x\nabla_y - \nabla_y\nabla_x)\phi_{n,n+1}(\mathbf{r}) = \mathbf{2\pi}\sum_{\mathbf{n}}(|\,\delta(\mathbf{r}-\mathbf{r_{n\nu}}) - \delta(\mathbf{r}-\mathbf{r_{n\nu}})\,|) \tag{11}$$

The equations (9) are a system of the sine-Gordon equations in two dimensions coupled in the perpendicular direction c in a discrete way.

1.4 THE DISTANCE s IS COMPARABLE WITH THE PENETRATION DEPTH λ_{ab}

When the distance s is comparable with the penetration depth λ_{ab} the mutual interlayer inductance $L_{n,n}$ is nonzero and much larger than the other inductancies $L_{n,m}$

$$L_{n,n} = \frac{\lambda_{ab}}{s} \tag{12}$$

For the phase difference $\phi_{m,m+1}$ the equations have in this case the form

$$-L_{n,n}\nabla^2\phi_{n,n+1} + \frac{1}{\lambda_J^2}\sin(\phi_{n,n+1}) = 0 \tag{13}$$

where the lower order terms in $\frac{L_{n,m}}{L_{n,n}}$ are neglected.
Denote by

$$\phi = \phi_{n,n+1} \tag{14}$$

an angle variable. We assume that this angle is the same for every plane, which may be done due to expected homogeneity in the sense of the same states in all layers. Let us denote further

$$\lambda_p = \lambda_J . L_{n,n} \tag{15}$$

The two-dimensional sine-Gordon equation (13) takes the form

$$\Delta\phi = \frac{1}{\lambda_p^2}\sin(\phi) \tag{16}$$

It is this form of the two-dimensional sine-Gordon equation which were studied in [5].

1.5 THE DISTANCE s COMPARABLE WITH THE PENETRATION DEPTH λ_{ab} - INFLUENCE OF NEXT LAYERS

When the distance s is comparable with the penetration depth λ_{ab} we will study the influence of neighbouring layers. The interlayer inductance for neighbouring layers $L_{n,n\pm1}$ is nonzero

$$L_{n,n\pm1} = \frac{\lambda_{ab}}{s}(1 - \frac{s}{\lambda_{ab}}) \tag{17}$$

The equation for the phase differences $\phi_{m,m+1}$ has now the form

$$-L_{n,n}\nabla^2\phi_{n,n+1} - L_{n,n+1}\nabla^2\phi_{n+1,n+2} - L_{n-1,n}\nabla^2\phi_{n-1,n} + \frac{1}{\lambda_J^2}\sin(\phi_{n,n+1}) = 0 \tag{18}$$

Let us denote further

$$\lambda_p = \lambda_J . \sqrt{L_{n,n}} \tag{19}$$

The two-dimensional sine-Gordon equation (18) takes the form

$$-\nabla^2\phi_{n,n+1} - l\nabla^2\phi_{n+1,n+2} - l\nabla^2\phi_{n-1,n} + \frac{1}{\lambda_p^2}\sin(\phi_{n,n+1}) = 0 \tag{20}$$

where the parameter l is small and is defined as

$$l = (1 - \frac{s}{\lambda_{ab}})^1 \tag{21}$$

In this form of the two-dimensional sine-Gordon equation three neighbouring planes are coupled explicitly.

The solution to the equation (20) has the form

$$\phi = \phi_{n,n+1} \tag{22}$$

for every n, which again may be done due to expected homogeneity in the sense mentioned above of states in planes in this limit. The equation (20) takes the form

$$-\nabla^2\phi + \frac{1}{\lambda_p'^2}\sin(\phi) = 0 \tag{23}$$

where we further define the renormalized Josephson length

$$\lambda_p' = \lambda_J\sqrt{1+2l} \tag{24}$$

which is larger than λ_p due to the fact that l is positive. This homogeneous solution has the characteristic length λ_p' larger than the one-layer length λ_p due to the inter-layer coupling via interlayer inductance.

The free energy of the single vortex in a given layer and the same homogeneous state in all layers depends linearly on L, the system linear dimension

$$2^{\frac{3}{2}}N\pi J2^2\frac{L}{\lambda_p'} - 2k_BTN\ln(\frac{L}{a_0}) \tag{25}$$

where N is the number of layers, the cutoff constant a_0 is a constant of the order of lattice constant. We see that if the coupling between layers is taken into account then the free energy is lower due to smaller energy of the vortex. The interlayer coupling decreases the free energy and stabilizes the state with vortices the cores of which are localizes at the same line in the direction perpendicular to the planes. Taking into account higher order inductance coupling it is expected that further lowering of the free energy will be found and will stabilise the vortex state.

Let us consider the solution to the equation (20) in which the sign of the angle ϕ changes in neighbouring layers, e.i. it is shifted by the angle π

$$\phi_{n,n+1} = (-1)^n\phi \tag{26}$$

for every n. The equation (20) takes the form

$$-\nabla^2\phi + \frac{1}{\lambda_p''^2}\sin(\phi) = 0 \tag{27}$$

where we define now the Josephson characteristic length

$$\lambda_p'' = \lambda_J\sqrt{1-2l} \tag{28}$$

which is smaller than λ_p, we assume l to be small parameter. This solution in which the sign of angle ϕ changes in neighbouring layers has the length λ_p'' smaller than the length λ_p' and smaller than the original in-plane Josephson length λ_p. The free energy of the single vortex in a given layer and the homogeneous in above sense state with a vortex corresponding to the minus sign in neighbouring layers depends again linearly on L, the system linear dimension

$$2^{\frac{3}{2}} N\pi J 2^2 \frac{L}{\lambda_p''} - 2k_B T N \ln(\frac{L}{a_0}) \tag{29}$$

where again N is the number of layers, the cutoff constant a_0 is a constant of the order of the lattice constant.

We see that if the inductance coupling l between layers is taken into account then the free energy of the state in which the sign of the angle ϕ changes in neighbouring layers is higher with respect to the state in which the sign of the angle ϕ does not change in neighbouring layers. The free energy of the state in which the sign of the angle ϕ changes in neighbouring layers is also higher with respect to the state in which the layers are not coupled at all. The interlayer coupling in this case increases the free energy and the vortex state is less stable now. Taking into account higher order coupling in this case we expect to increse further the free energy.

The small parameter l may be taken as an expansion parameter when solving the equation (20). One finds in that case that the homogeneous state with a vortex in one layer and other vortices in other layer with their core in the same point in the plane /a line of vortices/ is more stable. Single-vortex state in one layer and the same state in all layers give lower free energy, one then can ask at which value of the parameter l the transition to the vortex state occurs? And if the many-vortex states are realized, at which distance between vortices the state becomes more preferable than the normal state, or than the homogeneous state without vortices. The free energy of the single vortex in a given layer and the same homogeneous state with a vortex in neighbouring layers depends linearly on L, the system linear dimension and the critical value of the parameter l. At given temperature it is given by the condition that the free energy with the vortex at the same place in every layer is the same as the free energy of the superconducting state without vortex

$$F(vortex) = 2^{\frac{3}{2}} N\pi J q^2 \frac{L}{\lambda_p''} - 2k_B T N \ln(\frac{L}{a_0}) = F(hom) = 0 \tag{30}$$

For a given temperature and other parameters of the system increasing the parameter l decreases the energy in the free energy, it may happen that the vortex state become the most stable state. For a given temperature and other parameters of the system increasing the parameter l decreases the energy in the free energy, and it is clear that the transition to the superconducting state occurs at higher temperature than for the state without vortex. The transition to the single-vortex state with the inter-layer coupling occurs at higher temperature than for the state in which the single-vortex superconducting state does not take into account the inter-layer inductance coupling.

If the density of superconducting pairs depends on temperature, we introduce $\alpha = \alpha_0(T - T_c)$, T_c is the bulk transition temperature, β is a constant, $F(hom)$ is given from (5) for $\mid \Psi \mid^2 = \frac{|\alpha|}{\beta}$. The Lawrence-Doniach functional for this constant order parameter Φ gives

$$F(\Phi_{hom}) = -\frac{H_c^2 s L^2 N \alpha^2}{8\pi\beta} \tag{31}$$

where H_c is the bulk critical magnetic field, the constant J and the bulk critical magnetic field value H_c are related by

$$J = \frac{H_c^2 s \zeta_{ab}^2 \alpha^2}{4\pi\beta} \tag{32}$$

here we consider the constant amplitude approximation $| \Psi |^2 = \frac{|\alpha|}{\beta}$ instead of $| \Psi |^2 = 1$, thus temperature is near the transition temperature. The gradient term in the phase in this case gives also contribution to the free energy. It is of the order $\frac{4s}{\lambda}$ as a contribution to the bulk critical temperature, which we denote T_c . Note that the case with non-zero magnetic field may be easily described and will be considered elsevier.

The normal state is given by $\Psi = 0$ and its free energy is 0. The superconducting state with the homogeneous superconducting order parameterthe without vortices is given by $| \Psi |^2 = \frac{|\alpha|}{\beta}$ and its free energy is given by

$$F(hom) = -\frac{NsL^2\zeta_{ab}^2 H_c^2}{8\pi} = -\frac{JNL^2\alpha^2}{2\beta\zeta_{ab}^2} \tag{33}$$

The vortex state with the N_v vortices cores of which are localized in the square /for simplicity/ lattice has the free energy

$$F(N_v - vortex) = N_v N[2^{\frac{3}{2}}\pi J q^2 \frac{L_v}{\lambda_p'} - 2k_B T \ln(\frac{L_v}{a_0})] \tag{34}$$

It is necessary to calculate the inter-vortex distance $L_v = \frac{L}{N_v}$. Minimizing the free energy $F(N_v - vortex)$ with respect to L_v we have found that the minimal value L_v is

$$L_v = \frac{\sqrt{2}\lambda_p' k_B T}{\pi J q^2} ln(\frac{\sqrt{2}\lambda_p' k_B T}{\pi J q^2 a_0}) \tag{35}$$

The inter-vortex distance /vortex lattice constant/ is increasing with temperature T and decrease with temperature T depending whether it is above or below the critical temperature T_c^v given by

$$T_c^v = \frac{\pi J q^2 a_0}{\sqrt{2}\lambda_p' k_B e} \tag{36}$$

The free energy for the N_v vortex lattice state has the form

$$F(N_v - vortex) = NL^2[\frac{2\pi^2 J^2 q^4 ln(\frac{\sqrt{2}\lambda_p' k_B T}{\pi J q^2 a_0})}{\lambda_p'^2 k_B T} - \frac{\pi^2 J^2 q^4}{\lambda_p'^2 k_B T ln^2(\frac{\sqrt{2}\lambda_p' k_B T}{\pi J q^2 a_0})}] \tag{37}$$

The free energy of the N_v - vortex state should be compared with the free energy $F(N) = 0$ of the normal state, and with the free energy of the homogeneous superconducting state without vortices given by

$$F(sc) = -\frac{JNL^2\alpha^2}{2\beta\zeta_{ab}^2} \tag{38}$$

The transition at the transition temperature T_{cv} from the normal phase to the N_v - vortex superconducting state occurs, it is of the first order. Depending on the parameters the transition temperature T_{cv} may be higher for this transition than in the case of the second order phase transition from the normal phase to the hogheneous superconducting phase without vortices

$$T_{cv} = \frac{\pi J q^2 a_0}{\sqrt{2}\lambda_p' k_B} \tag{39}$$

Note that the transition temperature T_{cv} would be equal zero for zero topological charge q. Note further that the transition temperature T_{cv} increases when the interlayer inductance increases. Also note that the external magnetic field is zero. The case of non-zero magnetic field will not be treated here in this paper.

2 TWO-DIMENSIONAL SINE-GORDON VORTEX SOLUTIONS

2.1 SINGLE-VORTEX CONFIGURATIONS

In [5] we have found the single vortex configuration of the order parameter, it has the form

$$\phi = \pm \tan^{-1}\left(\frac{\sqrt{\frac{a}{(2-a)}}\sinh\left(\sqrt{\frac{2-a}{a}}\frac{(x-x_0)}{\lambda}\right)}{\sinh\left(\sqrt{\frac{a}{2}}\frac{(y-y_0)}{\lambda}\right)}\right) \tag{40}$$

here λ equals λ_p. The vorticity of this vortex is plus minus four, in the limit of $\lambda >> L$ we obtain usual isotropic vortex configuration. The configuration (40) breaks the rotational symmetry and thus besides two Goldstone modes corresponding to the breaking translational symmetry in two dimensions, we expext that a third Goldstone mode appears which corresponds to breaking the rotational symmetry. The free energy of the configuration given by (40) is linear in L. In the limit of $\lambda >> L$ where an usual isotropic vortex appears the logarithmic dependence of the free energy of the configuration is found.

2.2 MULTIVORTEX CONFIGURATIONS

Multi - (Resonant-Soliton) -Soliton solutions and vortex-like solutions in two and three dimension for the sine-Gordon equation were studied in [18]. For the two - and three - dimensional sine-Gordon equations there exist exact multi - (resonant-soliton) - soliton solutions and vortex-like solutions, in addition to exact multisoliton and resonant - soliton solutions.

In [19] it was shown that the quasi-vortex type solutions from [5] can be derived from the multiple soliton solutions by the proper procedure. Thus it follows from that there exist multiple vortex-like solutions.

2.3 STATIC LATTICE OF VORTICES

Other new types of static and dynamic two-dimensional solutions, multi-vortex solutions, space periodic vortex chains and lattices and breather solutions were studied in [20]. The boundary problem for two-dimensional sine-Gordon equation by the inverse scattering problem was studied in [25]. In Josephson junctions vortex configurations were studied [5]. The single vortex-type solution has the form

$$\phi = \pm \tan^{-1}\left(\sqrt{\frac{a}{2-a}} \sinh\left(\sqrt{\frac{2-a}{2}} \frac{(x-x_0)}{\lambda_p}\right) \sinh^{-1}\left(\sqrt{\frac{a}{2}} \frac{(y-y_0)}{\lambda_p}\right)\right) \tag{41}$$

Its vorticity q is ± 4, in the limit of $\lambda_p \gg L$ the vortex has isotropic form

$$\phi = \tan^{-1}\left(\frac{(x-x_0)}{(y-y_0)}\right) \tag{42}$$

The free energy of the single vortex depends linearly on L, the system linear dimension

$$2^{\frac{3}{2}} \pi J 2^2 \frac{L}{\lambda} - 2k_B T \ln\left(\frac{L}{a_0}\right) \tag{43}$$

where the cutoff constant a_0 is a constant of the order of lattice constant. x_0 and y_0 are coordinates of the center of the vortex.

Relation between vortex solutions of the two-dimensional sine-Gordon equation and solitons of the same equations were studied in [19]. Single vortex solution [5] can be derived from the known multiple soliton solutions by a proper procedure. This fact shows possibility to find multi-vortex solutions of the two-dimensional sine-Gordon equations due to existence of multisoliton solutions.

Vortex-type lattice solutions found by Hirota method and by Backlund transformation represent alternating vortices with $q = \pm 4$ and form a tetragonal or square lattice. Interaction of the spin-wave and vortex lattice are similar to the soliton-wave interaction.

Vortex-antivortex pair with $q = \pm 4$ is given by

$$\phi = \tan^{-1}\left(\frac{1 - x + \sinh(y)\exp(-x)}{\sinh(y) - (1+x)\exp(-x)}\right) \tag{44}$$

A vortex array was found with topological charges $q = \pm 4$. Numerical singular solutions of the elliptic sine-Gordon equation [23] were studied. Vortex solution as a semi-infinite 2π - kink starting from a singular point of the vortex type was found. Ring-like 2π - kinks with a common center in the singular point was found. The singular point with a linear region of the logarithmic spiral was found. Exact solutions of the sine-Gordon equations as rational-exponential solutions are described in [26]. Exact rational-exponential solutions of the two-dimensional sine-Gordon equations are constructed by a method based on the formal perturbation method. These solutions represent the nonlinear superposition of two interacting 2π kinks, at the point of intersection we have a vortex with topological charge plus and minus 4. Numerical analysis shows that the minimum distance between centers of vortices with opposite topological charges in dipole state exists and is given by approximately 1.1 in lambda units. These two vortices cannot annihilate in spite of zero total topological charge of the dipole.

2.4 DYNAMIC PROPERTIES OF TWO-DIMENSIONAL SINE-GORDON VORTICES

The two-dimensional sine-Gordon equation with time dependence was studied in [24], and it was found that a particular dynamical pattern in a perturbed two -dimensional sine - Gordon systems exists. The stability of vortex-like solutions were checked.

The main topological invariant of the vortex-like solutions is the total length of the $\pm 2\pi$ wave fronts entering as elementary kink-like patterns in the constitution of the whole configuration - this corresponds to the energy conservation law. In the paper [24] it is also discussed a possible observation of the vortex-like configurations in Josephson junctions.

In [18] the time dependent sine-Gordon equation without damping was studied. The procedure like that used in [19] leads to a two-soliton solution, thus intersection of these solitons corresponds with a moving vortex. On the other hand the author shows that three or higher number soliton solutions are known to exist only with limited parameter range. Thus time dependence without damping does not lead to vortex-lattice in general, this may exist only in a limited range of parameters.

In [24] it was shown that a particular dynamical pattern in a perturbed or not two-dimensional sine-Gordon system exists. Its stability with respect to perturbations and in time was studied. Their main topological invariant is the total length of the kink wave fronts forming the pattern and the invariant is consistent with the conservation of energy in all situations.

3 SUMMARY

In this paper we considered vortex states in high-temperature superconductors. Multilayered systems of superconducing planes coupled by Josephson effect are realised in the high-temperature superconductors. Layered superconductors of the high temperature type are coupled due to intrinsic Josephson effect. The Lawrence - Doniach free energy functional describes the order parameters in all layers, in the n-th layer the order parameter is dependent on the two-dimesional position $\mathbf{R} = (\mathbf{r}, z).r << 1$ is the Josephson coupling constant between neighbouring layers. The coherence length in the c-direction and the penetration depths define the anisotropy ratio. We assume that the interlayer coupling is weak, $\zeta_c << s$, and this corresponds with layered materials in which the in-plane distance is shorter than the inter-layer distance. We assume further that the constant amplitude approximation $|\Psi|^2 = 1$ holds and, when temprature is near the transition temperature the amplitude has the form $|\Psi|^2 = \frac{|\alpha|}{\beta}$, where the $\phi_n(\mathbf{r})$ is the phase of the order parameter.

Lawrence-Doniach functional for the order parameter leads to a set of coupled sine-Gordon equations for difference of phases in the neighbouring planes. The inductance coupling between planes exponentially decreases. Thus we consider firstly ground state and other order parameter configurations of the in-layer type. An homogeneous state with a vortex of the same type and position in every layer is one type of the state. In this case the order parameter phases are either the same either differing by π. The gauge invariant difference of phases between neighbouring planes is found of the soliton and vortex type /single, dipole, lattice/. Coupling between planes leads to coupling of these solitons and vortices. The interlayer Josephson current has its corresponding forms, which may be easily found. The mutual interlayer inductance for nearest neighbouring planes is a small parameter which leads to renormalized coupling of solitons and vorices in neighbouring planes. The boundary conditions for the Lagrange - Euler equations are known.

When the interlayer distance s is comparable with the in-plane penetration depth, the in-layer inductance is much larger than the interlayer inductancies. We concentrated on the vortex-like solutions of the phase differences. The equation for the phase difference leads to single vortex configurations with a given vorticity. Vortex-antivortex pairs may exist which do not annihilate due to existence of a minimal distance between them. Multivortex configurations which form different linear chains and lattices may exist, too. One may expect that the lattice of vortices has the same symmetry as that of the in-layer plane. Stability of the vortex-like states with respect to small pertubations was studied in general. Dynamic properties of two-dimensional sine-Gordon vortices show that, while the forces leading to movement without damping lead to destruction of the lattices, under some conditions, the damping terms lead to stabilisation of vortex lattices. A study of the two-dimensional sine-Gordon equation with time dependence has shown a particular dynamical pattern in a perturbed two -dimensional sine - Gordon systems, the stability of vortex-like solutions were checked. The main topological invariant of the vortex - like solutions corresponds to the energy conservation law. The Josephson current depends on the lattice of the vortices in the layered material and the type of the superconductivity, either s or d. Interpretation of static and dynamical properties of the two-dimensional vortex solutions is still an open area.

3.1 ACKNOWLEDGMENTS

The author wishes to expres his sincere thanks for discusions about the topological defects to V. Dvořák, V. Janovec, J. Holakovský, and M. Glogarová. The author would like to express his thanks for invitation to present this paper on the Workshop to the organizers of the Second International Workshop on Electron Correlations and Materials Properties, Rhodos, Greece, 25th Jun - 29th Jun 2001, especialy his sincere thanks he expresses to Dr. A. Gonis. The paper was supported by the grant VEGA No. 1/7069/20, Commenius University Bratislava.

References

[1] Hudak O., Phil Magazine B (2001), in print

[2] Hudak O., Phys.Lett. **119** 2 (1986) 89 - 91

[3] Hudak O., in Novel Superconductivity, eds. S.A. Wolf and V.Z. Kresin, Proc. Internatonal Workshop on Novel mechanisms of Superconductivity, 1987, Berkeley, California, Plenum Press,, New York, 1977, 223 - 232

[4] Hudak O., JETP Letters **42** (1985) 300

[5] Hudak O., Phys.Lett. **89A** (1982) 245 - 248

[6] Moler K.A., Kirtley J.R., Liang Ruixing et al., Physica C 341-348 (2000) 977-980

[7] Kirtley J.R., Moler K.A., Villard G. and Maignan A., Phys.Rev.Lett. 81 (1998) 2140-2143

[8] Tsvetkov A.A., van der Marel D., Moler K.A. et al., 395 Nature (1998) 360-362

[9] Moler K.A., Kirtley J.R., Hinks D. et al., Science 279 (1998) 119-1196

[10] Lawrence W.E. and Doniach S., in Proceedings of the 12th International Conference on Low temperature Physics, Kyoto 1970, ed. by Kanda E., Keigaku, Tokyo, 1970

[11] Bulaevski L.N., Tr.J. of Physics, **20** (1996)594 - 626

[12] Mertens F.G., Bishop A.R., Goueva M.E. and Wysin G.M., J. de Physique, Coll.8, Suppl.12, **49** (1988) C8-1385 – C8-1386

[13] Pereira A.R. and Pires A.S.T., Phys.Rev. **B51** 2 (1995) 996 - 1002

[14] Hudak O., in Proc. European meeting on Crystal Growth 82, Materials for Electronics, Prague, August 23-27, 1982

[15] Dikande A.M. and Kofane T.C., J.Phys., Condens. Matter **7** (1995) L141 - L146

[16] Borisov A.B. and Kiseliev V.V., Sol.St. Communications **59** 7 (1986) 445 - 448

[17] Borisov A.B. and Kiseliev V.V., Physica **D31** (1988) 49 - 64

[18] Takeno S., Prog.Theor.Phys. **68** 3 (1982) 992 - 995

[19] Nakamura A., J.Phys.Soc.Japan **32** 6 (1983) 1918 - 1920

[20] Borisov A.B., Tankeyev A.G., Shagalov A.G.

[21] Bezmaternih G.V., Phzs.Lett. **A 149** (1990) 492-495 and Bezmaternih G.V., Phys.Lett. **111A** 1,2 (1983) 15 - 18

[22] Borisov A.B. and Taluc G.G., Fiz.Met.met. (in russian) **34** (1991) 34 - 43

[23] Shagalov A.G., Phys.Lett. **A165** (1992) 412 - 416

[24] Turlot E. et al., Phys.Rev. **B42** 11 (1990) 8418 - 8425

[25] Borisov A.B., Phys.Lett. **143** 1,2 (1990) 52 - 56

[26] Goueva M.E., Wysin G.M., Bishop A.R. and Mertens F.G., J.Phys.: Condens.Matter **1**(1989) 4387-4398

SPIN PERMUTATION AT WORK IN THE BCS HAMILTONIAN

Jacob Szeftel*

Owing to conservation of the total projected spin in the elementary two-body scattering process, two electron spin transpositions are shown to secure an exact block-diagonal splitting of the whole BCS Hamiltonian matrix H. Because the transpositions do not commute with the Hamiltonian due to the Pauli principle, the Hilbert space does not supply any representation of the group of transpositions. Likewise every other eigenstate of H is found to be an (odd) eigenstate of a spin transposition operator while the remaining eigenstates of H are not eigenstates of the spin transpositions. Each block-diagonal submatrix is characterized by its own signature indicative of whether all of its eigenstates are eigenstate of each spin transposition operator or not. The submatrix sizes range from a lowest value (= 1 at half-filling) up to the largest value d_M. The groundstate is included in the largest submatrix or in a smaller one according to whether the number of electron pairs is odd or even. The ratio $\frac{d_0}{d_M}$ where d_0 designates the dimension of the BCS Hilbert space grows exponentially with the electron concentration. This result holds for any dimension of space, sign of the two-body interaction and electron filling. The whole spectrum of eigenvalues has been obtained for all possible values of the total momentum of the eigenstates of H up to $d_0 \sim 10^5$. Possible extensions towards much higher d_0 values are outlined.

I. INTRODUCTION

The long-standing interest for the Hubbard model stems from the belief that understanding its basic properties would be helpful in unraveling the paramount role of electron correlations in magnetism and superconductivity. Although the Hubbard Hamiltonian can be diagonalized [1] only in space dimension $D = 1$ for an infinite crystal, the Bethe wave-function mediated solution does not provide the corresponding two-electron correlation functions. Besides even though the numerous schemes used for $D > 1$ may happen to approximate well the groundstate energy, the associ-

*Université Paris 7, Laboratoire de Physique Théorique de la Matière Condensée, case 7020, 2 Place Jussieu, 75251 Paris Cedex 05, France.

ated correlation data remain unreliable so that exact results are known for very small clusters [2] only. A tremendous progress was achieved in the many-electron problem by BCS [3] who explained superconductivity by resorting to a variational state characterized by off-diagonal long-range order [4]. The impressive success of the BCS picture in accounting for all experimental observations in superconducting materials is all the more amazing since the BCS Hamiltonian is obtained by truncating the attractive Hubbard Hamiltonian in reciprocal space so that Fourier-transforming the BCS Hamiltonian back to real space gives rise to four site coupling terms, not present in any realistic two-body electron potential such as the Hubbard coupling. Anyhow it emphasizes the close relationship between the BCS and Hubbard models [5]. Our diagonalizations carried out on small clusters have indeed confirmed that both spectra of eigenvalues were similar and even partially identical in some particular cases such as the η pairing state [4] or the system of four electrons hopping on four sites. This conclusion has been reached elsewhere [6] by comparing the groundstate energy of the $D = 1$ Hubbard model with that of the BCS variational state. As the BCS model, albeit simpler than the Hubbard one, is a genuine many-body problem, solving it should pave the way for a better understanding of the many-electron physics, as emphasized by other authors [7] too.

The same physical quantities are conserved in each two-electron scattering process in both Hubbard and BCS models, namely the vanishing projected spin of the electron pair and its momentum K. While K-conservation has been studied elsewhere [5], this work is dedicated to unveiling the consequences of spin conservation in the BCS model. This will be shown here to give rise to a splitting of the Hamiltonian matrix into block-diagonal submatrices, the sizes of which turn out to be much smaller than the dimension of the BCS Hilbert space. As a striking peculiarity, although owing to spin conservation, the two electrons of the pair involved in the scattering process may exchange their respective spin, the spin transposition operators do not commute with the BCS Hamiltonian so that the eigenstates of the BCS Hamiltonian partition onto *2 kinds of eigenstates*, respectively being or *not* eigenstates of the spin transposition operators. This noteworthy property which differs markedly from the conclusions of usual group theory ensues solely from the Pauli principle.

The outline is as follows: the BCS Hamiltonian H is recalled in section I and the special procedure implemented to construct the basis of the Hilbert space is presented in section II; the proof of the block-diagonalization of H is detailed in section III and the power of this method is exemplified in section IV; the main results and outlooks are given in the conclusion.

II. THE BCS MODEL

The dimension of space being irrelevant to spin conservation, let us consider for simplicity a one-dimensional crystal containing N sites and $2n$ electrons. The Pauli principle requires that $n \leq N$ and n is taken not to be a divisor of N. The electrons populate a single band where the one-electron energy reads $E(k) = \cos(k)$ and $k = 0, .., N-1$ is a vector of the Brillouin zone. K-conservation is more conveniently [5] taken advantage of by working in the reciprocal space. It is then helpful to introduce the pair creation and annihilation operators

$$b^\dagger(k,k') = c^\dagger_{k,+}c^\dagger_{k',-} \quad , \quad b(k,k') = c_{k',-}c_{k,+} \quad . \tag{1}$$

The operator $c_{k,\sigma}^{(\dagger)}$ is a one-electron annihilation (creation) operator in the Bloch state $k, \sigma = \pm$ where the subscripts + or − refer to the two possible directions of a spin $\frac{1}{2}$. As the pair momentum $K = k + k'$ is constant in the Hilbert space S_K, the pair operators $b^{(\dagger)}(k, k')$ will be written $b_k^{(\dagger)}$. S_K is spanned by $2n$ electron Slater determinants of vanishing total projected spin and total momentum [8] equal to nK. A typical one reads

$$\phi = \prod_{j=1}^{n} b_{k_j}^{\dagger}|0\rangle \quad , \tag{2}$$

where $|0\rangle$ designates the no-electron state.

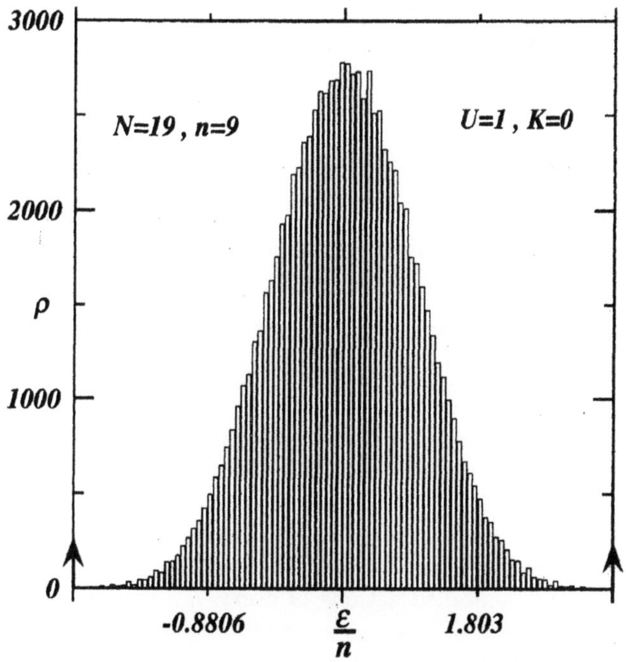

FIG. 1. Plot of the density of states of the BCS eigenspectrum versus the energy per electron pair

Given K in the Brillouin zone, the BCS Hamiltonian reads $H = H_D + H_K$ where the diagonal (H_D) and off-diagonal (H_K) parts are written respectively

$$\begin{aligned} H_D &= \sum_{k,\sigma} E(k) c_{k,\sigma}^{\dagger} c_{k,\sigma} + \frac{U}{N} \sum_{k,k'} c_{k,+}^{\dagger} c_{k,+} c_{k',-}^{\dagger} c_{k',-} \quad , \\ H_K &= \frac{U}{N} \sum_{k,k' \neq k} b_k^{\dagger} b_{k'} \quad , \end{aligned} \tag{3}$$

where $U \in \mathbb{R}$ assigns the interelectron coupling strength and $U > 0$ ($U < 0$) corresponds to a repulsive (attractive) model. The Hubbard Hamiltonian reads $H_D + \sum_K H_K$ where the sum, carried out over all possible K in the Brillouin zone, produces a mutual cancellation of the spurious four site coupling terms mentioned above. As a typical feature of H and in marked contrast with the Hubbard model where every electron pair keeps getting broken by the two-body scattering, the pairs get never dissociated by the BCS coupling so that they behave like true particles, having the characteristic features of hard-core bosons

$$b_k^{(\dagger)} b_{k'\neq k}^{(\dagger)} = b_{k'}^{(\dagger)} b_k^{(\dagger)} \quad , \quad (b_k^\dagger)^2 = (b_k)^2 = 0 \quad , \quad b_k^\dagger b_k + b_k b_k^\dagger = 1 \quad .$$

Furthermore it is inferred from replacing electrons by holes that there is a one to one correspondence between the eigenvalues $\varepsilon(\pm U, y)$ of H, associated with $\pm U$, ($y = \frac{2n}{N}$ stands for the electron concentration)

$$\frac{\varepsilon(U,y) + \varepsilon(-U, 2-y)}{N} = U(1-2y) \quad . \tag{4}$$

Eq.4, valid in any space dimension, is redolent of similar properties of the Hubbard model [1].

III. SPIN CONSERVATION ORDERED BASIS

In order to build up the basis of S_K of dimension [8] $d_0 = \binom{N}{n}$, we introduce the group G of n transpositions T. Each $T_{m=1,..n} \in G$ maps $\phi \in S_K$ in Eq.2 onto $T_m\phi$ so that ϕ and $T_m\phi$ differ by one pair only, namely the mth one reading $b_{k_m}^\dagger$ and $b_{K-k_m}^\dagger$ for ϕ and $T_m\phi$, respectively. The pairs $b_{k_m}^\dagger$ and $b_{K-k_m}^\dagger$ are actually inferred from each other by exchanging the spins of their constituting electrons. As any pair $b_k^\dagger$ may show up at all possible values $m = 1, ..n$ in the various vectors of S_K involving this pair, the k_m value and the pair ordering index m turn out to be unrelated whence T_m cannot be expressed as $b_{k_m}^\dagger b_{K-k_m}$. Actually this pair ordering procedure is irrelevant to the definition of ϕ as far as the pairs are hard-core bosons whereby each of $n!$ of permutations designates the same and one vector of S_K. However it is by all means required for the combined actions of every T_m and H to be unambiguously and consistently defined for all $\phi \in S_K$, as shown below.

Every ϕ is either of the singular (ϕ_s) or regular (ϕ_r) type with respect to $T_m \in G$:

- as a characteristic feature, ϕ_s involves at least two pairs $b_{k_i}^\dagger$ and $b_{k_j}^\dagger$ such that $k_i + k_j = K$. The Pauli principle implies that $T_i\phi_s = T_j\phi_s = 0$. As such very states prevent S_K from being a representation of G, they will be called singular for T_i, T_j
- every $\phi_r \in S_K$, not being singular, will be referred to hereafter as regular for T_m. Its typical property $T_m\phi_r \neq 0$ implies $T_m^2\phi_r = \phi_r$ so that the subset of S_K spanned by all ϕ_r's does provide a representation of G

Let us now select arbitrarily a vector $\phi_1 \in S_K$ and work out the subset $\{T\phi_1\} \subset S_K$ comprising any $\prod_m T_m\phi_1$ where $\prod_m T_m$ designates all possible products of transpositions. The dimension of the subspace $\{T\phi_r\}$ is generally equal to 2^n. But it reduces to 2^{n-1} if the mth pair $b_{k_m}^+$ of ϕ_r is such that $2k_m = K \bmod(N)$ because this entails

$T_m\phi_r = \phi_r$. The dimension of $\{T\phi_s\}$ is equal to 2^{n-2p} if ϕ_s is singular with respect to p of distinct pairs of transpositions. Then we let H in Eq.3 operate on ϕ_1 so as to generate the subset $\{H\phi_1\}$

$$(H - H_D)\,\phi_1 = \frac{U}{N} \sum_{\phi_j \in \{H\phi_1\}} \phi_j \quad ,$$

where every vector of $\{T\phi_1\}$ is to be discarded from the sum. Due to Eq.3, the two Slater determinants associated with ϕ_1 and any $\phi_j \in \{H\phi_1\}$ comprise the same $(n-1)$ pairs, ordered in the same way. Finally ϕ_1 and ϕ_j differ by a single pair associated with the same pair ordering index taken between 1 and n. This warrants consistent matrix elements $\langle\phi|H|\phi'\rangle$ for $\phi \in S_K$ and $\phi' \in \{H\phi\}$ with respect to every $T_m \in G$. The same procedure is pursued on each $\phi_j \in \{T\phi_1\}$ to yield that piece of the basis of S_K

$$\{\widetilde{HT}\phi_1\} = \{T\phi_1\} \oplus \{H\phi_1\} \oplus_j \{H\phi_j\} \quad ,$$

and iterated further on every vector of $\{\widetilde{HT}\phi_1\}$ till the whole orthonormal basis of S_K is secured as $\oplus_{p=1,2...}\{\left(\widetilde{HT}\right)^p \phi_1\}$. Once ϕ_1 has been chosen, the ordering procedure gives rise to a unique basis.

From a practical point of view, the basis can be viewed as a family tree where $\left(\widetilde{HT}\right)^p \phi_1$ gives rise to the pth generation and $\phi' \in \{H\phi\}$ is a son of ϕ. Then any new vector ϕ' cannot be added to the basis of S_K until it has been checked to fulfil the Pauli principle and not to have been already generated. As for this latter point, much computing time is saved by realizing that ϕ' need be compared only with those already available vectors ϕ'' such that $\{T\phi'\}$ and $\{T\phi''\}$ have same dimension and ϕ'' has not been produced prior to the grandfather of ϕ'.

IV. THE IRREDUCIBLE CLASSES

To demonstrate the exact block-diagonalization of H into irreducible classes of eigenvectors, the first step consists of working out $[T_m, H]\phi_r$ and $[T_m, H]\phi_s$ where $[T_m, H] = T_mH - HT_m$ and ϕ_r, ϕ_s are regular or singular with respect to $T_m \in G$. All along the proof we shall take for specificity $K = 0, m = 1$ and write T and $\phi = |i, I\rangle$ with $I = \{i_2, i_3..i_n\}$ instead of T_1 and $\phi = b_i^\dagger \prod_{j=2}^n b_{i_j}^\dagger |0\rangle$, respectively. Accordingly $T|i,I\rangle$ will be denoted $|-i, I\rangle$. Then it comes for $\phi_r = |i, I\rangle$ with $-i \notin I$ as expected for a regular state

$$\begin{aligned} TH|i,I\rangle &= H_D|-i,I\rangle + \tfrac{U}{N}\left(\sum_{j\notin I}|-j,I\rangle + \sum_{i\notin J}|-i,J\rangle\right), \\ HT|i,I\rangle &= H_D|-i,I\rangle + \tfrac{U}{N}\left(\sum|-j,I\rangle + \sum|-i,J\rangle\right) \quad , \end{aligned}$$

where I, J differ by a single pair. All constraints conditioning the sums stem from the Pauli principle. We have written explicitly only those contributing to

$$[T, H]\phi_r = -\frac{U}{N}\sum_r \phi_s' \quad ,$$

where $\sum_r \phi'_s$ includes every singular vector of $\{H\phi_r\}$ with respect to $T_1, T_{q\neq 1}$. This implies

$$T[T,H]\phi_r = 0 \quad , \quad [T,H]T\phi_r = [T,H]\phi_r \quad . \tag{5}$$

A similar calculation yields for ϕ_s

$$\left.\begin{array}{l} H\phi_s = H_D\phi_s + \frac{U}{N}\sum_s(\phi'_r + \phi'_s) \\ [T,H]\phi_s = TH\phi_s \end{array}\right\} \Rightarrow \\ T[T,H]\phi_s = (H - H_D)\,\phi_s - \frac{U}{N}\sum_s \phi'_s \quad . \tag{6}$$

where $\sum_s \phi'_r, \sum_s \phi'_s$ involve respectively every regular and singular vector of $\{H\phi_s\}$ with respect to $T_1, T_{q\neq 1}$. Eqs.5,6 bring evidence that the Pauli principle is responsible via the singular vectors ϕ_s, ϕ'_s for every transposition $T \in G$ not commuting with H despite spin conservation.

Let ψ, ε be any eigenvector and eigenvalue of H such that $(H - \varepsilon)\,\psi = 0$. The eigenvector ψ is expanded as

$$\psi = \chi_r + \chi_s, \quad \chi_r = \sum x_r\phi_r, \quad \chi_s = \sum x_s\phi_s \quad , \tag{7}$$

where $x_r, x_s \in \mathbb{R}$ and χ_r, χ_s are seen to contribute respectively the regular and singular parts to ψ. It is inferred from calculating $[T,H]\psi$ in two different manners

$$[T,H]\chi_r + [T,H]\chi_s = (\varepsilon - H)\,T\chi_r \quad . \tag{8}$$

Multiplying further both sides of Eq.8 by T while taking advantage of Eqs.5,6 leads finally to the main result

$$(H_D - \varepsilon)\,\chi_s + \frac{U}{N}\chi'_s = [T,H]\chi_r \quad , \tag{9}$$

where χ'_s includes all singular vectors belonging to $\{H\chi_s\}$. Consistently both sides of Eq.9 involve singular vectors only. Eq.9 implies that the eigenvectors of H are of two kinds according to whether $\chi_s = 0$ or $\chi_s \neq 0$.

In case $\chi_s = 0$ ($\Rightarrow \chi'_s = 0$), it comes $\psi = \chi_r$ (see Eq.7) so that $[T,H]\psi = 0$. Consequently ψ shows up an eigenvector common to H and T. Projecting ψ onto $\phi_r \oplus T\phi_r$ yields $x\phi_r + x_T T\phi_r$ with $x, x_T \in \mathbb{R}$. Then projecting $[T,H]\psi = 0$ onto $\phi_r \oplus T\phi_r$ entails via Eq.5 that $x = -x_T$ so that every $\chi_s = 0$ eigenvector fulfils $T\psi = -\psi$ and all such eigenvectors make up an odd irreducible representation of T.

As implied by Eq.9, a $\chi_s \neq 0$ eigenvector ψ of H cannot be an eigenvector of T because of $[T,H]\psi \neq 0$. Nevertheless the χ_r contribution to ψ (see Eq.7) is indeed an even eigenvector of T, that is $T\chi_r = \chi_r$ and the χ_r's comprise an even irreducible representation of T (but the associated $\chi_s \neq 0$ eigenvectors of H do not !). For a proof consider again the projection $x\phi_r + x_T T\phi_r$ of ψ onto $\phi_r \oplus T\phi_r$. Then select any $\chi_s = 0$ eigenvector ψ'. As all eigenvectors of H are orthogonal, projecting $\langle\psi|\psi'\rangle = 0$ onto $\phi_r \oplus T\phi_r$, while taking into account that ψ' is odd with respect to T, leads to the searched result $x = x_T$. Additional relations between the x_r, x_s coefficients with respect to the Schrödinger equation follow from Eq.9.

Each irreducible class of eigenvectors is characterized by its signature $\{s_{m=1,..n} = \pm 1\}$ indicative of whether every eigenvector belonging to this class is of the $\chi_s = 0$

($s_m = -1$) or the $\chi_s \neq 0$ type ($s_m = 1$) with respect to $T_m \in G$. Thus there are 2^n distinct classes. All χ_r's associated with eigenvectors pertaining to the same class fulfil $T_{m=1,..n}\chi_r = s_m\chi_r$ as assigned by the class signature. The class of largest dimension has a completely even signature $\{s_{m=1,..n} = 1\}$ whereas that of smallest dimension has a completely odd one $\{s_{m=1,..n} = -1\}$. Consequently this latter class comprises all eigenvectors of H which are also (odd) eigenvectors of every $T_m \in G$. Besides in case of half-filling ($N = 2n + 1$) its dimension equals unity. All eigenvectors of H are nondegenerate because the electron pairs are distinct particles so that the T_m's commute together. By contrast with the two-spin transpositions, the total electron spin commutes with H but this is of little practical interest because it becomes dauntingly difficult to work out the associated irreducible representations [9] for systems comprising more than 4 electrons.

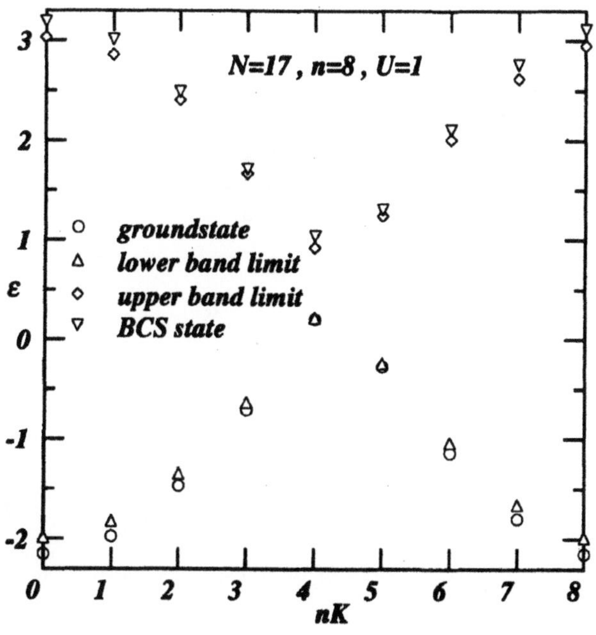

FIG. 2. Plots of BCS dispersion data

V. DISCUSSION

The $N = 5, n = 2$ case is the simplest one which can be used as a benchmark to check the validity of the conclusions inferred from Eq.9. G has 2 elements, S_K has dimension $d_0 = 10$ and its basis is made up of 2 ϕ_s's and 8 ϕ_r's. The 10×10 H matrix splits into one 5×5, two 2×2 and one 1×1 submatrices, associated with irreducible classes of signatures $\{1, 1\}$, $\{1, -1\}$, $\{-1, 1\}$ and $\{-1, -1\}$, respectively.

Diagonalizing these four submatrices on the one hand and the H matrix on the other hand yields indeed the same eigenspectrum. As a by-product the validity of Eq.9 has been confirmed for all five of $\chi_s \neq 0$ eigenvectors. As no Bethe wave-function based solution is available for the BCS model, the consistency of the present analysis has been checked by extending at half-filling and $U = 1$ the comparison between the eigenspectra pertaining to all irreducible submatrices and H, respectively, up to $N = 11$ and all K's, achieving an agreement always better than 10^{-13} over several thousands of eigenvalues. The results are summarized in Table I where d_M, d_g designate the dimensions of the largest irreducible class and that of the class containing the groundstate, respectively.

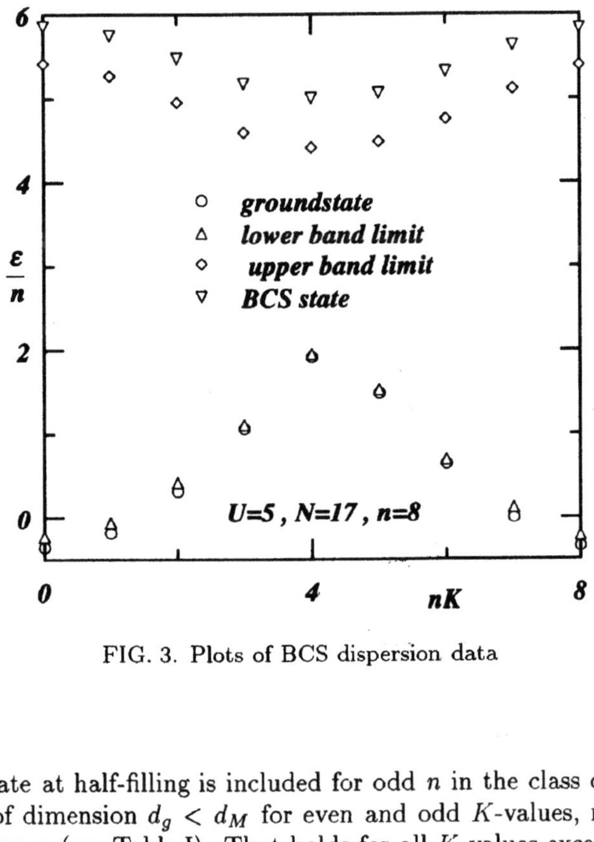

FIG. 3. Plots of BCS dispersion data

The groundstate at half-filling is included for odd n in the class of dimension d_M or in the class of dimension $d_g < d_M$ for even and odd K-values, respectively, and vice-versa for even n (see Table I). That holds for all K-values except for $K = n$, in which case the groundstate is included in the completely odd class of dimension unity for any n. The quantities of interest grow exponentially with respect to n

$$d_0 = 0.7 * 3.69^n, \quad \frac{d_0}{d_M} = 1.13 * 1.34^n, \quad \frac{d_0}{d_g} = 2.73 * 1.36^n \quad .$$

It is noteworthy that, were it not for the Pauli principle, the ratios $\frac{d_0}{d_M}, \frac{d_0}{d_g}$ would rather behave like 2^n because all subspaces $\{T_m\phi\}$ would have same dimension 2^n.

As a matter of fact there is rather at half filling a single $\{T_m\phi\}$ having dimension 2^n, namely that stemming from $\phi = \prod_{j=1}^{\frac{N-1}{2}} b_j^\dagger|0\rangle$ for $K = 0$. The computation of the eigenspectrum of large matrices is hampered by two setbacks. Not only does the memory space needed to store the matrix elements of H scale with d_0^2, which already outstrips by far the capacity of state of art computers in the $N = 19, n = 9$ case ($\approx 10^{11}$ bytes) but also the resolving power of two close eigenvalues decreases steeply with the matrix size. Both drawbacks are easily dodged here because the largest matrix size to be coped with is $d_M \ll d_0$ and any two close eigenvalues are likely to belong to two different irreducible classes which are thence diagonalized independently. Even though d_M increases beyond the realm of modern computers, it still remains possible to achieve at least the lowest eigenvalue of those very large classes by implementing the Lanczos or Monte-Carlo algorithms inside each corresponding submatrix separately.

The density of states attached to the whole eigenspectrum has been plotted in Fig.1 for the $U = 1, N = 19, n = 9, K = 0$ case. The prominent features are the two gaps opening below and above the continuum and isolating the groundstate and the most excited one, pinpointed by the arrows. The former gap is reminiscent of a similar property in the Hubbard model [1] while the latter should be compared with the BCS gap [10]. The groundstate and the highest excited state are actually a singlet and a resolved doublet for odd n and vice-versa for even n. The dispersion data have been depicted in Figs.2,3 from the Brillouin zone center ($nK = 0$) to the edge ($nK = 8$) to illustrate the U dependence.

VI. CONCLUSION

We have shown that owing to two-electron spin conservation, the BCS Hamiltonian H splits into block-diagonal, irreducible submatrices which are distinguished by their respective signature indicating how the eigenvectors transform under spin transposition. This conclusion holds regardless of space dimension D, electron concentration or crystal size. Due to the spin transpositions not commuting with H, the eigenvectors of H may not be (odd) eigenvectors of the spin transpositions. Since this work supplies all eigenvalues but also all eigenvectors, it enables one to compute the exact correlation functions for the groundstate and all excited states. As the ratio of the BCS Hilbert space dimension to that of the largest submatrice increases exponentially with the electron concentration, long-range order can henceforth be investigated in much larger crystals than done so far. Likewise the powerful numerical tools dedicated to finding out the groundstate energy of huge matrices could be applied to each big submatrix, separately. As the salient resemblence between the BCS and Hubbard models has been supported by strong evidence [3,5,6], it is planned to take advantage of spin conservation to achieve block-diagonalization of the Hubbard matrix, which would secure access to much larger cluster sizes than the very limited ones studied so far.

I thank Michel Caffarel and Ernest Ilisca for invaluable encouragement and criticism.

TABLE I. All symbols are defined in the text

N	5	7	9	11	13	15	17	19
n	2	3	4	5	6	7	8	9
d_0	10	35	126	462	1716	6435	24310	92378
d_M	5	13	35	96	267	750	2123	6046
$\frac{d_0}{d_M}$	2	2.69	3.6	4.81	6.43	8.58	11.45	15.28
d_g	2	13	13	96	96	750	750	6046
$\frac{d_0}{d_g}$	5	2.69	9.69	4.81	17.87	8.58	32.41	15.28

References

[1] E.H. Lieb and F.Y. Wu, Phys. Rev.Lett., 20, 1445 (1968)
[2] M. Caffarel and R. Mosseri, Phys.Rev.B, 57,12651 (1998)
[3] J. Bardeen, L.N. Cooper and J.R. Schrieffer, Phys. Rev.,108, 1175 (1957)
[4] C.N. Yang, Phys. Rev. Lett., 63, 2144 (1989)
[5] J. Szeftel and A. Khater, Phys.Rev.B, 54,13581 (1996)
[6] A. Kocharian, C. Yang and Y.L. Chiang, Phys.Rev.B, 59, 7458 (1999)
[7] H.R. Krishnamurthy and B. Sriram Shastry, Phys. Rev. Lett., 84, 4918 (2000)
[8] this expression of d_0 holds only for n not being a divisor of N, which ensures furthermore that any two Hilbert spaces $S_K, S_{K'}$ are disjoint for $K \neq K'$
[9] R. Pauncz, Spin Eigenfunctions (Plenum Press, New-York,1979)
[10] for $U > 0$ the BCS state approximates the most excited state instead of the groundstate in case of $U < 0$ (see Eq.4)

Part 3

Ab initio Studies of Correlation Effects

ELECTRON CORRELATION FROM MOLECULES TO MATERIALS

Rodney J. Bartlett*

1 INTRODUCTION

In the treatment of electron correlation, ***ab initio*** quantum chemical methods occupy a central role. However, when compared with the approaches used in some areas of physics, particularly with reference to strong correlation, there seems to be important differences. In this chapter, I plan to discuss the different philosophy, access where some limitations are, and suggest some ways to overcome these limitations to the mutual benefit of all aspects of electronic structure theory.

Electron correlation from the viewpoint of quantum chemistry introduces two critical approximations: the basis set and the extent of the configuration space that can be included in the correlated treatment. Coupled cluster theory[1, 2, 3, 4] has greatly aided in solving the latter problem, but the basis set problem persists. Today, the CC approach can provide energies ranging from 0.001 to 100 eV for a wealth of chemical and spectroscopic problems.[5, 6] However, there are still situations that are not accessible. Some of these occur when stronger correlation effects are encountered, while others require eliminating the basis set error. Some ideas are discussed. The relationship of coupled-cluster theory to density functional theory is also considered, providing an *ab initio* DFT framework that should offer converging exchange-correlation potentials and resultant approximations for electronic properties. In addition, coupled-cluster theory offers the natural vehicle for transferring first-principle electronic structure information into materials modeling, as it provides a rigorous basis for

*Rodney J. Bartlett, Quantum Theory Project, Departments of Chemistry and Physics, University of Florida, PO Box 118435, Gainesville, FL 32611-8435, Fax: 352-392-8722, Tel: 352-392-1597, E-mail: bartlett@qtp.ufl.edu.

low-rank Hamiltonians that can be rapidly solved for quite complicated problems.

2 THEORY

Consider the Hamiltonian,

$$H = \sum_i h(i) + \sum_{i,j} 1/r_{ij} \tag{1}$$

$$H = \sum_{p,q} h_{pq} p^\dagger q + \frac{1}{4} \sum_{p,q,r,s} \langle pq||rs\rangle p^\dagger q^\dagger sr \tag{2}$$

where the second-quantized form might also be written in normal order as

$$H = \sum_p f_{pp}\{p^\dagger p\} + \sum_{p \neq q} f_{pq}\{p^\dagger q\} + \frac{1}{4} \sum_{p,q,r,s} \langle pq||rs\rangle\{p^\dagger q^\dagger sr\} + \langle 0|H|0\rangle \tag{3}$$

Here, the Fock operator has its usual form, f=h+J-K, where in terms of the occupied molecular orbitals, J-K= $\sum_i \int d2\varphi_i(2)(1 - P_{12})\varphi_i(2)/r_{12}$. The f operator arises from the normal ordering procedure, and does not imply that Eq. (3) depends upon Hartree-Fock theory, though for that case and canonical orbitals the second term vanishes, and $f_{pp} = \epsilon_p$. For other choices of orbitals, the off-diagonal f_{pq} are required.

This 'matrix element' form of the Hamiltonian is particularly appropriate as our starting point as all answers can be no better than the basis functions used to represent H. In this form, the orbitals are expected to be molecular orbitals of some kind, obtained from the solution to an independent particle equation,

$$h^{eff}|p\rangle = \epsilon_p|p\rangle \tag{4}$$

of which Hartree-Fock is one choice. As the solutions to an eigenvalue problem, they are orthogonal or can be chosen to be, $\langle p|q\rangle = \delta_{pq}$. However, since the orbitals are the solution to some mean field equation, the actual computational basis set is $|\chi_1, \chi_2, \cdots \chi_m\rangle = |\chi\rangle$, so the orbital basis is $|\varphi\rangle = |\chi\rangle\mathbf{D}$, where $\mathbf{D}$ defines the MO coefficients. Today, almost invariably, this basis is chosen to be composed of Gaussian orbitals. In some atomic calculations and some diatomic molecule calculations, Slater orbitals are still used, but for polyatomic molecules, the many centered integrals are far more difficult for functions of the form exp(-αr) than for those with an exp(-αr^2) dependence. The angular and remaining radial part of the Gaussian is usually taken as $x^p y^q z^r$ meaning that we use Cartesian Gaussians, although, of course, the spherical combinations can be made as well, eg, to reduce the d orbitals to five and f orbitals to seven.

As square integrable atomic functions centered on the different atoms in a molecule, Gaussian functions are appropriate in many ways for molecular studies. Typically, they are contracted so that linear combinations of many Gaussians assume the role of the atomic orbitals in a calculation. This enables a better description of the atomic orbitals without introducing the additional linear variational parameters into the calculation that would be required if 'primitive' Gaussians were the basis. As the number of distinct basis functions is the rate determining step in the calculations, and since the dependence of *ab initio* correlated calculations depend non-linearly on the number of functions, typically as m^5-m^8, today, the number, m, is a very important one. The Gaussian basis functions are not orthogonal, $\langle\chi|\chi\rangle = S$, and if they are required to be located on each atomic center without 'tails' on other centers, they cannot be orthogonalized. Hence, if we want to fully exploit the localization of the Gaussian orbitals to obtain lower scaling of calculations, it would be nice to rewrite our second-quantized Hamiltonian in terms of the Gaussians, but the non-orthogonality would then introduce overlap integrals every time a contraction is made, making the calculations difficult; although some clever procedures are being developed that make the expressions tractable. [7, 8] Gaussians also offer ways to construct a complete set, although in practice, we are far from complete. The error introduced in any calculation is referred to as the basis set error, and today, we pragmatically extrapolate calculations in a consistent series of basis sets to overcome this error. [6] This basis set error is the principal remaining one in quantum chemistry.

The other principal error in *ab initio* correlated methods is the truncation of the basis of n-particle configurations that are responsible for electron correlation. Once the mean field problem is solved, we have a set of occupied and unoccupied spin-orbitals, designated as i,j,k,l... and a,b,c,d... while p,q,r,s... indicate any kind of orbital. The treatment of electron correlation requires that in addition to our mean-field solution, $\Phi_0 = |0\rangle$, which is a determinant of the n-occupied spin-orbitals, that we also consider all possible up to n-fold excitations where we successively replace 1 to n occupied orbitals by 1 to n of the unoccupied ones, to create single, $a^\dagger i|0\rangle = |_i^a\rangle$, double, $a^\dagger i b^\dagger j|0\rangle = |_{ij}^{ab}\rangle$, etc. excitations. If we can use all possible excitations for n electrons, we have the full CI solution,

$$\Psi = |0\rangle + \sum_{i,a} |_i^a\rangle C_i^a + \sum_{i<j,a<b} |_{ij}^{ab}\rangle C_{ij}^{ab} + \ldots + \sum_{i<j<\ldots<n,a<b<\ldots} |_{ijk\ldots}^{abc\ldots}\rangle C_{ijk\ldots}^{abc\ldots} \quad (5)$$

$$\Psi = (1 + C_1 + C_2 + \ldots + C_n)|0\rangle \quad (6)$$

This solution is the exact one in the underlying Gaussian basis set. It is variational, giving the best possible upper bound to the energy; invariant to any transformation in the basis set, and behaves properly with the number of electrons (i.e. is size-extensive). Its solution scales as $\sim m^n$ and is prohibitive except for some comparatively small examples, like H_2O in a DZP basis.[9] With the core electrons frozen, we have 8 fold excitations, and the DZP basis means 4s,2p, and 1d contracted Gaussian functions on O and 2s1p on H, for a total of 25 functions. A similar full CI for N_2 with the core frozen requires 10 fold excitations and 30 functions, and is the largest reported to date.[10] Such benchmark calculations are quite useful to calibrate more widely applicable methods, like CC theory.

The fundamental idea of CC theory is the exponential ansatz,

$$\Psi = \exp(T)|0\rangle = (1 + T + T^2/2 + \ldots)|0\rangle \tag{7}$$

where the operator

$$T = T_1 + T_2 + T_3 + \ldots + T_n \tag{8}$$

and

$$T_1 = \sum_{i,a} t_i^a \{a^\dagger i\} \tag{9}$$

$$T_2 = \sum_{i<j,a<b} t_{ij}^{ab} \{a^\dagger i b^\dagger j\} \tag{10}$$

$$T_3 = \sum_{i<j<k,a<b<c} t_{ijk}^{abc} \{a^\dagger i b^\dagger j c^\dagger k\} \tag{11}$$

Inserting the CC wavefunction into the Schroedinger equation, we have the CC equations for the energy and for the $\{t_{ij\ldots}^{ab\ldots}\}$ amplitudes,

$$E = \langle 0|\overline{H}|0\rangle \tag{12}$$

$$\langle {}_{ij\ldots}^{ab\ldots}|\overline{H}|0\rangle = 0 \tag{13}$$

where $\overline{H} = e^{-T}He^{T}$. This similarity transformed Hamiltonian has a central role in CC theory, [4] and will serve as the connecting link to excited states and materials modeling as discussed later.

The cluster expansion introduces products of lower-order operators for any choice of T. Hence, relating the cluster operators to the CI operators, we have

$$C_1 = T_1 \tag{14}$$

$$C_2 = T_2 + T_1^2/2 \tag{15}$$

$$C_3 = T_3 + T_2T_1 + T_1^3/3! \tag{16}$$

$$C_4 = T_4 + T_2^2/2 + T_2T_1^2/2 + T_3T_1 + T_1^4/4! \tag{17}$$

$$C_5 = T_5 + T_2T_3 + T_1T_4 + (T_2)^2T_1/2 + (T_1)^3T_2/3! + (T_1)^2T_3/2 \tag{18}$$

$$\vdots \tag{19}$$

For $T=T_1+T_2$, we have the CCSD wavefunction and the products of T_1 and T_2 introduce additional triple and quadruple excitations that would not be in CISD calculations. Furthermore, by virtue of the exponential form, CC theory has the very important property of size-extensivity. That is for A and B being two non-interacting units, $\exp(T_A+T_B)|0_A0_B\rangle = \exp(T_A)|0_A\rangle \exp(T_B)|0_B\rangle$, yet this essential property is not satisfied by a truncated CI like CISD, because the products of double excitations with double excitations, eg, are not included. Hence, for even two non-interacting He atoms and four electrons, CISD cannot give the right answer while CCSD does; and it does so for any number of separated pairs of electrons. Similarly, CCSDT would provide exact results for any number of non-interacting Li atoms, etc. The 'extensive' property is absolutely essential if we are to apply a correlated approach to polymers, surfaces, or crystals, so CC theory, unlike CI, is applicable (see Ref. [11] for the first application of CCSD to a polymer.) Since our Hamiltonian only has two-particle interactions, it also makes sense physically that products of T_2 (which are disconnected) will be far more important than (connected) T_4; and similarly, $T_2^3/3! >> T_6$. These disconnected terms arise earlier in perturbation theory, too, as all CI quadruple excitation contributions in fourth order come from $T_2^2/2$, with T_4 first contributing in fifth-order. Figure 1 shows the convergence of CC theory to the full CI limit, compared to truncated CI and many-body perturbation theory (MBPT) of various orders. The much more rapid convergence of CC theory with excitation level is apparent. Truncated CI suffers from retention of unlinked diagrams, which is another manifestation of its inextensivity. MBPT eliminates the unlinked diagrams, and consequently shows improved convergence, but CC theory further improves upon MBPT via its infinite sum of selected diagrams, rather than just those through a given order. It is also important that CC theory is independent of a choice of unperturbed Hamiltonian, unlike perturbation theory, which enables CC theory to avoid the divergence often associated with order-by-order perturbation theory.

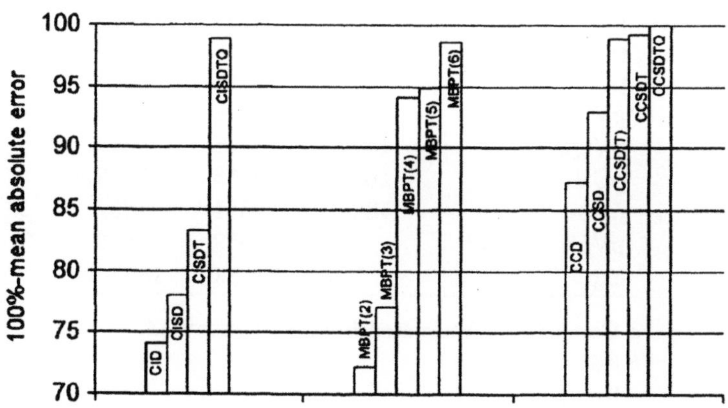

Figure 1. Mean absolute errors for a series of small molecules (BH, HF, H_2O at three geometries) with the CI, MBPT and CC approximations compared to full CI results.

The very rapid convergence of CC theory to the full CI limit frequently means that the correlation problem for small molecules can be essentially solved (some caveats are discussed below), leaving the Gaussian basis set as the remaining source of error. Many illustrations of this fact are in the literature. For example, despite these limitations, using quite large basis sets, typically cc-pVTZ and higher (30 functions per C,N,O,F,B and 14 functions per H), the average errors are competitive with experiment. [3] As just one example, Table 1 shows the convergence of the vibrational frequency of N_2 and C_2 at the full CCSDT level including a factorized approximation for connected quadruple excitations.[12, 13] We have shown that to achieve ≈1 cm^{-1} accuracy in its vibrational frequency, we also require the effects of pentuple excitations, T_5. [15] Furthermore, we have recently reported the initial implementation of the full CCSDTQP method, [16] which now represents the state-of-the art in CC theory.

Table 1. Computed CCSDT(Q_f) harmonic frequencies (cm^{-1}) of the N_2 and C_2 molecules with various correlation consistent basis sets.

Basis set	No. of b.f.	N_2	C_2
pVDZ	28	2325	1821
pVDZ+	46	2304	1807
pCVDZ	36	2327	1820
pCVDZ+	54	2309	1810
pVTZ	60	2337	1838
pVTZ+	92	2330	1834
pCVTZ	86	2345	1849
pCVTZ+	118	2342	1844
pVQZ	110	2348	1849
pVQZ+	160	2346	1848
pCVQZ	168	2356	1859
pV5Z	182	2351	1852
pV5Z + core[a]		2359	1862
exp		2358.6[b]	1855.0[c]

a Core effect is estimated from pCVQZ-pVQZ.
b Ref. [17].
c Ref. [18].

For excited states, the equation-of-motion (EOM-CC) [4] allows us to readily describe excited, electron-attached, and ionized states, providing we have a good single reference state description. That is, for one state, whether the ground state or not, we require a good CC description, and then using a second solution to the Schroedinger equation, we can construct the EOM equations.

$$
\begin{aligned}
H\Psi_{CC} &= E_{CC}\Psi_{CC} && (20)\\
\Psi_K &= \widehat{R}_K\Psi_{CC} && (21)\\
\widehat{R}_K &= r_0 + \sum_{i,a} r_i^a a^\dagger i + \sum_{i<j,a<b} r_{ij}^{ab} a^\dagger i b^\dagger j + \ldots && (22)\\
H\widehat{R}_K|\Psi_{CC}\rangle &= E_K\widehat{R}_K|\Psi_{CC}\rangle && (23)\\
\left[\overline{H},\widehat{R}_K\right]|0\rangle &= \omega_K\widehat{R}_K|0\rangle && (24)
\end{aligned}
$$

We introduce the excited state eigenfunction, Ψ_K, which is composed of excitations with amplitudes r_i^a, r_{ij}^{ab}, etc. The amplitudes are determined from an equation derived from subtracting the energy of the reference CC state from that for the excited state, $\omega_K = (E_K - E_0)$. The $\overline{H}$ contains all the information from T.

In matrix form, the EOM equations become

$$
\begin{aligned}
\overline{\mathbf{H}}\mathbf{R}_K &= \omega_k\mathbf{R}_K && (25)\\
\mathbf{L}_K\overline{\mathbf{H}} &= \omega_K\mathbf{L}_K && (26)
\end{aligned}
$$

where since the $\overline{\mathbf{H}}$ matrix is not Hermitian, we have both a left and a right eigenvector, with the same eigenvalue. Both eigenvectors are required for properties like transition moments.

From simply modifying the $\widehat{R}_K$ operator to result in a net ionization, i.e. by letting one electron go into the continuum, or to use j or $ja^\dagger i$ type operators, or $a^\dagger$ or $a^\dagger b^\dagger j$ to introduce electron attachment; we can tailor the operator to the problem at hand. Double ionizations and double electron attachments are similarly accessible. The correspondence between a series of Rydberg states and its corresponding ionization continuum is apparent. Such EOM-CC methods greatly extend the application of CC theory, as the excited states are obtained from an eigenvalue problem. So in the event of degeneracy or quasi-degeneracy in the excited state, the corresponding coefficients in the eigenvector are free to take whatever value they should. This is not the case in the reference CC state, since there is some bias introduced by the choice of the single determinant reference function until the full CI is reached. The state-of-the art in EOM-CC is currently EOM-CCSDT.[19] Analytical gradients [4] for standard CC [20] and EOM-CC [21] have been implemented to permit the forces on atoms to be evaluated simultaneously with the energies. Without such tools, we cannot readily move atoms to generate potential energy surfaces, and identify equilibrium and transition state geometries, or provide vibrational frequencies and intensities.

Table 2. Computed energies (Hartree) of the NH_3 molecule with various correlation consistent basis sets.

Basis set	SCF	CCSD(T) valence electrons	CCSD(T) all electrons
aug-cc-pVDZ	-56.205591	-56.425519	-56.428088
aug-cc-pVTZ	-56.220807	-56.480561	
aug-cc-pVQZ	-56.224504		-56.497865
aug-cc-pV5Z	-56.224617		
cc-pCVTZ			-56.527170
cc-pCVQZ			-56.550957
limit	-56.226[a]		-56.563269[b]

a Ref. [22].
b Total CCSD(T)-R12 energy. [23].

To return to the other limitation in quantum chemistry, the basis set problem, in Table 2 are shown the total energies for the NH_3 molecule as a function of larger and larger correlation-consistent basis sets. The Hartree-Fock energy rapidly converges to a mH with the cc-pVQZ basis. This basis consists of 218 functions yet going to the next basis, aug-cc-pV5Z of 367 functions, only achieves a 0.1 mH improvement. The correlation energy is still more slowly converging reaching ≈12 mH from the exact, all-electron non-relativistic value which is well approximated by CCSD(T)-R12 where two-electron interactions are included explicitly. The solutions to a one-particle equation like the HF equations, can be rapidly obtained in a basis, while the correlation part, being a function of two particles, requires an expansion in products of one particle functions; and such $[\varphi_p(1)\varphi_q(2) - \varphi_p(2)\varphi_q(1)]$ products, where the 1 and 2 represent the coordinates of the two particles, is a poor approximation. In fact, this is the most fundamental error left in quantum chemistry. Any such wavefunction that is only a function of products of one particle functions, as in the full CI solution above, clearly fails to satisfy the correlation cusp condition,

$$\left.\frac{\partial\Psi}{\partial r_{12}}\right|_0 = \eta\Psi(0) \tag{27}$$

where $\eta > 0$.

In a theory where we permit explicit, linear dependence on two particle coordinates, we can satisfy the cusp condition. However, its satisfaction is less important than the other numerical effects of r_{12}. For example, Gaussian geminals depend upon $\exp(-\alpha r_{12}^2)$ and they can provide extremely accurate answers though still not satisfy the cusp condition. For that matter, Gaussian orbitals, unlike Slater orbitals, cannot exactly satisfy the electron-nuclear cusp, $-1/(r_i - R_\alpha)$, on nucleus α, but since many Gaussian functions are located on each nucleus, this is not generally a problem except perhaps for some potentially extremely accurate determinations of hyperfine constants. However, the significant numerical effect of explicit two-particle functions is what keeps *ab initio* correlated methods from reaching their full potential. Recent linear r_{12} methods in CC theory have been presented [23, 24], showing that essentially exact results can be obtained from such a theory, as above for NH_3, but the demands on the calculation and basis are still severe, and no such theory is yet suitable for the general treatment of potential energy surfaces as required by chemistry, primarily because there is no analytical gradient method as has been developed for the usual CC theory.[20]

To correct the basis set problem in a way that would be as easily applicable as the normal CC theory, we require an alternative treatment of the short-range correlation. If we do a very accurate calculation of the correlation hole as a function of basis set, we should be able to see why the standard correlated theory is inferior. To the contrary, the so-called on top correlation in DFT is said to be essentially correct, even at the level of the local density approximation. [25] Hence, it should be possible to introduce such a basis set correcting potential into the second-quantized Hamiltonian to overcome the basis set problem. This is one of our objectives. [26]

Another area where there might be an interconnection pertains to strong-correlations. In DFT, it is straightforward to define strong-correlation effects via an adiabatic perturbation theory that introduces an order parameter in front of the two-particle interaction. [27] By simply asking what the solution is when this parameter approaches infinity, we have a limit defined. Just as we are interested in introducing cusp conditions into quantum chemistry to alleviate some basis set problems, we would be equally interested to arrange for our methods to incorporate, semi-analytically, these kind of effects that are also frequently outside the scope of applicable CC theory.

CC theory based upon a single reference function rapidly converges to full CI as shown in Fig. 1. However, there are situations involving quasi-degeneracy, with bond breaking being one that is very important for chemistry, [28] and others that might include extensive amounts of quasi-degeneracy or collective effects that are not as amenable to single reference CC theory. With the exception of EOM-CC, the traditional multi-reference CC analogs are less developed and more difficult, and also introduce 'active orbital' choices and other complications that make them less desirable. [29] Chemists tend to identify the effects of electron correlation that encourage multi-reference descriptions 'non-dynamic' to distinguish them from 'dynamic' correlation that primarily keeps electrons apart and is introduced by normal single reference CC theory. Sometimes, 'static' correlation is used to refer to cases where several determinants must be included in a reference simply because of symmetry considerations. Obviously, this is an approximate seperation of correlation effects since all types are in any full CI solution. However, short of full CI, such 'non-dynamic' effects could use a boost from graphing some kind of strong correlation DFT effects onto CC theory, e.g. as one might hope to do with short range correlation.

Both the above requires trying to take something that density functional theory does right, to correct something that is difficult for wavefunction theory to do. This raises the question of the interconnections between DFT and wavefunction theory. The enormous advantage of the latter is the systematic convergence with basis set and correlation leading to the now well-known paradigm of size-extensive results in electronic structure theory, MBPT(2) < CCD < CCSD < CCSD(T) < CCSDT < CCSDT(Q_f) < CCSDTQ < Full CI. This makes it possible for wavefunction theory to be ***predictive.*** DFT is not, as we do not know if LDA will provide worse answers than a gradient corrected functional like BLYP or PBE, or a hybrid functional like B3LYP, for any given problem. This raises the question of whether there is an *ab initio* analog of DFT? We recognize that DFT is a first-principle method in the sense that it has a rigorous basis. But in practice, it either takes a functional from a known model problem (LDA), or enforces a very few of the many known consistency conditions to guide the development of a functional; or adopts a semi-empirical approach where parameters are fit to a test suite of problems to get the best answers. Our *ab initio* DFT approach[?, 30, 31] is different. Like wavefunction theory, we insist that we will work in a basis set; second, that our approach is parameter free; and third, that we exploit properties of wavefunction theory so that we know we are using expressions that can, in principal, be made exact.

We know that a converged Kohn-Sham result must give the exact density for the (interacting) problem. This density corresponds to

$$\rho(1) = \sum_i \varphi_i(1)\varphi_i^*(1) \tag{28}$$

and the orbitals must be solutions to the problem,

$$h^{KS}(1)\varphi_p = \epsilon_p\varphi_p(1) \tag{29}$$

$$\Phi_{KS} = \mathcal{A}[\varphi_1(1)\varphi_2(2)...\varphi_n(n)] \tag{30}$$

where

$$h^{KS} = t + v + J(\rho) + V_{XC}(\rho) \tag{31}$$

Hence, the expedient of insisting that the density ρ does not change from its converged KS value, means that in any order of perturbation theory, that $\rho^{(k)} = 0$. Using coupled-cluster theory, we can write the density in the form,

$$\rho(1) = \frac{\langle 0|\exp(T^\dagger)\widehat{\delta}\exp(T)|0\rangle}{\langle 0|\exp(T^\dagger)\exp(T)|0\rangle} = \langle 0|[\exp(T^\dagger)\widehat{\delta}\exp(T)]_C|0\rangle \tag{32}$$

where the subscript indicates only connected terms and $\widehat{\delta}$ indicates the Dirac delta function. Actually, for CC theory a superior way to introduce the density is to use the CC functional [4] in terms of the Λ, deexcitation operator,

$$E = \langle 0|(1+\Lambda)\overline{H}|0\rangle \tag{33}$$

which means the density is given by

$$\rho(1) = \langle 0|(1+\Lambda)\exp(-T)\widehat{\delta}\exp(T)|0\rangle \tag{34}$$

where Λ satisfies the complementary equation,

$$\langle 0|\overline{H}\Lambda - E\Lambda|_{ij..}^{ab..}\rangle = 0 \qquad (35)$$

Since, $\Lambda = T^\dagger$ through linear terms, for low orders the two approaches are equivalent, but the Λ based functional has the advantage that the density (or any other expectation value) is given in closed form. The connected form does not have this property. For the purposes of this chapter, this distinction is not important. What is important, is that we have a CC way to rigorously define the density to any order in perturbation theory. Imposing the condition that $\rho^{(k)} = 0$, provides a series of equations whose solution has to give the rigorous form of V_{XC}.

Without going through details, [30] the vanishing first-order density provides the condition that

$$\sum_{i,a} \frac{\langle i|\widehat{\delta}|a\rangle\langle a|K + V_{XC}^{(1)}|i\rangle}{\epsilon_i - \epsilon_a} = 0 \qquad (36)$$

The limitation to first-order tells us that $V_{XC}^{(1)} = V_X$, and this equation might be recognized as the exchange only optimized effective potential (OEP) equation. Hence, knowing integrals of the non-local exchange operator, K, we can solve for the local, multiplicative, V_X, in a basis set. [31] This is usually done by introducing the response function,

$$X(1,2) = \sum_{i,a} \frac{\varphi_i^*(1)\varphi_a(1)\varphi_i(2)\varphi_a^*(2)}{\epsilon_i - \epsilon_a} \qquad (37)$$

from which some manipulation leads to the matrix form

$$\mathbf{X}\mathbf{V}_X = -\mathbf{W} \qquad (38)$$

where $\mathbf{W}$ is closely related to $\mathbf{K}$. This may be solved using single value decomposition to provide the matrix representation of V_X. Illustrations of V_X compared to conventionally used exchange potentials like B88 [32] or LDA are shown in Fig. 2. The important element is that the V_X obtained from the above procedure gives the quantum mechanical effects missing from the other potentials. Furthermore, as a consequence, this potential properly handles the self-interaction missing from conventional DFT; has the formally correct long range (-1/r) behavior; and at least in a complete basis set, should satisfy the exact condition that for the homo, h,

$$\langle h|K|h\rangle = \langle h|V_X|h\rangle \qquad (39)$$

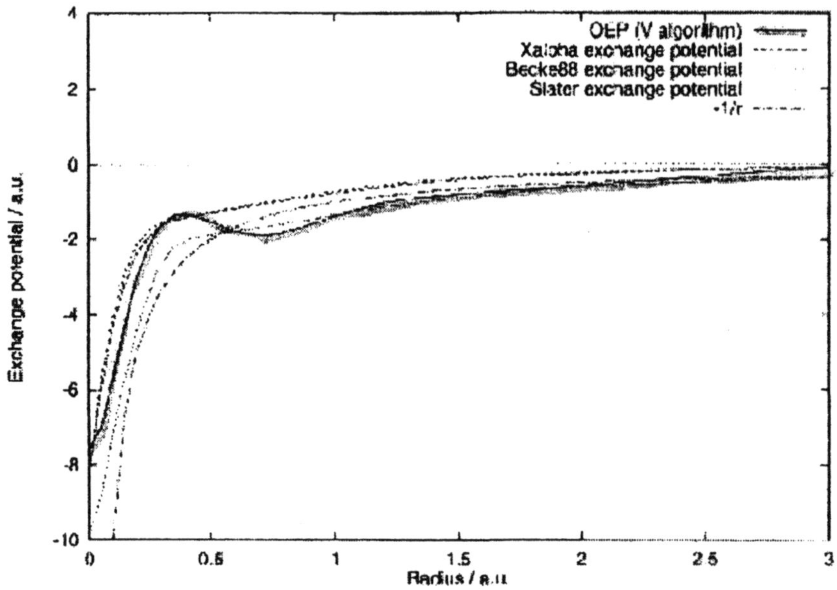

Figure 2. Comparison of exchange potentials for neon.

The latter greatly helps to enable ab initio DFT to give much better approximations to the exact ionization potential of the homo. Alternatively, $V_X(1) = \delta E_X/\delta\rho(1)$.

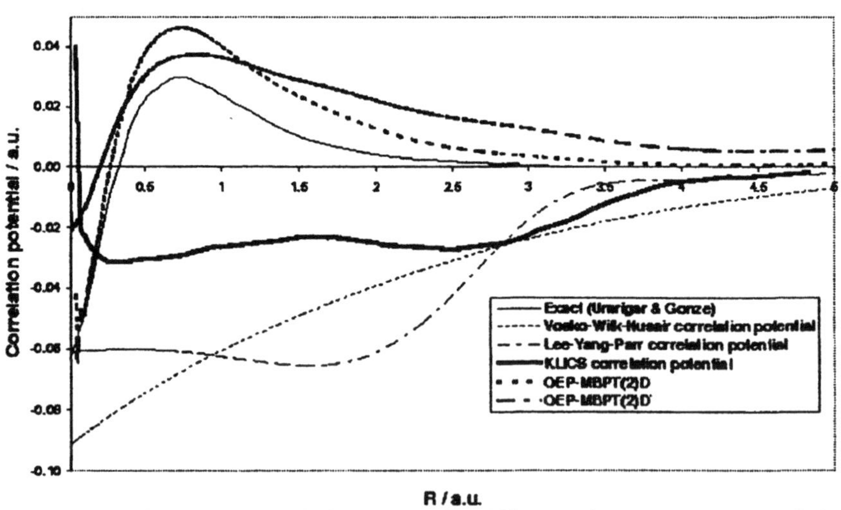

Figure 3. Comparison of the exact (QMC) correlation potential of the helium atom with correlation potentials obtained from VNW, LYP, KLICS and OEP-MBPT(2) functionals. The exact (QMC) correlation potential is from Ref. [33]. The usual choices: LYP, VWN and KLICS are below the correct results.

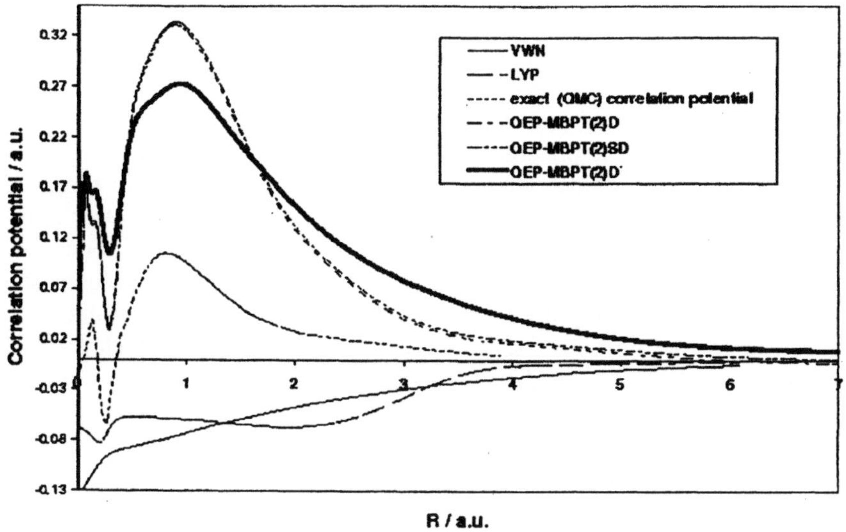

Figure 4. Comparison of the exact (QMC) correlation potential of the neon atom with correlation potentials obtained from the VWN, LYP and OEP-MBPT(2) functionals. The exact (QMC) correlation potential is from Ref. [33]. The LYP and VWN are below, contrary to the correct results.

Continuing to the next order, we obtain the first approximation to the correlation potential, $V_{XC}^{(2)} = V_C^{(2)} = \delta E_2/\delta\rho(1)$ from insisting that $\rho^{(2)} = 0$. In this case, E_2 is the second-order MBPT approximation to the energy, taking into account that the KS orbitals introduce off-diagonal Fock operators in the second-quantized Hamiltonian which arise in second-order due to mixing between Φ_{KS} with single excitations.

$$\mathbf{X}\left(\mathbf{V}_c^{(2)}\right) = -\mathbf{M} \tag{40}$$

In Eqn. (40), **M** represents several terms that are products of one and two electron integrals divided by orbital energy differences (i.e. diagrams). Again I will forego the details, as they are discussed in more detail elsewhere, [30] but the results from this procedure can be compared with the standard ones (LYP, VWN, etc.), and those obtained from quantum Monte Carlo results for He and Ne [33] in Fig. 3 and 4. Unlike all of the standard correlation potentials, our ab initio DFT $V_C^{(2)}$ have the correct sign as a function of r and go to the correct limit. Continuing to infinite order, ie CC theory, leads to the trivial satisfaction of all the order-by-order equations becoming equivalent to the procedure for inverting a known density to extract the corresponding V_{XC} discussed by Zhao, Morrison, and Parr. [34]

To conclude this survey of the correlation problem for molecules and how elements of CC can contribute, consider a problem that is very important in current materials simulations: ie, the problem of 'predictive' simulations. Since quantum chemistry is built upon the converging paradigm of electronic structure methods above, which when used with converging basis sets provide predictive results; an essential component of predictive simulations is that the potentials of interaction be similarly predictive. However, this is never the case. Almost invariably, simulations of materials, which require very large scale molecular dynamics to describe the phenomena of interest, force the use of simple, classical potentials of interaction. These potentials are most frequently simply pair potentials, although three-body ones are used as well. Even the latter imposes a significant cost in the rapid determination of forces for tens of thousand up to millions of atoms. Though such classical potentials can be tailored to describe some of the bulk properties of a system, the ultimate results for any phenomena can be no better than the potentials of interaction chosen. This is the great weakness of 'predictive' simulations. For some smaller examples, quantum mechanical forces can be used, most frequently from DFT. However, these simulations are limited to <200 atoms, far less than what is required for many problems. The largest quasi-quantum mechanical simulations, involving several hundred atoms, have used the tight-binding model, but that is far too crude to be predictive, as it does not have any self-consistency or charge transfer built into it, making it impossible to break bonds correctly, eg.

Our partial solution to this problem again uses the underlying framework of CC theory. The $\overline{H}$ operator rigorously satisfies the role of the Hamiltonians normally used in semi-empirical theory. That is, it satisfies the equations,

$$\overline{H}|0\rangle = E|0\rangle \tag{41}$$

$$\langle^{ab..}_{ij..}|\overline{H}\widehat{R}_K|0\rangle = \omega_K\langle^{ab..}_{ij..}|R_K|0\rangle \tag{42}$$

for the energy of the correlated ground state and for all associated excited, ionized, electron attached, etc. states. That means that $\overline{H}$ has a single determinant as its eigenfunction and its projection onto excited determinants provides the other states. Now if we can replace $\overline{H}$ by an operator that has a far lower rank than the *ab initio* $\overline{H}$, but one that can retain the accuracy of CC theory, very large molecules can be described, and, indeed, large scale quantum simulations would be possible. The type of quantity of interest is:

$$\overline{H} \approx \overline{H_T} = \sum_{\alpha<\beta} \bar{h}_{\alpha\beta}\alpha^\dagger\beta + \sum_{\alpha<\beta<\gamma<\delta} \overline{\langle\alpha\beta|\gamma\delta\rangle}\alpha^\dagger\beta^\dagger\delta\gamma + \sum_{\alpha<\beta<\gamma<\delta<\epsilon<\zeta} \overline{\langle\alpha\beta\gamma|\delta\epsilon\zeta\rangle}\alpha^\dagger\beta^\dagger\gamma^\dagger\zeta\epsilon\delta + \ldots \tag{43}$$

where the new, transfer Hamiltonian, $\overline{H}_T$, is limited to some suitable p functions whose number p<<<m, which was the number of functions in the original basis set. We also pay the price that now we have three and four-body operators (not shown) in $\overline{H}_T$.

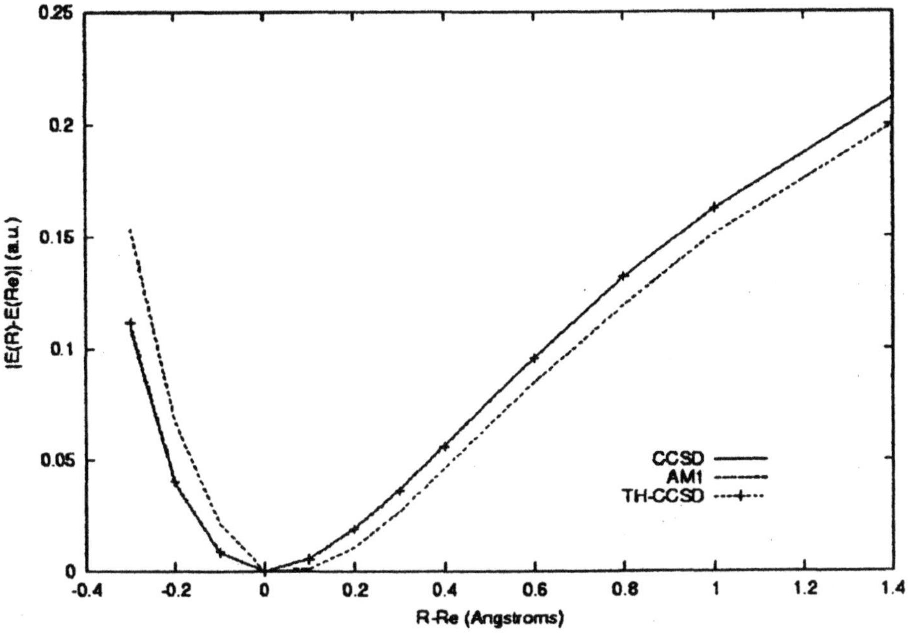

Figure 5. Comparison of PES from standard semi-empirical theory (AM1), CCSD and $\overline{H}_T$ (TH-CCSD) for breaking the Si-O bond in pyrosilicic acid.

The assumption of being able to define an $\overline{H}_T$ underlies all widely used semi-empirical theories, like complete neglect of differential overlap CNDO, and its variants like 'intermediate'(INDO) and neglect of diatomic differential overlap (NDDO) and others like the Austin Method 1 (AM1), also an NDDO scheme, etc. Instead of actual evaluation of the matrix elements in $\overline{H}_T$, such semi-empirical methods attempt to use parameters to facilitate the form of $\overline{H}_T$, invariably truncated to two-body terms. However, in practice, no set of parameters have ever been able to describe both ground state energetics and excited states as in electronic spectroscopy. AM1 for example, is expected to do fairly well for molecular structures and comparative energies on potential energy surfaces, but spectroscopy requires INDO/S where the S indicates spectroscopic parameters, and INDO with those parameters is a disaster for structures and energetics. $\overline{H}_T$ in principle is exactly what we would want, as it gives CC quality results for energetics and spectra.

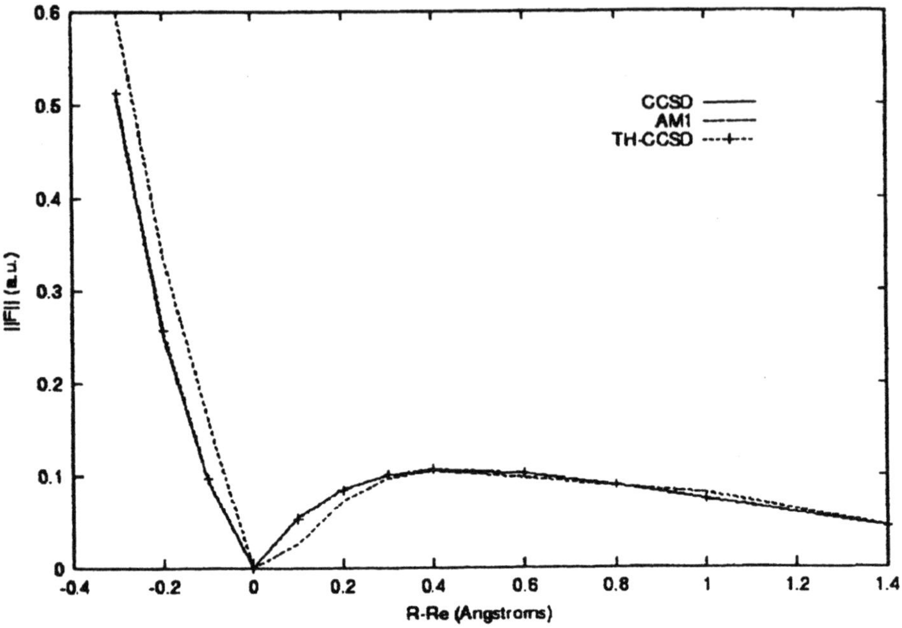

Figure 6. Comparison of forces from standard semi-empirical theory (AM1), CCSD and $\overline{H}_T$ (TH-CCSD) for breaking the Si-O bond in pyrosilicic acid.

If we can represent $\overline{H}_T$ in a greatly reduced rank from its purely *ab initio* form, guided by semi-empricial theory, eg, and without need for the three and four body operators; then we can hope to retain the accuracy of CC theory in a framework that permits ~1000 atoms to be described with QM forces. Recently, we have investigated this procedure by doing CC calculations for clusters of SiO_2 to describe its fracture under strain. The CC results properly break the single bonds involved, providing accurate potential energy curves. At each point on the curve, we also compute the forces on the atoms using analytical gradient theories. Given these high-level ab initio curves, we fit a Hamiltonian of the NDDO form, see Figs. 5 and 6, which means it formally has only four functions per Si or O atom, and only two center interactions involving those atoms need to be in the $\overline{H}_T$, compared to the exorbitant number of terms that would be in the ab initio form. This fit shows that for both forces and energies, CC quality results are obtained. Then we depend upon the fact that Si-O, Si-Si, and O-O interactions that are more than three or four neighbors apart fall off suitably rapidly that only results for clusters of moderate size need to be evaluated by CC theory. We use various means like genetic algorithms to define the NDDO parameters in $\overline{H}_T$ that reproduces the CC forces on the atoms involved yet with a drastically lower rank hamiltonian. Consequently, we can use that Hamiltonian to rapidly generate the QM forces, fully self-consistently, with all appropriate charge transfer for bond breaking, for $\approx$1000 atoms.[35]

3 ACKNOWLEDGMENTS

The author thanks Drs. Mikael Ciffan (ARO), Tony Gonis, and Nick Kioussis for the opportunity to attend this stimulating meeting. This work was supported by the NSF-DMR under grant No. 9980015.

References

[1] J. Čížek, *Adv. Chem. Phys.* **14**, 35 (1969).

[2] J. Paldus, J. Čížek and I. Shavitt, *Phys. Rev. A* **5**, 50 (1972).

[3] R.J. Bartlett, *J. Phys. Chem.* **93**, 1697 (1989).

[4] R.J. Bartlett, Coupled-cluster theory: An overview of recent developments, in *Modern Electronic Structure Theory, Part I*, edited by D. R. Yarkony. World Scientific Publishing Co., Singapore, (1995).

[5] T.H. Dunning Jr., *J. Phys. Chem. A* **104**, 9062 (2000).

[6] K.L. Bak, J. Gauss, P. Jørgensen, J. Olsen, T. Helgaker and J.F. Stanton, *J. Chem. Phys.* **114**, 6548 (2001).

[7] M. Schutz and H.J. Werner, *J. Chem. Phys.* **114**, 661 (2001).

[8] G.E. Scuseria and P.Y. Ayala, *J. Chem. Phys.* **111**, 8330 (1999).

[9] C.W. Bauschlicher Jr. and P.R. Taylor, *J. Chem. Phys.* **85**, 2779 (1986).

[10] H. Larsen, J. Olsen, P. Jorgensen and O. Christiansen, *J. Chem. Phys.* **113**, 6677 (2000).

[11] S. Hirata, I. Grabowski, M. Tobita and R.J. Bartlett, *Chem. Phys. Lett.* **345**, 475 (2001).

[12] S.A. Kucharski and R.J. Bartlett, *J. Chem. Phys.* **108**, 5243 (1998).

[13] S. Kucharski and R.J. Bartlett, *J. Chem. Phys.* **108**, 9221 (1998).

[14] M. Musial, S.A. Kucharski and R.J. Bartlett, *J. Mol. Struct. (Theochem)* **547**, 269 (2001).

[15] M. Musial, S.A. Kucharski and R.J. Bartlett, J. Chem. Phys. (in press).

[16] K.P. Huber and G. Herzberg, in *NIST Chemistry WebBook, NIST Standard Reference Database Number 69*, edited by W. Mallard and P. Linstrom (National Institue of Standards and Technology, Gaithersburg, MD, 1998), http://webbook.nist.gov.

[17] M. Douay, R. Nietmann and P.F. Bernath, *J. Mol Spectrosc.* **131**, 250 (1988).

[18] S.A. Kucharski, M. Wloch, M. Musial and R.J. Bartlett, *J. Chem. Phys.* **115**, 8263 (2001).

[19] E.A. Salter, G.W. Trucks and R.J. Bartlett, *J. Chem. Phys.* **90**, 1752 (1989).

[20] J.F. Stanton and J. Gauss, *J. Chem. Phys.* **100**, 4695 (1994).

[21] J.A. Pople and J.S. Binkley, *Mol. Phys.* **29**, 599 (1975).

[22] J. Noga, P. Valiron and W. Klopper, *J. Chem. Phys.* **115**, 2022 (2001).

[23] J. Noga, P. Valiron and W. Klopper, *J. Chem. Phys.* **115**, 5690 (2001).

[24] K. Burke, J.P. Perdew and M. Ernzerhof, *J. Chem. Phys.* **109**, 3760 (1998).

[25] T. Henderson and R.J. Bartlett, (to be published).

[26] M. Seidl, J.P. Perdew and S. Kurth, *Phys. Rev. A* **62**, 2502 (2000).

[27] R.J. Bartlett and J.F. Stanton, in *Reviews in Computational Chemistry*, edited by K.B. Lipkowitz and D.B. Boyd, page 65 (VCH Publishers, New York, 1994).

[28] P. Piecuch, S.A. Kucharski and R.J. Bartlett, *J. Chem. Phys.* **110**, 6103 (1999).

[29] R.J. Bartlett, in *Chemistry for the 21st Century*, edited by E. Keinan and I. Schechter, page 271 (Wiley-VCH, Weinheim, 2000).

[30] I. Grabowski, S. Hirata, S. Ivanov and R.J. Bartlett, J. Chem. Phys. (in press).

[31] S. Hirata, S. Ivanov, I. Grabowski, R.J. Bartlett, K. Burke and J.D. Talman, *J. Chem. Phys.* **115**, 1635 (2001).

[32] A.D. Becke, *Phys. Rev. A* **38**, 3098 (1988).

[33] C. Filippi, C.J. Umrigar and X. Gonze, *Phys. Rev. A* **54**, 4810 (1996).

[34] Q.S. Zhao, R.C. Morrison and R.G. Parr, *Phys. Rev. A* **50**, 2138 (1994).

[35] C.E. Taylor, M.G. Cory and R.J. Bartlett, J. Phys. Chem. A (submitted).

CAN DENSITY FUNCTIONAL THEORY DESCRIBE STRONGLY CORRELATED ELECTRONIC SYSTEMS?

John P. Perdew

Department of Physics and Quantum Theory Group
Tulane University
New Orleans, LA 70118

Abstract

Summarized here are the ideas of Kohn-Sham spin density functional theory for the many-electron ground state, the effects of exchange and correlation, and the limits of weak and strong correlation. In particular, the suppression of electron number fluctuations in one-electron regions due to correlation is discussed. While the standard local spin density and generalized gradient approximations (GGA's) for the exchange-correlation energy can work for certain cases of strong correlation, they fail dramatically for others. It is suggested that the new self-correlation-free meta-GGA's and self-interaction-free hyper-GGA's might yield a more reliable description of strong correlation.

1. INTRODUCTION

At the first International Workshop on Electron Correlations and Material Properties (Crete 1998), a lively discussion led to the conclusion that Kohn-Sham density functional theory in the standard local spin density and generalized gradient approximations is a theory of "moderately correlated" ground states, and does not describe "strong correlation" because of self-interaction error.

After brief summaries of Kohn-Sham density functional theory (section 2), of exchange and correlation (section 3), and of strong correlation (section 4), section 5 presents several standard and relatively new density functional approximations for the exchange-correlation energy, and section 6 discusses the extent to which they may or may not describe strongly-correlated systems.

Electron Correlations and Materials Properties 2
Edited by Gonis *et al.*, Kluwer Academic/Plenum Publishers, 2002

2. EXACT KOHN-SHAM SPIN DENSITY FUNCTIONAL THEORY

Kohn and Sham [1,2] proved that one can find the exact ground-state energy E and electron spin densities $n_\uparrow(\mathbf{r})$ and $n_\downarrow(\mathbf{r})$ for N electrons in an external spin-dependent potential $v^\sigma(\mathbf{r})$ from solution of the selfconsistent single-particle equations

$$\left[-\frac{1}{2}\nabla^2 + v^\sigma(\mathbf{r}) + \int d^3\mathbf{r}' \frac{n(\mathbf{r}')}{\mid \mathbf{r}-\mathbf{r}' \mid} + v^\sigma_{xc}(\mathbf{r})\right] \psi_{\alpha\sigma}(\mathbf{r}) = \varepsilon_{\alpha\sigma}\psi_{\alpha\sigma}(\mathbf{r}), \quad (1)$$

where

$$n_\sigma(\mathbf{r}) = \sum_\alpha \theta(\mu - \varepsilon_{\alpha\sigma}) \mid \psi_{\alpha\sigma}(\mathbf{r}) \mid^2, \quad (2)$$

$$E = \sum_{\alpha\sigma} \theta(\mu - \varepsilon_{\alpha\sigma}) < \psi_{\alpha\sigma} \mid -\frac{1}{2}\nabla^2 \mid \psi_{\alpha\sigma} > + \sum_\sigma \int d^3r\, v_\sigma(\mathbf{r}) n_\sigma(\mathbf{r})$$
$$+\frac{1}{2}\int d^3r \int d^3r' \frac{n(\mathbf{r})n(\mathbf{r}')}{\mid \mathbf{r}-\mathbf{r}' \mid} + E_{xc}[n_\uparrow, n_\downarrow], \quad (3)$$

and

$$v^\sigma_{xc}(\mathbf{r}) = \frac{\delta E_{xc}}{\delta n_\sigma(\mathbf{r})}. \quad (4)$$

Unless otherwise stated, we use atomic units in which $\hbar = m = e^2 = 1$. Of course $n(\mathbf{r}) = n_\uparrow(\mathbf{r}) + n_\downarrow(\mathbf{r})$. The chemical potential μ is to be adjusted so that

$$\int d^3r n(\mathbf{r}) = N. \quad (5)$$

Eqs. (1) - (5) are just the Hartree equations [3] (without self-interaction correction), corrected by the exchange-correlation energy $E_{xc}[n_\uparrow, n_\downarrow]$ and its functional derivative of Eq. (4). The theorem tells us that there is an exact spin-density functional for E_{xc}, but not what it is, so approximations are necessary in practice.

The theorem is rigorous, and applies as well to strongly-correlated systems as to any others. If we knew the exact $E_{xc}[n_\uparrow, n_\downarrow]$, we could predict the exact ground-state spin densities and total energy or ground-state energy differences (but not other properties such as the quasi-particle spectrum). Within the Born-Oppenheimer approximation for the nuclear motion, we could also predict the equilibrium positions and vibrational frequencies of the nuclei, and the response of the system to certain external perturbations such as static potentials or pressure.

Useful insights into the exchange-correlation energy are provided by assuming an adiabatic connection [4,5] between the interacting and non-interacting ground states of the same density. Given fixed spin densities $n_\uparrow(\mathbf{r})$ and $n_\downarrow(\mathbf{r})$, define a Hamiltonian depending upon a coupling constant λ which has those spin densities in its ground state:

$$\hat{H}_\lambda = \sum_{i=1}^{N} \left[-\frac{1}{2}\nabla_i^2 + \sum_\sigma v_\lambda^\sigma(\mathbf{r}_i)\delta_{\sigma,\hat{s}_z^i} \right] + \lambda \frac{1}{2} \sum_i \sum_{j \neq i} \frac{1}{\mid \mathbf{r}_i - \mathbf{r}_j \mid}. \tag{6}$$

The ground-state eigenvalue problem

$$\hat{H}_\lambda \Psi_\lambda = E_\lambda \Psi_\lambda \tag{7}$$

reduces for $\lambda = 0$ to the non-interacting or Kohn-Sham problem (Ψ_0 = Slater determinant of Kohn-Sham orbitals $\psi_{\alpha\sigma}(\mathbf{r})$,

$$E_0 = \sum_{\alpha\sigma} \varepsilon_{\alpha\sigma}\theta(\mu - \varepsilon_{\alpha\sigma})),$$

and for $\lambda = 1$ to the physical interacting system of interest (Ψ_1 = physical correlated wavefunction, $E_1 = E$ = physical ground-state energy). Now it can be proved that

$$E_{xc}[n_\uparrow, n_\downarrow] = \int_0^1 d\lambda\, W_\lambda[n_\uparrow, n_\downarrow], \tag{8}$$

where

$$W_\lambda\,[n_\uparrow, n_\downarrow] = < \Psi_\lambda \mid \frac{1}{2} \sum_i \sum_{j \neq i} \frac{1}{\mid \mathbf{r}_i - \mathbf{r}_j \mid} \mid \Psi_\lambda > - \frac{1}{2} \int d^3r \int d^3r' \, \frac{n(\mathbf{r})n(\mathbf{r}')}{\mid \mathbf{r} - \mathbf{r}' \mid}. \tag{9}$$

The spin density and the pair density are positive expectation values

$$n_\sigma(\mathbf{r}) = N \int d^3r_2 \cdots d^3r_N \sum_{\sigma_2 \cdots \sigma_N} \mid \Psi_\lambda(\mathbf{r}, \sigma, \mathbf{r}_2, \sigma_2, \cdots \mathbf{r}_N, \sigma_N) \mid^2, \tag{10}$$

$$\rho_2^\lambda(\mathbf{r},\mathbf{r}') = N(N-1)\int d^3r_3\cdots d^3r_N \sum_{\sigma_1\cdots\sigma_N} \mid \Psi_\lambda(\mathbf{r},\sigma_1,\mathbf{r}',\sigma_2,\cdots,\mathbf{r}_N,\sigma_N)\mid^2, \tag{11}$$

such that

$$W_\lambda[n_\uparrow,n_\downarrow] = \frac{1}{2}\int d^3r \int d^3r' \frac{\rho_2^\lambda(\mathbf{r},\mathbf{r}')}{\mid \mathbf{r}-\mathbf{r}'\mid} - \frac{1}{2}\int d^3r \int d^3r' \frac{n(\mathbf{r})n(\mathbf{r}')}{\mid \mathbf{r}-\mathbf{r}'\mid}. \tag{12}$$

From Eqs. (10) and (11), we see immediately that

$$\int d^3r' \rho_2^\lambda(\mathbf{r},\mathbf{r}') = (N-1)\, n(\mathbf{r}). \tag{13}$$

We can define $n_{xc}^\lambda(\mathbf{r},\mathbf{r}')$ as the correction to the Hartree approximation for the pair density $\rho_2^\lambda(\mathbf{r},\mathbf{r}')$:

$$\rho_2^\lambda(\mathbf{r},\mathbf{r}') = n(\mathbf{r})[n(\mathbf{r}') + n_{xc}^\lambda(\mathbf{r},\mathbf{r}')]. \tag{14}$$

Then clearly

$$E_{xc}[n_\uparrow,n_\downarrow] = \int_0^1 d\lambda \frac{1}{2}\int d^3r \int d^3r' \frac{n(\mathbf{r})n_{xc}^\lambda(\mathbf{r},\mathbf{r}')}{\mid \mathbf{r}-\mathbf{r}'\mid}, \tag{15}$$

where

$$\int d^3r' n_{xc}^\lambda(\mathbf{r},\mathbf{r}') = -1. \tag{16}$$

We interpret $n_{xc}^\lambda(\mathbf{r},\mathbf{r}')$ as the density at $\mathbf{r}'$ of the exchange-correlation hole around an electron at $\mathbf{r}$, for coupling strength λ.

3. EXCHANGE AND CORRELATION

The Hartree approximation ($\rho_2^{H,\lambda}(\mathbf{r},\mathbf{r}') = n(\mathbf{r})n(\mathbf{r}')$, $E_{xc}^H = 0$) violates the particle conservation sum rule (16). Exchange and correlation, the correction to the Hartree approximation, describes the "tendency of electrons not to bump into other electrons or into themselves" [6].

An electron travelling through the density $n(\mathbf{r})$ is much like a shopper moving through a mall [6]. Just as the shopper is surrounded by his or her "personal space," from which other shoppers are largely excluded, the electron is surrounded by its own exchange-correlation hole, from which other electrons are largely excluded.

The personal space of one shopper in an otherwise-empty mall is the entire mall, and the exchange-correlation hole of one electron in a one-electron system is clearly $n_{xc}^\lambda = n_x + n_c^\lambda$ where

$$n_x(\mathbf{r},\mathbf{r}') = -n(\mathbf{r}'), \quad n_c^\lambda(\mathbf{r},\mathbf{r}') = 0 \quad (N=1), \tag{17}$$

$$E_x[n_\uparrow, 0] = -\frac{1}{2}\int d^3r \int d^3r' \frac{n(\mathbf{r})n(\mathbf{r}')}{|\mathbf{r}-\mathbf{r}'|}, \quad E_c[n_\uparrow, 0] = 0 \quad (N=1). \tag{18}$$

Eqs. (17) and (18) remain true whether our electron is a fermion (as it really is), or a boson, or a classical particle: Exchange and correlation must provide a self-interaction correction to the Hartree approximation in any one-electron system. Of course, Hartree was aware of this, and introduced an orbital-dependent self-interaction correction to the Hartree approximation [3], but at the cost that his one-electron potential became orbital-dependent (and thus not a Kohn-Sham potential).

For real electrons, which are fermions, the exchange hole

$$n_x(\mathbf{r}, \mathbf{r}') = n_{xc}^{\lambda=0}(\mathbf{r}, \mathbf{r}') \tag{19}$$

includes not only the self-interaction correction but also the effect of the Pauli exclusion principle, which keeps two electrons with the same z-component of spin from coming together at the same point in space. The correlation hole

$$n_c^\lambda(\mathbf{r}, \mathbf{r}') = n_{xc}^\lambda(\mathbf{r}, \mathbf{r}') - n_x(\mathbf{r}, \mathbf{r}') \tag{20}$$

includes the effect of the Coulomb repulsion, which reduces the probability of any two electrons being close together.

For given ground-state spin densities $n_\uparrow(\mathbf{r})$ and $n_\downarrow(\mathbf{r})$, the weak-correlation limit is

$$\lambda \to 0. \tag{21}$$

For a given *shape* of the spin densities, the weak correlation limit is achieved for $\lambda = 1$ by uniform density scaling [7] to the high-density limit:

$$n_\sigma(\mathbf{r}) \to \gamma^3 n_\sigma(\gamma\mathbf{r}) \; and \; \gamma \to \infty. \tag{22}$$

In this limit, the exchange-only Kohn-Sham theory, which replaces $E_{xc}[n_\uparrow, n_\downarrow]$ by the exact $E_x[n_\uparrow, n_\downarrow]$ and thus resembles "Hartree-Fock theory with Kohn-Sham orbitals", becomes exact. If the ground-state of $\hat{H}_0$ is non-degenerate, i.e., if the non-interacting wavefunction is a single Slater determinant, then the pair density approaches its Hartree-Fock limit [8]

$$\rho_2^\lambda(\mathbf{r}, \mathbf{r}') \to \rho_2^{\lambda=0}(\mathbf{r}, \mathbf{r}') \quad (\lambda \to 0). \tag{23}$$

In this limit, the "on-top" ($\mathbf{r}' = \mathbf{r}$) pair density becomes

$$\rho_2^{\lambda=0}(\mathbf{r},\mathbf{r}) = n^2(\mathbf{r}) - n_\uparrow^2(\mathbf{r}) - n_\downarrow^2(\mathbf{r}). \tag{24}$$

The right-hand side of Eq. (24) varies from 0 for a fully-polarized density ($n_\uparrow = n$, $n_\downarrow = 0$) to $n^2/2$ for an unpolarized one ($n_\uparrow = n_\downarrow = n/2$).

We can also write

$$\hat{H}_\lambda = \hat{H}_0 + (\hat{H}_\lambda - \hat{H}_0), \tag{25}$$

where $(\hat{H}_\lambda - \hat{H}_0)$ is a weak perturbation in the weak correlation limit. If the ground state of $\hat{H}_0$ is non-degenerate, we expect to be able to evaluate the short-range (small $| \mathbf{r}' - \mathbf{r} |$) part of the pair density $\rho_2^\lambda(\mathbf{r},\mathbf{r}')$ by low-order perturbation theory [9]. In infinite metallic systems, such as the uniform electron gas, the long-range part of $\rho_2^\lambda(\mathbf{r},\mathbf{r}')$ is inaccessible to finite-order perturbation theory, even when $\lambda \to 0$, because of the long range of the Coulomb interaction[10].

The strong-correlation limit [11,12] is

$$\lambda \to \infty \tag{26}$$

(which is outside the range of the coupling constant integration of Eq. (15)), or for $\lambda = 1$ the low-density limit

$$n_\sigma(\mathbf{r}) \to \gamma^3 n_\sigma(\gamma\mathbf{r})\, and\, \gamma \to 0. \tag{27}$$

In this limit, even the short-range part of the pair density $\rho_2^\lambda(\mathbf{r},\mathbf{r}')$ is inaccessible to finite-order perturbation theory, but we know the limit for the on-top pair density:

$$\rho_2^\lambda(\mathbf{r},\mathbf{r}) \to 0 \quad (\lambda \to \infty). \tag{28}$$

There is another way to think about strong correlation which is worth mentioning. If Ω is any volume fragment of a system, one can prove [13,14] that

$$\int_\Omega d^3r \int_\Omega d^3r' \rho_2^\lambda(\mathbf{r},\mathbf{r}) =< \hat{N}_\Omega > [< \hat{N}_\Omega > -1] \\ +\{< \hat{N}_\Omega^2 >_\lambda - < \hat{N}_\Omega >^2\}, \tag{29}$$

where

$$\hat{N}_\Omega = \int_\Omega d^3r \hat{n}(\mathbf{r}) = \int_\Omega d^3r \sum_{i=1}^{N} \delta\mathbf{r} - \mathbf{r}_i) \tag{30}$$

is the electron number operator for volume fragment Ω. If Ω is the whole system, then the system is in an eigenstate of $\hat{N}_\Omega$ with eigenvalue N, and the right-hand side of Eq. (29) becomes N(N-1), in agreement with Eq (13).

If the volume fragment Ω is a one-electron region of space in the weak sense, or more precisely if

$$\int_\Omega d^3r n(\mathbf{r}) = 1, \tag{31}$$

then Eq. (29) becomes

$$\int_\Omega d^3r \int_\Omega d^3r' \rho_2^\lambda(\mathbf{r},\mathbf{r}') = < \hat{N}_\Omega^2 >_\lambda - < \hat{N}_\Omega >^2, \tag{32}$$

the mean square fluctuation of electron number in Ω. Strong correlation means a strong suppression of the short-range part of $\rho_2^\lambda(\mathbf{r},\mathbf{r}')$, and thus a strong suppression of electron number fluctuations in a one-electron region of space [14].

In summary, there is a competition between the kinetic energy and the Coulomb repulsion of a system of electrons. When the kinetic energy dominates, the system is weakly correlated and its wavefunction is close to that of a non-interacting electron system, which minimizes the kinetic energy for a given density. When the Coulomb repulsion dominates, the system is strongly correlated and its wavefunction is close to one which minimizes the Coulomb repulsion for a given density. This minimization is achieved by digging a deep exchange-correlation hole around the electron and so suppressing the electron number fluctuations in one-electron regions that are characteristic of the non-interacting wavefunction.

4. EXAMPLES OF STRONG CORRELATION

Having defined strong correlation at the end of section 2, we consider some examples in model or real systems.

A uniform electron density $n(\mathbf{r}) = n = 3/4\pi r_s^3$ (neutralized by a uniform position background) in one, two, or three dimensions can be uniformly scaled to the low density limit $r_s \to \infty$ (Eq. (27)), where it becomes strongly correlated. The electrons then avoid one another as much as they can while forming the uniform density, which turns out to be almost but not quite as much as they can avoid one another by forming a Wigner lattice. For a discussion of the strong suppression of electron number fluctuations in this limit, see Ref. [14]. If we want to predict the r_s at which the ground state density first becomes non-uniform (a charge density wave on its way to becoming a Wigner crystal), then we need an accurate theory of strongly-correlated electrons.

Hydrogen atoms on a lattice are weakly correlated in the limit of small lattice constant (monatomic metallic hydrogen), but strongly correlated in the limit of large lattice constant (Mott-Hubbard insulator). At small lattice constant, where we have a broad valence band of Kohn-Sham orbital energies, the system wavefunction is close to the non-interacting one which minimizes the kinetic energy for a given density, and there are significant fluctuations in the number of electrons within a Wigner-Seitz cell around a nucleus. At large lattice constant, where we have a narrow band of Kohn-Sham orbitals, the system wavefunction approaches the strongly-correlated limit in which one and only one electron is found in each Wigner-Seitz cell. In this limit, the system is nothing more than a collection of hydrogen atoms.

The site-localization of electrons, as described above for an expanded lattice of hydrogen atoms, can also occur in real systems at equilibrium lattice constants. It is found in some materials containing transition-metal or rare-earth ions with partially-filled, narrow *d* or *f* Kohn-Sham bands [15-23]. The *d* or *f* electrons may localize, and the fluctuations of their electron numbers on each site may be suppressed, in order to minimize the Coulomb repulsion between electrons. Standard density functionals like the local spin density (LSD) or generalized gradient approximations (GGA) can fail for the spin magnetic moments and lattice constants of these strongly-correlated solids. Methods such as "LSD+U" and "dynamical mean field theory" [23-25] have been developed to deal with these problems, and also with the quasi-particle spectrum.

As a specific example, consider Ce in the fcc structure [18]. The atomic configuration is $(4f)^1(5d)^1(6s)^2$. At room temperature and pressure, the electronic equilibrium state is γ-Ce, which has localized *f* electrons with disordered local magnetic moments. Under a pressure of 8 kbar, there is a transition to α-Ce, which has delocalized *f* eletrons and a volume per atom that is 15% less than γ-Ce. Calculations within the local spin density approximation find only α-Ce, while calculations with a self-interaction correction also find the strongly-correlated γ-Ce.

A key observation is that electrons in both the weakly-correlated and strongly-correlated limits can often be viewed as noninteracting in a sense. The weakly-interacting electrons are in delocalized (e.g., Bloch) orbitals, and the strongly-interacting electrons are in localized (e.g., Wannier) orbitals. Within Hartree-Fock theory, these descriptions are unitarily equivalent for a filled band, but not for a partly-filled band. Within Kohn-Sham theory, only the delocalized orbitals are solutions of the Kohn-Sham equation (1).

Thus, a necessary requirement for a density functional $E_{xc}[n_\uparrow, n_\downarrow]$ that can describe the strong-interaction limit is that it must satisfy at least approximately the self-interaction-free property of Eq. (17), in order to describe correctly the energy of a localized electron in a many-electron system.

5. LADDER OF DENSITY FUNCTIONAL APPROXIMATIONS

It is the existence of usefully accurate approximations to $E_{xc}[n_\uparrow, n_\downarrow]$ that has made density functional theory the cornerstone of most electronic structure calculations in condensed matter physics and quantum chemistry.

Density functional approximations can typically be cast in the form

$$E_{xc}^{approx}[n_\uparrow, n_\downarrow] = \int d^3r n(\mathbf{r}) \varepsilon_{xc}^{approx}(\mathbf{r}). \tag{33}$$

A natural hierarchy of approximations can be built around the use of increasingly complex ingredients to construct $\varepsilon_{xc}^{approx}(\mathbf{r})$. By using more complex ingredients, we can satisfy more known exact constraints on $E_{xc}[n_\uparrow, n_\downarrow]$ and achieve greater accuracy. In our own functionals, we strive to minimize or eliminate empirical parameters, but we are otherwise pragmatic: From the possible nonempirical functionals which satisfy the exact constraints appropriate to a given level of the hierarchy, we choose the one that works best in practice.

Below we will review the Jacob's Ladder hierarchy [26]. The first two rungs on the ladder (local spin density and generalized gradient approximations) can be regarded as controlled extrapolations away from the limit of slowly-varying densities. These functionals are essentially in final form, and are widely used today. The second two rungs of the ladder (meta- and hyper-generalized gradient approximations) can be regarded as controlled interpolations between the slowly-varying and one-electron limits. These functionals are still under development, and are not yet in wide use.

We will also review the self-interaction correction to the local spin density approximation, which is not properly a rung of the ladder but is relevant to our discussion of strong correlation.

5.1. LOCAL SPIN DENSITY (LSD) APPROXIMATION

The only ingredients of LSD [1,27] are the local spin densities $n_\uparrow(\mathbf{r})$ and $n_\downarrow(\mathbf{r})$:

$$E_{xc}^{LSD}[n_\uparrow, n_\downarrow] = \int d^3r\, n(\mathbf{r}) \varepsilon_{xc}^{unif}(n_\uparrow(\mathbf{r}), n_\downarrow(\mathbf{r})), \tag{34}$$

where $\varepsilon_{xc}^{unif}(n_\uparrow, n_\downarrow)$ is the known [28] exchange-correlation energy per electron of an electron gas with uniform spin densities $n_\uparrow \, and \, n_\downarrow$. By construction, LSD is exact for a uniform or slowly-varying density. But LSD also has "hidden assets" inherited from its underlying uniform gas which make it more widely useful. For example, the LSD exchange-correlation hole satisfies [4,5] the sum rule of Eq. (16). The LSD on-top ($\mathbf{r}' = \mathbf{r}$) hole is also exact at the exchange-only level (Eq. (24)) [8] and very accurate for correlation in many moderately-correlated systems [29].

5.2. SELF-INTERACTION CORRECTION TO LSD

LSD is not self-interaction-free, i.e., does not satisfy Eq. (18). Perdew and Zunger [30] proposed an orbital-by-orbital self-interaction correction (SIC)

$$\begin{aligned} E_{xc}^{SIC} &= E_{xc}^{LSD}[n_\uparrow, n_\downarrow] - \sum_{\alpha\sigma} \{ \frac{1}{2} \int d^3r \int d^3r' \frac{n_{\alpha\sigma}(\mathbf{r}) n_{\alpha\sigma}(\mathbf{r}')}{|\mathbf{r}' - \mathbf{r}|} \\ &\quad + E_{xc}^{LSD}[n_{\alpha\sigma}, 0] \}, \end{aligned} \tag{35}$$

where $n_{\alpha\sigma}(\mathbf{r}) = f_{\alpha\sigma} \mid \psi_{\alpha\sigma}(\mathbf{r}) \mid^2$ is the density of an orbital with occupation number $f_{\alpha\sigma}$. The corresponding exchange-correlation potential

$$\begin{aligned} v_{xc}^{SIC,\alpha\sigma}(\mathbf{r}) &= v_{xc}^{LSD,\sigma}(n_\uparrow(\mathbf{r}), n_\downarrow(\mathbf{r})) - \int d^3r' \frac{n_{\alpha\sigma}(\mathbf{r}')}{|\mathbf{r} - \mathbf{r}'|} \\ &\quad - v_{xc}^{LSD,\uparrow}(n_{\alpha\sigma}(\mathbf{r}), 0) \end{aligned} \tag{36}$$

is orbital-dependent, and thus not a proper Kohn-Sham potential. Many studies [16-21] by Svane, Temmerman and others have shown that SIC gives a much better description than LSD of strongly-correlated electrons, which energetically prefer to be in site-localized orbitals.

5.3. GENERALIZED GRADIENT APPROXIMATION (GGA)

The GGA form [31-35] is

$$E_{xc}^{GGA}[n_\uparrow, n_\downarrow] = \int d^3r \, n \varepsilon_{xc}^{GGA}(n_\uparrow, n_\downarrow, \nabla n_\uparrow, \nabla n_\downarrow). \tag{37}$$

The function ε_{xc}^{GGA} can be found by cutting off the spurious long-range parts of the second-order gradient expansion of the exchange-correlation hole, or by imposing exact constraints, or by fitting to experimental data. The "hidden assets" of LSD are largely retained in GGA. The gradient corrections significantly reduce the LSD overbinding of atoms in molecules and solids [36], but a self-interaction error persists.

5.4. META-GENERALIZED GRADIENT APPROXIMATION (MGGA)

The MGGA form [36-43] is

$$E_{xc}^{MGGA}[n_\uparrow, n_\downarrow] = \int d^3r n \varepsilon_{xc}^{MGGA}(n_\uparrow, n_\downarrow, \nabla n_\uparrow, \nabla n_\downarrow, \nabla^2 n_\uparrow, \nabla^2 n_\downarrow, \tau_\uparrow, \tau_\downarrow), \quad (38)$$

where the orbital kinetic energy density

$$\tau_\sigma(\mathbf{r}) = \frac{1}{2}\sum_\alpha \theta(\mu - \varepsilon_{\alpha\sigma}) \mid \nabla\psi_{\alpha\sigma}(\mathbf{r}) \mid^2 \quad (39)$$

requires only the *occupied* Kohn-Sham orbitals, fully-nonlocal functionals of the density which are readily available in any Kohn-Sham calculation.

Part of the appeal of $\tau_\sigma(\mathbf{r})$ is that it can be used to eliminate the self-correlation error[42]: We can make $E_c^{MGGA} = 0$ for any one-electron density, which can be recognized not by Eq. (31) but by the stronger conditions

$$\left|\frac{n_\uparrow(\mathbf{r}) - n_\downarrow(\mathbf{r})}{n_\uparrow(\mathbf{r}) + n_\downarrow(\mathbf{r})}\right| = 1 \quad (40)$$

and

$$\tau(\mathbf{r}) = \tau_\uparrow(\mathbf{r}) + \tau_\downarrow(\mathbf{r}) = \frac{1}{8}\frac{\mid \nabla n(\mathbf{r}) \mid^2}{n(\mathbf{r})}. \quad (41)$$

The meta-GGA of Perdew, Kurth, Zupan and Blaha [43] (PKZB) achieves freedom from self-correlation error while preserving the correct formal properties of the GGA and adding others. The self-correlation error of LSD and GGA can be especially troublesome in the strong-interaction limit. The integrand of Eq. (8) is [44]

$$W_\lambda[n_\uparrow, n_\downarrow] = \frac{d}{d\lambda}\{\lambda^2 E_{xc}[n_{\uparrow,\frac{1}{\lambda}}, n_{\downarrow,\frac{1}{\lambda}}]\}, \quad (42)$$

where

$$n_{\sigma,\gamma}(\mathbf{r}) = \gamma^3 n_\sigma(\gamma\mathbf{r}). \quad (43)$$

Table 1: The strong-interaction limit of $E_{xc}[n_\uparrow, n_\downarrow]$, W_∞ of Eq. (42), for the atoms H and He. LSD and GGA make nearly the same self-correlation error, which is eliminated in meta-GGA. (From Ref. [45].)

	W_∞ *(hartree)*			
Atom	*LSD*	*GGA*	*MGGA*	*Exact*
H	-0.421	-0.417	-0.308	-0.3125
He	-1.720	-1.689	-1.502	-1.500

Table 1 shows that, for the atoms H and He, W_∞ is much closer to exact values in MGGA than in LSD or GGA. The same conclusion appears to be true for multi-electron atomic densities as well [45].

Because the exact exchange energy of a one-electron system (Eq. (18)) is a fully nonlocal functional of the density, we cannot eliminate the self-exchange error within the meta-GGA form, but only at the higher hyper-GGA level.

5.5. HYPER-GENERALIZED GRADIENT APPROXIMATION (HGGA)

The HGGA form [26] is

$$E_{xc}^{HGGA}[n_\uparrow, n_\downarrow] = \int d^3r n(\mathbf{r})\{\varepsilon_x^{exact}(\mathbf{r})$$
$$+\varepsilon_c^{HGGA}(n_\uparrow, n_\downarrow, \nabla n_\uparrow, \nabla n_\downarrow, \tau_\uparrow, \tau_\downarrow, \varepsilon_x^{exact})\}, \qquad (44)$$

where $\varepsilon_x^{exact}(\mathbf{r})$ is the exact exchange energy density, i.e., the Fock integral of the occupied Kohn-Sham orbitals. Because exchange is treated in an exact and fully nonlocal way in Eq. (44), a compatible ε_c^{HGGA} must also be fully nonlocal, e.g., through a dependence upon $\varepsilon_x^{exact}(\mathbf{r})$.

A possible form [26] is

$$\begin{aligned}\varepsilon_c^{HGGA} &= \varepsilon_c^{MGGA} \\ &+ (\varepsilon_x^{exact} - \varepsilon_x^{MGGA})[\exp\{-\nu\frac{\varepsilon_c^{MGGA}}{\varepsilon_x^{MGGA}}\} - 1], \end{aligned}\qquad (45)$$

where ν is a positive number. Eq. (45) is presented, not as a practical functional but as an example of how additional exact constraints can be met by using $\varepsilon_x^{exact}(\mathbf{r})$: In the one-electron limit, where $\varepsilon_c^{MGGA} = 0$, we find $\varepsilon_c^{HGGA} = 0$, and in the high-density limit, where $\varepsilon_c^{MGGA}/\varepsilon_x^{MGGA} \to 0$, we find

$$\frac{E_{xc}^{HGGA}}{E_x^{exact}} \to 1 \quad (\gamma \to \infty). \tag{46}$$

So, a hyper-GGA can be fully self-interaction free, and also fully correct in the high-density or weakly-correlated limit.

6. DISCUSSIONS AND CONCLUSIONS

The exact Kohn-Sham density functional theory of section 2 is exact for the ground-state spin densities $n_\uparrow(\mathbf{r})$ and $n_\downarrow(\mathbf{r})$ and for the total energy E, no matter how weakly or strongly correlated the electrons are.

The four approximate functionals on Jacob's Ladder are also exact for uniform or slowly-varying electron densities, no matter how weakly or strongly correlated the electrons are. When we look instead at rapidly-varying densities, like the densities of the hydrogen and helium atoms in Table I, we can find that LSD and GGA fail in the strong-interaction limit due to self-correlation errors, while meta-GGA and hyper-GGA still work.

For other rapidly-varying densities, those of molecules, the LSD, GGA and meta-GGA functionals can also fail in the *weakly*-correlated limit; see the exchange-only atomization energies in Table III of Ref. [36]. For the valence electron densities of atoms, molecules, and solids, these functionals still tend to work for "moderate correlation" that is not too different from the correlation in the underlying uniform electron gas.

LSD and higher-level functionals can sometimes simulate strong correlation through symmetry breaking. Examples are the singlet ground states of the molecules Cr_2 (at equilibrium bond length) and H_2 (at a stretched bond length). Although the exact ground states have $n_\uparrow(\mathbf{r}) = n_\downarrow(\mathbf{r})$, the LSD and other functionals concentrate spin-up density on one nucleus and spin-down density on the other, creating an on-top pair density (Eq. (24)) and energy appropriate to a strongly-correlated system [46].

We know that LSD and GGA do not *reliably* describe strongly-correlated electrons with realistic densities. We also know that the self-interaction correction to LSD, or the "LSD+U" method, can more reliably describe such electrons. It is an interesting open question whether the self-correlation-free meta-GGA or the self-interaction-free hyper-GGA will be able to describe such systems reliably, without the introduction of an orbital-dependent potential that can produce site-localized orbitals in periodic solids.

ACKNOWLEDGEMENTS This article draws upon the work of many recent collaborators including Michael Seidl, Stefan Kurth, Karla Schmidt, Paul Ziesche, and Jianmin Tao. Work supported in part by the U.S. National Science Foundation under Grant No. DMR98-10620.

References

[1] W. Kohn and L.J. Sham, Phys. Rev. 140, A1133 (1965).

[2] W. Kohn, Rev. Mod. Phys. 71, 1253 (1999).

[3] D.R. Hartree, Proc. Camb. Phil. Soc. 24, 89, 111, 426 (1928).

[4] D.C. Langreth and J.P. Perdew, Solid State Commun. 17, 1425 (1975).

[5] O. Gunnarsson and B.I. Lundqvist, Phys. Rev. B13, 4274 (1976).

[6] H.T. Tran and J.P. Perdew, unpublished.

[7] M. Levy and J.P. Perdew, Phys. Rev. A32, 2010 (1985).

[8] T. Ziegler, A. Rauk, and E.J. Baerends, Theoret. Chim. Acta 43, 261 (1977).

[9] V.A. Rassolov, J.A. Pople, and M.A. Ratner, Phys. Rev. B62, 2232 (2000).

[10] D. Pines and P. Nozières, *The Theory of Quantum Liquids* (Benjamin, NY, 1966).

[11] M. Levy and J.P. Perdew, Phys. Rev. B48, 11638 (1993).

[12] M. Seidl, J.P. Perdew, and M. Levy, Phys. Rev. A59, 51 (1999).

[13] J.P. Perdew, in *Density Functional Methods in Physics*, edited by R.M. Dreizler and J. da Providencia (Plenum, NY, 1985).

[14] P. Ziesche, J. Tao, M. Seidl, and J.P. Perdew, Int. J. Quantum Chem. 77, 819 (2000).

[15] P. Dufek, P. Blaha, V. Sliwko, and K. Schwarz, Phys. Rev. B49, 10170 (1994).

[16] A. Svane and O. Gunnarsson, Phys. Rev. Lett. 65, 1148 (1990).

[17] A. Svane, Phys. Rev. Lett. 68, 1900 (1992).

[18] A. Svane, Phys. Rev. Lett. 72, 1248 (1994).

[19] Z. Szotek, W.M. Temmerman, and H. Winter, Phys. Rev. B47, 4029 (1993).

[20] P. Strange, A. Svane, W.M. Temmerman, Z. Szotek, and H. Winter, Nature 399, 756 (1999)

[21] W.M. Temmerman, H. Winter, Z. Szotek, and A. Svane, Phys. Rev. Lett. 86, 2435 (2001).

[22] J.-T. Hoeft, M. Kittel, M. Polcik, S. Bao, R.L. Toomes, J.-H. Kang, D.P. Woodruff, M. Pascal, and C.L.A. Lamont, Phys. Rev. Lett. 87, 86101 (2001).

[23] V.I. Anisimov, F. Aryasetiawan, and A.I. Lichtenstein, J. Phys.: Condens. Matter 9, 767 (1997).

[24] V.I. Anisimov, I.V. Solovyev, M.A. Korotin, M.T. Czyżyk, and G.A. Sawatzky, Phys. Rev. B48, 16929 (1993).

[25] A. Georges, G. Kotliar, W. Krauth, and M.J. Rozenberg, Rev. Mod. Phys. 68, 13 (1996).

[26] J.P. Perdew and K. Schmidt, in *Density Functional Theory and Its Application to Materials*, edited by V. E. Van Doren, K. Van Alsenoy, and P. Geerlings (American Institute of Physics, 2001).

[27] U. von Barth and L. Hedin, J. Phys. C: Solid State Phys. 5, 1629 (1972).

[28] J.P. Perdew and Y. Wang, Phys. Rev. B45, 13244 (1992).

[29] K. Burke, J.P. Perdew, and M. Ernzerhof, J. Chem. Phys. 109, 3760 (1998).

[30] J.P. Perdew and A. Zunger, Phys. Rev. B23, 5048 (1981).

[31] D.C. Langreth and M.J. Mehl, Phys. Rev. B28, 1809 (1983).

[32] J.P. Perdew and Y. Wang, Phys. Rev. B33, 8800 (1986).

[33] A.D. Becke, Phys. Rev. A38, 3098 (1988).

[34] J.P. Perdew, K. Burke, and M. Ernzerhof, Phys. Rev. Lett. 77, 3865 (1996).

[35] J.P. Perdew, K. Burke, and Y. Wang, Phys. Rev. B54, 16533 (1996).

[36] S. Kurth, J.P. Perdew, and P. Blaha, Int. J. Quantum Chem. 75, 889 (1999).

[37] R. Colle and O. Salvetti, Theoret. Chim. Acta 37, 329 (1975).

[38] J.P. Perdew, Phys. Rev. Lett. 55, 1665 (1985).

[39] A.D. Becke and M.R. Roussel, Phys. Rev. A 39, 3761 (1989).

[40] E.I. Proynov, S. Sirois, and D.R. Salahub, Int. J. Quantum Chem. 64, 427 (1998).

[41] T. Van Voorhis and G.E. Scuseria, J. Chem. Phys. 109, 400 (1998).

[42] A.D. Becke, J. Chem. Phys. 109, 2092 (1998).

[43] J.P. Perdew, S. Kurth, A. Zupan, and P. Blaha, Phys. Rev. Lett. 82, 2544 (1999).

[44] A. Goerling and M. Levy, Phys. Rev. B47, 13105 (1993), Phys. Rev. A52, 4493 (1995).

[45] M. Seidl, J.P. Perdew, and S. Kurth, Phys. Rev. A62, 12502 (2000).

[46] J.P. Perdew, A. Savin, and K. Burke, Phys. Rev. A51, 4531 (1995).

EXCHANGE-CORRELATION ENERGY AS A FUNCTION OF THE ORBITAL OCCUPANCIES

P. Pou, R. Oszwaldowski, R. Pérez, F. Flores and J. Ortega *

1. INTRODUCTION

Electronic structure methods comprise both first-principles total energy methods and many-body theory (MBT). These two approaches have taken different hamiltonians as their paradigms (the nearly-free electron hamiltonian in the case of the total energy methods and the Hubbard hamiltonian for MBT) and have developed as two separate fields. This is also reflected in the basis sets used in these calculations: typically, first-principles methods are implemented using a plane wave basis [1, 2], while MBT calculations are naturally performed using atomic-like orbitals. In an attempt to bring together these two different methodologies, a novel scheme for the description of electron exchange-correlation effects has been developed: the Orbital-Occupancy (OO) approach [3, 4]. Basically, the OO-approach is a DFT-like method, suitable for highly correlated systems, in which the role that the electron density $\rho(\bar{r})$ plays in standard DFT methods (LDA,GGA) is assumed by the set of the orbital occupation numbers, $\{n_{\mu\sigma}\}$:

$$\rho(\bar{r}) \Longrightarrow \{n_{\mu\sigma}\} \tag{1}$$

($n_{\mu\sigma} =< \hat{n}_{\mu\sigma} >$). Then, the total energy E is obtained as a *function* of the orbital occupancies:

$$E(\{n_{\mu\sigma}\}) = E^H(\{n_{\mu\sigma}\}) + E^{XC}(\{n_{\mu\sigma}\}). \tag{2}$$

The Hartree-like contribution E^H, as a function of the $\{n_{\mu\sigma}\}$, is easily obtained [5] and, as usual, the difficult part resides in determining the exchange-correlation term $E^{XC}(\{n_{\mu\sigma}\})$.

*Departamento de Física Teórica de la Materia Condensada, Universidad Autónoma de Madrid, E-28049 Madrid, Spain.

Electron Correlations and Materials Properties 2
Edited by Gonis *et al.*, Kluwer Academic/Plenum Publishers, 2002

In the OO-method a non-orthogonal basis set of optimized atomic-like orbitals $\psi_{\mu\sigma}$ is usually assumed, and the orbital occupation numbers $\{n_{\mu\sigma}\}$ are obtained using the orthonormal Löwdin orbitals, $\phi_{\mu\sigma}$:

$$\phi_{\mu\sigma} = \sum_{\nu} (S^{-\frac{1}{2}})_{\mu\nu} \psi_{\nu\sigma}; \tag{3}$$

$S_{\mu\nu}$ is the overlap matrix $< \psi_{\mu\sigma}|\psi_{\nu\sigma} >$. Once E^{XC} is known as a function of the $\{n_{\mu\sigma}\}$ a Kohn-Sham-like procedure can be introduced [3]. Thus, the $\{n_{\mu\sigma}\}$ can be determined selfconsistently via the corresponding one-electron effective hamiltonian $\hat{H}_{eff}$

$$\hat{H}_{eff}|\varphi_\alpha >= \epsilon_\alpha|\varphi_\alpha > \tag{4}$$

$$n_{\mu\sigma} = \sum_{\alpha\equiv occ.} | < \phi_{\mu\sigma}|\varphi_\alpha > |^2, \tag{5}$$

since $\hat{H}_{eff}$ contains Hartree and exchange-correlation contributions which, in turn, depend on the $\{n_{\mu\sigma}\}$. For example, the exchange-correlation potentials are

$$V_{\mu\sigma}^{XC} = \frac{\partial E^{XC}[\{n_{\nu\sigma}\}]}{\partial n_{\mu\sigma}}. \tag{6}$$

Notice that orbital-dependent potentials are obtained in this way.

The purpose of this paper is to present in a concise way the basic ideas behind the OO-approach; most of the many details con be found in reference [4]. In the next section we summarize how E^{XC} is obtained as a function of the $\{n_{\mu\sigma}\}$, giving special attention to the intra-atomic correlation energy. In section 3 some results of the application of the OO-approach are presented. First, we discuss the case of GaN. For this material it is essential to include explicitly the Ga–$3d$ states in the calculations, and the proper description of these states requires the inclusion of self–interaction corrections [6]. Finally, the case of Fe is used as an example to show how a self-energy approach can be alternatively applied, within the same approximations, when a MBT solution is required for the material under study.

2. EXCHANGE–CORRELATION ENERGY

In order to determine $E^{XC}(\{n_{i\sigma}\})$, we first analyze in this section the generalized Hubbard hamiltonian $\hat{H}_{GH}$: this hamiltonian is used to identify the main E^{XC}-terms. One of these terms, the intra-atomic correlation energy, $E^{C,I}$, is then discussed in some detail.

2.1 Generalized Hubbard Hamiltonian.

The generalized Hubbard hamiltonian

$$\begin{aligned}\hat{H}_{GH} &= \sum_{\nu\sigma} \epsilon_\nu \hat{n}_{\nu\sigma} + \sum_{(\nu,\mu),\sigma} T^{\sigma}_{\nu\mu}(\hat{c}^{\dagger}_{\nu\sigma}\hat{c}_{\mu\sigma} + \hat{c}^{\dagger}_{\mu\sigma}\hat{c}_{\nu\sigma}) \\ &+ \frac{1}{2}\sum_{\nu,\mu\neq\nu,\sigma} [J_{\nu\mu}\hat{n}_{\nu\sigma}\hat{n}_{\mu\bar{\sigma}} + (J_{\nu\mu} - J_{x,\nu\mu})\hat{n}_{\nu\sigma}\hat{n}_{\mu\sigma}] \\ &+ \sum_{\nu} U_\nu \hat{n}_{\nu\uparrow}\hat{n}_{\nu\downarrow}\end{aligned} \tag{7}$$

is an approximate hamiltonian that includes the most important contributions to the electron-electron interaction of the system. In particular, it contains the following electron-electron interactions:

$$U_\nu = \int \phi^2_\nu(\bar{r}) \frac{1}{\mid \bar{r} - \bar{r}' \mid} \phi^2_\nu(\bar{r}') d\bar{r} d\bar{r}'$$

$$J_{\nu\mu} = \int \phi^2_\nu(\bar{r}) \frac{1}{\mid \bar{r} - \bar{r}' \mid} \phi^2_\mu(\bar{r}') d\bar{r} d\bar{r}'$$

$$J_{x,\nu\mu} = \int \phi_\nu(\bar{r})\phi_\mu(\bar{r}) \frac{1}{\mid \bar{r} - \bar{r}' \mid} \phi_\nu(\bar{r}')\phi_\mu(\bar{r}') d\bar{r} d\bar{r}'$$

as well as other electron-electron interactions included in the hoppings $T^{\sigma}_{\nu\mu}$ (see section V.A in reference [4]). A detailed analysis of this hamiltonian [4] leads to the following main contributions:

$$E^{XC} = E^{SIC} + E^{X} + E^{C,I} + E^{J_x}. \tag{8}$$

The first term is a self–interaction correction

$$E^{SIC} = -\frac{1}{2}\sum_{\mu\sigma} U_\mu n_{\mu\sigma} n_{\mu\sigma} \tag{9}$$

due to the cancellation of U_μ and $J_{x,\mu\mu}$ (already taken into account in equation 7). The second contribution is an exchange term

$$E^{X} = -\frac{1}{2}\sum_{\mu\sigma} J_\mu n_{\mu\sigma}(1 - n_{\mu\sigma}). \tag{10}$$

J_μ represents the interaction between charge n_μ and its exchange hole $(1 - n_{\mu\sigma})$. In practice, if the interatomic correlation energy is considered [4], J_μ is equal to the nearest-neighbour Coulomb interaction J^{NN}_μ. The next term is the intra-atomic correlation energy $E^{C,I}$, which takes into account that a fraction of the exchange-correlation hole $(1 - n_{\mu\sigma})$ is actually in the same atom as the orbital $\mu\sigma$. This term is discussed in the next subsection. Finally, E^{J_x} is the mean-field contribution associated with the exchange integral J_x'(see equation 7):

$$E^{J_x} = -\frac{1}{2} \sum_{\nu,\mu\neq\nu,\sigma} J_{x,\nu\mu} n_{\nu\sigma} n_{\mu\sigma}. \tag{11}$$

Some of the exchange terms not considered in the generalized Hubbard hamiltonian can be important to obtain correctly the details of the total energy. These terms can be easily added perturbatively.

2.2 Intra-atomic Correlation Energy

The analysis of the generalized Hubbard hamiltonian suggests that the intra-atomic correlation energy $E^{C,I}$ can be described by the following Hubbard hamiltonian [4]:

$$\begin{aligned} \hat{H}_H &= \sum_{\nu\sigma} \tilde{E}^{\sigma}_{i\nu} \hat{n}_{\nu\sigma} + \sum_{(\nu,\mu),\sigma} T^{\sigma}_{\nu\mu} (\hat{c}^{\dagger}_{\nu\sigma} \hat{c}_{\mu\sigma} + \hat{c}^{\dagger}_{\mu\sigma} \hat{c}_{\nu\sigma}) \\ &+ \sum_{\nu} \tilde{U}_i \hat{n}_{\nu\uparrow} \hat{n}_{\nu\downarrow} \end{aligned} \tag{12}$$

with an effective $\tilde{U}_i$ per site i given by

$$\tilde{U}_i = < U_i > - J_i^{NN}. \tag{13}$$

J_i^{NN} is a nearest-neighbour Coulomb interaction (see above), and $< U_i >$ is an average U per atomic site.

$E^{C,I}$ is obtained using an interpolation between two limits for which the solution to hamiltonian 12 can be obtained analytically.

(a) Atomic limit : $\tilde{U}_i/T \to \infty$. In this limit, the correlation energy corresponding to hamiltonian 12 is given by

$$E^{C,I} = -\frac{1}{2} \sum_{\mu\sigma} \tilde{U}_i n_{\mu\sigma} (1 - n_{\mu\sigma}) + \frac{1}{2} \tilde{U}_i (\mathcal{N}_i - N_i)(1 + N_i - \mathcal{N}_i) \tag{14}$$

with $\mathcal{N}_i = \sum_{\mu\sigma} n_{\mu\sigma}$, and $N_i = \text{Int}\,[\mathcal{N}_i]$. The first term in equation 14 is associated with fluctuations of charge in each orbital, and the second with total fluctuations of charge in the atom.

(b) Second order perturbation theory limit: $\tilde{U}_i/T \to 0$. In this case, the correlation energy can be written as follows

$$E^{C,I} = -\frac{1}{2}\sum_{\mu\sigma\neq\nu\sigma'} \tilde{U}_i^2 \frac{n_{\mu\sigma}(1-n_{\mu\sigma})n_{\nu\sigma'}(1-n_{\nu\sigma'})}{W_{\mu\nu}};$$

$W_{\mu\nu}$ represents a mean excitation energy related with states $\mu\sigma$ and $\nu\sigma'$ (see section III.B in reference [4]).

Interpolation. These results suggest to use the following interpolation for the intra-atomic correlation energy:

$$E^{C,I} = -\frac{1}{2}\sum_{\mu\sigma} \tilde{U}_i n_{\mu\sigma}(1-n_{\mu\sigma})F(x) + ...(\mathcal{N}_i, N_i) \tag{15}$$

In this expression, the second contribution (not explicitly written) is related with the total fluctuation of charge in the atom, and is usually small. $F(x)$ can be interpreted as the fraction of the exchange-correlation hole located inside atom i. It is a function of x

$$x = \sum_{\nu\sigma'\neq\mu\sigma} (\tilde{U}_i/W_{\mu\nu})n_{\nu\sigma'}(1-n_{\nu\sigma'}). \tag{16}$$

$F(x)$ needs to fulfill the following conditions
$F(x) \to x, \qquad x \to 0$
$F(x) \to 1, \qquad x \to \infty$
in order to satisfy the above two limits. For intermediate values, $F(x)$ is obtained by fitting the results for simple models (see [4] for details).

2.3 Selfenergy for Hubbard Hamiltonians

We look for a MBT solution of the Hubbard Hamiltonian $\hat{H}_H$ by using Green-function techniques and introducing diagonal self-energies $\Sigma^{\sigma}_{i\alpha\alpha}(\omega)$. This implies neglecting off-diagonal contributions, as corresponds to the Dynamic Mean Field (DMF) approximation [7].

As in the case of the intra-atomic correlation energy, the calculation of the self-energy is based on an interpolative approach between the high and low correlation limits.

(a) Atomic limit : $\tilde{U}_i/T \to \infty$. In this limit, it is sufficient to consider fluctuations involving only one electron. The use of the equation of motion technique leads to the following three-pole expression for the Green's function:

$$G_{i,\mu\mu}^{(at)\sigma} \approx \frac{A_{N_i-1}^{\mu}}{\omega - \tilde{E}_{i\mu}^{\sigma} - \tilde{U}_i(N_i-1) + i0^+} + $$
$$+ \frac{A_{N_i}^{\mu}}{\omega - \tilde{E}_{i\mu}^{\sigma} - \tilde{U}_i N_i + i0^+} + \frac{A_{N_i+1}^{\mu}}{\omega - \tilde{E}_{i\mu}^{\sigma} - \tilde{U}_i(N_i+1) + i0^+} \tag{17}$$

where the weight factors, $A_{N_i-1}^{\sigma}$, $A_{N_i}^{\sigma}$ and $A_{N_i+1}^{\sigma}$ should satisfy sum rules involving the orbital occupancies $< \hat{n}_{i\beta\sigma'} >$ and the two-body correlation functions, $< \hat{n}_{i\mu\sigma}\hat{n}_{i\beta\sigma'} >$ [4].

From $G_{i\mu\mu}^{(at)\sigma}$, one can define an atomic self-energy:

$$\Sigma_{i\mu\mu}^{(at)\sigma} = \omega - \tilde{E}_{i\mu}^{\sigma H} - [G^{(at)\sigma}]_{i\mu\mu}^{-1} \tag{18}$$

where $\tilde{E}_{i\mu}^{\sigma H}$ is the Hartree level.

(b) Second order perturbation theory limit: $\tilde{U}_i/T \to 0$. Conventional perturbation theory yields the second-order selfenergy $\Sigma_{i,\mu\mu}^{\sigma(2)}(\omega)$:

$$\Sigma_{i\mu\mu}^{\sigma(2)} = \tilde{U}_i^2 \sum_{\beta\sigma' \neq \mu\sigma} \int_{-\infty}^{\infty} d\epsilon_1 d\epsilon_2 d\epsilon_3 \frac{\rho_{i\mu\sigma}(\epsilon_1)\rho_{i\beta\sigma'}(\epsilon_2)\rho_{i\beta\sigma'}(\epsilon_3)}{\omega + \epsilon_2 - \epsilon_1 - \epsilon_3 + 0^+}$$
$$\times [f_1 f_3(1-f_2) + (1-f_1)(1-f_3)f_2] \tag{19}$$

where $f_i = f(\epsilon_i)$ denotes the Fermi distribution function, and $\rho_{i\beta\sigma'}(\omega)$ are effective densities of states obtained using effective levels, $E_{i\mu}^{\sigma,eff}$, instead of $E_{i\mu}^{\sigma,H} + \Sigma_{i\mu\mu}^{\sigma}$. The effective levels, $E_{i\mu}^{\sigma,eff}$, in the i-site are introduced to fulfill charge consistency conditions: the orbital occupancies $n_{i\mu\sigma}$ has to be the same in the exact problem and in the effective one-electron solution.

Interpolation and Selfconsistency conditions. An interpolative self-energy can be defined replacing $(\omega - \tilde{E}_{i\mu}^{\sigma,eff})$ in the atomic selfenergy by $\tilde{U}_i^2 \sum_{\beta\sigma' \neq \mu\sigma} n_{i\beta\sigma'}(1 - n_{i\beta\sigma'})/\Sigma_{i\mu\mu}^{\sigma(2)}(\omega)$.

The final step in order to determine $\Sigma_{i\mu\mu}^{\sigma}(\omega)$ is to calculate $n_{i\beta\sigma}$, the correlation functions $< \hat{n}_{i\mu\sigma}\hat{n}_{i\beta\sigma'} >$ and the effective levels, $E_{i\mu}^{\sigma,eff}$, selfconsistently. Notice that, again, $E_{i\mu}^{\sigma,eff}$ should be determined to give the same local charge, $n_{i\mu\sigma}$, as the exact problem. In the case of the interpolative self-energy this leads to:

$$E_{i\mu}^{\sigma,eff} = \tilde{E}_{i\mu}^{\sigma,H} + \frac{\partial E^{C,I}[\{n_{i\mu\sigma}\}]}{\partial n_{i\mu\sigma}}, \tag{20}$$

This shows that the effective hamiltonian we have to use to calculate the effective density of states, $\rho_{i\mu\sigma}(\omega)$, is nothing else that the OO-solution of the Hubbard hamiltonian $\hat{H}_H$.

3. RESULTS

3.1 Self-interaction corrections in the $Ga - 3d$ bands of GaN

LDA or GGA plane-wave calculations for *GaN* show the need to include the *core Ga*-3*d* states explicitly in the calculations in order to obtain the lattice parameter correctly. When this is done, the *Ga*-3*d* bands appear as overlapping the *N*-2*s* bands, in contradiction with the experimental evidence which shows the *Ga*-3*d* states well below the *N*-2*s* bands. This discrepancy is due to the localized character of the *Ga*-3*d* states. The imperfect cancellation of Hartree contributions by the XC terms in the standard LDA or GGA functionals pushes up the electronic levels, the error being larger for the more localised states.

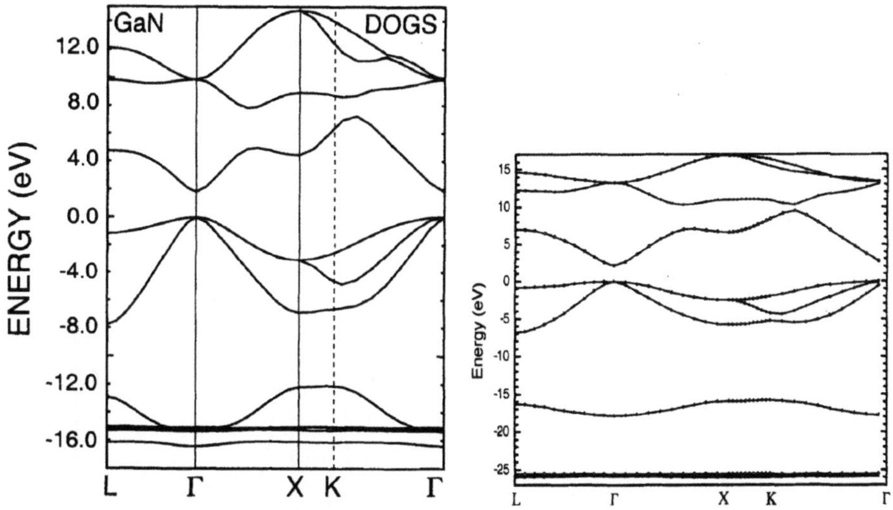

Figure 1. Band structure for GaN calculated with an LDA functional (left) and the OO-approach (right). Selfinteraction corrections, properly included in the OO-approach, provide a good description of the $Ga - 3d$ bands.

Our method provides a good description of the structural properties of GaN (comparable to that of the standard functionals with the same basis) and corrects the problem in the description of the localised states. In the bandstructure shown in figure 1 the *Ga*-3*d* states appear several eV below the $N - 2s$ bands, in good agreement with the experiment.

Notice that in the OO-method a self-interaction correction term appears naturally (see equation 9).

3.2 Quasiparticle spectrum of Iron

Finally, we present an example of the many-body solution that can be also obtained following the ideas discussed in section 2. Experiments for the excitation spectra of $3d$ metals show significant qualitative differences with the results of DFT calculations.

We have calculated the quasiparticle spectrum of Fe using the interpolative self–energy, $\Sigma^{\sigma}_{\mu\mu}(\omega)$ discussed in section 2.3.

The local orbital basis have been taken from a LDA calculation [8]. The bare interactions calculated for the $3d$ orbitals give 0.83 eV for the intraatomic exchange U^x_d and 22.2 eV for the coulomb interaction U_d. Once relaxation and screening effects are introduced the coulomb interaction reduces to $\tilde{U}_d = 4.0$ eV. The effective interaction one has to consider for the MBT solution is the average between $\tilde{U}_d$ and $\tilde{U}_d - U^x_d$, $\langle \tilde{U}_d \rangle = 3.6$ eV.

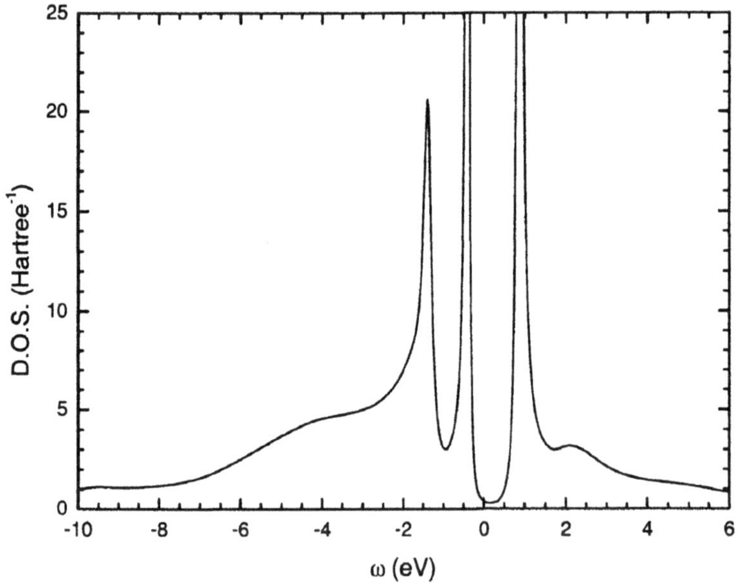

Figure 2. Quasiparticle spectrum for Fe at $\bar{k} = P$.

Figure 2 shows the quasiparticle spectrum for Fe at $\bar{k} = P$. It compares very well with the three peaks observed in the experiment, with a remarkable quantitative agreement for the two occupied peaks at -1.5 eV and 0.5 eV. There is a small difference for the peak observed at 1.6 eV with inverse photoemission, that appears at 0.9 eV in our calculations.

References

[1] J. Ihm, A. Zunger and M.L. Cohen, *J. Phys. C* **12**, 4409 (1979).

[2] M.C. Payne et al. , *Rev. Mod. Phys.* **64**, 1045 (1992).

[3] F.J. García-Vidal et al. , *Phys. Rev. B* **50**, 10537 (1994).

[4] P. Pou et al., *Phys. Rev. B* **62**, 4309 (2000).

[5] A.A. Demkov, J. Ortega, O.F. Sankey and M.P. Grumbach *Phys. Rev. B* **52**, 1618 (1995).

[6] D. Vogel, P. Krüger and J. Pollman, *Phys. Rev. B* **55**, 12836 (1996).

[7] A. Georges, G. Kotliar, W. Krauth, and M.J. Rozenberg, *Rev. Mod. Phys.* **68**, 13 (1996).

[8] J. P. Lewis *et al*, *Phys. Rev. B* (in press).

BSSE-CORRECTED PERTURBATION THEORIES OF INTERMOLECULAR INTERACTIONS

Ágnes Vibók
Institute of Theoretical Physics, University of Debrecen
H-4010 Debrecen, P.O. Box 5, Hungary
vibok@cseles.atomki.hu

Gábor J. Halász
Institute of Mathematics and Informatics, University of Debrecen
H-4010 Debrecen, P.O. Box 12, Hungary
halasz@math.klte.hu

István Mayer
Institute of Chemistry, Chemical Research Center, Hungarian Academy of Sciences
H-1525 Budapest, P.O.Box 17, Hungary
mayer@chemres.hu

1. Introduction

Intermolecular interactions play very important role in chemistry and physics. Several specific quantum chemical methods have been developed for calculating them. The problem of basis set superposition error (BSSE) occurring in these calculations has also been recognized for many years. It is due to the use of finite basis sets: neither the energy of the whole interacting complex ("supermolecule") nor the energies of the indioovidual constituting molecules ("monomers") are exact. Calculating the interaction energy as the difference of these quantities, one observes a systematic imbalance, usually leading to too deep minima on the potential surfaces. This is connected with a better description – in the energetic sense – of the individual constituting molecules within the complex, when the extended basis of the whole supermolecule becomes available, than that in the free monomer case. Different approaches have been proposed in the literature to account for this BSSE effect both at

Electron Correlations and Materials Properties 2
Edited by Gonis *et al.*, Kluwer Academic/Plenum Publishers, 2002

the SCF and at the correlated levels of theory. These methods can approximately be divided into two groups. The first, most commonly used approach is the Boys-Bernardi counterpoise (CP) correction scheme[1] in which the supermolecule energy is used without any correction, while the monomer energies are recomputed by using the full dimer basis set for every geometrical arrangement of the complex, so they become distance dependent. We shall discuss that this treatment involves a tacit assumption about the *additivity* of the BSSE effects. In the other approach, one assumes that the free monomer energies are correct *for the given basis set*, and the BSSE has to be eliminated in some manner from the supermolecule calculation. Either one omits some terms from the supermolecule energy expression as being due to BSSE[2, 3] or one tries to get rid of it when determining the wavefunction, as used in the so called "Chemical Hamiltonian Approach" (CHA).

In the CHA scheme (first introduced by Mayer[4]) one identifies and omits those terms of the Hamiltonian which are responsible for the BSSE. Using second-quantized formalism, the CHA lets us to present the Born-Oppenheimer Hamiltonian as a sum of intramolecular interactions, true physical interactions between the monomers and the "non-physical" terms causing BSSE. In this manner an efficient *a priori* BSSE correction method can be constructed. As BSSE is not a physical quantity, the "remaining" CHA-Hamiltonian is not Hermitian*. Using this new BSSE-free Hamiltonian and applying the method of momenta, appropriate SCF-type equations have been derived to calculate BSSE-free single determinant wavefunctions[5]. (Slightly different CHA–SCF equations giving practically the same numerical results have been obtained by applying an analogous treatment directly to the Fockian[6, 7].) Both the numerical calculations and the analytical considerations indicated, that the energy should be calculated as a standard expectation value of the full conventional Born-Oppenheimer Hamiltonian with the BSSE-free CHA wavefunction, so only the wavefunction but not the energy should be calculated by using the CHA Hamiltonian, making the treatment somewhat more involved[7]–[9]. (This scheme of energy calculation is usually indicated by the expression "CHA with conventional energy", CHA/CE.)

Several comparative studies have been performed at the SCF level[6, 10] and showed that for not too small basis sets the CHA and CP results are usually very close to each other. For medium-size basis sets some

*It should be emphasized that the BSSE is clearly an artificial "mathematical effect" which is solely due to the incomplete description of the individual monomers, so there is no reason to assume that it could correspond to any Hermitian operator.

"overcorrecting" behavior of the CP scheme can usually be observed, leading to curves with slightly too shallow minima which are also somewhat shifted to larger intermolecular separations. It was also found that the difference between the *a posteriori* CP and the *a priori* CHA curves diminishes much faster when the basis set increases than the BSSE disappears from the uncorrected SCF results.

As it is known the electron correlation plays very important role for the intermolecular interactions in the hydrogen bonded and van der Waals systems[11]–[13], and its proper treatment is essential. For large enough basis sets, the BSSE effects at the Hartree-Fock level are much smaller than those at the correlated level. In practice, one can closely approach the basis set limit of the SCF interaction energy by using basis sets for which the BSSE content of the correlation energy is still very large and is comparable with the actual value of the intermolecular interaction energy. (At the same time, the use of much larger basis sets might be affordable for SCF calculations but not for the correlated ones.) So, it is very important to deal properly with the BSSE problem at the correlated level and different approaches have been developed to account for it[11]–[13]. Several attempts have also been made to find an appropriate simultaneous treatment of electron correlation and the BSSE problem by using the CHA philosophy[14]–[22]. (These included the development of the CHA–DFT[19]–[21] theory, too.) The pivot four-electron full CI calculations[16] showed that the CHA scheme is working well at the electron correlated level of theory, too. In practice one needs a method based, e.g., on perturbation theory (PT) which is applicable also for larger systems for which full CI is not feasible. Our attempts to develop such a theory remained unsuccessful until it was found that the non-Hermitian CHA Hamiltonian must not be partitioned as a Hermitian zero-order Hamiltonian and a non-Hermitian perturbation: the separation of BSSE causes non-Hermiticity and this is an effect already manifesting at the single determinant level, so it must be taken into account in the unperturbed Hamiltonian. Another difficulty was connected with the CHA/CE framework in which the wavefunction is determined by using the (non-Hermitian) BSSE-free CHA Hamiltonian, while the energy is calculated as the expectation value of the Hermitian (conventional) Hamiltonian. It was not quite trivial to determine how one should calculate the energy *in a given order* of PT. This problem was solved by introducing the proper generalized Hylleraas functional[23] (see Section 4 below).

Based on these considerations, two different types of second-order perturbation theory (CHA–PT2, CHA–MP2) have been developed[17,18,22,24]. In the first case one starts from the orbitals and orbital energies obtained by solving the free monomer SCF problems, defines the unperturbed Hamiltonian as the sum of the monomer Fock operators by taking properly into account the intermolecular overlap, calculates the first-order wavefunction by using the BSSE-free CHA Hamiltonian and then determines the second-order energy contribution by using the generalized Hylleraas functional. (We shall note that in this case intermolecular overlap alone causes the unperturbed Hamiltonian to be non-Hermitian.) In the second approach (CHA–MP2) the unperturbed problem is defined by the solutions of the BSSE-free CHA supermolecule SCF equations and the MP2 formalism is generalized in the CHA framework. These two perturbation theories differ from each other only by the definition of the unperturbed problem.

In this work our aim is to give a brief summary of the two conceptually different BSSE correction schemes – the *a posteriori* Boys-Bernardi method and the *a priori* Chemical Hamiltonian Approach – and then to describe the presently available CHA-based perturbational electron correlation methods (CHA–PT2 and CHA–MP2). Sample calculations for hydrogen bonded systems will be presented in order to assess the possibilities of the *a priori* BSSE-free CHA formalism at the electron correlation level, by comparing the results obtained by the two versions of the CHA-based second-order PT with those of the *a posteriori* BSSE-corrected MP2–CP method.

2. The Boys-Bernardi counterpoise (CP) correction method

The traditional definition of the interaction energy $\Delta E_{AB}^{unc.}$ between the "monomers" A and B is the difference of the supermolecule energy and of the sum of the free monomer energies, each calculated in its own basis set:

$$\Delta E_{AB}^{unc.} = E_{AB}(AB) - E_A(A) - E_B(B), \tag{1}$$

where $E_{AB}(AB)$, $E_A(A)$ and $E_B(B)$ denote the total energy of the AB "supermolecule" and of the monomers A and B, respectively. The notations in the parentheses indicate that basis sets corresponding to the (sub)system A, B and AB, respectively, were used.

The above "uncorrected" definition would be the only correct one if we were able to perform the calculations of the free monomers and of the supermolecule exactly or almost exactly; this would require the use of (nearly) complete basis sets, which is usually far to be feasible. Of course, the above formula would give a correct result also if the error of the energy would be *constant*, that is, if the error in the energy of the supermolecule would equal the sum of the errors in the free monomer energies. The success of practically all quantum chemical calculations relies on such error compensations. The problem is extremely acute if intermolecular interactions are considered, as one is looking for rather small energy differences.

As noted in the Introduction, the use of incomplete monomer basis sets has the consequence that the individual monomers can utilize to some extent the basis orbitals of the partner molecule for improving (lowering) their *internal* energies within the complex (supermolecule). This means that the total energy of the complex is also somewhat lowered by an "effect" which does not correspond to any physical interaction, and one obtains slightly overestimated interaction energies, *i.e.*, a BSSE. Thus BSSE measures the extent to which the error of the internal monomer energies varies during the complex formation. (BSSE should be distinguished from those basis set errors, which are due to the insufficiency of the given basis set to account for some physical interactions.)

The counterpoise correction scheme represents an effective method to get a simple and usually rather good estimate for the *change* of the error in monomer energies taking place during the complex formation. As the reason of BSSE is that monomer A uses also the orbitals of monomer B within the complex (and *vice versa*), we can get an approximation of how much the presence of the orbitals of B can lower the energy of A by performing a calculation for the latter in the full supermolecule basis, i.e. by adding the orbitals – but not the nuclei and electrons – of monomer B (so called "ghost orbitals calculation"). Denoting by $E_A(AB)$ the energy of monomer A in the whole basis of the supermolecule AB, one gets the energy difference

$$\delta E_A(AB) = E_A(AB) - E_A(A). \tag{2}$$

A similar calculation should be performed for monomer B, giving

$$\delta E_B(AB) = E_B(AB) - E_B(B). \tag{3}$$

Assuming that the energy lowering corresponding to the individual monomers *within the supermolecule* is approximately the same as in these "ghost orbitals" calculations, we can correct the total energy $E_{AB}(AB)$ by the sum of these energy lowerings. By this ***additivity assumption*** one obtains the corrected total energy as

$$\begin{aligned} E_{AB}^{CP} &= E_{AB}(AB) - \delta E_A(AB) - \delta E_B(AB) \qquad (4) \\ &= E_{AB}(AB) - E_A(AB) + E_A(A) - E_B(AB) + E_B(B). \end{aligned}$$

Replacing $\Delta E_{AB}^{unc.}$ in formula (1) by E_{AB}^{CP}, one arrives at the CP corrected interaction energy

$$\Delta E_{AB}^{CP} = E_{AB}(AB) - E_A(AB) - A_B(AB). \tag{5}$$

Equ. (5) differs from (1), because here the monomer energies $E_A(AB)$ and $E_B(AB)$ are calculated in the same basis set as that used for the supermolecule AB. In agreement of this, the formula (5) is usually interpreted in terms that the description of the monomers is adjusted to the dimer problem so the energies (and other properties) of the "free" monomers become distance-dependent. This is usually justified by stating that this interaction energy is the "correct one", because "all quantities are computed by using the same basis set". Our derivation and the usual interpretation represent two different looks on *the same* result, so one might consider their difference immaterial; there are, however, more complex cases for which our treatment works but the standard interpretation seems inapplicable[14].

The simplest case in which the standard treatment of CP corrections meets a difficulty is encountered if the internal geometries of the monomers are allowed to change during the dimer formation (so called "monomer relaxation")[25]. To take into account this effect, a more complicated scheme is needed for calculating the appropriate BSSE-corrected interaction energy:

$$\begin{aligned} \Delta E_{AB}^{CP,rel.} &= E_{AB}^{rel.}(AB) - E_{A}^{rel.}(AB) - E_{B}^{rel.}(AB) \\ &+ [E_{A}^{rel.}(A) - E_A(A)] + [E_{B}^{rel.}(B) - E_B(B)], \end{aligned} \tag{6}$$

where *rel.* indicates the relaxed internal monomer energies (corresponding to the actual monomer geometries in the complex) and the differences $[E_{A}^{rel.}(A) - E_A(A)]$ and $[E_{B}^{rel.}(B) - E_B(B)]$ are the relaxation energies for monomers A and B, respectively. Mayer and Surján showed[25] that the relaxation energy cannot be consistently calculated by using the supermolecule basis, only the free monomer one. As equation (6) shows, in order to calculate the CP corrected interaction energy of relaxed monomers, one needs to perform five different calculations – in addition to the free monomer energy calculations performed in the monomer basis set – for obtaining each point on the interaction energy surface†.

The CP method performs rather well in most cases, and it has the advantage that it can be applied at any level of theory, usually without the need in any significant programming work.

†The terms of (6) can be regroupped as

$$\Delta E_{AB}^{CP,rel.} = E_{AB}^{rel.}(A) - E_A(A) - E_B(B) - [E_{A}^{rel.}(AB) - E_{A}^{rel.}(A)] - [E_{B}^{rel.}(AB) - E_{B}^{rel.}(B)]$$

in which the uncorrected interaction energy is corrected with the actual BSSE content of the monomer energies in their relaxed geometries.

3. The Chemical Hamiltonian Approach (CHA)

As it was mentioned, the aim is the alternative CHA method as applied to the problem of intermolecular interactions consists in *a priori* exclusion of the BSSE by modifying the supermolecule calculations. That means that the supermolecule calculation is *forced* to keep consistency with the treatment of the individual monomers, performed in their own basis set. For that reason, one needs to *identify* and *exclude* those terms which are responsible for the BSSE.

Let us start from the one-electron part of the intramolecular Hamiltonian of monomer A. It contains the sum of the kinetic energy operator and the potential energy of the interaction with the nuclei in the same monomer:

$$\hat{h}_A = \hat{T} + \hat{U}_A = -\frac{1}{2}\Delta + \hat{U}_A. \tag{7}$$

If $\hat{h}_A$ acts on some orbital φ_i^A belonging to the monomer A (this is marked by the superscript A), then, in general, the resulting function $\hat{h}_A \varphi_i^A$ will have components both in the subspace spanned by the basis orbitals of monomer A and in the orthogonal complement to that subspace. Using the Dirac's "bra" and "ket" formalism, this can be expressed in the form of an identity:

$$\hat{h}^A|\varphi_i^A> \equiv \hat{P}^A\hat{h}^A|\varphi_i^A> + (1-\hat{P}^A)\hat{h}^A|\varphi_i^A>, \tag{8}$$

where $\hat{P}^A$ is the projector on the subspace of the molecular basis of monomer A; it can be written in the general case when the basis functions on A are not orthonormalized ($S_{\mu\nu} =< \varphi_\mu \mid \varphi_\nu >=< \mu \mid \nu > \neq \delta_{\mu\nu}$) as

$$\hat{P}^A = \sum_{\mu,\nu \in A} |\mu > S^{-1}_{(A)\mu\nu} < \nu|. \tag{9}$$

Here and further on, $S^{-1}_{(A)\mu\nu}$ and $S^{-1}_{\mu\nu}$ are elements of the inverse overlap matrices for molecule A and the whole supermolecule, respectively, and the notations like $\mu \in A$ indicate that the summation is restricted to the orbitals of monomer A. Now, it is easy to understand that only the first term on the right-hand-side of equation (8) is actually included in the calculations of the free monomer, because it is within the space spanned by the monomer basis. The second term gives the projection on the orthogonal complement of the molecular basis; it decreases if the molecular basis increases and vanishes in the limit of the infinite basis. However, for finite basis sets, this term enters the supermolecule calculations and is responsible for BSSE. (The same term causes the energy lowering in the "ghost orbitals" calculations, too.) Accordingly, for keeping consistency with the free monomer calculations and avoiding the appearance of BSSE in incomplete basis sets, one has to drop all such terms which are connected to the orthogonal complement of the monomer basis. This means one has to use the following substitutions:

$$\hat{h}^A|\varphi_i^A> \Rightarrow \hat{P}^A\hat{h}^A|\varphi_i^A> . \qquad (10)$$

Similar considerations apply in the two-electron case, too, leading to the substitutions:

$$\frac{1}{r_{12}}\varphi_i^A(1)\varphi_j^A(2) \Rightarrow \hat{P}^A(1)\hat{P}^A(2)\frac{1}{r_{12}}\varphi_i^A(1)\varphi_j^A(2), \qquad (11)$$

where the indexes of the electrons are indicated in parentheses both for the projection operators and for the functions. The substitutions of (10) and (11) lead to the respective substitutions for the one and two-electron integrals:

$$< \chi_\nu|\hat{h}^A|\chi_\mu > \Rightarrow \sum_{\lambda,\rho\in A} S_{\nu\lambda}S^{-1}_{(A)\lambda\rho} < \chi_\rho|\hat{h}^A|\chi_\mu > \quad (\mu \in A) \qquad (12)$$

$$[\gamma\vartheta|\mu\nu] \Rightarrow \sum_{\kappa,\lambda,\rho,\tau\in A} S_{\gamma\rho}S_{\vartheta\lambda}S^{-1}_{(A)\kappa\rho}S^{-1}_{(A)\lambda\tau}[\kappa\tau|\mu\nu] \quad (\mu,\nu \in A). \qquad (13)$$

We use the [12—12] convention for the two-electron integrals, i.e.,

$$[\mu\rho|\nu\tau] = \int\int \chi_\mu^\star(1)\chi_\rho^\star(2)\frac{1}{r_{12}}\chi_\nu(1)\chi_\tau(2)dv_1dv_2. \qquad (14)$$

Introducing (12) and (13) in the one and two-electron integrals which occur in the second quantized form of the usual Born-Oppenheimer Hamiltonian, we arrive to the BSSE-free Hamiltonian $\hat{H}_{CHA}$ of the so-called *Chemical Hamiltonian Approach* (CHA). Two explicit forms of $\hat{H}_{CHA}$ will be used in the next sections.

The expressions (12) and (13) represent the essence of the CHA scheme in solving the BSSE problem. It can also be seen by inspecting the formulae that these substitutions leave unchanged all the intramonomer integrals, only the integrals which contain effects of intermolecular type are modified.

Based on this CHA scheme, two different perturbation theories have been developed (we shall denote them CHA-PT2 and CHA-MP2) for taking into account the electron correlation effects in the intermolecular interactions.

4. The CHA based perturbation theories

Mayer showed[23] that the proper second-order energy can be obtained as follows. At first one has to calculate the first-order CHA wavefunction χ by using the non-Hermitian CHA Hamiltonian partitioned as

$\hat{H}_{CHA} = \hat{H}^0 + \hat{V}_{CHA}$. The Hermitian *conventional* Hamiltonian should also be presented as a sum of the same non-Hermitian unperturbed Hamiltonian $\hat{H}^0$ as used in determining the first-order CHA wavefunction and of some (also non-Hermitian) perturbation $\hat{V}$. Then the second-order energy contribution is given by the generalized Hylleraas functional J_2.

As noted in the Introduction, two different second-order BSSE-free perturbational schemes have been developed[17, 18, 22, 24]; they differ in the selection of the zero-order (unperturbed) problem. Both methods start from the same second-order BSSE-free intermolecular energy formula:

$$E^{(2)} = \frac{< \Psi_0|\hat{H}|\Psi_0 >}{< \Psi_0|\Psi_0 >} + J_2 \tag{15}$$

although the operators and wavefunctions entering it are not the same. The first term in the right-hand-side of (15) is the expectation value of the total Hamiltonian over the unperturbed wavefunction, thus represents the sum of zero and first-order energy terms of the PT. The wavefunction used is the antisymmetrized product of the free monomer SCF wavefunctions in one perturbational scheme and the BSSE-free CHA-SCF wavefunction of the supermolecule in the another one. Accordingly, the first term of (15) represents in these two cases the so called "Heitler-London" energy and the BSSE-free CHA-SCF supermolecule energy, respectively. The second term in (15) is the Hylleraas-functional, giving the second-order BSSE-free energy contribution. In both theories the unperturbed Hamiltonian is non-Hermitian; the expression for J_2 functional for such a case has been given in paper[23]:

$$\begin{aligned} J_2 &= \frac{1}{< \Psi_0|\Psi_0 >}\Big[< \chi|\hat{V} - E_1|\Psi_0 > + < \Psi_0|\hat{V}^{\dagger} - E_1^*|\chi > \\ &+ Re(< \chi|\hat{H}_0 - E_0|\chi >)\Big] \end{aligned} \tag{16}$$

In these formulae Ψ_0 and E_0 are the zero-order ground state wavefunction and energy, respectively, $\hat{V}$ is the perturbation operator and $|\chi >$ is the first-order wavefunction. E_1 is defined as $E_1 =< \Psi_0|\hat{V}|\Psi_0 >$, as usual. (As it is easy to see, for Hermitian $\hat{H}_0$ and $\hat{V}$ and for normalized Ψ_0 the equation (15) trivially reduces to the usual second-order Hylleraas-functional.)

5. The CHA–PT2 formalism

Using the well-known second quantized formalism[26]–[28] the usual Born-Oppenheimer Hamiltonian can be written as

$$\hat{H} = \sum_{i,k} < \tilde{k}|\hat{h}|i > \hat{k}^{+}\hat{\tilde{i}}^{-} + \sum_{i,j,k,l} [\tilde{i}\tilde{j}|kl]\hat{i}^{+}\hat{j}^{+}\hat{\tilde{l}}^{-}\hat{\tilde{k}}^{-}, \tag{17}$$

where $\hat{h}$ is the one-electron Hamiltonian and the one- and two-electron integrals are defined over the spin-orbitals. (All integrations include also summations over the spins.) Symbols with a "$\sim$" indicate that the integrals contain the corresponding biorthogonal spin-orbitals

$$\tilde{\varphi}_i = \sum_k (\mathbf{S}^{-1})_{ki}\varphi_k, \tag{18}$$

where

$$S_{ip} = <\varphi_i|\varphi_p> \neq \delta_{ip}, \quad \text{and} \tag{19}$$

$$\hat{\tilde{i}}^{-} = (\hat{\tilde{i}}^{+})^{\dagger}. \tag{20}$$

These creation and "effective" annihilation operators $\hat{i}^{+}$ and $\hat{\tilde{i}}^{-}$ are connected to the original and biorthogonal orbitals, respectively, and have defined[4] as to satisfy the standard fermion anticommutation rules despite of the overlap:

$$\{\hat{i}^{+};\hat{\tilde{p}}^{-}\} = \hat{i}^{+}\hat{\tilde{p}}^{-} + \hat{\tilde{p}}^{-}\hat{i}^{+} = \delta_{ip}. \tag{21}$$

Introducing the relationship (12) and (13) into the Born-Oppenheimer Hamiltonian (17), one obtains its BSSE-free CHA counterpart:

$$\begin{aligned}
\hat{H}_{CHA} &= \sum_{i,k\in A} < k|\hat{h}_A|i > \hat{k}^{+}\hat{\tilde{i}}^{-} + \frac{1}{2}\sum_{i,j,k,l\in A} [ij|kl]\hat{i}^{+}\hat{j}^{+}\hat{\tilde{l}}^{-}\hat{\tilde{k}}^{-} \\
&+ \sum_{i,k\in B} < k|\hat{h}_B|i > \hat{k}^{+}\hat{\tilde{i}}^{-} + \frac{1}{2}\sum_{i,j,k,l\in B} [ij|kl]\hat{i}^{+}\hat{j}^{+}\hat{\tilde{l}}^{-}\hat{\tilde{k}}^{-} \\
&- \sum_{a\in A}\sum_{i\in B}\sum_{k} < \tilde{k}|\frac{Z_a}{r_a}|i > \hat{k}^{+}\hat{\tilde{i}}^{-} \\
&- \sum_{b\in B}\sum_{i\in A}\sum_{k} < \tilde{k}|\frac{Z_b}{r_b}|i > \hat{k}^{+}\hat{\tilde{i}}^{-} \\
&+ \sum_{k\in A}\sum_{l\in B}\sum_{i,j} [\tilde{i}\tilde{j}|kl]\hat{i}^{+}\hat{j}^{+}\hat{\tilde{l}}^{-}\hat{\tilde{k}}^{-},
\end{aligned} \tag{22}$$

where $\hat{h}_A$ and $\hat{h}_B$ are the *intra*molecular parts of the one-electron Hamiltonian Equ. (7), corresponding to molecule A and B, respectively.

The first two sums of equation (22) represent the effective intramolecular Hamiltonians for monomer A: it acts on a monomer wavefunction in the whole non-orthogonal supermolecule basis exactly as the original molecular Hamiltonian does it in the monomer basis. The next two terms represent the similar effective Hamiltonian of monomer B. These effective intramonomer Hamiltonians contain integrals over the original basis function (and not over their biorthogonal counterparts); this is the consequence of BSSE being excluded from the CHA Hamiltonian $\hat{H}_{CHA}$. (Integrals over biorthogonal orbitals remain in the terms describing intermolecular interactions.)

In the CHA–PT2 scheme one starts from the canonic SCF orbitals and orbital energies of the unperturbed free monomers and defines the zero-order effective one-electron Hamiltonian as the sum of the "effective" monomer Hartree-Fock operators

$$\hat{H}_0 = \hat{F}_A + \hat{F}_B \tag{23}$$

where

$$\hat{F}_A = \sum_{i \in A} \varepsilon_i \hat{i}^+ \hat{\tilde{i}}^- \quad \text{and} \quad \hat{F}_B = \sum_{i \in B} \varepsilon_i \hat{i}^+ \hat{\tilde{i}}^- \tag{24}$$

with ε_i being the respective orbital energies. The unperturbed ground state wavefunction is chosen as the antisymmetrized product of the monomer ground state wavefunctions

$$|\Psi_0> = \hat{i}^+_{A_1} \hat{i}^+_{A_2} \ldots \hat{i}^+_{A_n} \hat{i}^+_{B_1} \ldots \hat{i}^+_{B_m} |vac> \tag{25}$$

n and m being the number of electrons on molecules A and B, respectively. Then the perturbation can be written as:

$$\hat{V}_{CHA} = \hat{H}_{CHA} - \hat{H}_0 \, . \tag{26}$$

In this method, the first term on the right-hand-side of (15) is gives the energy corresponding to the wavefunction obtained by the simple antisymmetrization of the unperturbed monomer wavefunctions; it is sometimes called the "Heitler-London energy". It can be calculated by performing an orthogonalization of the original occupied monomer orbitals in the supermolecule basis. (For an explicit analytical expression of this energy and its decomposition into different physical components we refer to the papers[29].) Using the CHA Hamiltonian (22) and a biorthogonal formalism[30], an explicit expression for the first-order wavefunction has been obtained[17]:

$$|\chi> = \; - \; \frac{1}{4} \sum_{i,j \in A}^{occ} \sum_{p,q \in A}^{virt} \frac{[pq||ij]}{\varepsilon_p + \varepsilon_q - \varepsilon_i - \varepsilon_j} |\Psi_{ij}^{pq}>$$

$$
\begin{aligned}
&- \sum_{i\in A}^{occ}\sum_{p}^{virt} \frac{<\tilde{p}|\hat{V}_B^{aux}|i>}{\varepsilon_p - \varepsilon_i}|\Psi_i^p> \\
&- \frac{1}{4}\sum_{i\in A}^{occ}\sum_{j\in B}^{occ}\sum_{p,q}^{virt} \frac{[\tilde{p}\tilde{q}||ij]}{\varepsilon_p + \varepsilon_q - \varepsilon_i - \varepsilon_j}|\Psi_{ij}^{pq}> \\
&+ (A \Leftrightarrow B).
\end{aligned} \tag{27}
$$

Here the notation $(A \Leftrightarrow B)$ means that all terms with A and B interchanged should be added. The auxiliary operator $\hat{V}_B^{aux}$ is defined through its matrix elements as:

$$
<\tilde{p}|\hat{V}_B^{aux}|i> = <\tilde{p}|\sum_{a\in B}\frac{-Z_a}{r_a}|i> + \sum_{j\in B}^{occ}[\tilde{p}\tilde{j}||ij]. \tag{28}
$$

The first sum in Equ. (27) coincides with the first-order wavefunction describing the intramonomer correlation in the free-monomer; due to the non-orthogonality, however, the monomer correlation energies are not *strictly* recovered when the wavefunction (27) is substituted into the Hylleraas functional (15).

The non-orthogonality of the monomer molecular orbitals φ_i and φ_j makes practically infeasible to evaluate explicitly the matrix elements in the expression of the generalized Hylleraas functional (15) directly in their terms. One can circumvent this difficulty by introducing an auxiliary orthonormalized set of spin-orbitals $\{\vartheta_i\}$. These orbitals span the same space as the original ones $\{\varphi_i\}$ and are defined in such manner that first the occupied orbitals of both monomers are orthonormalized among others and then the virtual orbitals are orthogonalized to the occupied subspace. Then the original and the orthogonalized orbitals are connected with the linear transformations

$$
\varphi_p = \sum_l K_{lp}\vartheta_l, \tag{29}
$$

$$
\vartheta_p = \sum_l (\mathbf{K}^{-1})_{lp}\varphi_l. \tag{30}
$$

$\mathbf{K}$ is the transformation matrix of this orthogonalization. Using these relationships, after somewhat lengthy derivation explicit expression can be obtained for J_2 in terms of the *spatial orbitals*:

$$
\begin{aligned}
J_2 &= 2\sum_{k}^{occ}\left\{a(k|k) + \sum_{l<k}^{occ}[2u(kl|kl) - u(kl|lk)]\right\} \\
&\times 2\sum_{k}^{occ}\left\{d(k|k) + \sum_{l<k}^{occ}[2v(kl|kl) - v(kl|lk)\,]\right\}
\end{aligned}
$$

$$
\begin{aligned}
&+ \quad 4\sum_{k}^{occ}\sum_{t}^{virt}\left\{\left[a(t|k)+\sum_{l}^{occ}(\ 2u(tl|kl)-u(tl|lk))\right]\right. \qquad (31)\\
&\quad\times\left.\left[2<\vartheta_t|\hat{h}|\vartheta_k>+d(t|k)+\sum_{l}^{occ}(2v(tl|kl)-v(tl|lk))\right]\right\}\\
&+ \quad 4\sum_{k<l}^{occ}\sum_{t<r}^{virt}\left\{\left[2\ u(tr|kl)-u(tr|lk)\right]\right.\\
&\quad\times\left.\left[4\ [\vartheta_t\vartheta_r|\vartheta_k\vartheta_l]-2[\vartheta_t\vartheta_r|\vartheta_l\vartheta_k]+2v(tr|kl)-v(tr|lk)\right]\right\}.
\end{aligned}
$$

Here the different auxiliary quantities are defined as follows:

$$
a(t|k) = \sum_{i\in A}^{occ}\sum_{p}^{virt} K_{tp}\frac{-<\tilde{p}|\hat{V}_B^{aux}|i>}{\varepsilon_p-\varepsilon_i}K_{ik}^{-1}+(A\Leftrightarrow B) \qquad (32)
$$

$$
d(t|k) = \sum_{i\in A}^{occ}\sum_{p}^{virt} -K_{tp}<\tilde{p}|\hat{V}_B^{aux}|i>K_{ik}^{-1}+(A\Leftrightarrow B) \qquad (33)
$$

$$
u(tr|kl) = b(tr|kl)+c(tr|kl) \qquad (34)
$$

$$
v(tr|kl) = e(tr|kl)+f(tr|kl) \qquad (35)
$$

with

$$
\begin{aligned}
b(tr|kl) &= \sum_{i,j\in A}^{occ}\sum_{p,q\in A}^{virt} K_{tp}K_{rq}\frac{-[pq|ij]}{\varepsilon_p+\varepsilon_q-\varepsilon_i-\varepsilon_j}K_{jl}^{-1}K_{ik}^{-1} \qquad (36)\\
&+ (A\Rightarrow B)
\end{aligned}
$$

$$
\begin{aligned}
c(tr|kl) &= \sum_{i\in A}^{occ}\sum_{j\in B}^{occ}\sum_{p,q}^{virt} K_{tp}K_{rq}\frac{-[\tilde{p}\tilde{q}|ij]}{\varepsilon_p+\varepsilon_q-\varepsilon_i-\varepsilon_j}K_{jl}^{-1}K_{ik}^{-1} \qquad (37)\\
&+ (A\Leftrightarrow B)
\end{aligned}
$$

$$
e(tr|kl) = \sum_{i,j\in A}^{occ}\sum_{p,q\in A}^{virt} -K_{tp}K_{rq}[pq|ij]K_{jl}^{-1}K_{ik}^{-1}+(A\Rightarrow B) \qquad (38)
$$

$$
f(tr|kl) = \sum_{i\in A}^{occ}\sum_{j\in B}^{occ}\sum_{p,q}^{virt} -K_{tp}K_{rq}[\tilde{p}\tilde{q}|ij]K_{jl}^{-1}K_{ik}^{-1}+(A\Leftrightarrow B). \qquad (39)
$$

The notations $A \Leftrightarrow B$ ($A \Rightarrow B$) indicate that the terms with A interchanged (replaced) with B should be added to the respective expression.

These expressions represent our working formulae[22] for the case of CHA–PT2 calculations.

6. The CHA–MP2 method

In the CHA–MP2 scheme we start again from the CHA Hamiltonian[4] written down in terms of integrals over molecular orbitals[18] transformed with the use of transformations like (12), (13). As a first step one has to solve the BSSE-free CHA–SCF equations and to choose the zero-order unperturbed Hamiltonian as

$$\hat{H}_0 = \sum_i \varepsilon_i \varepsilon_i \hat{i}^+ \hat{\tilde{i}}^- \tag{40}$$

where $\hat{i}^+$ and $\hat{\tilde{i}}^-$ are the creation and "effective" annihilation operators corresponding to the non-orthogonal (occupied and virtual) CHA canonic supermolecular orbitals and ε_i is the respective eigenvalue of the CHA–SCF equations. To calculate the first-order wavefunction, one again introduces an appropriate biorthogonal basis set, defined by the left eigenvectors of the CHA Fockian. Then, the perturbation may again be defined as:

$$\hat{V}_{CHA} = \hat{H}_{CHA} - \hat{H}_0\,. \tag{41}$$

Only the orbitals and orbital energies on which $\hat{H}_0$ is built up differ from the previous case. One has to emphasize again that this $\hat{H}_0$ is not Hermitian. As already noted, the first term of (15) representing the sum of zero and first-order energy contributions coincides in this case with the BSSE-free CHA-SCF energy.

As the orbitals used are solution of the CHA–SCF equations, an analogue of the Brillouin-theorem holds and the first-order wavefunction contains no singly excited configurations, only doubly excited ones:

$$|\Psi_1> = -\sum_{i<j}^{occ}\sum_{a<b}^{virt} \frac{\{\tilde{a}\tilde{b}||ij\}}{\varepsilon_a - \varepsilon_b - \varepsilon_i - \varepsilon_j}|\Psi_{ij}^{ab}>\,. \tag{42}$$

Here the notation $\{\tilde{a}\tilde{b}||ij\}$ denotes the "CHA transformed" counterpart of the integral $[\tilde{a}\tilde{b}||ij]$ which can be obtained by the technique described in papers[16, 18].

Using the orthonormalized auxiliary spin-orbitals $\{\vartheta_i\}$ defined in a manner similarly to that discussed above, one can again derive the formula for J_2:

$$\begin{aligned} J_2 &= Re\left\{\sum_{k,l}^{occ} t(kl||kl)\sum_{i,j}^{occ} y(ij||ij)\right. \\ &+ \sum_r^{virt}\sum_k^{occ}\left[\sum_l^{occ} t(lr||lk)\sum_j^{occ} y(jr||jk) - 2 < \vartheta_r|\hat{F}|\vartheta_k >\right] \end{aligned} \tag{43}$$

$$- \sum_{k<l}^{occ} \sum_{p<q}^{virt} t(pq||kl) \left[2[pq||kl] - y(pq||kl) \right] \Bigg\}$$

where the coefficients $t(pq||kl)$ and $y(pq||kl)$ are defined as

$$t(pq||kl) = \sum_{i,j}^{occ} \sum_{a,b}^{virt} K_{pa} K_{qb} \frac{\{\tilde{a}\tilde{b}||ij\}}{\varepsilon_a + \varepsilon_b - \varepsilon_i - \varepsilon_j} K_{ik}^{-1} K_{jl}^{-1} \tag{44}$$

$$y(pq||kl) = \sum_{i,j}^{occ} \sum_{a,b}^{virt} K_{pa} K_{qb} \{\tilde{a}\tilde{b}||ij\} K_{ik}^{-1} K_{jl}^{-1} . \tag{45}$$

Despite the seemingly simpler PT expressions, this method is more expensive but has a better performance than the CHA-PT2 scheme discussed above. Its advantages (and disadvantages) are obviously connected with the fact that it is based on the converged BSSE-free SCF orbitals, reflecting all the polarization and charge transfer effects exactly, while CHA-PT2 contains them only up to first-order in the wavefunction and second-order in the energy. Accordingly, the application of this CHA-MP2 method requires the CHA–SCF equations to be solved at every geometry of the complex studied, which is a straightforward but relatively complex and costly procedure‡. (In the CHA-PT2 scheme one always uses the orbitals calculated for the *free* monomers.) Another complication in the CHA-MP2 formalism is that the eigenvalues of the non-Hermitian CHA Fockian can be complex. (A number of such cases has really been encountered.) The resulting value of the second-order energy contribution is always real automatically so, in principle, one could use complex computer arithmetics throughout to get the real second-order energy, but this would reduce dramatically the effectivity of calculations. Instead, we have rewritten the formulae in such a manner as to collect separately those terms the imaginary parts of which cancel in the final expression; this required quite lengthy derivations but the application of the resulting formulae needs a negligible computational effort[18].

7. Applications

Many different van der Waals and hydrogen bonded systems have already been studied by using the above PT schemes[17, 18, 22, 24]. Here we

‡For use with CHA–MP2 one has to solve the more complicated "original" CHA–SCF equations (usually denoted CHA/CE[6, 8]) and not the much simpler heuristic CHA/F[7, 10] ones, because the Brillouin-type theorem does not strictly hold for the latter.

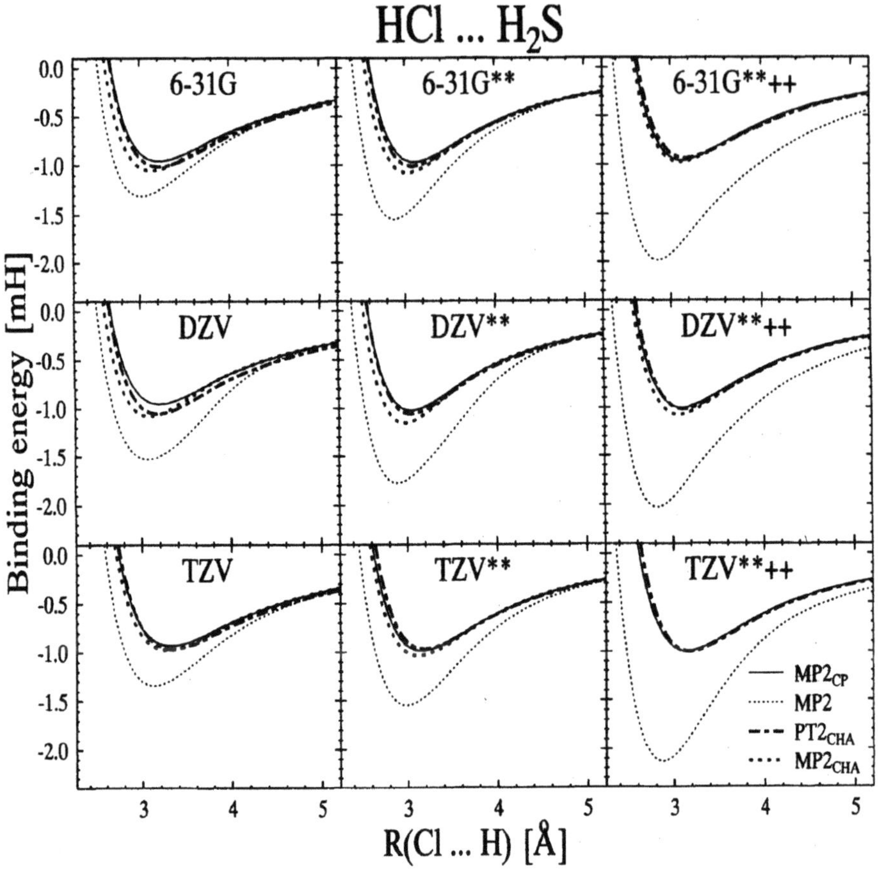

Figure 1. Potential curves of the HCl ... H_2S system calculated in nine different basis sets as functions of the hydrogen bond length. The curves display four different approximations to the interaction energy: MP2, standard uncorrected second-order perturbation theory; MP2-CP, counterpoise corrected MP2 method , CHA-PT2 and CHA-MP2, *a priori* BSSE-free second-order perturbation methods based on the "Chemical Hamiltonian Approach".

show the results for four different hydrogen bonded dimers, HCl ... H_2S, $H_3N \ldots H_3N$, $H_2O \ldots H_3N$ and HF ... HF, for which we have published previously only SCF results[6, 10]. These systems represent the whole spectrum of hydrogen bonds, ranging from the very weak interactions to the very strong ones. We used the same nine different basis sets (from 6–31G to TZV**++) as in the work[10] and simplified geometries as reported in paper[6]. During the studies of these systems, the intermolecular distance R has been varied in a wide range, keeping fixed the

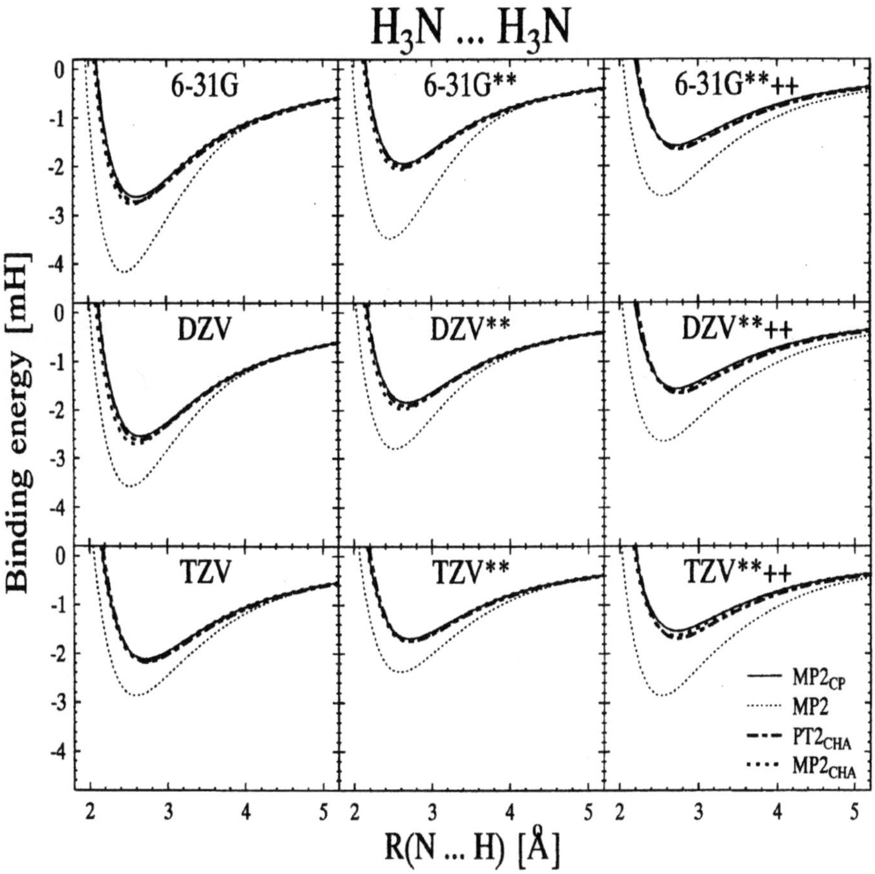

Figure 2. Potential curves of the $H_3N \ldots H_3N$ system calculated in nine different basis sets as functions of the hydrogen bond length. The notations of the individual curves are the same as for Fig. 1.

internal geometry as well as the relative orientation of the monomers. These simplifications seemed reasonable because we are interested in comparison of the different methods and not in the fine details of the structure of the actual systems.

The results are presented in Figures 1-4. Comparing the results obtained with the uncorrected (MP2) and corrected (CHA–PT2, CHA–MP2 and MP2–CP) perturbational methods, the following general remarks can be made. First, the amount of the BSSE content in the uncorrected energy is very large in all cases considered. (This is characteristic for the electron correlation; for the largest basis sets considered, the BSSE effects observed at the SCF level are rather small.) Furthermore,

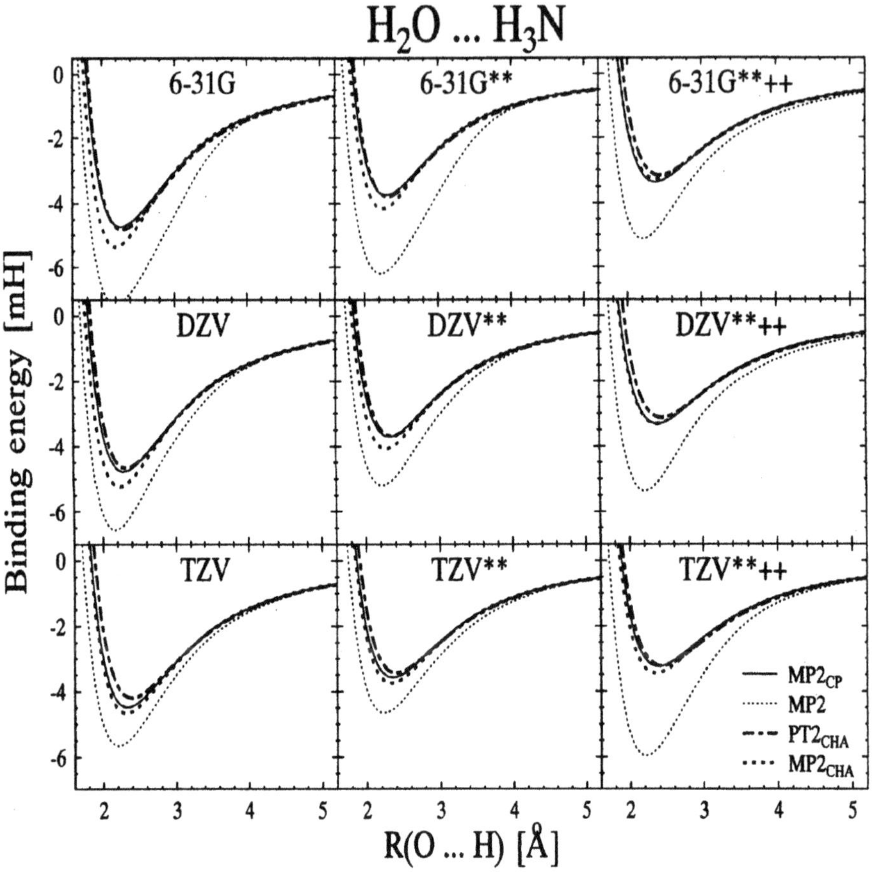

Figure 3. Potential curves of the $H_2O \ldots H_3N$ system calculated in nine different basis sets as functions of the hydrogen bond length. The notations of the individual curves are the same as for Fig. 1.

in practically all cases the counterpoise corrected (MP2–CP) results are very close to those given by the CHA-based perturbation theories, and the differences reduce as the basis set improves. This is especially the case for the CHA–MP2 method utilizing the supermolecule CHA–SCF orbitals.

Considering the CHA–PT2 scheme, it may be noted that the minima obtained with this method are somewhat less deep and slightly shifted to the larger intermolecular separations, as compared with the CHA–MP2 and the MP2–CP ones. This effect is often not very evident, it is most pronounced for the system $H_2O \ldots H_3N$. That observation is in contrast to the general trend that – at different levels of the theory – the CHA

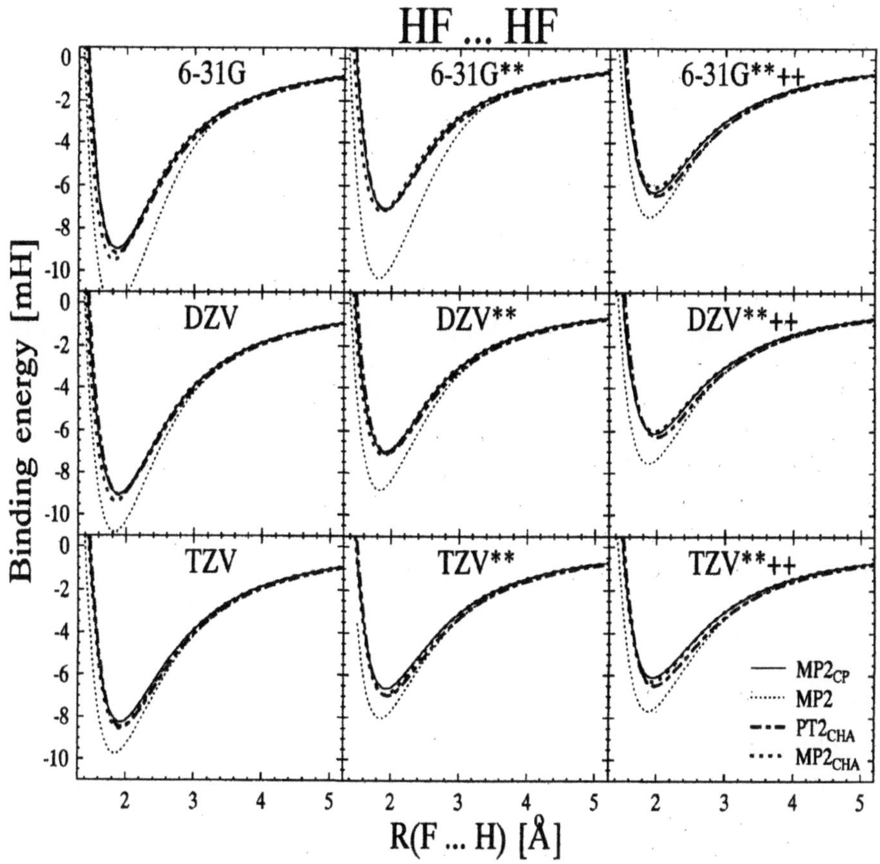

Figure 4. Potential curves of the HF ... HF system calculated in nine different basis sets as functions of the hydrogen bond length. The notations of the individual curves are the same as for Fig. 1.

curves usually lie somewhat below the CP corrected ones and predict slightly shorter equilibrium intermolecular separations. This behavior of the CHA–PT method is a consequence of the fact, that in this theory the unperturbed Hamiltonian was chosen as the sum of the monomer Fockians, so that polarization and delocalization effects are taken into account up to second-order only, while these effects appear up to infinite order in the different supermolecule-based calculations.

We have to emphasize that the importance of electron correlation in intermolecular interactions is varying very much depending on the systems studied. Electron correlation effects play a decisive role for the true van der Waals systems (like rare gas dimers) which are not bonded at all at the SCF level of the theory, while its importance is less for the hydrogen bonded systems discussed here. Obviously, their role is larger in the case of weakly bonded situations (see the work of Mayer and Valiron[18] for an elaborated example) while in the strongly bonded complexes the electrostatic and charge transfer effects dominate.

As it can be seen from the figures, the BSSE content of the results obtained by the 6–31G and 6–31G** basis sets is extremely large for HF ... HF, but it is much smaller with the better basis sets (Fig. 4). An interesting, but not unusual, observation can be made by inspecting the curves (Figs. 1 and 2) for HCl ... H_2S and H_3N ... H_3N): the better basis sets exhibit larger BSSE. (The same holds for the TZV**++ basis in the case of H_2O ... H_3N.) This is connected with the known maximum character of BSSE: for small basis sets there is no flexibility for BSSE-caused energy lowerings, then BSSE increases with increasing basis sets and is reduced only after the saturation starts. This maximum behavior is usually not observed at the SCF level for which saturation occurs at much smaller basis sets than in the correlated case.

8. Conclusion

This paper gives a short general description of the Boys-Bernardi counterpoise corrected scheme and of the Chemical Hamiltonian Approach as are applied to the BSSE problem of intermolecular interactions. The *a priori* corrected CHA–PT2 and CHA–MP2 BSSE-free second-order perturbation methods are summarized and compared with the *a posteriori* CP method at the electron correlated level on the example of several hydrogen-bonded model systems. The results show that at the correlated level the basis set superposition error is very large in the uncorrected interaction energy if medium-size basis sets are used, so its removal is imperative. The differences of the two CHA-based perturbational theories discussed are connected solely with the different one-electron orbital sets used for building up the unperturbed Hamiltonian.

Acknowledgments

This work was supported by the grants OTKA No. T25369 and T29716. Á. V. also gratefully acknowledges the grant "Széchenyi" from the Hungarian Ministry of Culture. The authors also thank Dr. Pierre Valiron (Grenoble) for the long-term cooperation in the subject.

References

[1] H. B. Jansen and P. Ross, *Chem. Phys. Lett.* **3**, 140 (1969).
S. B. Boys and F. Bernardi, *Mol. Phys.* **19**, 553 (1970).

[2] P. W. Fowler and A. D. Buckingham, *Mol.Phys.* **50**, 1349 (1983).

[3] M. M. Szczesniak and S. Scheiner, *Chem. Phys. Lett.* **131**, 230 (1986).

[4] I. Mayer, *Intern. J. Quantum Chem.* **23**, 341 (1983).
I. Mayer, in *"Modelling of structure and properties of molecules"* (ed. Z.B.Maksić), Ellis Horwood, Chicester, UK, 1987, p. 145.

[5] I. Mayer and Á. Vibók, *Chem. Phys. Lett.* **136**, 115 (1987).
I. Mayer and Á. Vibók, *Chem. Phys. Lett.* **140**, 558 (1987).
Á. Vibók and I. Mayer, *J. Mol. Struct. (Theochem)*, **170**, 9 (1988).

[6] Á. Vibók and I. Mayer, *Intern. J. Quantum Chem.* **43**, 801 (1992).

[7] I. Mayer, P.R. Surján, and Á. Vibók, *Int. J. Quantum Chem.* Quantum Chem. Symp. **23**, 281 (1989).

[8] I. Mayer and L. Turi, *J. Mol. Struct. (Theochem)*, **227**, 43 (1991).

[9] I. Mayer and P.R. Surján, *Int. J. Quantum Chem.* **36**, 225 (1989).

[10] P. Valiron, Á. Vibók, and I. Mayer, *J. Comp. Chem.* **14**, 401 (1993).

[11] Fu-Ming Tao, *J. Chem. Phys.* **100**, 3645 (1993).

[12] Juan J. Novoa, Marc Planas, and Myung-Hwan Whangbo, *Chem. Phys. Lett.* **225**, 240 (1994).

[13] Ernest R. Davidson and Subhas J. Chakravarty, *Chem. Phys. Lett.* **217**, 48 (1994).

[14] I. Mayer, *Int. J. Quant. Chem.* **70**, 41 (1998).

[15] J. Noga and Á. Vibók, *Chem. Phys. Lett.* **180**, 114 (1991).

[16] I. Mayer, Á. Vibók, and P. Valiron, *Chem. Phys. Lett.* **224**, 166 (1994).

[17] I. Mayer and Á. Vibók, *Mol.Phys.* **92**, 503 (1997).

[18] I. Mayer and P. Valiron, *J. Chem. Phys.* **109**, 3360 (1998).

[19] M. Kieninger, S. Suhai, and I. Mayer, *Chem. Phys. Lett.* **230**, 485 (1994).

[20] B. Paizs and S. Suhai, *J. Comput. Chem.* **18**, 694 (1997).

[21] B. Paizs and S. Suhai, *J. Comput. Chem.* **19**, 575 (1998).

[22] Á. Vibók, G. Halász, and I. Mayer, *Mol.Phys.* **93**, 873 (1998).

[23] I. Mayer, *Mol. Phys.* **89**, 515 (1996).

[24] G. Halász, Á. Vibók, and I. Mayer, *J. Comput. Chem.* **20**, 274 (1999).

[25] I. Mayer and P. R. Surján, *Chem. Phys. Lett.* **191**, 497 (1992).

[26] H. C. Longuet-Higgins, p. 105 in "Quantum Theory of Atoms, Molecules and the Solid State" (ed. P.-O. Löwdin) Academic Press, New York, 1966.

[27] P. R. Surján, "Second Quantized Approach To Quantum Chemistry" Springer, Berlin, 1989.

[28] 1723 (1994).

[29] A. Hamza and I. Mayer, *Int. J. Quantum Chem.* **82**, 53, 105, (2001).

[30] Á. Vibók and I. Mayer, *Acta Phys. Hung.* **68**, 241 (1990).

WHAT IS TIME-DEPENDENT DENSITY FUNCTIONAL THEORY? SUCCESSES AND CHALLENGES

Neepa T. Maitra, Adam Wasserman, and Kieron Burke
Departments of Chemistry and Physics, Rutgers University, 610 Taylor Road, NJ08904, USA

Abstract We discuss ongoing projects in ground-state density functional theory (DFT) before introducing some basic concepts in time-dependent DFT (TDDFT). The accuracy of simple approximations to transition frequencies and oscillator strengths is analyzed, developing scattering theory within TDDFT is discussed, and the importance of memory in fully time-dependent calculations is emphasized.

1. Introduction: ground-state DFT

Since its birth in the 1960's (Hohenberg and Kohn, 1964) and especially with the developments in recent years (Kohn, 1999), density functional theory (DFT) has been enormously successful for calculating the electronic structure and properties of ground-state systems. Modern DFT is in principle an exact theory, built on the firm foundation of the Hohenberg-Kohn theorem (1964): the ground-state density $n(\mathbf{r})$ of an interacting N-electron system in a static external potential completely determines all of its properties (e.g. energy, bond lengths, excited states). From a practical point of view, the power of DFT lies in the mapping of the interacting system to a fictitious non-interacting system of the same ground-state density, the Kohn-Sham system. The Kohn-Sham orbitals experience a one-body effective potential functional $v_S[n](\mathbf{r})$ and so the calculation reduces to solving self-consistently N one-particle Schrödinger equations, finally producing the exact density of the true interacting system. In practical calculations, the exact Kohn-Sham potential is unknown and approximations must be made. Popular approximate functionals include the local spin-density approximation (Kohn and Sham, 1965) (particularly ubiquitous in solid-state calculations), generalized gradient approximations such as PBE (Perdew, Burke, and Ernzerhof, 1996) (becoming increasingly popular in quantum chemistry), and empirical hybrids such as B3LYP (Becke, 1998, 1993; Lee,Yang, and Parr, 1988) (perhaps the most used in quantum chemistry).

In the Rutgers DFT group, there are many projects underway to investigate and further improve the functionals used in ground-state DFT. We mention here a few of these projects. Consider the strictly correlated limit: studying what happens when electron-electron repulsion becomes infinite should help guide the construction of new functionals (Magyar, Terilla, Burke, 2001). We are calculating unambiguously the exchange-correlation energy density (Burke, Cruz and Lam, 1998) in molecules in order to visualize bond formation as well as to study functionals. We are studying properties of the non-interacting kinetic energy density (Larkin et al., 2001), to investigate the validity of the von Weisacker functional and to shed some light on the construction of a good kinetic energy functional. We are developing a new tool to study systems in which the spins are not equivalent, such as magnetic systems: by independently scaling the spin densities, one can analyze the behaviour of the correlation energy for various polarizations and test whether approximate functionals follow this behaviour (Whittingham, Magyar, and Burke, 2001). The problem of degenerate ground-states is being investigated through a superhamiltonian approach (Katriel, Zahariev and Burke, 2001).

2. What is TDDFT?

The time-dependent Schrödinger equation,

$$\hat{H}(t)\Psi(t) = i\dot{\Psi}(t), \tag{1}$$

where the dot indicates a time derivative, gives the solution to the many-body wavefunction for a given initial state $\Psi(0)$. The Hamiltonian $\hat{H}$ is the sum of the kinetic energy operator, the Coulomb repulsion, and an external time-varying one-body potential, $v_{\text{ext}}(\mathbf{r}, t)$. For more than very few electrons, the two-body Coulomb potential makes this a nightmare to solve.

The success of DFT methods for ground-state properties encouraged a search for a time-dependent theory of a similar flavor (TDDFT). This came to fruition with the Runge-Gross theorem (1984): for a given initial state, there is a unique mapping between the evolving density and the time-dependent potential. A non-interacting system of the same density may be defined, whose evolution is described by time-dependent Kohn-Sham equations where the orbitals evolve under a time-dependent one-body potential functional $v_S[n;\Phi(0)](\mathbf{r}t)$:

$$\left\{-\frac{1}{2}\nabla^2 + v_S[n;\Phi(0)](\mathbf{r},t)\right\}\phi_i(\mathbf{r}t) = i\dot{\phi}_i(\mathbf{r}t) \tag{2}$$

$$n(\mathbf{r}t) = \sum_{i=1}^{N} |\phi_i(\mathbf{r}t)|^2 \tag{3}$$

$$v_S[n;\Phi(0)](\mathbf{r}t) = v_{\rm ext}[n;\Psi(0)](\mathbf{r}t) + \int d^3r' \frac{n(\mathbf{r}'t)}{|\mathbf{r}-\mathbf{r}'|} + v_{\rm XC}[n;\Psi(0),\Phi(0)](\mathbf{r}t) \tag{4}$$

As in the ground-state case, this fictitious system of non-interacting electrons is far faster to solve than the full interacting problem, and, in principle, yields everything we would want to know about the exact interacting problem. Any quantity that depends only on the time-dependent density is given exactly, while exact formulas for other expectation values can be written down. In any practical calculation, we must make approximations for the exchange-correlation potential. The most popular in use today is the adiabatic local density approximation (ALDA), which uses the potential for a uniform gas of electrons of density $n(\mathbf{r}t)$:

$$v_{\rm XC}^{\rm ALDA}(\mathbf{r}t) = v_{\rm XC}^{\rm unif}(n(\mathbf{r}t)), \tag{5}$$

as was used in the historic calculations of Zangwill and Soven (1980), and Ando (1977). This approximation should work well for a system beginning in its ground state which varies slowly in time and space. It is on shaky ground in the general case however: in particular, because it has no memory. We shall come back to this point shortly, and demonstrate a drastic consequence of this lack of memory.

3. Applications of TDDFT

There is an exponentially increasing use of TDDFT in chemistry and in physics as described in the recent reviews of Burke and Gross (1998), and Maitra et al. (2001). Most of the calculations are in the linear response regime, which is discussed in the next section. For example, there are many calculations of electronic excitations in atoms and molecules (Appel, Burke, and Gross (2001)) and this is now built into many commercial chemical codes such as Gaussian, where ALDA is used. There have been the first few optical response calculations in solids (Bertsch et al., 2000; Kootstra, de Boeij, and Snijders, 2000). Within TDDFT, one is now able to calculate van der Waal's forces (Andersson, Langreth, and Lunqvist, 1996), whose origin lies in small time-dependent fluctuations;

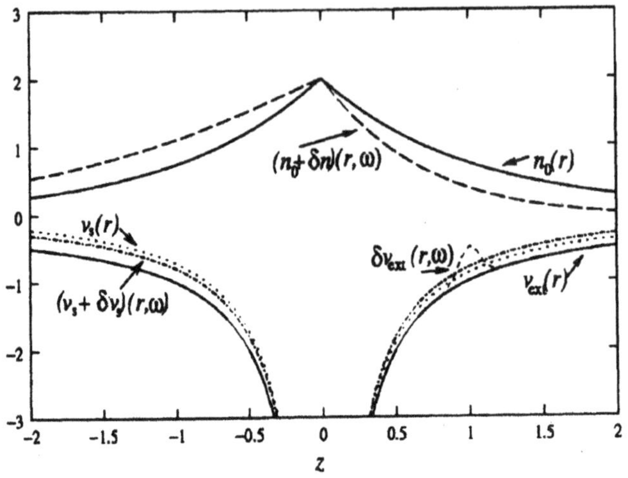

Figure 1. How TDDFT response works: A ground state density is perturbed by a frequency dependent change in the external potential. The change in the KS potential is defined to be that which causes the KS system to produce the same density change as the real system.

this is of great interest in biochemistry. Two years ago, Sundholm (1999) calculated the spectrum of chlorophyll A. One can improve on *ground-state* energies, including more non-local effects, through the adiabatic-connection-fluctuation-dissipation formula (Kurth and Perdew, 1999).

Far beyond linear response, TDDFT appears to be the only feasible approach to studying atoms in intense laser fields, and during energetic collisions (Burke and Gross, 1998). We are building a fully time-dependent Kohn-Sham code to calculate transport through molecular wires.

4. Linear response theory

Most of the applications of TDDFT have been in the linear response regime (Burke and Gross, 1998), where one turns on a small time-dependent perturbation on a system initially in its ground-state. This is illustrated in Fig. 1, where the perturbation is shown in the frequency domain. The change in the density may be written in terms of the susceptibility: for either the non-interacting Kohn-Sham or the physical system:

$$\delta n(\mathbf{r}\omega) = \int d^3r' \chi[n_0](\mathbf{r}\mathbf{r}'\omega)\delta v_{\rm ext}(\mathbf{r}'\omega) \quad (6)$$

$$= \int d^3r' \chi_s[n_0](\mathbf{r}\mathbf{r}'\omega)\delta v_{\rm S}(\mathbf{r}'\omega) \quad (7)$$

where the susceptibility $\chi[n_0](\mathbf{rr}'\omega)$ is the response of the interacting system to a small change in the external potential. The second line follows since the density change is the same for the interacting system and the non-interacting system; $\chi_s[n_0](\mathbf{rr}'\omega)$ is the Kohn-Sham susceptibility. It is convenient to write the change in the Kohn-Sham potential as

$$\delta v_{\rm S}(\mathbf{r}\omega) = \delta v_{\rm ext}(\mathbf{r}\omega) + \int d^3r' \left(\frac{1}{|\mathbf{r}-\mathbf{r}'|} + f_{\rm XC}[n_0](\mathbf{rr}'\omega)\right) \delta n(\mathbf{r}'\omega) \quad (8)$$

where $f_{\rm XC}[n_0](\mathbf{rr}'\omega)$ is known as the exchange-correlation kernel. In the time-domain,

$$f_{\rm XC}[n_0](\mathbf{rr}', t-t') = \frac{\delta v_{\rm XC}[n_0](\mathbf{r}t)}{\delta n(\mathbf{r}'t')}. \quad (9)$$

These equations lead to the Dyson-type response equation

$$\chi(\mathbf{rr}'\omega) = \chi_{\rm S}(\mathbf{rr}'\omega) + \int d^3r_1 \int d^3r_2 \chi_{\rm S}(\omega)(\mathbf{rr}_1\omega) f_{\rm HXC}[n_0](\mathbf{r}_1\mathbf{r}_2\omega)\chi(\mathbf{r}_2\mathbf{r}'\omega),$$

where $f_{\rm HXC} = 1/|\mathbf{r}-\mathbf{r}'| + f_{\rm XC}$. There are two key ingredients to composing the interacting response: first, the Kohn-Sham susceptibility, which may be written as a sum over Kohn-Sham orbitals:

$$\chi_{\rm S}(\mathbf{r},\mathbf{r}',\omega) = \sum_{jk}(f_k - f_j)\frac{\phi_j(\mathbf{r})\phi_k^*(\mathbf{r})\phi_j^*(\mathbf{r}')\phi_k(\mathbf{r}')}{\omega - (\epsilon_j - \epsilon_k) + i0_+} \quad (10)$$

where the f_k are occupation numbers; and second, the exchange-correlation kernel, Eq. (9). Excitation energies lie at the poles of the true response function (Petersilka, Gossmann and Gross, 1996), and oscillator strengths are related to the pole strengths (Casida, 1996).

Both of the two ingredients mentioned above involve approximations: in obtaining the ground-state Kohn-Sham potential whose orbitals to use in Eq. (10), and in approximating $f_{\rm XC}$. Which of these has the greater influence?

Petersilka, Gross and Burke (2000) found that an accurate *ground-state* Kohn-Sham potential is a vital starting point for these calculations. For atoms and molecules, KS eigenvalues and orbitals of valence states are well approximated with standard LDA and GGA potentials, but Rydberg states require the slow $-1/r$ decay of $v_{\rm XC}(r)$ far from a Coulomb system. This can be achieved via an exact exchange calculation or some grafting procedure for the long-range tail. In that article, the case of singlet excitations in Be was also studied. This case dramatically illustrates that the approximation for the ground-state potential used in calculating the Kohn-Sham response χ_s has a much bigger impact on the positions of the higher-lying transitions than the approximation used for f_{XC}.

The other key ingredient is the XC kernel, about which relatively little is known. This is almost always approximated by its ground-state limit, $\omega \to 0$. Again, the most popular approximation for the f_{XC} is the ALDA, completely local in space and in time:

$$f_{XC}^{ALDA} = \frac{d^2 e_{XC}^{unif}}{dn^2}(n(\mathbf{r}t))\delta^{(3)}(\mathbf{r}-\mathbf{r}')\delta(t-t') \tag{11}$$

Memory dependence gives the exact $f_{XC}[n_0](\mathbf{r}\mathbf{r}'\omega)$ a frequency-dependence, neglected in any adiabatic approximation. Within this approximation, Filippi, Umrigar and Gonze (1997) found that the exact triplet and singlet excitation tends to sandwich the exact Kohn-Sham level. The triplet-singlet splitting is pretty well approximated by using f_{XC} in the ALDA approximation, but then the mean of the two is shifted. This motivated Burke, Petersilka, and Gross (2000) to develop a hybrid f_{XC} functional.

When are the bare Kohn-Sham transitions good approximations? Casida (1996) transformed the response equation Eq. (9) to a linear response problem. Using q as a double-index, we consider the basis $\Phi_q(\mathbf{r}) = \phi_i(\mathbf{r})\phi_a(\mathbf{r})$, where i labels an occupied state and a an unoccupied one, we obtain

$$\sum_{q'} \tilde{\Omega}_{qq'} v_{q'} = \Omega v_q \tag{12}$$

where $\Omega = \omega^2$, the square of the true transition frequency, and

$$\tilde{\Omega}_{qq'} = \delta_{qq'}\Omega_q + 2\sqrt{\omega_q \omega_{q'}}\langle q|f_{HXC}|q'\rangle, \tag{13}$$

with $\Omega_q = \omega_q^2$ being the square of the KS transition frequency. The square arises naturally when both forward and back transitions are treated on an equal footing.

If we then do perturbation theory in the *distance from the diagonal* of the matrix (Appel, Burke, and Gross (2001)), we obtain, for transition frequencies that are only weakly shifted from their Kohn-Sham values,

$$\omega = \omega_q + \langle q|f_{HXC}|q\rangle \left(1 + \sum_{q \neq q'} \frac{\langle q'|f_{HXC}|q\rangle}{\omega_q - \omega_q'}\right) \tag{14}$$

Neglecting the sum over off-diagonal elements on the right-hand side constitutes the "single pole approximation" (SPA) (Petersilka, Gossmann, and Gross, 1996). For the oscillator strengths, we find

$$f = f_q \left(1 + \frac{\langle q'|f_{HXC}|q\rangle}{\omega_q - \omega_q'}\sqrt{\frac{f_{q'}}{f_q}}\right). \tag{15}$$

The precise expressions for these corrections are given in the article cited just above.

Ignoring the corrections leads to the single-pole approximation of Petersilka, Gossmann and Gross (1996). Now we can see precisely when this is a good approximation. It requires the poles to be decoupled, in the sense that the the ratio of the matrix elements of the Hartree-XC kernel to the frequency separation must be small. Hence it is accurate in the limit of weak Coulomb interaction, but also in the limit of well-separated poles. Note also that, when it is accurate, it provides a simple formula for correcting KS transition frequencies, without solving further sets of equations. Finally, note that a signal for this SPA to be valid is that the KS oscillator strengths are accurate. In such circumstances, the corrections to the KS transition frequencies yield expectation values of the unknown XC kernel.

The single-pole approximation has proved to be very successful in predicting excitation energies for finite systems, such as atoms and molecules, when the bound states are well-separated and the off-diagonal elements are truly small. For solids however, the poles are no longer separated and this analysis cannot be used. There are subtleties involved in extracting the long wavelength limit of an extended system (de Boeij et al. , 2001).

5. Density Functional Scattering Theory

The Kohn-Sham potential for an N-electron system has proven to be a very accurate approximation of the effective optical potential felt by an electron in a Rydberg state orbiting in the field of the daughter (N-1)-electron system (Al-Sharif, Resta, and Umrigar, 1998). The natural question arises if the N-electron ground-state Kohn-Sham potential also produces accurate *continuum* states for an electron scattering off the corresponding (N-1)-electron system. In particular, when an electron collides at low energies with a neutral (N-1)-electron system, is it possible to get scattering information from the corresponding N-electron ground-state Kohn-Sham potential? The photoresponse problem has already been addressed by Zangwill and Soven (1980), but the electronic scattering one has not been formulated yet within the TDDFT framework.

Consider the example (in one dimension) of one electron scattering off a negative delta function potential $V(x) = -Z\delta(x)$ of strength Z. This potential admits one bound state, and its energy is $-Z^2/2$. The transmission coefficient must be singular at that energy, and should therefore be of the form $T(\varepsilon) = f(\varepsilon)/(\varepsilon + Z^2/2)$, where $f(\varepsilon)$ must go to zero as $\varepsilon \to 0$ and to ε as $\varepsilon \to \infty$. The simplest guess $f(\varepsilon) = \varepsilon$ provides in fact the exact answer: $T(\varepsilon) = \varepsilon/(\varepsilon + Z^2/2)$. Notice how knowledge of the ground-state energy for an electron *bound* in an external potential was enough in this case to easily guess the transmission probability (1d-cross-section) for an electron *scattering* off that potential.

We can also obtain the same answer from the susceptibility, which for one electron in one dimension (arbitrary $V(x)$) is given by:

$$\chi(x, x', \varepsilon) = \sqrt{n(x)n(x')}[g^>(x, x', \varepsilon + \varepsilon_0) + g^{>*}(x, x', -\varepsilon + \varepsilon_0)], \quad (16)$$

where $n(x)$ is the ground-state density, ε_0 is the ground-state energy, and $g^>$ is the outgoing Green's function satisfying:

$$\left(-\frac{1}{2}\frac{\partial^2}{\partial x^2} + V(x) - i\frac{\partial}{\partial t}\right) g^>(x, x'; t - t') = -i\delta(x - x')\delta(t - t'). \quad (17)$$

Eq. (16) can be proven in the following way: due to an external time-dependent perturbation $\delta v(x, t)$, the ground state $\phi_0(x)e^{-i\varepsilon_0 t}$ changes by an amount determined by the Green's function:

$$\delta\phi_0(x, t)e^{-i\varepsilon_0 t} = \int dx' \int dt' \phi_0(x')e^{-i\varepsilon_0 t'} g^>(x, x'; t - t')\delta v(x', t'), \quad (18)$$

and since for one electron $\delta n(x, t) = \phi_0(x, t)\delta\phi_0^*(x, t) + \phi_0^*(x, t)\delta\phi_0(x, t)$, taking the Fourier transform and using the definition of the susceptibility (eq. (7)), we get eq. (16).

Since at large distances only the continuum states contribute to the susceptibility (the bound states have decayed), it is possible to get the transmission amplitude $t(\varepsilon)$ for one electron scattering off any potential $V(x)$ by extracting the oscillating part of the susceptibility at large distances:

$$t(\varepsilon) = \lim_{x \to -\infty} \left(\frac{i\sqrt{2\varepsilon}}{n(x)} \chi^{osc}(x, -x; \varepsilon - \varepsilon_0)\right). \quad (19)$$

Eq. (19) can be readily tested for $V(x) = -Z\delta(x)$. In this case, the ground-state density is $n(x) = Ze^{-2Z|x|}$ and:

$$g^>(x, x', \varepsilon) = \frac{1}{i\sqrt{2\varepsilon}}\left(e^{i\sqrt{2\varepsilon}|x - x'|} - \frac{Ze^{i\sqrt{2\varepsilon}(|x| + |x'|)}}{i\sqrt{2\varepsilon} + Z}\right). \quad (20)$$

Using eqns. (16) and (19) we get (see also Fig. 2):

$$T(\varepsilon) = |t(\varepsilon)|^2 = \frac{\varepsilon}{\varepsilon + Z^2/2}. \quad (21)$$

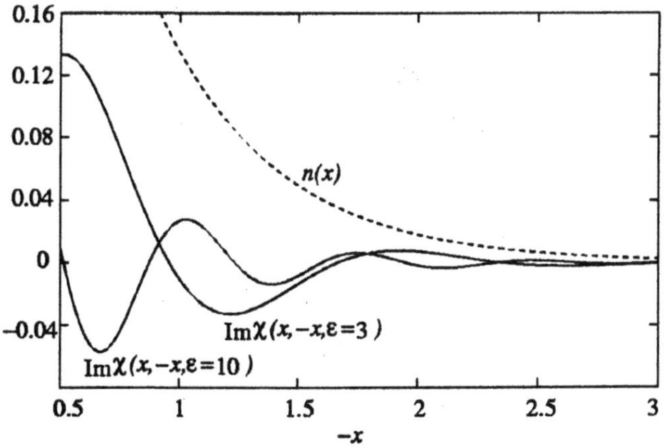

Figure 2. Although $\chi(x, -x, \varepsilon)$ decays exponentially (because the density decays exponentially), it oscillates at large distances, and the amplitude of these oscillations yields the transmission coefficient. $Z = 1$ in this plot.

The Rutgers group is exploring how general this prescription is. Can we always extract cross-sections from the Kohn-Sham susceptibility? Is it possible to relate the Kohn-Sham scattering amplitudes to the true amplitudes using the linear response formalism presented in the preceding section? These are some of the challenges that TDDFT will face in the near future.

6. Memory in TDDFT

There are two sources of memory in TDDFT. Functionals depend, in principle, on both the history of the density and on the initial-state of the system. Often the term memory is used to refer to the former, but we prefer to call it history-dependence to distinguish it from the other memory effect, initial-state dependence.

Very little is known about initial-state dependence and it is completely ignored in all functionals used today. Recently, we have begun to investigate the importance of initial-state dependence (Maitra and Burke, 2001). We found that for one electron, there is in fact *no* initial-state dependence: the time-evolving density uniquely specifies the time-dependent potential. Any attempt to find an alternative initial wavefunction that evolves with the same density as another, in a different potential, results in unphysical potentials which diverge at large distances. We also note that there is no initial-state dependence for the case of one electron scattering off a (many-electron) target (see

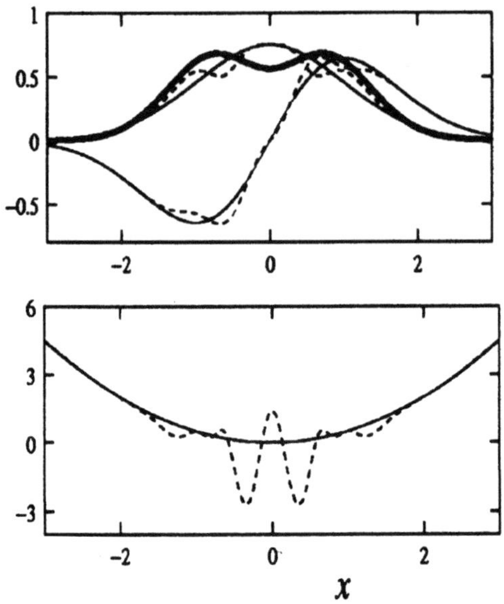

Figure 3. Initial-state dependence: the solid lines in the top panel are the two orbitals of one wavefunction, which happens to be a stationary state of the harmonic oscillator shown below as a solid line. The dashed lines are the orbitals of an alternative initial wavefunction which evolve with the same density (thick solid line in top figure) in the potential which at time 0 is shown as the dashed line in the bottom figure.

Sec. 5). Assuming the target to be initially in its ground-state reduces the initial-state dependence problem to that of one electron, for which knowledge of the density is enough to pin down the potential (and the wavefunction).

For more than one electron, however, the initial-state question is alive and kicking. Using van Leeuwen's construction (1999), we constructed an example for two non-interacting electrons, where two different initial wavefunctions were made to evolve in two different potentials with the same density. An example is given in Fig. 3, where the simple case of two non-interacting electrons in one dimension are considered. If we consider the density in the top panel to be that of an interacting system of two electrons, then the difference between the two potentials in the lower panel corresponds to the difference in the exchange-correlation potential for the two different choices of initial KS state.

The dependence of functionals on the history of the density has been dramatically illustrated in a recent work of Hessler, Maitra and Burke (2001). There we studied exactly numerically, time-dependent Hooke's atom: two interacting electrons in a harmonic well with a time-varying force constant. We began in the ground-state. We could compare the exact correlation energy with an exact adiabatic approximation (i.e. using the exact ground-state functional). We found major qualitative differences in the behaviour of the correlation energy, including positivity of E_C which can never occur in any adiabatic approximation due to the ground-state variational principle. In particular, we found very strong history effects: there were two time slices in which the density evolved essentially identically but in which the correlation potential looked completely different. The density at time t and right around time t is not enough to specify the correlation potential; v_C is a highly non-local functional in time.

In fact the two memory effects are intimately connected. Maitra, Burke and Woodward (2001) derive a memory formula:

$$v_{XC}[n_{t'}; \Psi(t'), \Phi(t')](\mathbf{r}t) = v_{XC}[n_0; \Psi(0), \Phi(0)](\mathbf{r}t), \quad t > t' > 0 \qquad (22)$$

where $n_{t'}(\mathbf{r}t) = n_0(\mathbf{r}t)$, $t > t'$ and is undefined for $t < t'$. This follows from a very simple statement in quantum mechanics, saying that it doesn't matter what we consider to be the "initial" time. We just need to adjust the arguments of the functionals consistently. But it has profound consequences for TDDFT, providing a stringent test of any functional with memory. Notice that ALDA satisfies it trivially, since it has no memory of any sort, but any of the recent functionals which attempt to capture some of the memory (Vignale and Kohn, 1996; Vignale, Ullrich and Conti, 1997; Dobson, Bünner and Gross, 1997) will not. These have history dependence but no initial-state dependence. By functional differentiation of Eq. (22), one obtains a relation expressing the exchange-correlation kernel f_{XC} entirely in terms of initial-state dependence. The memory formula translates the two memory effects into each other; indeed, if one can find a "pseudo-prehistory" (Maitra, Burke and Woodward, 2001) during which the initial state evolves into a non-degenerate ground-state, initial-state dependence can be completely absorbed into a history effect. This is discussed in much detail in the article above.

7. Excited-state densities in TDDFT

We now encounter a conundrum: what are excited state densities in TDDFT, and what is the nature of the Kohn-Sham state when an interacting system is in an excited state? Let us consider the specific case of a single singlet excitation in He. Suppose we find some $v_{ext}(\mathbf{r}t)$ which starts with the He atom in the ground state $1s^2$, with wavefunction $\Psi(0)$ and brings it in a finite time to the first excited singlet state, 1s2p, along a history $n_0(\mathbf{r}t)$. What is an exact Kohn-Sham potential for this problem? Since the Kohn-Sham system begins in a single Slater determinant, with one occupied spatial orbital, it will remain in a single (time-dependent) orbital throughout the history that brings it to the excited state. This means that the excited state density for 1s2p state is represented in the KS system by *one* orbital! Now, consider the case that v_S becomes static once the interacting excited state is reached: the density and $v_{ext}(\mathbf{r}t)$ become static, and the KS state is an eigenstate of the static $v_S(\mathbf{r})$. Clearly, this one-orbital KS eigenstate would be rather unpleasant to deal with, being of a completely different nature to the two-orbital 1s2p state it is representing. On the other hand, with v_{XC} ultra-nonlocal in time, consider the case that the KS wavefunction and potential continue to evolve even as the density remains static. In either situation, it is hard to imagine ALDA being a useful approximation.

8. Summary

TDDFT is a promising theory, generalizing ground-state DFT to include time-dependent systems. Most applications are so far in the linear response regime where, for example, it is being used to get electronic excitations of molecules. For systems with well-separated transition frequencies, the single-pole approximation is accurate for transition frequencies and the Kohn-Sham oscillator strengths are good. We discuss a new idea for developing scattering theory within TDDFT, based on properties of response functions. However TDDFT is not limited to linear response.

Much has yet to be learnt about properties of functionals, one of which is memory. Functionals depend in principle on both the initial-state of the system as well as the entire history of the density. Until very recently little was known about the two effects. Recently it has been shown that initial-state dependence can be significant, as can the dependence on the history. The two memory effects are inextricably intertwined and often initial-state dependence can be completely absorbed into a history dependence. Finally, we discussed an important challenge to TDDFT: representing certain excited states fully within TDDFT (not just in the linear response regime).

Acknowledgments

We gratefully acknowledge support from the National Science Foundation grant no. CHE-9875091 and the Petroleum Research Fund.

References

A. I. Al-Sharif, R. Resta, C.J. Umrigar, *Evidence of physical reality in the Kohn-Sham potential: The case of atomic Ne*, Phys. Rev. A **57**, 2466 (1998).

Y. Andersson, D.C. Langreth, and B.I. Lunqvist, *van der Waals Interactions in Density-Functional Theory*, Phys. Rev. Lett. **76**, 102 (1996).

T. Ando, *Inter-subband optical absorption in space-charge layers on semiconductor surfaces.*, Z. Phys. B. **26**, 263 (1977).

H. Appel, K. Burke, and E.K.U. Gross, *Oscillator strengths in density functional theory*, in preparation.

A.D. Becke, *Density-functional exchange-energy approximation with correct asymptotic behavior*, Phys. Rev. A **38**, 3098 (1988).

A.D. Becke, *Density-functional thermochemistry. III. The role of exact exchange*, J. Chem. Phys. **98**, 5648 (1993).

G.F. Bertsch, J.-I. Iwata, A. Rubio, and K. Yabana, *Real-space, real-time method for the dielectric function*, Phys. Rev. B **61**, 7998 (2000).

P. L. de Boeij, F. Kootstra, J. A. Berger, R. van Leeuwen, and J. G. Snijders, *Current density functional theory for optical spectra: A polarization functional*, J. Chem. Phys. **115**, 1995 (2001).

K. Burke, F.G. Cruz, and K.C. Lam, *Unambiguous exchange-correlation energy density*, J. Chem. Phys. **109**, 8161 (1998).

K. Burke and E.K.U. Gross, *A guided tour of time-dependent density functional theory*, in *Density functionals: Theory and applications*, ed. D. Joubert (Springer, Berlin, 1998).

K. Burke, M. Petersilka, and E.K.U. Gross, *A hybrid functional for the exchange-correlation kernel in time-dependent density functional theory*, in *Recent advances in density functional methods, vol. III*, ed. P. Fantucci and A. Bencini (World Scientific Press, 2000).

M.E. Casida, *Time-dependent density functional response theory of molecular systems: Theory, computational methods, and functionals*, in *Recent developments and applications in density functional theory*, ed. J.M. Seminario (Elsevier, Amsterdam, 1996).

J.F. Dobson, M. Bünner, and E.K.U. Gross, *Time-dependent density functional theory beyond linear response: An exchange-correlation potential with memory*, Phys. Rev. Lett. **79**, 1905 (1997).

C. Filippi, C.J. Umrigar, and X. Gonze, *Excitation energies from density functional perturbation theory*, J. Chem. Phys., **107**, 9994 (1997).

P. Hessler, N.T. Maitra, and K. Burke, *Dynamic effects in time-dependent density functional theory* submitted (2001).

P. Hohenberg and W. Kohn, *Inhomogeneous electron gas*, Phys. Rev. **136**, B 864 (1964).

J. Katriel, F. Zahariev, and K. Burke, *Symmetry and degeneracy in density functional theory* to appear in Int. J. Quantum Chem. (2001).

W. Kohn, *Nobel Lecture: Electronic structure of matter - wave functions and density functionals*, Rev. Mod. Phys. **71**, 1253 (1999).

W. Kohn and L.J. Sham, *Self-consistent equations including exchange and correlation effects*, Phys. Rev. **140**, A 1133 (1965).

F. Kootstra, P.L. de Boeij, and J.G. Snijders, *Efficient real-space approach to time-dependent density functional theory for the dielectric response of nonmetallic crystals*, J. Chem. Phys. **112**, 6517 (2000).

S. Kurth and J.P. Perdew, *Density-functional correction of random-phase-approximation correlation with results for jellium surface energies*, Phys. Rev. B **59**, 10461 (1999).

J. Larkin, C. Bock, N.T. Maitra, and K. Burke, *A new tool for studying the kinetic energy density functional*, in prep., Fall 2001.

C. Lee, W. Yang, and R.G. Parr, *Development of the Colle-Salvetti correlation-energy formula into a functional of the electron density*, Phys. Rev. B **37**, 785 (1988).

R. van Leeuwen, *Mapping from Densities to Potentials in Time-Dependent Density-Functional Theory*, Phys. Rev. Lett. **82**, 3863, (1999).

R. Magyar,B. Terilla, and K. Burke, *Extrapolating the adiabatic connection*, in preparation for Chem. Phys. Letts., Fall 2001.

N.T. Maitra and K. Burke, *Demonstration of initial-state dependence in time-dependent density functional theory*, Phys. Rev. A **63**, 042501 (2001); (E) **64**, 039901 (2001).

N.T. Maitra, K. Burke, H. Appel, E.K.U. Gross, and R. van Leeuwen, *Ten topical questions in time-dependent density functional theory*, to appear in *Reviews in Modern Quantum Chemistry: A Celebration of the Contributions of R.G. Parr*, ed. K.D. Sen (World Scientific, 2001).

N.T. Maitra, K. Burke, and C. Woodward, *Memory in time-dependent density functional theory*, submitted (2001).

J.P. Perdew, K. Burke, and M. Ernzerhof, *Generalized gradient approximation made simple*, Phys. Rev. Lett. **77**, 3865 (1996); **78**, 1396 (1997) (E).

M. Petersilka, E.K.U. Gross, and K. Burke, *Excitation energies from time-dependent density functional theory using exact and approximate functionals*, Int. J. Quantum Chem. **80**, 534 (2000).

M. Petersilka, U.J. Gossmann, and E.K.U. Gross, *Excitation energies from time-dependent density-functional theory*, Phys. Rev. Lett. **76**, 1212 (1996).

E. Runge and E.K.U. Gross, *Density-functional theory for time-dependent systems*, Phys. Rev. Lett. **52**, 997 (1984).

D. Sundholm, *Density functional theory calculations of the visible spectrum of chlorophyll a*, Chem. Phys. Lett., **302**, 480 (1999).

G. Vignale and W. Kohn, *Current-Dependent Exchange-Correlation Potential for Dynamical Linear Response Theory*, Phys. Rev. Lett. **77**, 2037 (1996).

G. Vignale, C.A. Ullrich, and S. Conti, *Time-dependent density functional theory beyond the adiabatic local density approximation*, Phys. Rev. Lett. **79**, 4878 (1997).

T. Whittingham, R.J. Magyar, and K. Burke, *Scaling one spin density at a time*, in prep, Fall 2001.

A. Zangwill and P Soven, *Density-functional approach to local-field effects in finite systems: Photoabsorption in the rare gases*, Phys. Rev. A **21**, 1561 (1980).

ENSEMBLE HARTREE-FOCK AND KOHN-SHAM SCHEMES FOR EXCITED STATES
The optimised effective potential method

N.I. Gidopoulos, P. Papaconstantinou, and E.K.U. Gross*

1. INTRODUCTION

Ground state Hartree Fock (HF) theory not only has proven a valuable tool to calculate the ground state properties of electronic systems, but has also served as a starting point for more sophisticated approaches which deal with correlation, like configuration interaction methods and density functional theory (DFT). Ground state HF theory is based on the Rayleigh-Ritz variational principle, which states that, the expectation value of the Hamiltonian of the N-electron system (atomic units)

$$H = \sum_i \left[-\frac{\nabla^2}{2} + U_{ext}(\mathbf{r}) + \frac{1}{2}\sum_{j \neq i} \frac{1}{r_{ij}} \right]$$

attains its minimum value for the true ground state energy E_0:

$$\langle \Psi | H | \Psi \rangle \geq E_0 \qquad (1)$$

where, Ψ is any normalised N-electron antisymmetric state. Let us take N' and N'' the number of spin up and spin down electrons in the ground state, $N=N'+N''$. In the (unrestricted) HF approximation, one substitutes a trial N-particle Slater determinant

$$\Phi_0 = \frac{1}{\sqrt{N!}} \det\left[\varphi_1^{\uparrow}, \ldots, \varphi_{N'}^{\uparrow}; \varphi_1^{\downarrow}, \ldots, \varphi_{N''}^{\downarrow}\right]$$

* N.I. Gidopoulos, Isis facility, Rutherford Appleton Laboratory, Chilton, Didcot, Oxfordshire, OX11 0QX, UK, P. Papaconstantinou, Institute of Materials Science, NCSR Demokritos, GR 153 10 Ag. Paraskevi, Greece, E.K.U. Gross, Institut für Theoret. Physik, Universität Würzburg, Am Hubland, D97074 Würzburg, Germany.

Electron Correlations and Materials Properties 2
Edited by Gonis *et al.*, Kluwer Academic/Plenum Publishers, 2002

for Ψ and minimizes the expectation value in (1) with respect to the spin-orbitals φ_i^σ of Φ_0. One ends up with single particle Schrödinger-like equations which determine φ_i^σ:

$$h_{\Phi_0} \varphi_j^\sigma(\mathbf{r}) = \varepsilon_j^\sigma \varphi_j^\sigma(\mathbf{r})$$

where,

$$h_{\Phi_0} \varphi_j^\sigma(\mathbf{r}) = \left[-\frac{\nabla^2}{2} + U_{ext}(\mathbf{r}) + \sum_{s=\uparrow,\downarrow} \sum_{\varphi_n^s \in \Phi_0} \int d^3r' \frac{|\varphi_n^s(\mathbf{r}')|^2}{|\mathbf{r}-\mathbf{r}'|} \right] \varphi_j^\sigma(\mathbf{r})$$

$$- \sum_{\varphi_n^\sigma \in \Phi_0} \int d^3r' \frac{\rho_n^\sigma(\mathbf{r},\mathbf{r}')}{|\mathbf{r}-\mathbf{r}'|} \varphi_j^\sigma(\mathbf{r}')$$

In HF theory, one assigns the physical meaning of an independent electron to each occupied spin-orbital φ_n^σ.

The extension of the HF scheme to excited states is more subtle. For example, if one wants to obtain information about the first excited state, the Rayleigh-Ritz variational principle dictates that the trial Slater determinant to approximate the first excited state must be orthogonal to the exact (unknown) ground state. If, instead, one imposes the orthogonality constraint on the Slater determinant Φ_0 which approximately represents the ground state, then the initial error will multiply. Nevertheless, the physical meaning of excited electrons, is often assigned to the eigen-orbitals of Φ_0 which are not occupied in Φ_0. To ascertain that his is clearly problematic, consider that each unoccupied orbital lies in the field of N rather than N-1 electrons, causing the virtual orbitals of Φ_0 to be too diffuse. Various methods to overcome the problem have appeared,[1-3] within ground state HF theory.

An elegant method to study low excitations in HF theory is to abandon the Rayleigh-Ritz variational principle and use the ensemble variational principle (or subspace, or trace variational principle), first used in the foundation of DFT for excited states by Theophilou[4] and later by Gross, Oliveira and Kohn.[5-7] Considering for simplicity only the ground state and the first excited state Ψ_0, Ψ_1 of H, the minimum principle states that the trace $E_\omega[\Psi, \Psi']$

$$E_\omega[\Psi,\Psi'] = (1-\omega)\langle\Psi|H|\Psi\rangle + \omega\langle\Psi'|H|\Psi'\rangle$$

with Ψ, Ψ' orthonormal and $0 \leq \omega \leq 1/2$, attains its minimum value when $\Psi=\Psi_0$ and $\Psi'=\Psi_1$. This statement holds true for any value of ω in the interval $0 \leq \omega \leq 1/2$. Theophilou's subspace or equi-ensemble theory was restricted to the value $\omega=1/2$, where the weight of the excited state is maximized.

2. ENSEMBLE-HARTREE-FOCK EQUATIONS

We will consider for simplicity only the ground state and the lowest single particle excitation of an N-electron system described by H. We also take the highest occupied orbital spin up and non degenerate. The Slater determinant representing the lowest excitation will be

$$\Phi_1 = \frac{1}{\sqrt{N!}} \det\left[\varphi_1^{\uparrow}, \ldots, \varphi_{N'-1}^{\uparrow}, \varphi_{N'+1}^{\uparrow}; \varphi_1^{\downarrow}, \ldots, \varphi_{N''}^{\downarrow}\right]$$

Form the trace:

$$E_\omega = (1-\omega)\langle \Phi | H | \Phi \rangle + \omega \langle \Phi' | H | \Phi' \rangle$$

with $0 \le \omega \le 1/2$. The trace E_ω can be written in terms of the spin orbitals

$$\begin{aligned}
E_\omega = & -\frac{1}{2} \sum_{\sigma=\uparrow,\downarrow} \sum_{\varphi_n^\sigma \in \Phi_0} \int d^3r\, \varphi_n^{\sigma *}(\mathbf{r}) \nabla^2 \varphi_n^{\sigma}(\mathbf{r}) \\
& -\frac{\omega}{2} \int d^3r\, \varphi_{N'+1}^{\sigma *}(\mathbf{r}) \nabla^2 \varphi_{N'+1}^{\sigma *}(\mathbf{r}) + \frac{\omega}{2} \int d^3r\, \varphi_{N'}^{\sigma *}(\mathbf{r}) \nabla^2 \varphi_{N'}^{\sigma *}(\mathbf{r}) \\
& + \int d^3r\, U_{ext}(\mathbf{r}) \rho(\mathbf{r}) \\
& + \frac{1}{2} \iint d^3r\, d^3r' \frac{\rho(\mathbf{r})\rho(\mathbf{r}')}{|\mathbf{r}-\mathbf{r}'|} + \frac{\omega(1-\omega)}{2} \iint d^3r\, d^3r' \frac{g(\mathbf{r})g(\mathbf{r}')}{|\mathbf{r}-\mathbf{r}'|} \\
& - \frac{1}{2} \sum_{\sigma=\uparrow,\downarrow} \iint d^3r\, d^3r' \frac{|\rho^\sigma(\mathbf{r},\mathbf{r}')|^2}{|\mathbf{r}-\mathbf{r}'|} - \frac{\omega(1-\omega)}{2} \iint d^3r\, d^3r' \frac{|g(\mathbf{r},\mathbf{r}')|^2}{|\mathbf{r}-\mathbf{r}'|}
\end{aligned}$$

where,

$$\rho_n^\sigma(\mathbf{r},\mathbf{r}') = \varphi_n^\sigma(\mathbf{r})\, \varphi_n^{\sigma *}(\mathbf{r}')$$

$$g(\mathbf{r},\mathbf{r}') = \rho_{N'+1}^{\uparrow}(\mathbf{r},\mathbf{r}') - \rho_{N'}^{\uparrow}(\mathbf{r},\mathbf{r}')$$

$$g(\mathbf{r}) = g(\mathbf{r},\mathbf{r})$$

$$\rho^{\downarrow}(\mathbf{r},\mathbf{r}') = \sum_{\varphi_n^{\downarrow} \in \Phi_0} \rho_n^{\downarrow}(\mathbf{r},\mathbf{r}')$$

$$\rho^{\downarrow}(\mathbf{r},\mathbf{r}') = \sum_{\varphi_n^{\uparrow} \in \Phi_0} \rho_n^{\uparrow}(\mathbf{r},\mathbf{r}') + \omega\, g(\mathbf{r},\mathbf{r}')$$

$$\rho(\mathbf{r}) = \sum_{\sigma=\uparrow,\downarrow} \rho^\sigma(\mathbf{r},\mathbf{r}')$$

Based on the ensemble minimum principle, the ensemble-Hartree-Fock (eHF) equations for the ground and the lowest excited states are obtained by minimizing the trace with respect to the spin-orbitals.[8, 9] Observe that not all spin-orbitals appear symmetrically in E_ω. This implies that the single particle equations for the various orbitals of the same spin are inhomogeneous i.e., the equations have different potentials, and consequently nondiagonal Lagrange multipliers appear that cannot be chosen to vanish with a suitable unitary transformation of the minimizing orbitals.[9]

Since it is obviously desirable to overcome the difficulties associated with a scheme which involves different Hamiltonians for the various orbitals, we will choose to minimize the trace E_ω, requiring that the orbitals of the same spin form the lowest $N'+1$ and N'' single-particle eigenstates of some common (for each spin) local potential $V^\uparrow$ and $V^\downarrow$ to be determined. This constitutes the Optimised Effective Potential (OEP) method[10,11] for the eHF scheme which introduces an additional approximation to eHF.

In the following, we shall derive the equations determining $V^\uparrow$ and $V^\downarrow$. In the OEP picture, E_ω is a functional of the local potentials, $E_\omega = E_\omega[V^\uparrow,V^\downarrow]$. The functional derivative $\delta E_\omega[V^\uparrow,V^\downarrow]/\delta V^\sigma(\mathbf{r})$ is defined by the limit

$$\lim_{\lambda\to 0}\frac{E_\omega[V^\sigma+\lambda\,\delta V^\sigma,V^{\sigma'}]-E_\omega[V^\sigma,V^{\sigma'}]}{\lambda}=\int d^3r\,\delta V^\sigma(\mathbf{r})\frac{\delta E_\omega}{\delta V^\sigma(\mathbf{r})} \qquad (2)$$

where, $\sigma=\uparrow,\downarrow$ and $\sigma'=\downarrow,\uparrow$.

The functional derivative must be independent of the choice of δV^σ. Denote here also by φ_n^σ the single-particle eigenstates of V^σ

$$\left[-\frac{\nabla^2}{2}+V^\sigma(\mathbf{r})\right]\varphi_n^\sigma(\mathbf{r})=\varepsilon_n^\sigma\,\varphi_n^\sigma(\mathbf{r}) \qquad (3)$$

Then, for $0<\lambda\ll 1$, the eigenstates of $V^\sigma+\lambda\,\delta V^\sigma$ will be given by

$$\varphi_{n\lambda}^\sigma(\mathbf{r})=\varphi_n^\sigma(\mathbf{r})-\lambda\int d^3r'\,\delta V^\sigma(\mathbf{r}')\,G_n^\sigma(\mathbf{r},\mathbf{r}')\,\varphi_n^\sigma(\mathbf{r}')$$

where,

$$G_n^\sigma(\mathbf{r},\mathbf{r}')=\sum_{m\neq n}\frac{\rho_n^\sigma(\mathbf{r},\mathbf{r}')}{\varepsilon_m^\sigma-\varepsilon_n^\sigma}$$

Substituting the above in (2) and using Eq. (3) we can calculate the functional derivative:

$$\frac{\delta E_\omega}{\delta V^\downarrow(\mathbf{r})}=\int d^3r'\,F^\downarrow(\mathbf{r};\mathbf{r}',\mathbf{r}')\left[V^\downarrow(\mathbf{r}')-U_{ext}(\mathbf{r}')-\int\frac{d^3r''\rho(\mathbf{r}'')}{|\mathbf{r}'-\mathbf{r}''|}\right]$$
$$+\iint d^3r'd^3r''\frac{F^\downarrow(\mathbf{r};\mathbf{r}',\mathbf{r}'')\,\rho^\downarrow(\mathbf{r}'',\mathbf{r}')}{|\mathbf{r}'-\mathbf{r}''|}$$

$$\frac{\delta E_{\omega}}{\delta V^{\uparrow}(\mathbf{r})}=\int d^3r' F^{\uparrow}(\mathbf{r};\mathbf{r}',\mathbf{r}')\left[V^{\uparrow}(\mathbf{r}')-U_{ext}(\mathbf{r}')-\int\frac{d^3r''\rho(\mathbf{r}'')}{|\mathbf{r}'-\mathbf{r}''|}\right]$$
$$+\iint d^3r'd^3r''\frac{F^{\uparrow}(\mathbf{r};\mathbf{r}',\mathbf{r}'')\,\rho^{\uparrow}(\mathbf{r}'',\mathbf{r}')}{|\mathbf{r}'-\mathbf{r}''|}$$
$$-\omega(1-\omega)\int d^3r' B(\mathbf{r};\mathbf{r}',\mathbf{r}')\int\frac{d^3r''g(\mathbf{r}'')}{|\mathbf{r}'-\mathbf{r}''|}$$
$$+\omega(1-\omega)\iint d^3r'd^3r''\frac{B(\mathbf{r};\mathbf{r}',\mathbf{r}'')\,g(\mathbf{r}'',\mathbf{r}')}{|\mathbf{r}'-\mathbf{r}''|}$$

where,

$$B(\mathbf{r};\mathbf{r}',\mathbf{r}'')=G^{\uparrow}_{N'+1}(\mathbf{r}',\mathbf{r})\,\rho^{\uparrow}_{N'+1}(\mathbf{r},\mathbf{r}'')+G^{\uparrow}_{N'+1}(\mathbf{r},\mathbf{r}'')\,\rho^{\uparrow}_{N'+1}(\mathbf{r}',\mathbf{r})$$
$$-G^{\uparrow}_{N'}(\mathbf{r}',\mathbf{r})\,\rho^{\uparrow}_{N'}(\mathbf{r},\mathbf{r}'')-G^{\uparrow}_{N'}(\mathbf{r},\mathbf{r}'')\,\rho^{\uparrow}_{N'}(\mathbf{r}',\mathbf{r})$$

$$F^{\downarrow}(\mathbf{r};\mathbf{r}',\mathbf{r}'')=\sum_{n=1}^{N''}\left[G^{\downarrow}_{n}(\mathbf{r}',\mathbf{r})\,\rho^{\downarrow}_{n}(\mathbf{r},\mathbf{r}'')+G^{\downarrow}_{n}(\mathbf{r},\mathbf{r}'')\,\rho^{\downarrow}_{n}(\mathbf{r}',\mathbf{r})\right]$$

$$F^{\uparrow}(\mathbf{r};\mathbf{r}',\mathbf{r}'')=\sum_{n=1}^{N'}\left[G^{\uparrow}_{n}(\mathbf{r}',\mathbf{r})\,\rho^{\uparrow}_{n}(\mathbf{r},\mathbf{r}'')+G^{\uparrow}_{n}(\mathbf{r},\mathbf{r}'')\,\rho^{\uparrow}_{n}(\mathbf{r}',\mathbf{r})\right]+\omega B(\mathbf{r};\mathbf{r}',\mathbf{r}'')$$

$B(\mathbf{r};\mathbf{r}',\mathbf{r}'')$ and $F^{\sigma}(\mathbf{r};\mathbf{r}',\mathbf{r}'')$ are Hermitian with respect to the second pair of arguments: $[B(\mathbf{r};\mathbf{r}',\mathbf{r}'')]^*=B(\mathbf{r};\mathbf{r}'',\mathbf{r}')$ and $[F^{\sigma}(\mathbf{r};\mathbf{r}',\mathbf{r}'')]^*=F^{\sigma}(\mathbf{r};\mathbf{r}'',\mathbf{r}')$. Consequently $B(\mathbf{r};\mathbf{r}',\mathbf{r}')$ and $F^{\sigma}(\mathbf{r};\mathbf{r}',\mathbf{r}')$ are real.

The minimising potentials of E_{ω} are determined by

$$\frac{\delta E_{\omega}}{\delta V^{\sigma}(\mathbf{r})}=0,\quad \sigma=\uparrow,\downarrow \tag{4}$$

Finally, applying the popular mean field approximation by Krieger Li and Iafrate[12-15] (KLI) to deal with the OEP integral equations (4) we get

$$\rho^{\downarrow}(\mathbf{r})V^{\downarrow}_{x}(\mathbf{r})=\sum_{n=1}^{N''}\left|\varphi^{\downarrow}_{n}(\mathbf{r})\right|^2\int d^3r'\left|\varphi^{\downarrow}_{n}(\mathbf{r}')\right|^2V^{\downarrow}_{x}(\mathbf{r}')-\int\frac{d^3r'|\rho^{\downarrow}(\mathbf{r},\mathbf{r}')|^2}{|\mathbf{r}-\mathbf{r}'|}$$
$$+\sum_{n=1}^{N''}\left|\varphi^{\downarrow}_{n}(\mathbf{r})\right|^2\iint\frac{d^3r'd^3r''}{|\mathbf{r}'-\mathbf{r}''|}\rho^{\downarrow}_{n}(\mathbf{r}',\mathbf{r}'')\,\rho^{\downarrow}(\mathbf{r}'',\mathbf{r}')$$

$$\rho^{\uparrow}(\mathbf{r})V_x^{\uparrow}(\mathbf{r})=\sum_{n=1}^{N'}\left|\varphi_n^{\uparrow}(\mathbf{r})\right|^2\int d^3r'\left|\varphi_n^{\uparrow}(\mathbf{r}')\right|^2V_x^{\uparrow}(\mathbf{r}')-\int\frac{d^3r'|\rho^{\uparrow}(\mathbf{r},\mathbf{r}')|^2}{|\mathbf{r}-\mathbf{r}'|}$$

$$+\omega\left|\varphi_{N'+1}^{\uparrow}(\mathbf{r})\right|^2\int d^3r'\left|\varphi_{N'+1}^{\uparrow}(\mathbf{r}')\right|^2V_x^{\uparrow}(\mathbf{r}')$$

$$-\omega\left|\varphi_{N'}^{\uparrow}(\mathbf{r})\right|^2\int d^3r'\left|\varphi_{N'}^{\uparrow}(\mathbf{r}')\right|^2V_x^{\uparrow}(\mathbf{r}')$$

$$+\sum_{n=1}^{N'}\left|\varphi_n^{\uparrow}(\mathbf{r})\right|^2\iint\frac{d^3r'd^3r''}{|\mathbf{r}'-\mathbf{r}''|}\rho_n^{\uparrow}(\mathbf{r}',\mathbf{r}'')\rho^{\uparrow}(\mathbf{r}'',\mathbf{r}')$$

$$+\omega\left|\varphi_{N'+1}^{\uparrow}(\mathbf{r})\right|^2\iint\frac{d^3r'd^3r''}{|\mathbf{r}'-\mathbf{r}''|}\rho_{N'+1}^{\uparrow}(\mathbf{r}',\mathbf{r}'')\rho^{\uparrow}(\mathbf{r}'',\mathbf{r}')$$

$$-\omega\left|\varphi_{N'}^{\uparrow}(\mathbf{r})\right|^2\iint\frac{d^3r'd^3r''}{|\mathbf{r}'-\mathbf{r}''|}\rho_{N'}^{\uparrow}(\mathbf{r}',\mathbf{r}'')\rho^{\uparrow}(\mathbf{r}'',\mathbf{r}')$$

$$+\omega(1-\omega)\left\{g(\mathbf{r})\int\frac{d^3r'g(\mathbf{r}')}{|\mathbf{r}-\mathbf{r}'|}-\int\frac{d^3r'|g(\mathbf{r},\mathbf{r}')|^2}{|\mathbf{r}-\mathbf{r}'|}\right.$$

$$-\left|\varphi_{N'+1}^{\uparrow}(\mathbf{r})\right|^2\iint\frac{d^3r'd^3r''}{|\mathbf{r}'-\mathbf{r}''|}\left[\left|\varphi_{N'+1}^{\uparrow}(\mathbf{r}')\right|^2g(\mathbf{r}'')-\rho_{N'+1}^{\uparrow}(\mathbf{r}',\mathbf{r}'')g(\mathbf{r}'',\mathbf{r}')\right]$$

$$\left.+\left|\varphi_{N'}^{\uparrow}(\mathbf{r})\right|^2\iint\frac{d^3r'd^3r''}{|\mathbf{r}'-\mathbf{r}''|}\left[\left|\varphi_{N'}^{\uparrow}(\mathbf{r}')\right|^2g(\mathbf{r}'')-\rho_{N'}^{\uparrow}(\mathbf{r}',\mathbf{r}'')g(\mathbf{r}'',\mathbf{r}')\right]\right\}$$

where,

$$V_x^{\sigma}(\mathbf{r})=V^{\sigma}(\mathbf{r})-U_{ext}(\mathbf{r})-\int d^3r'\frac{\rho(\mathbf{r}')}{|\mathbf{r}-\mathbf{r}'|}$$

The KLI equations can be solved self-consistently. Initially, one ignores the terms on the rhs which include the unknown potential, solves for $V_x^{\uparrow,\downarrow}$ on the lhs and then iterates until self-consistency is reached.

We have calculated the total ensemble energies of various atoms and present the results in Table 1. An equi-ensemble (ω=1/2) has been chosen and the KLI approximation is employed in all cases. The ensemble configuration and dimensionality (number of states) is shown in the second and third columns. The states in each ensemble are proper eigenfunctions of the total spin[16] and consequently we had to use a different potential for each ensemble. These potentials were derived in a similar way to Eqs. (4). Details of the derivation will be published elsewhere.

The experimental data in the last column are taken from the NIST database for Atomic Spectroscopy.[17] The eHF results in the fourth column give upper bounds to the experimental values as expected.

Table 1. Ensemble total energies for various atoms in atomic units.

Atom	Ensemble	Dim	eHF	eKS	Exper.
He	$\{1s^2\}\,^1S$; $\{1s2s\}\,^1S$	2	-2.479	-2.516	-2.525
He	$\{1s2s\}\,^3S$; $\{1s3s\}\,^3S$	2	-2.117	-2.117	-2.123
He	$\{1s2s\}\,^3S$; $\{1s2p\}\,^3P$	2	-2.154	-2.154	-2.155
He	$\{1s2s\}\,^3S$; $\{1s2p\}\,^3P$	4	-2.143	-2.143	-2.144
Li	$\{1s^22s\}\,^2S$; $\{1s^22p\}\,^2P$	4	-7.385	-7.434	-7.427
Li	$\{1s^22s\}\,^2S$; $\{1s^23s\}\,^2S$	2	-7.364	-7.411	-7.416
C^{2+}	$\{1s^22s^2\}\,^1S$; $\{1s^22s2p\}\,^3P$	4	-36.258	-36.357	-36.355

2.1 The Ensemble-Kohn-Sham Scheme

The ensemble-Kohn-Sham (eKS) scheme proposed by Theophilou[4] and later by Gross, Oliveira and Kohn[5-7] was also based on the ensemble variational principle. The basic quantity in the formulation of the theory is the ensemble (average) density $\rho(\mathbf{r})$ and in the eKS scheme, the interacting system with ensemble density $\rho(\mathbf{r})$ and energy E_ω is mapped to a non-interacting one with the same ensemble density and energy.

A comparison of the eHF and the eKS theories reveals that the definition of the ensemble Hartree energy

$$\frac{1}{2}\iint d^3r d^3r' \frac{\rho(\mathbf{r})\,\rho(\mathbf{r}')}{|\mathbf{r}-\mathbf{r}'|}$$

contains an unphysical interaction between the orbitals $\varphi^{\dagger}_{N'+1}$ and $\varphi^{\dagger}_{N'}$, which, in the non-interacting picture, are never occupied in the same configuration. We have called this unphysical interaction the ghost-interaction.[18] Clearly the ensemble exchange and correlation energy functional must correct for the ghost interaction. The reader is also referred to the recent reformulation of the OEP method for ensembles of excited states by Nagy.[19]

In the fifth column in Table 1, we give the results of the eKS calculation, where the ghost interaction is corrected and the approximate (ground state) correlation functional of Colle and Salvetti[20, 21] is used to account for the remaining part of the correlation energy. The eHF results we present, using the OEP method, are identical to the results of an x-only eKS calculation.

REFERENCES

1. H.J. Silverstone and M.L. Yin, J. Chem. Phys. **49**, 2026 (1968).
2. W.J. Hunt and W.A. Goddard III, Chem. Phys. Lett. **6**, 414 (1969).
3. S. Huzinaga and C. Arnau Phys. Rev. A **1**, 1285 (1970).
4. A.K. Theophilou, J. Phys. C, **12**, 5419 (1979).
5. E.K.U. Gross, L.N. Oliveira and W. Kohn, Phys. Rev. A **37**, 2805 (1988).
6. E.K.U. Gross, L.N. Oliveira and W. Kohn, Phys. Rev. A **37**, 2809 (1988).
7. L.N. Oliveira, E.K.U. Gross and W. Kohn, Phys. Rev. A **37**, 2821 (1988).
8. A.K. Theophilou, *Recent Developments in Condensed Matter Physics*, Vol.2, edited by J.T. DeVreese, L.F. Lemmens, V.E. Doren and J. Van Royen, Plenum, New York, p.125 (1981).
9. N. Gidopoulos and A.Theophilou, Phil. Mag. B **69**, 1067 (1994).
10. R.T. Sharp and G.K. Horton, Phys. Rev. **90**, 317 (1953).
11. J.D. Talman and W.F. Shadwick, Phys. Rev. A **14**, 36 (1976).

12. J.B. Krieger, Y. Li and G.J. Iafrate, Phys. Rev. A **45**, 101 (1991).
13. J.B. Krieger, Y. Li and G.J. Iafrate, Phys. Rev. A **46**, 5453 (1992).
14. J.B. Krieger, Y. Li and G.J. Iafrate, Phys. Rev. A **47**, 165 (1993).
15. J.B. Krieger, Y. Li and G.J. Iafrate, in *Density Functional Theory*, edited by R.M. Dreizler and E.K.U. Gross, Plenum Press, New York, p.191, (1995).
16. A.K. Theophilou and P.G. Papaconstantinou, Phys. Rev. A **61**, 22502 (2000).
17. National Institute of Standards and Technology, U.S. Department of Commerce (1995).
18. N.I. Gidopoulos, P.G. Papaconstantinou and E.K.U. Gross, submitted to Phys. Rev. Lett. (2001).
19. Á. Nagy, J. Phys. B: At. Mol. Opt. Phys. **34**, 2363, (2001).
20. R. Colle and D. Salvetti, Theoret. Chim. Acta **37**, 329 (1975).
21. R. Colle and D. Salvetti, Theoret. Chim. Acta, **53**, 55 (1979).

DENSITY-MATRIX FUNCTIONAL THEORY AND THE HIGH-DENSITY ELECTRON GAS

P. Ziesche

Max Planck Institute for the
Physics of Complex Systems,
Nöthnitzer Straße 38,
D-01187 Dresden, Germany

1 INTRODUCTION

Often the ground state (GS) of the unpolarized uniform electron gas (EG) serves as a good model to describe the influence of the interelectron Coulomb repulsion on the energies and reduced densities [1]. Essential kinematic features of this quantum-many-body system are described by the electron density ρ, the momentum distribution $n(k)$, and the dimensionless pair distribution functions, or pair densities (PD), $g_{\uparrow\uparrow}$ and $g_{\uparrow\downarrow}$ for electron pairs with parallel and antiparallel spins, respectively. $n(k)$ satisfies $0 \leq n(k) \leq 1$ ($n(k)$ is non-idempotent), and the g's are probabilities, thus $g_{\uparrow\uparrow}, g_{\uparrow\downarrow} \geq 0$. ρ fixes the dimensionless density parameter r_s from $(4\pi r_s^3/3)a_{\rm B}^3 = 1/\rho$ and the Fermi wave number $k_{\rm F}$ from $k_{\rm F}^3 = 3\pi^2\rho$. The PD's $g_{\uparrow\uparrow}$ and $g_{\uparrow\downarrow}$ are each referred to as either exchange or Fermi hole, and correlation or Coulomb hole, respectively. They are functions of the interelectron distance $r_{12} = |\underline{r}_1 - \underline{r}_2|$. All the quantities of the EG depend parametrically on the density parameter r_s. The Coulomb repulsion between each electron pair causes the phenomenon denoted electron correlation [2–9], which shows up in the correlation 'tails' of $n(k)$, which means $n(k<1) \lessapprox 1$ for holes and $n(k>1) \gtrapprox 0$ for particles, where k is measured in units of $k_{\rm F}$. Naturally related to these tails is $z_{\rm F} < 1$, the quasiparticle weight or reduced jump discontinuity of $n(k)$ at $k = 1$, which causes the oscillatory long-range behavior of $g_{\uparrow\uparrow}$ and $g_{\uparrow\downarrow}$ for $r_{12} \to \infty$. Short-range (or dynamical) correlations show up in the curvature of $g_{\uparrow\uparrow}$ for zero-interelectron distance. We refer to this situation as the electrons being "in contact" with or "on-top" of each other. They also appear in the on-top value of $g_{\uparrow\downarrow}$, which determines simultaneously both the on-top slope of $g_{\uparrow\downarrow}$ from the coalescing cusp theorem [10], and the asymptotic behavior of $n(k)$ for $k \to \infty$. The spin-traced PD $g = (g_{\uparrow\uparrow} + g_{\uparrow\downarrow})/2$ and its parallel and antiparallel 'components' have been repeatedly studied over the years [11-35]. This is especially true for the on-top value of the Coulomb hole $g_{\uparrow\downarrow}$ [25-27, 34, 35] and for the on-top curvature of the Fermi hole $g_{\uparrow\uparrow}$ [17, 28, 34, 35], which are short-range correlation properties. Examples of studies of the Fermi and Coulomb holes in molecules are given in Ref. [36].

The PD $\rho_2 = \rho^2 g$ is discussed in terms of probabilities, average numbers of particle pairs, and fluctuations in Refs. [37-39]. Its relation to correlation and localization effects is discussed in Ref. [40]. Here we stress the importance of the cumulant expansion of higher-order reduced density matrices [41-45] in conjunction with the linked-diagram theorem of perturbation theory [46], where the Coulomb repulsion is treated as a perturbation. Although the Fermi-hole $g_{\uparrow\uparrow}$ is determined both from the correlated and thus non-idempotent momentum distribution $n(k)$, and from the cumulant PD $h_{\uparrow\uparrow}$, using the relation $g_{\uparrow\uparrow} = 1 - |f|^2 - h_{\uparrow\uparrow}$, where $f(k_F r_{12})$ is the Fourier transform of $n(k)$, the Coulomb hole $g_{\uparrow\downarrow}$ is simply determined from the cumulant PD $h_{\uparrow\downarrow}$ using $g_{\uparrow\downarrow} = 1 - h_{\uparrow\downarrow}$ [40-42]. From f and g follow t and v, respectively, the kinetic and interaction energies per particle. With the help of the virial theorem [47, 48] these energies t and v follow also from the r_s-dependence of $e = t + v$, the total energy per particle. All these GS properties have been calculated using perturbation theory (random phase approximation for e [49-52], for $n(k)$ [53, 54], for g [11, 12], and local field corrections beyond the random phase approximation, e.g. Refs. [13-20, 22-34, 55-62]). Other calculational schemes are the coupled-cluster [63], and quantum-Monte-Carlo [21, 64] methods.

New methods are appearing on the horizon: the density-matrix functional theory (DMFT) [9,65-78] and the contracted Schrödinger equation (CSE) approach [79-82], which may be added in future to the group of well-tried traditional (configurational interaction, coupled cluster [83], Møller-Plesset, quantum Monte Carlo, incremental [84], density functional and quasiparticle) methods. The DMFT rests upon the key quantities γ, the one-body reduced density matrix (1-matrix), which is for the EG simply $\gamma = \rho f$, and $\tilde{V}_{\text{int}}[\gamma]$, the interaction energy as a universal functional of this γ [9,65-78]. There are several ways to find proper DMFT functionals $\tilde{V}_{\text{int}}[\gamma]$: Goedecker and Umrigar generalized an expression known for 2-electron systems and calculated with it correlation energies of the isoelectronic series He(Z) and Be(Z). Another way makes use of the cumulant expansion, which gives e.g. the two-body reduced density matrix (2-matrix) as a sum of products of the 1-matrix and a non-reducible remainder, the cumulant 2-matrix χ, the contraction of which yields the 1-matrix. This cumulant expansion automatically leads to the interaction energy as a bilinear functional of this χ, hence $V_{\text{int}}[\chi]$. Appropriate functionals $\chi[\gamma]$ are suggested either from the CSE approach [81] or from perturbation theory [35], which then combined with $V_{\text{int}}[\chi]$, thus defining a 1-matrix functional $\tilde{V}_{\text{int}}[\gamma] = V_{\text{int}}[\chi[\gamma]]$, as required in the DMFT. With the CSE method Yasuda calculated correlation energies of several molecules and reaction energies of several reactions. J. Cioslowski posed the question, whether these new methods work also for extended systems with the uniform EG as the simplest test model. To find a first answer in Refs. [77, 78] the high-density EG has been studied with a generalized Goedecker-Umrigar functional $\tilde{V}_{\text{int}}[\chi]$ and with the Yasuda functional $\chi[\gamma]$.

In Sec. 2, the basic many-body notation and in Sec. 3 its specification for homogeneous systems and to the high-density EG are summarized. In Sec. 4, the DMFT is described and a generalized Goedecker-Umrigar functional is applied to the high-density EG. In Sec. 5, the Yasuda functional arising from the CSE approach is also applied to this model of a weakly correlated extended system.

2 THE SYSTEM AND ITS GROUND STATE

We consider the fermionic many-particle system described by the Hamiltonian

$$H = \sum_{i}^{N} (t_i + v_{\text{ext}}(\underline{r}_i)) + \frac{1}{2!}\sum_{i \neq j}^{N} v(i,j), \quad t_i = -\frac{\hbar^2}{2m}\frac{\partial^2}{\partial \underline{r}_i^2} \tag{2.1}$$

with $v(i,j) = v_{\text{int}}(r_{ij})$ and $r_{ij} = |\underline{r}_i - \underline{r}_j|$. Its GS energy is $E = \langle \Psi | H | \Psi \rangle / \langle \Psi | \Psi \rangle$ or

$$E = \int d1\; t_1 \gamma(1|1')\Big|_{1'=1} + \int d1\; \rho(1) v_{\text{ext}}(1) + \int \frac{d1 d2}{2!}\; \rho_2(1,2) v(1,2), \tag{2.2}$$

thus $E = T[\gamma] + V_{\text{ext}}[\rho] + V_{\text{int}}[\rho_2]$. $\Psi(1,\ldots,N)$ is the antisymmetric GS wave function. We use the short-hand notations $1 = (\underline{r}_1, \sigma_1)$ and $\int d1 = \sum_{\sigma_1} \int d^3 r_1$ and the normalization

$$\int \frac{d1 \ldots dN}{N!} |\Psi(1,\ldots,N)|^2 = 1. \tag{2.3}$$

Each particle configuration is correctly counted only once, just as in the partition function of classical equilibrium statistics, taking into account that the particles are identical. This introduces a factor of $1/N!$, as commonly occurs in quantum theory. The important quantum-kinematical quantities contained in E are the 1-matrix, $\gamma(1|1') = \gamma^*(1'|1)$, the 1-density, $\rho(1) = \gamma(1|1)$, and the PD, $\rho_2(1,2) = \rho_2(2,1)$, the contraction of which yields $\int d2\; \rho_2(1,2) = (N-1)\rho(1)$. $\rho(1)$ is normalized to N, the total number of particles, $\int d1\; \rho(1) = N$.

The density $\rho(1)$ enables us to calculate the external-potential and the Hartree 'components' of E,

$$V_{\text{ext}}[\rho] = \int d1\; \rho(1) v_{\text{ext}}(1), \quad V_{\text{H}}[\rho] = \frac{1}{2}\int d1 d2\; \rho(1)\rho(2) v(1,2). \tag{2.4}$$

From the 1-matrix $\gamma(1|1')$, we may calculate the kinetic and the Fock 'components' of E,

$$T[\gamma] = \int d1\; t_1 \gamma(1|1')\Big|_{1'=1}, \quad V_{\text{F}}[\gamma] = -\frac{1}{2}\int d1 d2\; \gamma(1|2)\gamma(2|1) v(1,2). \tag{2.5}$$

The diagonal of the Fourier transform of $\gamma(1|1')$ gives the momentum distribution $n(k)$.

From the PD $\rho_2(1,2)$, we may calculate V_{int}, the interaction 'component' of E, which is the last term of Eq. (2.2). It is normalized from

$$\int \frac{d1d2}{2!}\,\rho_2(1,2) = \binom{N}{2}, \tag{2.6}$$

where each 2-particle configuration is counted only once, and the rhs is the total number of particle pairs. It is the diagonal of the 2-matrix

$$\gamma(1|1',2|2') = \int \frac{d3\ldots dN}{(N-2)!}\Psi(1,2,3,\ldots)\Psi^*(1',2',3,\ldots). \tag{2.7}$$

Its cumulant expansion defines the cumulant 2-matrix, $\chi(1|1',2|2')$,

$$\gamma(1|1',2|2') = A\gamma(1|1')\gamma(2|2') - \chi(1|1',2|2'), \tag{2.8}$$

where A is the 'antisymmetrizer', defined by

$$A\gamma(1|1')\gamma(2|2') = \gamma(1|1')\gamma(2|2') - \gamma(1|2')\gamma(2|1'). \tag{2.9}$$

We note that both $\gamma(1|1',2|2')$ and $\chi(1|1',2|2')$ exhibit the Pauli principle property, e.g. $\gamma(1|1',2|2') = -\gamma(2|1',1|2')$. The sign in front of $\chi(1|1',2|2')$ is chosen such that it is positively normalized, as it is shown later in Eq. (2.12). From the 'contraction' property of the 2-matrix,

$$\int d2\;\gamma(1|1',2|2) = (N-1)\gamma(1|1'), \tag{2.10}$$

and the contraction property of the cumulant 2-matrix,

$$\int d2\;\chi(1|1',2|2) = \gamma(1|1') - \int d2\;\gamma(1|2)\gamma(2|1'), \tag{2.11}$$

the cumulant PD $u(1,2) = \chi(1|1,2|2)$ yields the normalization

$$\int d1d2\;u(1,2) = N - \int d1d2\;\gamma(1|2)\gamma(2|1) = \sum_k n_k(1-n_k), \tag{2.12}$$

which is related to the perfect screening sum rule. Because the momentum distribution is non-idempotent, i.e. $0 < n_k < 1$, the rhs of Eq. (2.12) is $< N$. When correlations are absent, $n_k(1-n_k) = 0$, in agreement with $\chi(1|1',2|2') = 0$ and $u(1,2) = 0$. With increasing correlations, the size-extensive normalization of $u(1,2)$ also increases. Note the invariance of the rhs under the replacement $n_k \to 1-n_k$ due to particle-hole symmetry [85]. Also note that the rhs contains the 2nd-order non-idempotency, which was recently introduced and discussed as a possible measure of the correlation strength [39, 40]. For the EG, an important feature of the cumulant 2-matrix χ is that it contains the cusp conditions that arise from the divergence of the Coulomb repulsion for zero interelectron distance [10]. Furthermore, from χ one obtains the PD in momentum space, allowing a fluctuation analysis as has been similarly done in position space [37–39] with the conclusion 'correlation suppresses or enhances fluctuations for repulsive or attractive interaction, respectively'. According to the cumulant expansion, Eqs. (2.8) and (2.9), the interaction energy $V_{\rm int}$ (the last term of Eq.(2.2)) has three 'components', the Hartree (H), Fock (F), and cumulant (c) components,

$$V_{\rm int}[\chi] = V_{\rm H}[\rho] + V_{\rm F}[\gamma] + V_{\rm c}[\chi], \quad V_{\rm c}[\chi] = -\int \frac{d1d2}{2!}\, \chi(1,2)v(1,2). \tag{2.13}$$

The last term is called the cumulant interaction energy.

So the most important quantity in the quantum-kinematical basis for calculating the energy E is the cumulant 2-matrix $\chi(1|1',2|2')$. From Eq. (2.10), it gives the 1-matrix $\gamma(1|2)$ and thus $\rho(1)$ and $n(k)$. From Eq. (2.8), it also gives the PD

$$\rho_2(1,2) = \rho(1)\rho(2) - \gamma(1|2)\gamma(2|1) - u(1,2), \tag{2.14}$$

which is the diagonal of Eq. (2.8). Note that the cumulant PD $u(1,2)$ is size-extensively normalized, as in Eq. (2.12).

The eigenfunctions and eigenvalues of $\gamma(1|1')$ are the natural spin-orbitals $\varphi_k(1)$ and the natural occupation numbers n_k. Thus, the spectral resolution

$$\gamma(1|1') = \sum_k \varphi_k(1) n_k \varphi_k^*(1') \tag{2.15}$$

holds. The eigenfunctions $\varphi_{k_1,k_2}(1,2)$ and eigenvalues n_{k_1,k_2} of the cumulant 2-matrix $\chi(1|1',2|2')$ may be referred to as natural 'geminals', or two-particle wave functions, and natural 'geminal occupation numbers',

$$\chi(1|1',2|2') = \frac{1}{2!}\sum_{k_1,k_2} \varphi_{k_1,k_2}(1,2) n_{k_1,k_2} \varphi^*_{k_1,k_2}(1',2'),$$

$$\int \frac{d1d2}{2!}\, |\varphi_{k_1,k_2}(1,2)|^2 = 1, \quad \sum_{k_1,k_2} n_{k_1,k_2} = \sum_k n_k(1-n_k). \tag{2.16}$$

The antisymmetric geminals $\varphi_{k_1,k_2}(1,2)$ are labeled with the geminal quantum numbers k_1, k_2, which means that in the configurational interaction expansion of $\varphi_{k_1,k_2}(1,2)$, the 2-particle Slater determinant of the natural orbitals $\varphi_{k_1}(1)$ and $\varphi_{k_2}(2)$ is assumed to be dominant. One may think of the natural geminals arising from an effective 2-particle Schrödinger equation with an effective 2-particle potential. similarly to Refs. [26, 34].

If the interparticle interaction $v_{\rm int}(r_{ij})$ is treated as a perturbation, then E or $e = E/N$, the GS energy per particle, is given by the linked 'vacuum' Feynman diagrams of Figs. 1a,b. The 1-matrix $\gamma(1|1')$ and the momentum distribution $n(k)$ are given by the diagrams of Fig. 1c, and the cumulant 2-matrix $\chi(1|1',2|2')$ is given by the diagrams of Fig. 1d,e. Note the relation between the cumulant expansion and the linked diagram theorem [46].

3 THE UNIFORM ELECTRON GAS

3.1 Homogeneous systems

For an unpolarized homogeneous system (e.g. the uniform EG) with $\rho(1) = \text{const}$ and $k_{\rm F}^3 = 3\pi^2\rho$, we may write $\rho(1) = \rho$ and

$$\gamma(1|2) = \delta_{\sigma_1,\sigma_2}\rho f(x), \quad f(x) = \int_0^\infty dk^3\, n(k)\frac{\sin kx}{kx}, \quad dk^3 = 3k^2dk, \tag{3.1}$$

with k in units of $k_{\rm F}$. $x = k_{\rm F}r_{12}$ is the interparticle distance r_{12} measured in units of $1/k_{\rm F}$. The momentum distribution $n(k)$ yields the kinetic energy per particle

$$t = \int_0^\infty dk^3 n(k)k^2\, \varepsilon_{\rm F},\ \varepsilon_{\rm F} = \frac{\hbar^2 k_{\rm F}^2}{2m},\ \int_0^\infty dk^3 n(k) = 1,\ dk^3 = 3k^2dk\,. \tag{3.2}$$

The quantity $f(x)$ in Eq. (3.1) is the dimensionless 1-matrix. The normalization of $n(k)$ requires $f(0) = 1$. In addition one finds

$$f(x \to 0) = 1 - \frac{< k^2 >}{3!}x^2 + \frac{< k^4 >}{5!}x^4 + \cdots, \tag{3.3}$$

when approaching the diagonal, where $< k^2 >= t/\varepsilon_{\rm F}$. For the EG, the dots may mean $O(x^6 \ln x)$, since $n(k \to \infty) \to 1/k^8$, thus the moment $< k^6 >$ does not exist. The jump discontinuity of $n(k)$ at the Fermi edge $k = 1$ is

$$z_{\rm F} = n(1-\delta) - n(1+\delta), \quad \delta \overset{>}{\to} 0, \tag{3.4}$$

which causes the long-range asymptotic oscillations

$$f(x \to \infty) \to -z_{\rm F}\frac{3\cos x}{x^2} + \cdots. \tag{3.5}$$

For the ideal Fermi gas, $n_0(k) = \Theta(1-k)$, $z_{\rm F} = 1$, and $f_0(x) = 3(\sin x - x\cos x)/x^3$, thus $f_0(x \to 0) = 1 - x^2/10 + \cdots$, and we note that $t_0/\varepsilon_{\rm F} = 3/5$.

The spin-traced 2-body quantites are

$$\rho_2(\underline{r}_1, \underline{r}_2) = \sum_{\sigma_1,\sigma_2} \rho_2(1,2) = \rho^2 g(x), \quad u(\underline{r}_1, \underline{r}_2) = \sum_{\sigma_1,\sigma_2} u(1,2) = \rho^2 h(x), \tag{3.6}$$

so that from Eq.(2.8) the dimensionless spin-traced PD is

$$g(x) = 1 - \frac{1}{2}|f(x)|^2 - h(x), \quad \alpha^3\int_0^\infty dx^3\, h(x) = \int_0^\infty dk^3\, n(k)[1-n(k)] < 1. \tag{3.7}$$

Distinguishing between electron pairs with parallel spins and others with antiparallel spins yields the (correlated) exchange or Fermi hole

$$g_{\uparrow\uparrow}(x) = 1-|f(x)|^2-h_{\uparrow\uparrow}(x), \quad \alpha^3\int_0^\infty dx^3\, h_{\uparrow\uparrow}(x) = 2\int_0^\infty dk^3\, n(k)[1-n(k)] < 1, \quad (3.8)$$

which is determined by the momentum distribution $n(k)$ and the spin-parallel cumulant PD $h_{\uparrow\uparrow}(x)$. We also have the correlation or Coulomb hole,

$$g_{\uparrow\downarrow}(x) = 1 - h_{\uparrow\downarrow}(x), \quad \int_0^\infty dx^3\, h_{\uparrow\downarrow}(x) = 0, \tag{3.9}$$

which is determined only by the spin-antiparallel cumulant PD $h_{\uparrow\downarrow}(x)$. The spin-traced pair quantities are [86]

$$g(x) = \frac{1}{2}[g_{\uparrow\uparrow}(x) + g_{\uparrow\downarrow}(x)], \quad h(x) = \frac{1}{2}[h_{\uparrow\uparrow}(x) + h_{\uparrow\downarrow}(x)]. \tag{3.10}$$

Note the on-top properties $h_{\uparrow\uparrow}(0) = 0$, due to the Pauli principle, and $h'_{\uparrow\uparrow}(0) = 0$.

Correlations appears in the on-top value $g_{\uparrow\downarrow}(0) = 1 - h_{\uparrow\downarrow}(0)$ and in the on-top curvature $g''_{\uparrow\uparrow}(0) = -2f''(0) - h''_{\uparrow\uparrow}(0)$. The 1-matrix contributes the amount

$$-2f''(0) = \frac{2}{3}\frac{t}{\varepsilon_F} \rightarrow \frac{2}{5} \quad \text{for} \quad t \rightarrow t_0 = \frac{3}{5}\varepsilon_F \tag{3.11}$$

to the on-top curvature of $g''_{\uparrow\uparrow}(x)$, and the amount

$$1 - |f(x \rightarrow \infty)|^2 \rightarrow 1 - z_F^2\,\frac{9}{2}\frac{\cos 2x}{x^4} \tag{3.12}$$

to the oscillatory long-range behavior [18, 29].

The Fock and the cumulant 'components' of the interaction energy per particle are

$$v_F[f(x)] + v_c[h(x)] = -\frac{1}{2}\alpha^3\int_0^\infty dx^3\left[\frac{1}{2}|f(x)|^2 + h(x)\right]v(x) \tag{3.13}$$

with $v(k_F r_{12}) = v_{\mathrm{int}}(r_{12})$.

The natural geminals of Eq. (2.16) now have the form,

$$\varphi^{S,T}_{\underline{K},L} \sim \mathrm{e}^{\mathrm{i}\underline{KR}}\varphi^{S,T}_{\underline{K},L}(\underline{r}), \tag{3.14}$$

as in Refs. [26, 34], where $\underline{R}$ and $\underline{r}$ are the centre of mass and the difference coordinates, respectively. Here $L = (l, m)$ are the angular momentum quantum numbers and we use the notation S = singlet, T = triplet.

3.2 Coulomb interaction

Here the particle interaction is specified as $v_{\rm int}(r_{12}) = \epsilon^2/r_{12}$, $\epsilon^2 = e^2/(4\pi\varepsilon_0)$. This quantum-mechanical many-body system is characterized by a homogeneous density ρ, a dimensionless density parameter r_s and the Fermi wave number $k_{\rm F}$:

$$\rho = \frac{3}{4\pi r_0^3}, \quad r_s = \frac{r_0}{a_{\rm B}}, \quad k_{\rm F}a_{\rm B} = \frac{1}{\alpha r_s}, \quad \alpha = (\frac{4}{9\pi})^{1/3}. \tag{3.15}$$

$a_{\rm B} = \hbar^2/m\epsilon^2$ is the Bohr radius, in the following the energies are measured in ryd $= \epsilon^2/2a_{\rm B} = (\alpha r_s)^2\varepsilon_{\rm F}$.

From the momentum distribution $n(k)$ follows the kinetic energy

$$t = \frac{1}{(\alpha r_s)^2}\int_0^\infty dk^3\, n(k)k^2, \; \int_0^\infty dk^3\, n(k) = 1, \; dk^3 = 3k^2dk. \tag{3.16}$$

Correlation (subscript corr) usually means the correction of the non-interacting or Hartree-Fock quantities (subscript 0) due to the interparticle Coulomb repulsion:

$$n(k) = n_0(k) + n_{\rm corr}(k), \quad t = t_0 + t_{\rm corr}, \quad t_{\rm corr} = \frac{1}{(\alpha r_s)^2}\int_0^\infty dk^3\, n_{\rm corr}(k)k^2 \tag{3.17}$$

with $n_0(k) = \Theta(1-k)$ and $t_0 = \frac{3}{5}\frac{1}{(\alpha r_s)^2}$.

From the PD $g(x)$ follows the interaction energy

$$v = -\frac{\alpha^2}{r_s}\int_0^\infty dx^3\, [1-g(x)]\frac{1}{x}, \; \alpha^3\int_0^\infty dx^3\, [1-g(x)] = 1, \; dx^3 = 3x^2dx. \tag{3.18}$$

The Hartree term $v_{\rm H}$ is compensated by $v_{\rm ext}$ (positive background compensation). In $v = v_{\rm F} + v_{\rm c}$ the Fock (or generalized exchange) energy $v_{\rm F}$ is usually split into v_0, its lowest-order contribution, and v_1, the corresponding remainder: $v_{\rm F} = v_0 + v_1$ with

$$v_0 = -\frac{\alpha^2}{r_s}\int_0^\infty dx^3\, \frac{1}{2}[f_0(x)]^2\frac{1}{x} = -\frac{3}{\pi\alpha r_s}\int_0^1 dk_1\int_0^1 dk_2\, k_1k_2\ln\frac{k_1+k_2}{|k_1-k_2|} = -\frac{3}{2\pi\alpha r_s}, \tag{3.19}$$

$$v_1 = -\frac{\alpha^2}{r_s}\int_0^\infty dx^3\, \frac{1}{2}\{[f(x)]^2 - [f_0(x)]^2\}\frac{1}{x}$$
$$= -\frac{3\cdot 2}{\pi\alpha r_s}\int_0^\infty dk_1\int_0^\infty dk_2\, n_{\rm corr}(k_1)n_0(k_2)k_1k_2\ln\frac{k_1+k_2}{|k_1-k_2|}. \tag{3.20}$$

v_1 and the cumulant interaction energy

$$v_{\rm c} = -\frac{\alpha^2}{r_s}\int_0^\infty dx^3\, h(x)\frac{1}{x}, \quad \alpha^3\int_0^\infty dx^3\, h(x) = 1 - \int_0^\infty dk^3\, [n(k)]^2 \tag{3.21}$$

define the correlation contribution to the interaction energy: $v = v_0 + v_{\rm corr}$, $v_{\rm corr} = v_1 + v_{\rm c}$. From $e = t + v$, the total energy per particle as a function of r_s, follow its kinetic and interaction 'components' according to the virial theorem [47]

$$t = -\frac{\partial}{\partial r_s}r_se, \quad v = \frac{1}{r_s}\frac{\partial}{\partial r_s}r_s^2e. \tag{3.22}$$

Relations between $n(k)$ and $g(x)$ on the basis of the virial theorem are discussed in Ref. [48].

The singularity of the Coulomb repulsion for zero interelectron distance causes the coalescing cusp theorems[10, 17, 19, 34]

$$h'_{\uparrow\downarrow}(0) = -\alpha r_s[1 - h_{\uparrow\downarrow}(0)], \quad h'''_{\uparrow\uparrow}(0) = \frac{3}{2}\alpha r_s h''_{\uparrow\uparrow}(0). \tag{3.23}$$

Whereas the momentum distribution influences the on-top curvature of $g_{\uparrow\uparrow}(x)$ and its oscillatory long-range behavior, the cusp theorems are contained only in the the cumulant PD's $h_{\uparrow\downarrow}(x)$ and $h_{\uparrow\uparrow}(x)$. The 1-particle quantities $n(k)$ or $f(x)$ do not contribute to the cusp theorems. Is this related to the fact that v_1 does not contribute to the logarithmic term of v_{corr} ?

3.3 High densities

In the high-density limit $r_s \to 0$ we have

$$e_{\text{corr}} = a \cdot \ln r_s + b + c + c' \cdot r_s \ln r_s + c'' \cdot r_s + O(r_s^2 \ln r_s),$$

$$a = \frac{2}{\pi^2}(1 - \ln 2),\ b = -0.1422,\ c = \frac{1}{3}\ln 2 - \frac{3}{2\pi^2}\zeta(3),\ c' = 0.0184,\ c'' = -0.020. \tag{3.24}$$

While a, the Macke number [49], and b, the Gell-Mann/Brueckner number [50], arise from

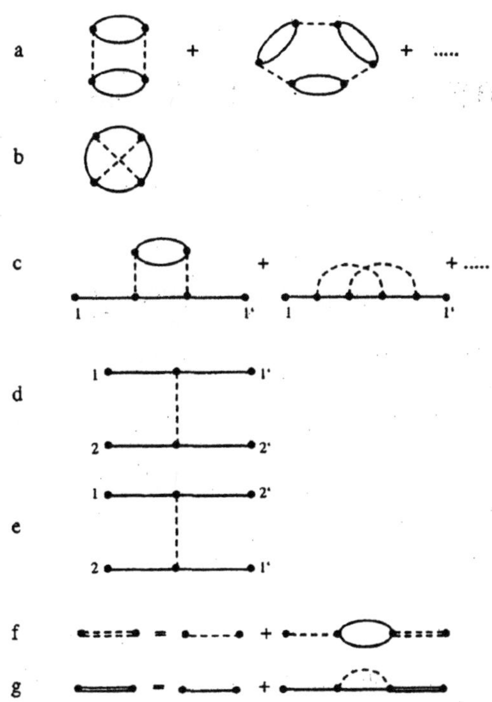

Fig. 1: Feynman diagrams of the uniform electron gas

1a,b: 'vacuum diagrams' for the ground state energy e

1c: diagrams for the 1-particle quantities $n(k)$ or $f(x)$

1d,e: diagrams for the cumulant 2-matrix $\chi_{11'22'}$

1f: effective interaction

1g: effective 1-particle propagator

the ring or RPA diagrams of Fig 1a, the coefficient c due to Onsager, Mittag, and Stephen [51] comes from the exchange diagram of Fig. 1b. c' is due to DuBois [55] and

Carr and Maradudin [56] and c'' is due to Endo et al. [57]. With the virial theorem (3.22) it follows that

$$t_{\text{corr}} = -(a\cdot\ln r_s + b + c) - a + O(r_s \ln r_s), \quad v_{\text{corr}} = 2(a\cdot\ln r_s + b + c) + a + O(r_s \ln r_s). \quad (3.25)$$

The logarithmic term of v_{corr} stems from the cumulant interaction energy v_c, not from v_1. The on-top PD $g(0) = g_{\uparrow\downarrow}(0)/2 = [1 - h_{\uparrow\downarrow}(0)]/2$ is determined by

$$h_{\uparrow\downarrow}(0) = 2d \cdot r_s + O(r_s^2 \ln r_s), \quad d = \frac{\alpha}{5\pi}(\pi^2 + 6\ln 2 - 3). \quad (3.26)$$

This behavior has its origin in the exchange diagram of Fig. 1d. The coefficient d is due to Geldart [12] and Kimball [18]. The on-top curvature of $g_{\uparrow\uparrow}(x)$ is determined by [19, 33]

$$-2f''(0) = \alpha^2 r_s^2 \frac{t}{3} = \frac{2}{5} + O(r_s^2 \ln r_s), \quad -h''_{\uparrow\uparrow}(0) = -0.844 \cdot \alpha^2 r_s + O(r_s^2 \ln r_s). \quad (3.27)$$

The jump discontinuity of $n(k)$ is described by [53–62, 77, 78]

$$z_{\text{F}} = \frac{1}{1 + 3.353 \cdot \frac{\alpha}{\pi^2} r_s + 0.003 \cdot r_s^2 \ln r_s + \cdots}, \quad (3.28)$$

where the second prefactor 0.003 is only an estimate. This behavior arises from the diagrams of Figs. 1c and 1g. The particle correlation tail has the asymptotic property $n(k \to \infty) = (2\alpha r_s/3\pi)^2 h_{\uparrow\downarrow}(0)/k^8$, as given in Refs. [60, 61].

4 DENSITY-MATRIX FUNCTIONAL THEORY

The key quantities of density functional theory (DFT) are the density ρ and the density functional $(T + V_{\text{F}} + V_{\text{c}})[\rho]$. The key quantities of DMFT are the 1-matrix γ and the 1-matrix functionals $\tilde{V}_{\text{int}}[\gamma]$ [9,65–78] or better $\tilde{V}_{\text{c}}[\gamma]$, which is the cumulant 'component' of the interaction energy: $\tilde{V}_{\text{int}}[\gamma] = V_{\text{H}}[\rho] + V_{\text{F}}[\gamma] + \tilde{V}_{\text{c}}[\gamma]$. Note that $V_{\text{H}}[\rho]$ and $V_{\text{F}}[\gamma]$, the Hartree and Fock 'components' of the interaction energy, are by definition functionals of the density ρ and the 1-matrix γ, respectively.

The Goedecker-Umrigar functional

$$(V_{\text{F}} + V_{\text{c}})[\gamma] = -\frac{1}{2}\sum_{k_1,k_2} \sqrt{n_1 n_2}\, v_{1221} - \frac{1}{2}\sum_{k_1} n_1(1 - n_1) v_{1111} \quad (4.1)$$

with $k_1 = (\underline{k}_1, s_1)$, $n_1 = n(k_1)$, and

$$v_{11'22'} = \int d1d2\; \varphi_1^*(1)\varphi_{1'}(1) v_{\text{int}}(r_{12}) \varphi_2^*(2) \varphi_{2'}(2) \quad (4.2)$$

generalizes an expression known for 2-particle systems and has been successfully applied to atoms. It correctly reproduces the different dependencies of electron correlation on the nuclear charge Z in the isoelectronic series He(Z) and Be(Z) [68, 69]. Note that the last term of Eq. (4.1) makes the functional $\tilde{V}_{\rm int}[\gamma] = (V_{\rm H} + V_{\rm F} + V_{\rm c})[\gamma]$ free of self interactions. This functional turns into the Hartree-Fock expression, if all occupation numbers are either 0 or 1 or if the square-root sign and the last term are deleted. It is not particle-hole symmetric. Other disadvantages are discussed in Refs. [72, 75].

In Ref. [77] the question was posed, whether the more general functional

$$(V_{\rm F} + V_{\rm c})[\gamma] = -\frac{1}{2}\sum_{k_1,k_2} F(n_1,n_2)v_{1221} - \frac{1}{2}\sum_{k_1}[n_1^2 - F(n_1,n_2)]v_{1111} \tag{4.3}$$

(with a symmetric function $F(n_1, n_2)$ and with the last term making $\tilde{V}_{\rm int}[\gamma]$ again free of self-interactions) can describe the high-density EG as a simple example of an extended system. For the uniform EG the functional (4.3) yields

$$e[n] = \frac{1}{(\alpha r_s)^2}\int_0^\infty dk^3\, n(k)k^2 - \frac{3}{\pi\alpha r_s}\int_0^\infty dk_1 \int_0^\infty dk_2\, F(n_1,n_2)k_1k_2 \ln\frac{k_1+k_2}{|k_1-k_2|}, \tag{4.4}$$

the total energy per particle in units of ryd as a functional of the momentum distribution $n(k)$. Minimizing $e[n]$ under a density constraint produces the Euler equation

$$\frac{1}{(\alpha r_s)^2}k_1^2 - \frac{2}{\pi\alpha r_s}\int_0^\infty dk_2\, \frac{\partial F(n_1,n_2)}{\partial n_1}\frac{k_2}{k_1}\ln\frac{k_1+k_2}{|k_1-k_2|} = \zeta, \tag{4.5}$$

the solution of which gives at first $n(k, r_s, \zeta)$. Finally, the normalization condition (3.16) makes $\zeta = -(r_s^4/3)\partial(e/r_s^3)/\partial r_s$, the chemical potential [87] to be a function of r_s. In the high-density or small-r_s limit it is (again in ryd)

$$\zeta = \varepsilon_{\rm F} - \frac{2}{\pi\alpha r_s} + a\cdot\ln r_s + \cdots, \quad \varepsilon_{\rm F} = \frac{1}{(\alpha r_s)^2}. \tag{4.6}$$

Unfortunately the known features of the high-density momentum distribution $n(k)$ [53, 54] with its strong occupancy limit $n(k\to 0)\lesssim 1$ and its weak occupancy limit $n(k\to\infty)\gtrsim 0$ and its jump discontinuity (3.28) are incompatible with Eq. (4.3). This is sketched in the following, further details are given in Ref. [77]. Namely, in the weak occupancy limit $k_{1,2}\to\infty$ it is

$$F(n_1\to 0, n_2\to 0) \to C\sqrt{n_1n_2}, \quad C\approx -0.64. \tag{4.7}$$

In the mixed occupancy limit $k_1\to\infty, k_2\to 0$ it is

$$F(n_1\to 0, n_2\to 1)\to \sqrt{n_1}(1 - C\sqrt{1-n_2}). \tag{4.8}$$

These are the constraints imposed by the large-k asymptotics of $n(k)$. Its small-k asymptotics or strong occupancy limit $k_{1,2}\to 0$ imposes the constraint

$$F(n_1\to 1, n_2\to 1)\to (1 - D\sqrt{1-n_1})(1 - D\sqrt{1-n_2}), \quad D\approx -0.32. \tag{4.9}$$

Finally the jump discontinuity of $n(k)$ at $k = 1$ imposes the constraint $D = 1$, thus contradicting the above given value of D. So, the large- and small-asymptotics of $n(k)$

rigorously determine the behavior of the function $F(n_1, n_2)$ for each of its arguments approaching either 0 or 1. Unfortunately, the resulting $F(n_1, n_2)$ does not give rise to $n(k)$ with a proper discontinuity as a solution of the Euler equation (4.5). Consequently, functionals of the form (4.1) cannot be exact for a high-density EG. One thus has to be on the lookout for better functionals. One alternative is described in the next Section.

5 CONTRACTED SCHRÖDINGER EQUATIONS

Recently the contracted-Schrödinger-equation (CSE) approach [79–82] has been used to derive an equation which determines the cumulant 2-matrix $\chi_{11'22'}$ as a functional of the 1-matrix $\gamma_{11'}$ [81]. Thereby the many-body Schrödinger equation $H\Psi = E\Psi$ is multiplied by $\Psi(1', 2', \cdots)$ and in the 1st step integrated over $2' = 2,\ 3' = 3, \cdots$. The resulting equation contains (besides E) the 1-matrix, the 2-matrix, and the 3-matrix. Therefore this equation is called 1-3 contracted Schrödinger equation. In the next step integration over $3' = 3,\ 4' = 4, \cdots$ yields the 2-4 contracted Schrödinger equation as the next member of a whole hierarchy, which may be compared with the Bogolyubov-Born-Green-Kirkwood-Yvon hierarchy in statistical physics. These equations simplify themselves if the cumulant expansion is applied and $E = \mathrm{Tr}((t+v_{\mathrm{ext}})\gamma+\frac{1}{2}v_{\mathrm{int}}(A\gamma\gamma-\chi))$ is used. Recently Yasuda truncated this hierarchy by setting the cumulant 3-matrix and all the higher cumulant matrices equal to zero [81]. With this assumption the 1-3 contracted Schrödinger equation is a functional relation between the (unknown) 1-matrix $\gamma_{11'}$ and the (also unknown) cumulant 2-matrix $\chi_{11'22'}$:

$$\tilde{\varepsilon}_{11'}n_1'(1-n_1) = (v\chi - nv\chi - v\chi n)_{11'} + (\chi\tilde{\varepsilon})_{11'}, \tag{5.1}$$

where

$$(v\chi)_{11'} = \sum_{k_2,k_3,k_4} v_{1234}\,\chi_{1'234} \quad \text{etc.} \quad (\chi\tilde{\varepsilon})_{11'} = \sum_{k_2,k_2'} \chi_{11'22'}\,\tilde{\varepsilon}_{2'2} \tag{5.2}$$

and $\tilde{\varepsilon}_{11'}$ are the natural matrix elements of the generalized HF Hamiltonian

$$h^{\mathrm{HF}}(1)\gamma(1|1') = (t_1 + v_{\mathrm{ext}}(\underline{r}_1))\gamma(1|1') + \int d2\, v_{12}[\gamma(1|1')\gamma(2|2) - \gamma(1|2)\gamma(2|1')]. \tag{5.3}$$

The diagonal of Eq. (5.1) is

$$\tilde{\varepsilon}_{11}n_1(1-n_1) = (1-2n_1)(v\chi)_{11} + (\chi\tilde{\varepsilon})_{11}. \tag{5.4}$$

(The similar equation (19) in Ref. [81] does not contain the last term, but instead an adjustable parameter.) In the next step Eqs. (5.1) and (5.4) are used to simplify the 2-4 contracted Schrödinger equation, which is another relation between γ and χ. The result is

$$\chi_{11'22'} = \frac{n_1(1-n_{1'})n_2(1-n_{2'}) + (1-n_1)n_{1'}(1-n_2)n_{2'}}{\varepsilon_1 + \varepsilon_{1'} + \varepsilon_2 + \varepsilon_{2'}}(v_{11'22'} - v_{12'21'}). \tag{5.5}$$

with

$$\varepsilon_1 = \frac{(1-2n_1)^2}{n_1(1-n_1)}(v\chi)_{11}. \tag{5.6}$$

Eqs. (5.5) and (5.6) implicitly define the cumulant 2-matrix as a functional of the 1-matrix. This functional is particle-hole symmetric. Thus the cumulant interaction energy V_c is also a particle-hole symmetric functional of the 1-matrix, which can be written as

$$V_c = -\frac{1}{2}\sum_{k_1}(v\chi)_{11} = -\frac{1}{2}\sum_{k_1}\frac{n_1(1-n_1)}{(1-2n_1)^2}\varepsilon_1. \tag{5.7}$$

This method has been applied successfully to molecules by self-consistently solving Eqs. (5.5) and (5.6) [81]. Its applicability to the high-density EG as an example of an extended system has also been shown [78]. Namely, with the high-density $n(k)$ of Refs. [53, 54] the self-consistent solution of Eqs. (5.5) and (5.6) yields the high-density limits of v_{corr} and $h_{\uparrow\downarrow}$, as described in Subsec. 3.3.

The functional (5.5) is deficient in that it does not provide the correct contraction property. So, from Eq. (5.5) it follows that

$$\sum_{k_2}\chi_{1122} = \sum_{k_2}\tilde{v}_{1122}\frac{n_2(1-n_2)}{\varepsilon_1+\varepsilon_2}n_1(1-n_1) = \tilde{n}_1(1-\tilde{n}_1) \tag{5.8}$$

defining the new occupation numbers $\tilde{n}_1$, where presumably $\sum_{k_1}\tilde{n}_1 \neq N$. So the approximations contained in Eq. (5.5) act as if particles are added to the system or removed from it. A similar functional obtained from perturbation theory has the same disadvantage [35]. So these are particle number non-conserving approximations. However, their advantage, they are constructed to be free of self interactions and they are particle-hole symmetric.

6 SUMMARY

In addition to the 1-matrix γ with its characteristic short-range curvature in finite and extended systems and long-range oscillations in extended systems, the most important 2-body quantity is the cumulant 2-matrix χ. When contracted it yields the 1-matrix. It is size-extensively normalized, can be represented only by linked diagrams, and contains the cusp conditions. It yields the cumulant interaction energy $V_c[\chi]$, which is by definition a simple linear functional of χ. The cumulant expansion makes the interaction energy $V_{\text{int}}[\chi] = V_H[\rho] + V_F[\gamma] + V_c[\chi]$ free of self interactions. The importance of this requirement for the conventional DFT, especially when applied to strongly correlated systems, has been stressed by J. P. Perdew at this conference. So, the hyper generalized gradient approximation provides a DFT functional, which corresponds to the cumulant expansion and is therefore free of self interactions. Calculational schemes for χ and its diagonalizing geminals and occupation numbers are highly desirable.

The DMFT with a universal, but sophisticated and to date unknown 1-matrix functional $\tilde{V}_c[\gamma]$ goes beyond the conventional DFT with the also universal, but only approximately known density functional $(T + V_F + V_c)[\rho]$. From a corresponding Euler equation, it yields not only the diagonal elements of the 1-matrix γ, but also its off-diagonal elements, from which the momentum distribution is obtained. A significant advantage of the DMFT is that it does not require a non-interacting reference system. Hence, it should apply both to weakly correlated as well as to strongly correlated systems.

Within the CSE approach or from perturbation theory, particle-hole symmetric functionals $\chi[\gamma]$ may be derived, and when combined with $V_c[\chi]$, provide particle-hole symmetric DMFT functionals $\tilde{V}_c[\gamma] = V_c[\chi[\gamma]]$.

ACKNOWLEDGMENTS

The author thanks R. J. Bartlett, K. Burke, P. Gori-Giorgi, E. K. U. Gross, J. P. Perdew, A. Savin, M. Seidl, V. H. Smith for stimulating discussions, R. Klemm and P. Chocian for critically reading the manuscript, and acknowledges P. Fulde for support of this work. Thanks to Mrs. Schuppe and E. Mrosan for technical help.

References

[1] With the 'poor' jellium model, pure electronic correlations are studied. Contrarily, in all real molecules, clusters, and solids the many-electron problem is intimately combined with the multi-centre problem. To explain or predict properties of molecules, clusters, and solids these two problems have to be solved simultaneously. For an overview of recent developments in solid state theory cf. the book *Theoretical Materials Science. Tracing the Electronic Origins of Materials Behavior* (MRS, Warrendale, 2000) by A. Gonis with the Preface by P. Ziesche. For the application of the jellium model to metal surfaces cf. A. Kiejna and K. F. Wojciechowski, *Metal Surface Electron Physics*, Pergamon, Kidlington (1996).

[2] E.R. Davidson, *Reduced Density Matrices in Quantum Chemistry*, Academic, New York (1976).

[3] P. Ziesche and G. Lehmann, ed., *Ergebnisse in der Elektronentheorie der Metalle*, Akademie- and Springer-Verlag, Berlin (1983).

[4] R. Erdahl and V. H. Smith, Jr., ed., *Density Matrices and Density Functionals*, Reidel, Dortrecht (1987).

[5] E. K. U. Gross and E. Runge, *Vielteilchentheorie*, Teubner, Stuttgart (1986);
E. K. U. Gross, E. Runge, and O. Heinonen, *Many-Particle Theory*, Hilger, Bristol (1991).

[6] P. Fulde, *Electron Correlations in Molecules and Solids*, Springer, Berlin (1991, 3rd enlarged ed. 1995).

[7] W. Kutzelnigg and P. von Herigonte, *Adv. Quant. Chem.* 36:185(1999).

[8] A. J. Coleman and V. I. Yukalov, *Reduced Density Matrices*, Springer, Berlin (2000).

[9] J. Cioslowski, ed., *Many-Electron Densities and Reduced Density Matrices*, Kluwer/Plenum, New York (2000).

[10] For the coalescing or electron-electron or correlation cusp theorem cf.
T. Kato, *Commun. Pure Appl. Math.* 10:51(1957),
J. H. Smith, Jr., *Chem. Phys. Lett.* 9:365(1971),
A. J. Thakkar, V. H. Smith, Jr., *Chem. Phys. Lett.* 42:476(1976),
W. Klopper and W. Kutzelnigg, *J. Chem Phys.* 94:2020(1991) and references cited therein,
J. Cioslowski, B. B. Stefanov, A. Tan, and C. J. Umrigar, *J. Chem. Phys.* 103:6093(1995),
and H. F. King, *Theor. Chim. Acta* 94:345(1996).

[11] P. Noziéres and D. Pines, *Phys. Rev.* 111:442(1958).

[12] D. J. W. Geldart, *Can. J. Phys.* 45:3139(1967).

[13] P. Vashishta and K. S. Singwi, *Phys. Rev.* B 6:875(1972).

[14] A. Holas, P. K. Aravind, and K. S. Singwi, *Phys. Rev.* B 20:4912(1979).

[15] P. K. Aravind, A. Holas, and K. S. Singwi, *Phys. Rev.* B 25:561(1982).

[16] J. T. Devreese, F. Brosens, and L. F. Lemmens, *Phys. Rev.* B 21:1349,1363(1980) and references cited therein; cf. also
H. Nachtegaele, F. Brosens, and J. T. Devreese, *Phys. Rev.* B 28:6064(1983).

[17] J. C. Kimball, *Phys. Rev.* A 7:1648(1973).

[18] J. C. Kimball, *Phys. Rev.* B 14:2371(1976).

[19] A. K. Rajagopal, J. C. Kimball, and M. Banerjee, *Phys. Rev.* B 18:2339(1978).

[20] V. Contini, G. Mazzone, and F. Sacchetti, *Phys. Rev.* B 33:712(1986).

[21] D. M. Ceperley and B. J. Alder, *Phys. Rev. Lett.* 45:566(1980).

[22] W. E. Pickett and J. Q. Broughton, *Phys. Rev.* B 48:14859(1993).

[23] C. F. Richardson and N. W. Ashcroft, *Phys. Rev.* B 50:8170(1994); cf. also
M. Lein, E. K. U. Gross, and J. P. Perdew, *Phys. Rev.* B 61:13431(2000).

[24] G. Ortiz, M. Harris, and P. Ballone, *Phys. Rev. Lett.* 82:5317(1999).

[25] H. Yasuhara, *Solid State Commun.* 11:1481(1972).

[26] A. W. Overhauser, *Can. J. Phys.* 73:683(1995).

[27] K. Burke, J. P. Perdew, and M. Ernzerhof, *J. Chem. Phys.* 109:3760(1998).

[28] J. F. Dobson, *J. Chem. Phys.* 94:4328(1991); *J. Phys.: Cond. Matt.* 4:7877(1992).

[29] J. P. Perdew and Y. Wang, *Phys. Rev.* B 46:12947(1992); 56:7018(1997).

[30] V. A. Rassolov, J. A. Pople, and M. A. Ratner, *Phys. Rev.* B 59:15625(1999).

[31] P. Gori-Giorgi, F. Sacchetti, and G. B. Bachelet, *Phys. Rev.* B 61:7353(2000).

[32] K. Schmidt, S. Kurth, J. Tao, and J. P. Perdew, *Phys. Rev.* B 62:2227(2000).

[33] V. A. Rassolov, J. A. Pople, and M. A. Ratner, *Phys. Rev.* B 62:2232(2000).

[34] P. Gori-Giorgi and J. P. Perdew, *Phys. Rev.* B 64:155102(2001).

[35] P. Ziesche, Int. J. Quantum Chem., submitted.

[36] E. J. Baerends and O. Gritsenko, *J. Phys. Chem.* A 101:5383(1997);
M. A. Buijse, electron correlation - Fermi and Coulomb holes, dynamical and nondynamical correlation, Thesis, Amsterdam (1991); cf. also
M. Slamet and V. Sahni, *Phys. Rev.* A 51:2815(1995) for holes in the He atom,
J. Wang and V. H. Smith, Jr., *Int. J. Quantum Chem.* 56:509(1995) for excited states in molecules,
J. Wang, A. N. Tripathi, and V. H. Smith, Jr., *J. Chem. Phys.* 97:9188(1992) for intra- and extracule PDs; cf. also the articles by
E. Valdemoro, by J. Cioslowski and by T. Koga in Ref. [9].

[37] P. Ziesche, Pair Densities, Particle Number Fluctuations, and a Generalized Density Functional Theory, in: A. Gonis, N. Kioussis, M. Ciftan (ed.), *Electron Correlations and Materials Properties*, Kluwer/Plenum, New York (1999), p. 361. Note that after Eq. (3.15) therein the Coulomb hole (C-hole) normalization C is erroneously considered as a measure for the correlation strength. Correlation indeed creates the Coulomb hole, but its normalization C is always zero irrespective of the correlation strength. Note also that correlation widens the Fermi hole (X-hole), namely its on-top or short-range region, while further away from an electron the probability of finding another electron is enhanced (correlation reshovels the pair density from the nearby region of an electron to the region of its 1st pair-density shell, where $g(x)$ has its 1st maximum), such that the Fermi-hole normalization X remains unchanged. So, $C = 0$ and $X = N$.

[38] P. Ziesche, J. Tao, M. Seidl, and J. P. Perdew, *Int. J. Quant. Chem.* 77:819(2000).

[39] R. A. Römer and P. Ziesche, *Phys. Rev.* B 62:15279(2000) and *J. Phys. A: Math. Gen.* 34:1485(2001) and references cited therein.

[40] P. Ziesche, *J. Mol. Struc.* (Theochem.) 527:35(2000) and references cited therein.

[41] P. Ziesche, *Solid State Commun.* 82:597(1992).

[42] P. Ziesche, Cumulant Expansions of Reduced Densities, Reduced Density Matrices, and Green's Functions, in: Ref. [9], p. 33.

[43] K. Kladko, P. Fulde, and D. A. Garanin, *Europhys. Lett.* 46:425(1999).

[44] W. Kutzelnigg and D. Mukherjee, *J. Chem. Phys.* 110:2800(1999).

[45] The relation between the concept of cumulants and the quantum chemical coupled-cluster method is described in Ref. [6]. Note the formal similarity of the coupled-cluster wave function $\Psi = e^T\Psi_0$ and the exponential relation between the generating functional of the reduced density matrices and the generating functional of the cumulant matrices.

[46] W. Macke and P. Ziesche, *Acta Phys. Hung.* 17:215(1964) with references therein to the original papers by Gell-Mann and Low (1951), Goldstone (1957), Hubbard (1957/58), Sucher (1957), Rodberg (1958), Bogolyubov and Shirkov (1957) and Bloch (1958). For the general linked-diagram theorem cf. P. Ziesche, *Commun. Math. Phys.* 5:191(1967).

[47] N. H. March, *Phys. Rev.* 110:604(1958).

[48] W. Macke and P. Ziesche, *Ann. Physik* (Leipzig) 13:26(1964).

[49] W. Macke, *Z. Naturforsch.* A 5:192(1950).

[50] M. Gell-Mann and K. Brueckner, *Phys. Rev.* 106:364(1957).

[51] L. Onsager, L. Mittag, and M. J. Stephen, *Ann. Physik* (Leipzig) 18:71(1966).

[52] G. G. Hoffman, *Phys. Rev.* B 45:8730(1992).

[53] E. Daniel and S. H. Vosko, *Phys. Rev.* 120:2041(1960).

[54] I. O. Kulik, *Z. Eksp. Teor. Fiz.* 40:1343(1961) [*Sov. Phys. JETP* 13:946(1961)].

[55] D. F. DuBois, *Ann. Phys.* (N.Y.) 7:174(1959), 8:24(1959).

[56] W. J. Carr, Jr., and A. A. Maradudin, *Phys. Rev.* 133:A371(1964).

[57] T. Endo, M. Horiuchi, Y. Takada, and H. Yasuhara, *Phys. Rev.* B 59:7367(1999).

[58] L. Hedin, *Phys. Rev.* 139:A796(1965).

[59] P. Ziesche, *Ann. Physik* (Leipzig) 21:80(1968).

[60] J. C. Kimball, *J. Phys.* A 8:1513(1975).

[61] H. Yasuhara and Y. Kawazoe, *Physica* A 85:416(1976).

[62] Y. Takada and H. Yasuhara, *Phys. Rev.* B 44:7879(1991).

[63] D. L. Freeman, *Phys. Rev.* B 15:5512(1977);
R. F. Bishop, K.H. Lührmann, *Phys. Rev.* B 18:3757(1978), B 26:5523(1982).

[64] G. Ortiz and P. Ballone, *Phys. Rev.* B 50:1391(1994); 56:9970(1997).

[65] T. L. Gilbert, *Phys. Rev.* B 12:2111(1975).

[66] M. Levy, *Proc. Natl. Acad. Sci.* 76:6062(1979); Correlation Energy Functionals of One-Matrices and Hartree-Fock Densities, in: Ref. [4], p. 479.

[67] G. Zumbach and K. Maschke, *J. Chem. Phys.* 82:5604(1985).

[68] S. Goedecker and C. J. Umrigar, *Phys. Rev. Lett.* 81:866(1998).

[69] S. Goedecker and C. J. Umrigar, Natural Orbital Functional Theory, in: Ref. [9], p. 165.

[70] J. Cioslowski and R. Lopez-Boada, *J. Chem. Phys.* 109:4156(1998).

[71] J. Cioslowski and R. Lopez-Boada, *Chem. Phys. Lett.* 307:445(1999).

[72] J. Cioslowski and K. Pernal, *J. Chem. Phys.* 111:3396(1999).

[73] J. Cioslowski and K. Pernal, *Phys. Rev.* A 61:34503(2000).

[74] A. Holas, *Phys. Rev.* A 59:396(1999).

[75] G. Csanyi and T. A. Arias, *Phys. Rev.* B 61:7348(2000).

[76] B. Barbiellini, *J. Phys. Chem. Solids* 61:341(2000).

[77] J. Cioslowski, P. Ziesche, and K. Pernal, *Phys. Rev.* B 63:205105(2001). Therein, in Eq. (27) the erroneous constant $5\pi^{-1}$ must be replaced by $(5\pi)^{-1}$, consequently in Eq. (28) the factor $10\pi^{-1}$ must be replaced by $2(5\pi)^{-1}$, in Eq. (32) $5\pi^{-1}$ by $(5\pi)^{-1}$, and in Eq. (45) $5\pi/3$ by $\pi/15$, so the correct value of the constant C should read -0.663788. In Eq. (64) replace f_F by k_F and in Ref. 12 replace L. M. Hag by L. Mittag. In footnote 14 read small-r_s.

[78] J. Cioslowski, P. Ziesche, and K. Pernal, J. Chem. Phys., in press.

[79] C. Valdemoro, *Adv. Quant. Chem.* 31:37(1999); in: *Progress in Theoretical Chemistry*, S. Wilson and A. Hernandez-Laguna, ed., Kluwer, Dordrecht (1993) and references cited therein.

[80] K. Yasuda and H. Nakatsuji, *Phys. Rev.* A 56:2648(1997);
K. Yasuda, *Phys. Rev.* A 59:4133(1999);
M. Ehara, *Chem. Phys. Lett.* 305:483(1999) and references cited therein.

[81] K. Yasuda, *Phys. Rev.* A 63:32517(2001).

[82] D. Mazziotti, *Phys. Rev.* A 60:4396(1999) and references cited therein.

[83] R. J. Bartlett, Coupled Cluster Theory: An Overview of Recent Developments, in: D. R. Yarkony, ed., *Modern Electronic Structure Theory*, World Scientific, Singapore (1995)

[84] P. Fulde, H. Stoll, and K. Kladko, *Chem. Phys. Lett.* 299:481(1999) and refs. cited therein.

[85] M. B. Ruskai, *J. Math. Phys.* 11:3218(1970). Generally, in the more mathematically oriented community studying the properties of reduced density matrices, a series of theorems has been derived [2, 4, 8] which merit to be used and exploited more in quantum chemistry and solid state theory.

[86] In Ref. [40] the factors 1/2 are erroneously incorporated in the definition of $g_{\uparrow\uparrow}(x)$ and $g_{\uparrow\downarrow}(x)$ such that therein $g_{\uparrow\uparrow}(\infty) = g_{\uparrow\downarrow}(\infty) = 1/2$ appears instead of 1.

[87] The chemical potential ζ is not to be confused with the electrochemical potential $\mu = \zeta - D$, where D is the dipole barrier of an semi-infinite EG, the work function of which is determined by μ. Note the relation $\zeta N = E + p\Omega$ with p, the pressure. In the equilibrium statistics it is known as Duhem-Gibbs relation. With the virial theorem $2T + V = 3p\Omega$ it holds $\zeta N = \frac{5}{3}T + \frac{4}{3}V$, or per particle, $\zeta = \frac{5}{3}t + \frac{4}{3}v$. This follows also from the operator identity

$$-\frac{r_s^4}{3}\frac{\partial}{\partial r_s}\frac{1}{r_s^3} = \frac{5}{3}\cdot(-1)\frac{\partial}{\partial r_s}r_s + \frac{4}{3}\cdot\frac{1}{r_s}\frac{\partial}{\partial r_s}r_s^2,$$

cf. Eq. (3.22).

Toward the construction of an exchange-correlation potential in electronic-structure calculations of two-particle states in solids

A. Gonis
Lawrence Livermore National Laboratory
Chemistry and Materials Science
PO Box 808, L-353, Livermore, CA 94551

Abstract

It is shown that the treatment of jellium within the two-particle picture, in which the states of an interacting many-electron system are expressed in terms of two-particle states, can be used to yield an exchange-correlation potential for two-state electronic strudture calculations in solids. This potential can be used in a generalization of the local-density approximation of density-functional theory to obtain the electronic structure of pair states. From these effective single particle states can be obtained in which the Coulomb interaction between electrons has been taken directly into account within a pair approximation.

1 Introduction

Both Paul Ziesche and I along with our collaborators have introduced [1, 2] and are pursuing the application of density functional theory (DFT) to units of more than one particle. We did this not to improve DFT, which as an exact theory is in no need of improvement. Rather, the attempt is to improve approximate applications of DFT [3, 4, 5], such as the local density approximation (LDA) [6], or the local spin density approximation (LSDA), by providing a possibly more accurate treatment of the electron-electron interaction in the calculation of the electronic ground states of solids. The basic motivation is the desire to provide a coherent and unified treatment of the two distinct but interrelated aspects of electronic structure calculations of solid materials; energetics and spectra. The present contribution is directed at the possible determination of an exchange-correlation functional for carrying out calculations that would accomplish this task.

Currently, energetic information and related properties such as ground-state energies and densities and equilibrium phases of elemental solids and alloys is determined within computational methodologies based on DFT. The tremendous success that these methodologies have enjoyed [7, 8, 9] in the prediction of energetic information cannot be denied, in spite of some failures in the case of specific systems (e.g. Fe). Where they have failed, the culprit is invariably identified as the poor treatmnet of electron-electron correlations afforded through LDA in the particular system. Also, DFT-based methods often fail to produce reliable spectra, their failure to predict the correct gap in semiconductors (and insulators) and other spectroscopic data being as well known and daunting as their success in predicting energetics.

Many-body theory [10, 11, 12] on the other hand has provided a powerful methodology for the calculation of electron self-energies and often accurate predictions of spectra, such as the XPS spectrum for Ni [13]. It can claim justifiably that it provides a much more accurate treatment of correlation than DFT-LDA. However, its determination of energetics is much more problematic. The difficulty here can be easily identified as the failure to take into account the presence of the Fermi sea and its interaction with the global potential in a material, something that DFT-LDA is explicitly geared to do. To calculate spectra it is often sufficient to calculate the self-energy that is greatly influenced by the local environment of a site in the material (or around an electron) [13]. By contrast, cohesive energies are strongly dependent on global properties, such as the distribution of atoms throughout the (infinite) lattice, an information well represented by the electron density in the material. Thus, a dichotomy has developed, DFT-LDA (and many related schemes) on the one hand and MBT on the other for the treatment of the two different aspects of the electronic structure of matter.

Applying DFT-based concepts to two-particle states can be viewed as an attempt to bridge the gap between these two different methodologies, DFT-LDA and MBT. One can view the ensuing formalisms in either of two ways: As incorporating MBT principles and techniques into DFT, or as providing MBT with a formal procedure for accounting for the presence of the Fermi sea. In either case, the developmental task before us is far from trivial. And once some basic formal aspects have been clarified, such that DFT can indeed be formulated in terms of two-particle densities, we must deal with the details of implementing these concepts. One important detail is the establishment of an exchange-correlation potential to be used in an LDA-like approach to electronic structure within a two-particle picture.

A two-particle application of DFT is of particular importance. In his development, Paul is exploiting the fact that the exact knowledge of the pair density, the diagonal part of the two-particle density matrix, provides an exact account of the internal potential energy of the system. Thus, in constructing an exchange-correlation potential one needs to worry only about the kinetic energy functional and its contribution to the exchange-correlation potential. Paul is describing his efforts in that direction elsewhere in this volume.

A rather different motivation underlies my approach. The objective is to reduce the treatment of the Coulomb interaction to that of an external field, like the nuclear field in a material. Eliminating even partly the distinction between local nuclear potentials and non-local inter-particle interactions may lead to improvements because the treatment of local fields is essentially routine. Then one has a hope that the vast experience that we have gained in the treatment of external potentials can be put to good use in calculations carried out in two-particle space. In this formulation, the Coulomb potential between the members of a pair appears indeed as an external potential and it should be treatable with a straightforward, although not necessarily simple, generalization of well-known procedures. To proceed along these lines one divides the system of interacting particles into distinct pairs so that any given particle is a member of only one pair. These pairs are then allowed to interact via a four-point interaction that involves the Coulomb repulsion of a particle in one pair and that in another. The problem reduces to that of interacting **particles** but with the meaning of the term particle changed to mean a two-particle unit. For the sake of clarity, I will use bold face letters to indicate that a **particle** is a unit of two electrons treated as a single entity in a space of six dimensions.

It is useful to set forth some basic notions to help in the conceptual development of the formalism. First, we note that the properly anti-symmetrized system wave function is anti-symmetric with respect to single-particle states, *not* single-particle coordinates. It is very easy to exchange only coordinates but not spins and get a nonsensical expression, so the only meaningful elements to exchange are state indices. Second, the wave function can be written as a combination of products of determinants, each of which contains determinants that are anti-symmetric with respect to their internal state indices. This is due to beautiful theorem by Laplace [14] according to which a determinant can be written as the sum of products of lower order determinants, chosen in all possible ways, and their minors. Of central importance to our case is that a determinant of even order (and as the number of electrons approaches infinity this condition is satisfied) can be written as a sum of products of second order determinants.

Taken together these two notions lead to a third one that is particularly significant. Namely, the system of N interacting particles defines a $3N$-dimensional space which in the single-particle picture is taken as the tensor sum of 3-dimensional spaces, and in the two-particle picture as the tensor sum of 6-dimensional spaces. Each of these latter spaces accommodates a **particle**. It must be kept in mind that each of these spaces is absolute and immutable. One is not allowed to interchange a particle (a 3-dimensional subspace) of one six-dimensional space with that of another any more than one is allowed to interchange the x coordinate of one particle with the y coordinate of another. Because of this, it is allowed to consider distinct pairs of particles making up the whole system. Taking the products of second-order determinants distributed over all two-particle spaces, antisymmetrizing across different determinants and summing up the products then satisfies the condition of anti-symmetry for the total wave function. In any one product no one single-particle state can appear in more than one determinant. It is easy to see that the Pauli exclusion principle is accounted for. Also, the two-particle spaces are now convenient for defining basis states that accommodate two particles, and can be made orthonormal and complete. Indeed, at this stage, there is no evident difference between a two-particle formalism and the traditional single-particle picture as each element of one transcribes naturally into a corresponding element of the other.

This sets the formal basis for the development of an exchange-correlation functional in two-particle space.

2 DFT and the LDA in two-particle space

2.1 Jellium for pairs

The formulation of DFT [3, 4, 5] can be extended without special effort to interacting particles of any dimensionality. Thus, it can be applied to interacting pairs in the sense described above. However, approximations based on DFT, such as the local density approximation (LDA) within the Kohn-Sham computational scheme [6], do not carry through by simply transcribing symbols and assigning new meaning to them. Consider, for example, the LDA within the single-particle picture. Here the exchange-correlation potential at a point $\mathbf{r}$ in a real material is taken to be that of an interacting, uniform gas (jellium) at the real density $n(\mathbf{r})$. This is found by parametrizing the difference of the exact kinetic and potential energies for jellium and their "classical" counterparts, the kinetic energy being taken to be that of plane waves and the potential energy that of two classical charge distributions, and differentiating with respect to n. The derivative evaluated at $n(\mathbf{r})$ defines the potential at $\mathbf{r}$ in carrying through the calculation of the electronic structure of a material.

It would be ideal to be able to apply this scheme to two-particle states. At each position $\mathbf{x} = (\mathbf{r}_1, \mathbf{r}_2)$ in the space characterized by a density $n(\mathbf{x})$ we'd assign an exchange-correlation potential associated with a jellium of **particles** of constant density $n(\mathbf{x})$ defined through the parametrization procedure described above for the case of single particles. The problem with this scheme is that two-particle jellium (or jellium$_2$) does not exist.

The value of $n(\mathbf{r}_1, \mathbf{r}_2)$ in ordinary jellium of density n depends on the separation $|\mathbf{r}_1 - \mathbf{r}_2|$. For the case of parallel spins, for example, it starts at 0, rises to a maximum at a separation around r_s (with r_s being the effective radius of the electron in the system) and oscillates with diminishing amplitude approaching the constant value of n^2 as

$|\mathbf{r}_1 - \mathbf{r}_2|$ approaches infinity. Of course, this exchange-correlation density is not exactly what we need. What we need is the density of a system of **particles** forming jellium$_2$ that see no external potential except that of a uniform background of charge. Remember that in our formulation of DFT in two-particle space, the members of a pair do not interact with one another. This interaction is treated exactly as an external field. Still, the resulting inter-pair interaction is not invariant with respect to arbitrary translations in two-particle space so that the Hamiltonian does not commute with the corresponding set of translation operators. It follows that even in this case, the density of these pairs is not invariant with respect to translations in the space and hence not a constant.

Jellium$_2$ is similar to ordinary jellium in the sense that the particles in both systems feel only their mutual interaction (a positive and uniform background charge would not affect the density). In contrast to ordinary jellium, however, the **particles** forming jellium$_2$ interact with a potential that does not commute with the translations in the space so that they do not form a constant density. The density of such pairs is not known, but can be calculated. It can easily be surmised to approach a constant value as the separation between the particles approaches infinity. As the separation vanishes, its behavior must be obtained numerically. Given a density n_2 characterizing a system of non-interacting **particles**, we can assess the positional dependence of $n(\mathbf{x})$ by means of exact numerical procedures, and parametrize it in terms of the distance $|\mathbf{r}_1 - \mathbf{r}_2|$. The task is to describe a methodology that would use such a calculation to derive an exchange and correlation potential.

2.2 Exchange and correlation potential in ordinary LDA

In ordinary implementations of the LDA, one identifies an energy difference $\Delta E[n]$ associated with jellium at density n. This difference in energy is made up by the difference in the exact kinetic energy and that of plane-wave states, and between the exact potential energy, involving the exact two-particle correlation function, and the classical interaction between two charge distributions,

$$\Delta E[n] = T_{\text{ex}}[n] - T_{\text{pw}}[n] + U_{\text{ex}}[n] - U_{\text{cl}}[n]. \tag{1}$$

Here, T_{ex} is the exact kinetic energy of the system, T_{pw} is that of plane-wave states for particles forming a system of density n,

$$U_{ex} = \int \frac{n(\mathbf{r}_1, \mathbf{r}_2)}{|\mathbf{r}_1 - \mathbf{r}_2|} \tag{2}$$

is the exact interaction energy involving the exact pair correlation function $n(\mathbf{r}_1, \mathbf{r}_2)$, and

$$U_{cl} = \int \frac{n(\mathbf{r}_1)n(\mathbf{r}_2)}{|\mathbf{r}_1 - \mathbf{r}_2|} \tag{3}$$

is the classical Coulomb interaction between two charge distributions. (For the explicit case of jellium, of course, these distributions are constant).

This difference in energy, $\Delta E[n]$, can be calculated numerically and parametrized as a function of n, as was done by Ceperley and Alder [15]. Differentiating with respect to n gives an approximate value for the exchange-correlation potential,

$$\mu_{xc}(\mathbf{r}) = \frac{\delta \Delta E[n]}{\delta n(\mathbf{r})}, \tag{4}$$

to be used in the calculation of electronic structure. The approximate nature of the potential derives from the fact that $\Delta E[n]$ is a functional and not a function of the density. In a realistic calculation, the kinetic energy is taken to be that of independent particles arranged in single-particle states as determined by the calculation, and U_{cl} is calculated with respect to the single-particle density of the system under consideration. The exchange-correlation potential assigned to point $\mathbf{r}$ is that corresponding to jellium of constant density $n(\mathbf{r})$ and determined as just described. Thus, jellium is treated exactly in the LDA. Any extension of these concepts to two-particle space should preserve this feature, namely the exact treatment of a physical system, such as jellium.

2.3 Exchange and correlation potential for two-particle states

The notion of deriving an exchange-correlation potential from the exact treatment of a selected system is a practically sound one.

Regardless of the difficulties enumerated above, it is still possible to obtain such potentials for two-particle states from the exact study of jellium$_2$. One can conceive two different ways of doing this, one seemingly more approximate than the other. However, both are equivalent in providing an exact treatment of jellium$_2$. We'll begin with the conceptually simpler approximation, only to show that it is in fact equivalent to a seemingly more rigorous procedure.

We consider a completely non-interacting collection of **particles** at some constant density n_2. For the sake of reference we will call this system jellium$_2^0$. When the inter-pair Coulomb interaction is turned on, jellium$_2^0$ evolves into jellium$_2$ whose density $n(\mathbf{x}) = n(\mathbf{r}_1, \mathbf{r}_2)$ approaches n_2 as the distance $|\mathbf{r}_1 - \mathbf{r}_2|$ approaches infinity. Total energies for jellium$_2$ can be obtained in two different ways: exactly, through numerical means, and approximately. In the approximate treatment the kinetic energy, T_s, is taken to be that of non-interacting pairs but with exchange included, and the potential energy is obtained using the classical expression,

$$U_{\text{cl}}^{(2)} = \int d\mathbf{x}d\mathbf{x}' n(\mathbf{x})U(\mathbf{x},\mathbf{x}')n(\mathbf{x}'), \tag{5}$$

where $U(\mathbf{x}, \mathbf{x}')$ is the Coulomb interaction across two different pairs. By contrast, the exact potential energy involves, $n(\mathbf{x}, \mathbf{x}')$, the **particle-particle** (or pair-pair) correlation function for jellium$_2$, and is given by the expression,

$$U_{\text{cl}}^{(2)} = \int d\mathbf{x}d\mathbf{x}' U(\mathbf{x},\mathbf{x}')n(\mathbf{x},\mathbf{x}'). \tag{6}$$

The resulting energy difference, $\Delta E[n_2] = E_{\text{ex}}[n(\mathbf{x})] - E_{\text{cl}}[n(\mathbf{x})]$, takes a form identical to that given in Eq. (1) but in terms of the newly defined quantities involving two-particle states. It can once again be parametrized in terms of the density of the non-interacting system (jellium$_2^0$), n_2. Its derivative with respect to n_2,

$$\mu_{\text{xc}}(\mathbf{x}) = \frac{\delta \Delta E[n]}{\delta n(\mathbf{x})}, \tag{7}$$

would yield an exchange-correlation potential. *This potential could then be assigned to a point* $\mathbf{x}$ *in the real system which is such that*

$n(\mathbf{r}_1)n(\mathbf{r}_2) = n_2$, or more precisely where $n(|\mathbf{r}_1 - \mathbf{r}_2|)$ in the real material equals the corresponding quantity for jellium$_2$.

It is easily seen that this is unique and also leads to an exact treatment of jellium$_2$. The exact aspects of the treatment derive from the fact that in such a system $n(\mathbf{r}_1) = n(\mathbf{r}_2) = \sqrt{n_2} = const.$ Therefore, identifying any unique aspect of the system connected with point $\mathbf{x}$, such as the density for the interparticle separation defined by $\mathbf{x}$, suffices to identify a proper local exchange-correlation potential. It is interesting to note that ordinary LDA also subscribes to such a description, but does so in a more or less trivial way. It is to be noted that if an external field $V(\mathbf{x}) = \frac{1}{|\mathbf{r}_1 - \mathbf{r}_2|}$ is added to jellium$_2$ the use of an exchange-correlation potential as just described in an electronic structure calculation would provide an *exact determination of two-particle states in ordinary jellium of density $n = \sqrt{n_2}$.*

It is interesting to proceed a bit further and consider a more rigorous approximation, one which would rely on the local density, $n(\mathbf{x})$, directly in assigning an exchange-correlation potential at $\mathbf{x}$. Suppose that the exact exchange-correlation densities of jellium$_2$ corresponding to two different n_2 of the non-interacting system, jellium$_2^0$, do not cross anywhere when plotted on the same graph against $|\mathbf{r}_1 - \mathbf{r}_2|$. As is shown below, this is indeed the case. It follows that we can assign to point $\mathbf{x}$ (more precisely to $|\mathbf{r}_1 - \mathbf{r}_2|$) the exchange-correlation potential that corresponds to a jellium$_2$ with the same $n(\mathbf{x})$, the same $|\mathbf{r}_1 - \mathbf{r}_2|$, as the real material. Once again, this would certainly lead to the exact treatment of jellium$_2$.

We have arrived at two different approximations that yield the same results for jellium$_2$ so that they must be equivalent at least as far as that system is concerned. One central condition to be placed on any LDA-like exchange-correlation potential is that it must provide an exact treatment of a physical system. In the present case two seemingly different approximations lead to such a treatment and we can conclude that they must be equivalent at least at the level of jellium$_2$.

However, these two approximations can be expected to yield different results in the calculation of the electronic structure of a realistic material. In such a calculation, one methodology would assign at point $\mathbf{x} = (\mathbf{r}_1, \mathbf{r}_2)$ the exchange-correlation potential associated with that jellium$_2$ that evolves out of jellium$_2^0$ at constant density $n_2 = n(\mathbf{r}_1)n(\mathbf{r}_2)$. The other would assign that potential that corresponds to jellium$_2$ that at $\mathbf{x}$ (actually the same inter-particle distance $|\mathbf{r}_1 - \mathbf{r}_2|$) has the same density as the real material. Of the two methodologies, the latter is evidently the more rigorous one.

It remains to show that a unique identification of a jellium for every $n(\mathbf{x}) = n(|\mathbf{r}_1 - \mathbf{r}_2|)$ is indeed possible given the quantity $n(|\mathbf{r}_1 - \mathbf{r}_2|)$. *This would be the case provided that the curves corresponding to the various pair densities at* $|\mathbf{r}_1 - \mathbf{r}_2|$ *for different* n *do no cross when plotted on the same scale. That this is indeed the case* has been verified [16] numerically using highly precise and accurate numerical representations of the pair densities of jellium at different n, or equivalently at different effective electron radii r_s. Paola Gori-Giorgi describes the methodology followed in these calculations in her contribution to this volume. As shown in the panels of Fig. 1, two such curves do not cross, either for two very different n, or when the jellium densities differ only infinitesimally.

In the panels of the figure we plot the quantity $n(|\mathbf{r}_1 - \mathbf{r}_2|) = n^2 g(|\mathbf{r}_1 - \mathbf{r}_2|)$, which defines g, versus the inter-particle separation $r = |\mathbf{r}_1 - \mathbf{r}_2|$ for pairs of anti-parallel spins at different jellium densities (or equivalently differen effective electron radii r_s). The upper two panels show that that two such curves do not cross. The bottom panel shows that the difference between the pair densities at n and at $n + \mathrm{d}n$ is always positive.

Now, given $n(\mathbf{x})$, one can identify uniquely that jellium (at some density n) that has the same density at the inter-particle distance defined by $\mathbf{x}$. That jellium is to be treated as a collection of pairs that fully interact with another, but whose individual pairs do not see the Coulomb repulsion between the particles forming the pair. That repulsion is taken into account directly in the calculation as an external field and should not be counted in the determination of V_{xc} that accounts for the interaction between units only. From such exact treatment (as outlined above) an exchange correlation energy and potential can be derived.

We now turn to formal considerations of the results just quoted.

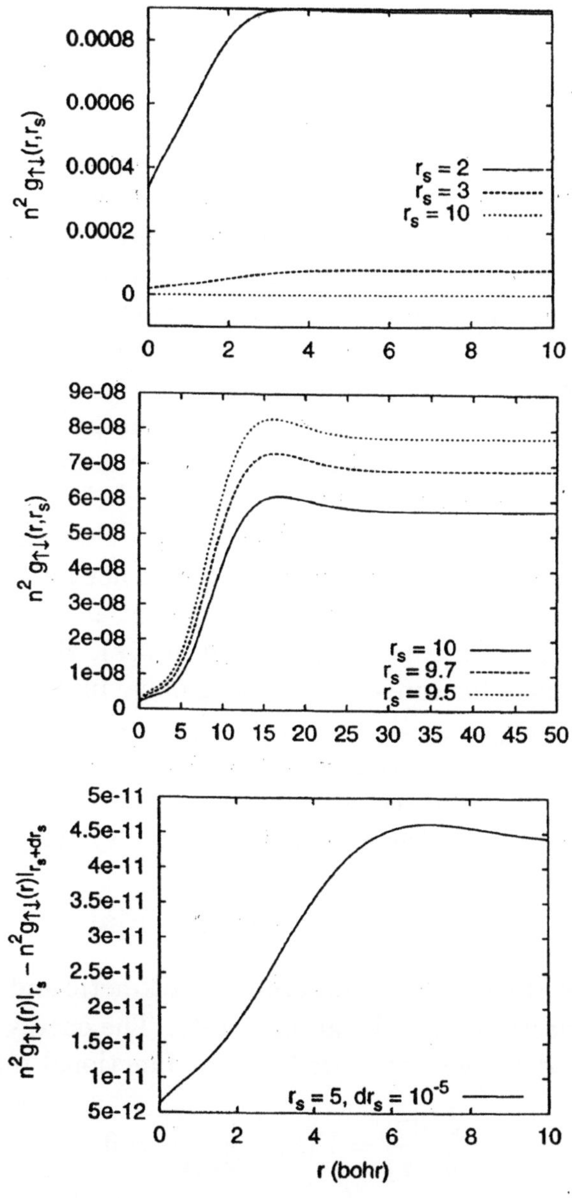

Figure 1: **Illustration that pair dnsities corresponding to different jellium single-particle densities do not cross as discussed in the text.**

2.4 Non-crossing of pair densities

As the calculations depicted in Fig. 1 illustrate, the pair densities corresponding to two different jellium densities do not cross. There is a fairly elegant proof that this is the case, which can be constructed in terms of the following proposition:

Consider a system in its ground state characterized by a single-particle density $n(\mathbf{r})$. *Let* $\delta V(\mathbf{r})$ *be an infinitesimal but everywhere positive change in the external field determining* $n(\mathbf{r})$. *The corresponding change* $\delta n(\mathbf{r})$ *is also positive for all* $\mathbf{r}$.

It is convenient to think of a system of interacting electrons and nuclei as consisting of an external positive charge distribution, $p(\mathbf{r})$ that produces a potential $V(\mathbf{r})$. This charge is in contact with an infinite reservoir of negative charge (electrons) that responds to the potential set up by $p(\mathbf{r})$ by providing enough electrons, with corresponding density $n(\mathbf{r})$, to render the system neutral. An infinitesimal everywhere positive change in the potential can be thought of as arising from a corresponding change in $p(\mathbf{r})$, although this construction is made for convenience's sake only. In general, it suffices to consider changes in the potential alone without specifying the physical reasons that may give rise to them. We now turn to the proof of the conjecture stated above.

The total energy in the ground state of the system is a functional of the density,

$$E[n] = \int V(\mathbf{r})n(\mathbf{r})d\mathbf{r} + F[n] \tag{8}$$

where $F[n]$ is the functional describing the kinetic and potential energy of the electron gas in its ground state. The energy is minimized by the exact density so that we have the functional relation,

$$\frac{\delta E[n]}{\delta n(\mathbf{r})} = V(\mathbf{r}) + \frac{\delta F[n]}{\delta n(\mathbf{r})} = 0. \tag{9}$$

¿From this follows that

$$V(\mathbf{r}) = -\frac{\delta F[n]}{\delta n(\mathbf{r})}. \tag{10}$$

One more functional differentiation yields,

$$\frac{\delta V(\mathbf{r})}{\delta n(\mathbf{r}')} = -\frac{\delta^2 F[n]}{\delta n(\mathbf{r})\delta n(\mathbf{r}')}. \tag{11}$$

Now, for an everywhere positive $\delta V(\mathbf{r})$, $\delta n(\mathbf{r})$ must be positive for some $\mathbf{r}'$ (otherwise charge neutrality would be violated). From the last expression it follows that the functional derivative $\frac{\delta V(\mathbf{r})}{\delta n(\mathbf{r}')}$ corresponding to these two points is positive. However, this derivative is independent of the order of differentiation, so that $\frac{\delta V(\mathbf{r}')}{\delta n(\mathbf{r})}$ is also positive. It follows that $\delta n(\mathbf{r})$ is positive everywhere in the volume.

This construction can be readily extended to arbitrary multiparticle densities through the simple replacement of the $\mathbf{r}$'s by the corresponding vectors $\mathbf{x}$ in the space under consideration.

In the case of jellium, a change in the electron density can be uniquely associated with a corresponding change in an external positive charge distribution. Furthermore, the infinitesimal changes used above can become finite because all ground-state densities n correspond to an infinite volume, thus preserving the one-to-one correspondence between points in the volumes of two different systems. (Note that in the case of a real system, two different densities will probably be associated with different volumes in the ground states of the corresponding systems, so that the proof above must be restricted to infinitesimal changes.) This shows that two different n-particle densities corresponding to two different jellium densities do not cross, providing a unique identification of a system for the calculation of an exchange-correlation potential for each $n(\mathbf{x})$ in a real system.

3 Discussion and Conclusions

The discussion in the body of the paper is motivated by the desire to provide a unique and consistent treatment of the two interrelated but distinct aspects of an electronic structure calculation; namely, energetics and spectra. Although developments have materialized (much too numerous to attempt even a cursory review here) that correct some of the deficiencies of the DFT-LDA approach, there exists no single methodology that treats spectra on an equal footing with the energetic aspects of an electronic structure calculation.

As mentioned in the introduction, the aim is to improve not DFT per se, but its approximate implementation with respect to the LDA (LSDA). Numerical calculations within a phenomenological tight-binding description of the Hamiltonian have given strong indications that the expectation for improvement is not altogether unfounded. At the same time, one should not expect the resulting computational methodology to cure all ills attending to the use of the LDA. For example, it should not be expected that excited states would be properly described through the LDA, regardless of the number of particles used in its implementation. At the same time, one can expect improvement in the determination of such features as characterize the XPS spectra of correlated systems, e.g., Ni.

We hope to proceed with work to determine the exchange-correlation potential for jellium_2. The results of this study will be communicated when they become available.

4 Acknowledgment

Discussions with John Perdew and Paola Gori-Giorgi are gratefully acknowledged. I am particularly indebted to Dr. Giorgi for providing camera-ready figures of her calculations of pair densities shown in Fig. 1. This work was performed under the auspices of the US Department of Energy under contract No. W-7405-ENG-48 With Lawrence Livermore National Laboratory.

References

[1] Paul Ziesche, Phys. Lett. **A195**, 213 (1994).

[2] A. Gonis, T. C. Schulthess, and P. E. A. Turchi, Phys. Rev. **B56**, 9335 (1997).

[3] P. Hohenberg and W. Kohn, Phys. Rev. **136** B864, (1964).

[4] R. G. Parr and C. Y. Yang, *Density Functional Theory of Atoms and Molecules* (Oxford University Press, Oxford, 1989).

[5] R. M. Dreitzler and E. K. U. Gross, *Density Functional Theory* (Springer-Verlag, Berlin, New York, 1990).

[6] W. Kohn and L. J. Sham, Phys. Rev. **140**, A1 133 (1965).

[7] V. L. Moruzzi, J. F. janak, and A. R. Williams, *Calculated Electronic Properties of Metals*, (Pergamon, 1978).

[8] J. S. Faulkner, in *Progress in Materials Science*, edited by J. W. Christian, P. Haasen, and T. B. Massalsky (Pergamon Press, New York, 1982), Nos. 1 and 2.

[9] A. Gonis, Theoretical Materials Science: Tracing the Electronic Origins of Materials Behavior (The Materials Research Society, Warrendale, PA, 2000).

[10] E. K. U. Gross and E. Runge, *Vielteilchentheorie*, (Teubner Texte, Stuttgart, 1986).

[11] T. Kato, T. Kobayashi, and M. Namiki, *Supplement of the Progress of Theoretical Physics*, No. **15**, (Phys. Soc. Japan, 1960), p. 3.

[12] Alexander L. Fetter and John Dirk Walecka, *Quantum Theory of Many-Particle Systems*, (McGraw-Hill, Inc., New York, NY, 1971).

[13] G. Treglia, F. Ducastelle, and D. Spanjaard, J. Physique **41**, 281 (1980).

[14] Thomas Muir, *A treatise on the Theory of Determinants*, (Dover, New York, 1960). Laplace's theorem can be stated as follows: If any m rows of a determinant be selected and every possible minor of the mth order be formed from them, and if each be multiplied by its complimentary and the sign + or - be affixed to the product according as the sum of the numbers indicating the rows and columns from which the minor is formed be even or odd, the aggregate of the products thus obtained is equal to the original determinant.

[15] D. M. Ceperley and B. J. Alder, Phys. Rev. Lett. **45**, 566 (1980).

[16] P. Gori-Giorgi, F. Sacchetti, and G.B. Bachelet, Phys. Rev. B 61, 7353 (2000), and private communication.

ELECTRON CORRELATIONS IN DISORDERED ALLOYS AND AT METALLIC SURFACES

Václav Drchal, Václav Janiš, and Josef Kudrnovský [1]

1. INTRODUCTION

The density functional theory (DFT)[1] supplemented by the local density approximation (LDA) or generalized gradient approximation (GGA) in which the electron-correlation part is treated on the basis of the electron gas model is a highly reliable method for evaluation of the ground state properties of molecules and solids[2]. However, the DFT fails in some cases, e.g., for excitation spectra of solids or in the evaluation of the gap in insulators and semiconductors and also for solids (such as lanthanides and actinides or transition-metal oxides) whose electronic structure is better described in terms of atomic-like electronic states rather than in terms of the electron gas model on which the LDA and GGA are based.

The electronic properties of strongly correlated materials which cannot be adequately described within the DFT are usually studied in the framework of simplified models like, e.g., the single-band Hubbard model. The strength of correlation effects is usually classified in terms of the ratio of the on-site Coulomb energy U to the bandwidth w. One can distinguish three regimes, namely, (i) the weak interaction case ($U/w < 1$, transition metals), (ii) the intermediate interaction case ($U/w \approx 1$, metal-insulator transition regime and Kondo systems), and (iii) the strong interaction case ($U/w > 1$, rare-earth systems and wide gap solids). It should be mentioned that a general approach bridging all the above cases is still missing even at the model level.

Recently some progress has been made in extending the validity of the DFT, for example the time-dependent LDA[3] can yield the excitation energies, and proper account of the self-interaction corrections (SIC) lead to a remarkable improvement

[1]Institute of Physics, Academy of Sciences of the Czech Republic, Na Slovance 2, CZ-182 21 Praha 8, Czech Republic

for rare-earths[4]. The so-called GW approximation (GWA)[5] was successfully applied to the gap problem in a number of semiconductors and insulators as well as to studies of photoemission spectra of metallic systems[6]. An alternative approach to the gap problem which is based on a straightforward generalization of the LDA is the LDA+U method[7] which also yielded band narrowing in the photoemission spectra of nickel[8], but it failed to produce the well-known satellite structure below the main peak[9]. The presence of the satellite in Ni was successfully explained in the framework of the T-matrix formalism developed on *ab initio* level (cf. Ref. 10). Another approach to the satellite and the band narrowing in nickel is based on a three-body scattering approximation that employs the Faddeev equations[11,12].

The dynamical mean-field theory (DMFT)[13] aims at accurate solution of the correlation problem on a single site which is coupled to the rest of the lattice in a selfconsistent way. The DMFT, besides many studies on the model level, was also applied to realistic calculations of correlated systems, e.g., δ-Pu[14].

The LDA++ method[15,16] introduced by Lichtenstein and Katsnelson extends the LDA+U method by accounting for dynamical electron correlations. The starting point of the LDA++ method is a multiband Hubbard Hamiltonian whose one-electron part is taken from the LDA Hamiltonian after the double counting corrections for the average Coulomb interactions in the LDA were subtracted. The LDA++ employs different many-body approximations, namely, the so-called Hubbard I solution[17] in the strong-interaction limit, the iterated perturbation theory (IPT) within the dynamical mean-field theory (DMFT)[13] for the intermediate-interaction case, and the fluctuation exchange approximation (FLEX)[18] in the weak-interaction limit.

A new method, denoted as LDA+DMFT, appeared very recently[19]. It follows the general scheme of the LDA+U, but the correction term is constructed from the DMFT selfenergy. Already the first results[20,21] show that it bears a great promise.

We have studied the electron correlations in $3d$ metals and their alloys[22,23] using the multiband Hamiltonian of Hubbard type derived from *ab initio* calculations and solved within the single-channel approximations of the many-body theory, in particular, employing the non-selfconsistent version of second order perturbation theory and including direct and exchange interactions U and J. In the present work we start from a multiband Hubbard Hamiltonian whose parameters are determined from a tight-binding linear muffin-tin orbital (TB-LMTO) Hamiltonian of the system (for a review of the TB-LMTO method and its applications to random surfaces and interfaces see Ref. 24). Our principal aim is the application to a broad class of materials including not only elemental solids, but also compounds, ordered and disordered alloys, as well as surfaces, interfaces, and multilayers. The formalism is thus based on Green functions which allow for configurational averaging in the case of alloys, and for using surface Green functions for surfaces and interfaces. We employ single-channel approximations (FLEX) of the canonical perturbation theory to solve the many-body part of the problem. The basic approximation we adopt is the assumption on the local (site-diagonal, or wave-vector independent) selfenergy which is reasonably well justified for transition metals and their alloys[25]. The local approximation for the selfenergy is also required from the formal point of view for the application of the theory to the case of random alloys treated within the coherent potential approximation (CPA)[26]. In contrast to the previous work[22,23] we now extend our study to infinite order in the particle-particle and particle-hole channels including both U and J. We solve the correlation problem selfconsistently within the DMFT.

2. MULTIBAND HUBBARD HAMILTONIAN

The electronic structure determined within the LDA is described by a TB-LMTO Hamiltonian with the overlap matrix equal to unity

$$H^{\rm LMTO} = \sum_{\mathbf{R}\lambda,\mathbf{R}'\lambda'} |\mathbf{R}\lambda\rangle\, H^{\rm LMTO}_{\mathbf{R}\lambda,\mathbf{R}'\lambda'}\, \langle \mathbf{R}'\lambda'|\,, \tag{1}$$

where

$$H^{\rm LMTO}_{\mathbf{R}\lambda,\mathbf{R}'\lambda'} = C_{\mathbf{R}\lambda\lambda'}\,\delta_{\mathbf{R}\mathbf{R}'} + \Delta^{1/2}_{\mathbf{R}\lambda}\, S^{\gamma}_{\mathbf{R}\lambda,\mathbf{R}'\lambda'}\, \Delta^{1/2}_{\mathbf{R}'\lambda'}\,. \tag{2}$$

Here, $\mathbf{R}$ is the site index, $\lambda = (L\sigma) = (\ell m\sigma)$ is the spinorbital index, $L = (\ell m)$ is the orbital index, σ is the z-projection of the spin, C, Δ, and γ are site-diagonal matrices of potential parameters, and S^{γ} is the matrix of structure constants in the orthogonal LMTO representation

$$S^{\gamma}_{\mathbf{R}\lambda,\mathbf{R}'\lambda'} = [S^0(1-\gamma S^0)^{-1}]_{\mathbf{R}\lambda,\mathbf{R}'\lambda'} = \left[S^{\beta}\left(1-(\gamma-\beta)S^{\beta}\right)^{-1}\right]_{\mathbf{R}\lambda,\mathbf{R}'\lambda'}. \tag{3}$$

Here, S^0 is a matrix of canonical structure constants, S^{β} is a matrix of screened structure constants, and β is a site-diagonal matrix of screening constants.

The parameters of a multiband Hubbard Hamiltonian in second quantization with creation ($a^{+}_{\mathbf{R}\lambda}$) and annihilation ($a_{\mathbf{R}\lambda}$) operators

$$\begin{aligned} H^{\rm Hubb} = & \sum_{\mathbf{R}\lambda,\mathbf{R}'\lambda'} t_{\mathbf{R}\lambda,\mathbf{R}'\lambda'}\, a^{+}_{\mathbf{R}\lambda}\, a_{\mathbf{R}'\lambda'} \\ & + \sum_{\mathbf{R},\lambda_1,\lambda_2\lambda_3\lambda_4} \langle \mathbf{R}\lambda_1, \mathbf{R}\lambda_2 |V| \mathbf{R}\lambda_3 \mathbf{R}\lambda_4\rangle\, a^{+}_{\mathbf{R}\lambda_1} a^{+}_{\mathbf{R}\lambda_2} a_{\mathbf{R}\lambda_4} a_{\mathbf{R}\lambda_3} \end{aligned} \tag{4}$$

are found from LDA calculations. The hopping integrals

$$t_{\mathbf{R}\lambda,\mathbf{R}'\lambda'} = \Delta^{1/2}_{\mathbf{R}\lambda}\, S^{\gamma}_{\mathbf{R}\lambda,\mathbf{R}'\lambda'}\, \Delta^{1/2}_{\mathbf{R}'\lambda'}\,, \quad \mathbf{R} \neq \mathbf{R}' \tag{5}$$

are identified with the site off-diagonal elements of (1), while the atomic levels $t_{\mathbf{R}\lambda,\mathbf{R}\lambda'} = \epsilon_{\mathbf{R}\lambda\lambda'} = \epsilon_{\mathbf{R}\lambda}\delta_{\lambda,\lambda'}$ are determined from the condition that the average occupancy $\langle n_{\mathbf{R}\lambda}\rangle$ of the state $|\mathbf{R}\lambda\rangle$ as calculated within the LDA and within the Hartree-Fock approximation for the multiband Hubbard Hamiltonian (4) are identical. It means

$$\epsilon_{\mathbf{R}\lambda} = C_{\mathbf{R}\lambda} - \Sigma^{\rm HFA}_{\mathbf{R}\lambda}\,. \tag{6}$$

Various methods to determine the interaction parameters $\langle \mathbf{R}\lambda_1 \mathbf{R}\lambda_2 | V | \mathbf{R}\lambda_3 \mathbf{R}\lambda_4 \rangle$ are currently in use. Usually the pair interactions among electrons in s and p states are neglected and only interactions between the d-states are considered. Very often all interactions except the direct term (the so-called Hubbard U) and the exchange term (J) are also neglected. Screening by s and p electrons leads to a strong suppression of U. This effect is properly included in the LDA calculations in which U is determined as a second derivative of E_{tot} with respect to occupancies of the interacting states λ, λ'. The exchange interaction J is not screened and thus values obtained from atomic-like calculations yield reasonable estimates.

The interaction part of the multi-band Hubbard-type Hamiltonian for one site (site index $\mathbf{R}$ is omitted here) reads

$$H^{\text{int}} = \sum_{i,j,k,l} \sum_{\sigma,\sigma'} \langle i\sigma j\sigma' | V | k\sigma l\sigma' \rangle \, a_{i\sigma}^{+} a_{j\sigma'}^{+} a_{l\sigma'} a_{k\sigma} \,, \tag{7}$$

where i, j, k, l denote the orbitals corresponding to complex spherical harmonics $Y_{\ell,m}$ (Condon-Shortley convention is assumed). For d-orbitals $\ell = 2$ and σ, σ' are spin variables ($\sigma, \sigma' = \uparrow, \downarrow$). The usual approximation for the interaction constants $\langle i\sigma j\sigma' | V | k\sigma l\sigma' \rangle$, which is adopted also here, consists in neglecting all terms but the direct and exchange ones, namely,

$$\langle i\sigma j\sigma' | V | k\sigma l\sigma' \rangle \approx \delta_{ik}\delta_{jl}\, U + \delta_{il}\delta_{jk}\, \delta_{\sigma\sigma'}\, J \,. \tag{8}$$

The numerical calculations can be simplified by the point group symmetry. For cubic lattices (point group O_h) the complex harmonics can be replaced by cubic harmonics. If we consider only the d-states, cubic harmonics are identical with the real spherical harmonics. This is no longer true for the f-states. The transformation relations read

$$Y_{22}^{s} = \frac{i}{\sqrt{2}}(Y_{2,-2} - Y_{2,2}) \,, \quad Y_{21}^{s} = \frac{i}{\sqrt{2}}(Y_{2,-1} + Y_{2,1}) \,, \quad Y_{21}^{c} = \frac{1}{\sqrt{2}}(Y_{2,-1} - Y_{2,1}) \tag{9}$$

for the basis functions of the t_{2g} representation and

$$Y_{22}^{c} = \frac{1}{\sqrt{2}}(Y_{2,-2} + Y_{2,2}) \,, \quad Y_{20} = Y_{2,0} \tag{10}$$

for the basis functions of the e_g representation. It is easy to show that (8) is invariant with respect to transformations (9) and (10), i.e.,

$$\langle a\sigma b\sigma' | V | c\sigma d\sigma' \rangle \approx \delta_{ac}\delta_{bd}\, U + \delta_{ad}\delta_{bc}\, \delta_{\sigma_a,\sigma_b}\, J \,, \tag{11}$$

where a, b, c, d denote the orbitals corresponding to cubic harmonics. Due to the cubic symmetry, the Green function and selfenergy are diagonal with respect to spinorbitals $\alpha = (a, \sigma_a)$, $\beta = (b, \sigma_b)$, etc., that correspond to cubic harmonics.

3. SOLUTION OF THE MANY-BODY PROBLEM

We determine the one-particle selfenergy within the fluctuation exchange approximation. Our formulation is different from, but equivalent to that given in the original paper by Bickers and Scalapino[18]. In order to simplify the notation, we introduce the quantity

$$v_{\alpha\beta} = (U - J\delta_{\sigma_\alpha,\sigma_\beta})\,(1 - \delta_{\alpha\beta})\,. \tag{12}$$

The first-order contribution corresponds to the Hartree-Fock approximation

$$\Sigma_\alpha^{\mathrm{HF}} = \sum_\beta v_{\alpha\beta}\,\langle n_\beta\rangle\,, \tag{13}$$

where $\langle n_\beta\rangle$ is the occupation number of the spinorbital β.

The second-order graphs (direct and exchange) give the contribution to the one-particle selfenergy

$$\Sigma_\alpha^{(2)}(E) = \sum_\beta v_{\alpha\beta}^2\, I_{\alpha\beta\beta}(E)\,, \tag{14}$$

where

$$I_{\alpha\beta\gamma}(E) = \int_{-\infty}^{\infty}\frac{\mathrm{d}\omega}{2\pi}\int_{-\infty}^{\infty}\frac{\mathrm{d}\omega'}{2\pi}\,G_\alpha(E-\omega)\,G_\beta(\omega+\omega')\,G_\gamma(\omega')\,. \tag{15}$$

The particle-particle channel includes the sum of ladder graphs. The two-particle propagator is defined as

$$\Psi_{\alpha\beta}(E) = i\int_{-\infty}^{\infty}\frac{\mathrm{d}\omega}{2\pi}\,G_\alpha(E-\omega)\,G_\beta(\omega)\,, \tag{16}$$

the particle-particle T-matrix that corresponds to an infinite sum of ladder graphs starting from the third-order graph is

$$T_{\alpha\beta}(E) = \frac{v_{\alpha\beta}^3\Psi_{\alpha\beta}^2(E)}{1 - v_{\alpha\beta}\Psi_{\alpha\beta}(E)}\,, \tag{17}$$

and the one-particle selfenergy

$$\Sigma_\alpha^{\mathrm{ee}}(E) = -i\sum_\beta\int_{-\infty}^{\infty}\frac{\mathrm{d}\omega}{2\pi}\,T_{\alpha\beta}(E+\omega)\,G_\beta(\omega)\,. \tag{18}$$

The particle-hole channel yields contributions of two different types, the first comes from the particle-hole T-matrix (random-phase approximation, RPA), while the second one corresponds to the renormalization of the pair interaction as in the GWA. We first define the particle-hole propagator

$$\Phi_{\alpha\beta}(E) = i\int_{-\infty}^{\infty}\frac{\mathrm{d}\omega}{2\pi}\,G_\alpha(E+\omega)\,G_\beta(\omega)\,, \tag{19}$$

and the particle-hole T-matrix of the first type that corresponds to an infinite sum of ladder graphs st^rting from the third order graph

$$T^{\mathrm{I}}_{\alpha\beta}(E) = \frac{v^3_{\alpha\beta}\Phi^2_{\alpha\beta}(E)}{1 - v_{\alpha\beta}\Phi_{\alpha\beta}(E)} \,. \tag{20}$$

The particle-hole T-matrix of the second type represents an infinite sum of graphs corresponding to a renormalization of pair interactions

$$\left[T^{\mathrm{II}}(E)\right]_{\alpha\beta} = \left[[1 - v\Phi(E)]^{-1} v\Phi(E) v\Phi(E) v\right]_{\alpha\beta} \,. \tag{21}$$

The quantities v and $\Phi(E)$ are matrices in spinorbital indices. The one-particle selfenergy then reads

$$\begin{aligned} \Sigma^{\mathrm{eh}}_{\alpha}(E) &= -i \int_{-\infty}^{\infty} \frac{\mathrm{d}\omega}{2\pi} \left[\sum_{\beta} T^{\mathrm{I}}_{\alpha\beta}(E-\omega)\, G_{\beta}(\omega) + T^{\mathrm{II}}_{\alpha\alpha}(E-\omega)\, G_{\alpha}(\omega)\right] \\ &\quad - \sum_{\beta(\beta\neq\alpha)} T^{\mathrm{II}}_{\alpha\beta}(0)\langle n_{\beta}\rangle \,. \end{aligned} \tag{22}$$

The full selfenergy within the FLEX approximation

$$\Sigma^{\mathrm{FLEX}}_{\alpha}(E) = \Sigma^{\mathrm{HF}}_{\alpha} + \Sigma^{(2)}_{\alpha}(E) + \Sigma^{\mathrm{ee}}_{\alpha}(E) + \Sigma^{\mathrm{eh}}_{\alpha}(E) \tag{23}$$

is a sum of all contributions (13), (14), (18), and (22).

We have so far used the causal quantities. In numerical calculations, however, it is advantageous to replace them by their retarded counterparts. This is particularly simple for one-particle quantities such as the Green function, $G(E)$, the selfenergy $\Sigma(E)$, and the integral $I(E)$ when we use the following generic notation

$$X(E) = \mathrm{Re}X(E) + i\,\mathrm{sgn}(E-\mu)\,\mathrm{Im}X(E) \tag{24}$$

valid at absolute zero. By inserting (24) for $G(E)$ into (15) we find equations for the retarded function $I(z)$ and the selfenergy $\Sigma(z)$ that are holomorphic in the upper half-plane of the complex energy z and vanishing when $z \to \infty$. It is then sufficient to calculate only their imaginary parts because the real parts can be found from a Kramers-Kronig relation

$$X(z) = \int_{-\infty}^{\infty} \frac{\mathrm{d}\omega}{\pi} \frac{\mathrm{Im}X(\omega)}{\omega - z} \,. \tag{25}$$

It allows us to determine the real and imaginary parts of $X(z)$ for complex z as well as for real values of the energy. For example, the retarded quantity $I(z)$ is then given by

$$\mathrm{Im}I_{\alpha,\beta,\gamma}(E) = \int_{0}^{E-\mu} \frac{\mathrm{d}\omega}{\pi} \int_{\mu-\omega}^{\mu} \frac{\mathrm{d}\omega'}{\pi}\, \mathrm{Im}G_{\alpha}(E-\omega)\, \mathrm{Im}G_{\beta}(\omega+\omega')\, \mathrm{Im}G_{\gamma}(\omega') \,, \tag{26}$$

where all Green functions are retarded.

The Green function G is calculated in a standard way within the TB-LMTO method. We first define an effective one-electron Hamiltonian

$$H^{\mathrm{eff}}_{\mathbf{R}\lambda,\mathbf{R}'\lambda'}(z) = H^{\mathrm{LMTO}}_{\mathbf{R}\lambda,\mathbf{R}'\lambda'} + \Sigma_{\mathbf{R},\lambda\lambda'}(z)\,\delta_{\mathbf{R}\mathbf{R}'}\,, \tag{27}$$

where z is a complex energy which in turn is used for the definition of the resolvent Green function of the interacting system analytic in both, upper and lower half-planes of complex energy

$$G(z) = [z - H^{\mathrm{eff}}(z)]^{-1}\,, \tag{28}$$

where the Green function and the effective Hamiltonian are matrices in the site-index as well as in the orbital momentum and spin indices, but are diagonal in the spin index for collinear magnetic structures.

If we know the selfenergy with proper analytic (Herglotz) properties we can determine all the quantities of interest. The density of states (DOS) and the Bloch spectral functions are defined for a translationally invariant system as

$$\rho(E) = -\frac{1}{\pi}\,\mathrm{Im}\sum_{\lambda} G_{\mathbf{R}\lambda,\mathbf{R}\lambda}(E+i0) \tag{29}$$

and

$$A(\mathbf{k},E) = -\frac{1}{\pi}\mathrm{Im}\sum_{\lambda} G_{\lambda\lambda}(\mathbf{k},E+i0)\,, \tag{30}$$

respectively, where $G_{\lambda\lambda'}(\mathbf{k},E+i0)$ is the lattice Fourier transform of $G_{\mathbf{R}\lambda,\mathbf{R}'\lambda'}(E+i0)$. The set of equations has to be completed by an equation for the chemical potential μ

$$\int_{-\infty}^{\mu} \mathrm{d}E\,\rho(E) = n\,, \tag{31}$$

where n is the number of electrons per lattice site.

The potential parameters $C_{\mathbf{R}\lambda}$ are replaced by

$$\tilde{C}_{\mathbf{R}\lambda}(z) = C_{\mathbf{R}L} + \Sigma_{\mathbf{R}\lambda}(z)\,. \tag{32}$$

The calculations are performed along the line $z = E + i\epsilon$ in the complex energy plane and then deconvoluted to the real axis[24]. The selfenergy in the complex energy plane is obtained from its imaginary part on the real axis determined above using Kramers-Kronig relation (25).

The quantities of the many-body theory have to be determined in a selfconsistent manner. This can be done at various levels of sophistication. In the simplest approach, only the Fermi energy μ is determined selfconsistently so as to preserve the number of particles n. The Green function and the selfenergy are then modified only by changes in μ. On the next level, one can renormalize the Green function by inserting the new selfenergy, and then repeating the calculations until the selfconsistency is achieved. This approach leads to renormalized Green functions, but it leaves the interaction vertices unchanged. Therefore, such an approach is not consistent as both, the Green functions and vertices should be renormalized. This can be done within the parquet theory which, however, goes beyond the FLEX approximation. An approximate, yet consistent, renormalization of propagators and vertices can be achieved by employing the DMFT which is exact in the limit of infinite spatial dimension ($d \to \infty$). We use the iterated perturbation method (IPT) of solving the DMFT equations (effective impurity problem)

$$\Sigma = \Sigma[\mathcal{G}_0] , \tag{33}$$

where $\mathcal{G}_0$ is an auxiliary Green function (bath, 'Weiss field') that represents the effective medium in which the selected site is embedded

$$[\mathcal{G}_0(z)]^{-1} = [G_{00}(z)]^{-1} + \Sigma(z) + (\mu_0 - \mu)\mathbf{1} . \tag{34}$$

Note that it is the idea of IPT[13] that $\mathcal{G}_0$ is used instead of G to determine the selfenergy Σ from (23). In (34), G_{00} is the on-site block of the Green function, μ_0 is the Fermi energy connected with the bath Green function $\mathcal{G}_0$, μ is the Fermi energy of the system (calculated from G_{00}), and $\mathbf{1}$ is a unit matrix.

4. DISORDER

In disordered alloys the lattice sites are randomly occupied by two or more atomic species. In such a case we have to solve the problem in two successive steps: we first evaluate the effect of electron correlations for one selected configuration of the system, and in the second step, we average over configurations. We wish to calculate the averaged or conditionally averaged one-particle Green function for a multi-band Hamiltonian of Hubbard type that represents a disordered alloy.

To make this problem tractable, one has to assume certain approximations. The first approximation we adopt is the terminal-point approximation[27]. All quantities in the many-body theory (like the selfenergy Σ, the Green functions, etc.) depend on the whole configuration of the alloy, however, it turns out that they depend mostly on the occupation of terminal lattice sites. For example, the element $G_{\mathbf{RR}'}$ of the Green function depends mostly on type of atoms at the lattice sites $\mathbf{R}$ and $\mathbf{R}'$. A further simplification is achieved if the one-particle selfenergy is assumed to be diagonal in the site representation, i.e., $\Sigma_{\mathbf{R}\lambda,\mathbf{R}'\lambda'}(z) = \delta_{\mathbf{R},\mathbf{R}'}\Sigma_{\mathbf{R},\lambda\lambda'}(z)$ which leads to a single terminal-point approximation

$$\Sigma_{\mathbf{R},\lambda\lambda'}(z) = \sum_Q \eta_{\mathbf{R}}^Q \, \Sigma_{\lambda\lambda'}^Q(z) , \tag{35}$$

where the occupation index $\eta_{\mathbf{R}}^Q = 1$ if site $\mathbf{R}$ is occupied by atom Q and $\eta_{\mathbf{R}}^Q = 0$ otherwise. The standard coherent potential approximation[26] can be then applied. Within the TB-LMTO scheme, the random site-diagonal selfenergy is simply added to the potential parameters $C_{\mathbf{R}\lambda}$

$$\widetilde{C}_{\mathbf{R},\lambda\lambda'}(z) = \sum_Q \eta_{\mathbf{R}}^Q \, \widetilde{C}_{\lambda\lambda'}^Q(z) , \quad \widetilde{C}_{\lambda\lambda'}^Q(z) = C_\lambda^Q(z)\,\delta_{\lambda\lambda'} + \Sigma_{\lambda\lambda'}^Q(z) , \tag{36}$$

similarly as in the non-random case (cf. also Eq.(32)). The configurational averaging is then performed in the same manner as in the standard TB-LMTO-CPA method[28].

5. SURFACES

At surfaces, the translational symmetry in the direction perpendicular to the sample surface is lost and only the translational symmetry parallel to the surface is preserved. It can be used to simplify the problem by using the two-dimensional Bloch transform. It is convenient to introduce the mixed basis, namely, $|\mathbf{R}\lambda\rangle = |\mathbf{R}_{\parallel}n\lambda\rangle$, where $\mathbf{R}_{\parallel}$ refers to the component parallel to the surface and n labels the sample layers. The problem then has to be solved in real space with respect to the layer index n. It can be shown[24] that introducing the surface Green functions allows to reduce the problem originally of infinite dimension with respect to the layer index n to a problem of finite dimension in the so-called intermediate region[24]. Vacuum is described within the TB-LMTO formalism in terms of empty spheres which allow to evaluate properly charge redistribution at the solid/vacuum interface which leads to a dipole barrier.

The selfenergy $\Sigma_{\mathbf{R}\lambda\lambda'} = \Sigma_{n\lambda\lambda'}$ is layer-dependent in the intermediate region, and it equals $\Sigma^{\rm bulk}_{\lambda\lambda'}$ in all bulk layers. Due to lowering of the symmetry at the surface, $\Sigma_{n\lambda\lambda'}$ is non-diagonal with respect to orbital indices L, L' but it is still diagonal with respect to spin indices σ, σ' (for collinear magnetic case). The values of $\Sigma_{n\lambda\lambda'}$ in the layer n depend on selfenergies in all other layers. Consequently, one has to solve a coupled system of DMFT equations for $\Sigma_{n\lambda\lambda'}$, $n = 1, ...N$, where N is the number of layers in the intermediate region.

One has to evaluate the new charge distribution at the surface selfconsistently to obtain correct values of effective levels $\tilde{C}_{n\lambda\lambda'}(z)$, i.e., one has to proceed similarly as in the DFT calculations for surfaces[24]. It should be finally mentioned that also the case of disordered surfaces can be addressed within the present computational scheme.

6. RESULTS AND DISCUSSION

We show the results for two systems as an illustration of the present theory. In the first case we have studied the bcc-based disordered Co_xFe_{1-x} alloy for which we have calculated the local magnetic moments within the non-selfconsistent second order perturbation theory (Fig. 1). We have assumed $U = 0.12$ Ry and $J = 0.09$ Ry for Co and $U = 0.12$ Ry and $J = 0.08$ Ry for Fe and used 1360 points in the irreducible wedge of the Brillouin zone. The calculated moments at Fe agree quite well with experimental data[29] in the whole concentration range in which the bcc structure is stable $(0 \leq x < 0.75)$, while the agreement for Co is worse. A further refinement of the computational procedure is needed.

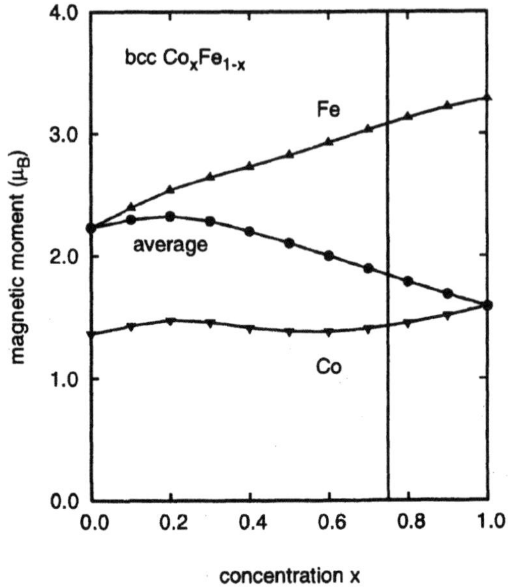

Figure 1. The local and average magnetic moments in the bcc-based disordered Co_xFe_{1-x} alloys calculated within the non-selfconsistent second order perturbation theory combined with the CPA. Vertical line at $x = 0.75$ shows the upper limit of stability of the bcc phase.

The other example is ferromagnetic fcc Ni. Using the electron-electron T-matrix approximation we have calculated spin-resolved densities of states within the DMFT. We have assumed $U = 0.18$ Ry and $J = 0.12$ Ry and used 2480 points in the irreducible wedge of the Brillouin zone. These parameters yield the magnetic moment (0.6 μ_B) in agreement with experimental value and the density of states (Fig. 2) closely resembles the photoemission spectra of Ni: one can observe a band narrowing and also the satellite at approximately 7 eV below the Fermi energy. A comparison with a similar calculation employing the full FLEX selfenergy (i.e., including also the electron-hole channel) has shown a little difference, because the e-h channel is not important for Ni which has a low concentration of holes. On the other hand, the satellite is missing in calculations based on the second order perturbation theory[23], which clearly shows the importance of infinite summation in the perturbation series.

The majority states exhibit strong correlation effects while the minority states are influenced only weakly. This is also confirmed by a large difference between the selfenergies for majority and minority states (Fig. 3). The effect is more pronounced than in the second-order theory[23] and it can be explained in a similar way: because U and J are comparable, the largest contribution to the selfenergy comes from states with opposite spin. The electrons in partially filled minority bands can give rise to an energy-dependent potential acting on the majority electrons. On the other hand, the almost filled majority bands which can be viewed approximately as closed shells, can yield merely a static (energy-independent) potential acting on the minority electrons. A rigorous proof of this statement follows from the retarded versions of the equations (14) - (18).

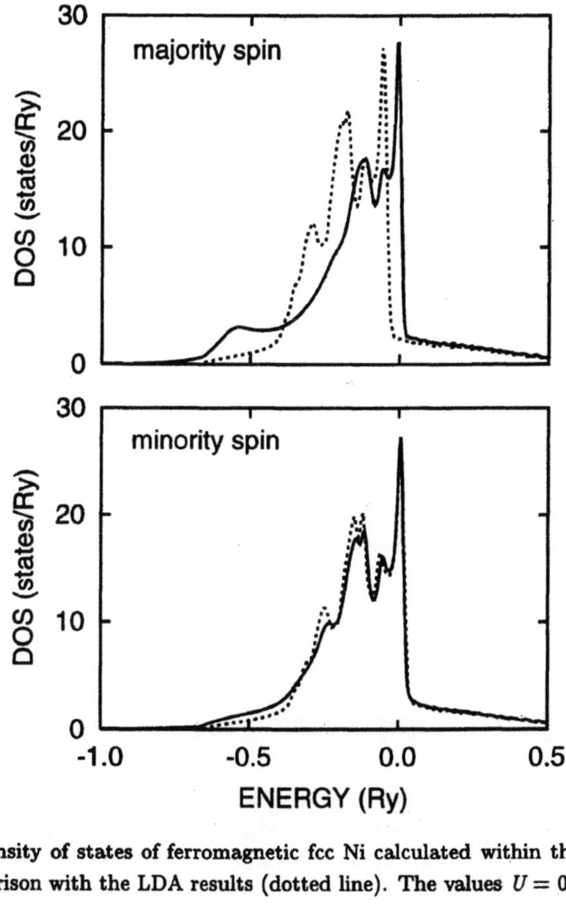

Figure 2. The density of states of ferromagnetic fcc Ni calculated within the TMA and DMFT (full line) in comparison with the LDA results (dotted line). The values $U = 0.18$ Ry and $J = 0.12$ Ry were assumed.

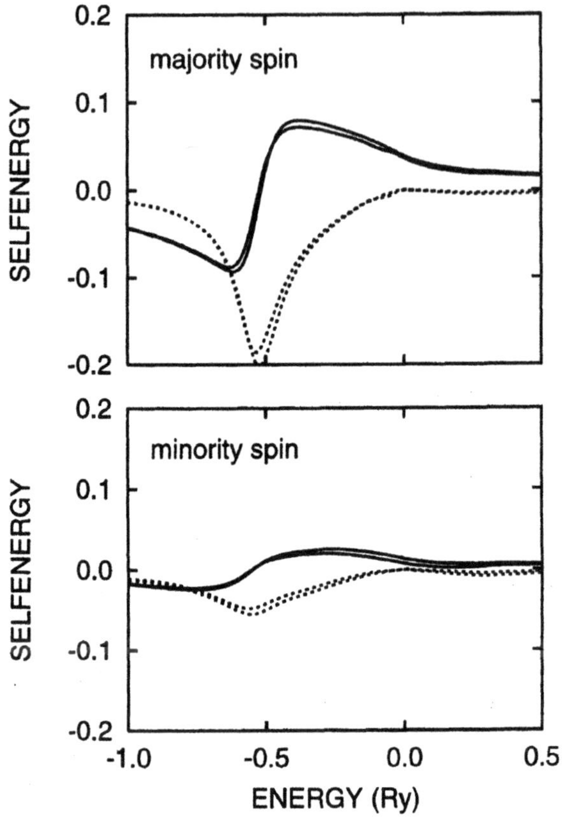

Figure 3. The selfenergies of ferromagnetic fcc Ni calculated within the TMA and DMFT ($U =$ 0.18 Ry and $J = 0.12$ Ry). The real parts (full lines) and imaginary parts (dotted lines) are shown for representations t_{2_g} and e_g that belong to d-states. The Hartree-Fock contributions are omitted in this figure.

7. CONCLUSIONS AND OUTLOOK

In summary, we have developed a scheme allowing to determine electronic properties of solids including the electron correlations. Our approach is based on multiband Hamiltonian of Hubbard type. Its one-electron part is derived from the TB-LMTO Hamiltonian. The TB-LMTO is a method based on the LDA that already describes the ground-state properties of solids reasonably well. The present formalism is particularly suitable for systems with a limited or lacking translational symmetry such as solid surfaces and disordered alloys.

The many-body problem is solved by using the FLEX approximation for the one-particle selfenergy. The Green functions and other quantities of many-body theory are determined selfconsistently by using the iterated perturbation method of solving the DMFT equations. In this way, an approximate, yet consistent renormalization of propagators and interaction vertices is achieved.

The basic approximation adopted in this paper is a local character of the self-energy. It is reasonably justified for transition metals and it is also required from formal reasons as it makes possible to employ the DMFT and to treat the disorder within the CPA.

We employ the retarded quantities evaluated along a line in the complex energy plane followed by an analytic deconvolution to the real axis. We then use Kramers-Kronig relations to reconstruct analytic functions from their imaginary parts. This formulation has certain advantages over the conventional approach using causal functions where the energy integrals are replaced by numerically less accurate finite sums over Matsubara frequencies.

We have applied the general scheme to transition metals and their alloys assuming the direct and exchange interactions only between the *d*-electrons. We have found a number of features in a reasonable agreement with the experiment such as the narrowing of the energy bands and the position of the satellite peak in fcc-Ni, the magnetic moments in fcc-Ni and in Co_xFe_{1-x} disordered alloys. This confirms the ability of the technique to investigate the effects of simultaneous presence of electron correlations and substitutional disorder.

We have shown that the present scheme can be extended also to surfaces. It opens the way to studies of electron correlations at solid surfaces and interfaces as well as in multilayers. The knowledge of the electron Green function allows one to evaluate not only the one-electron properties, but also the magnetic interactions in correlated systems by mapping their total energy onto an effective Heisenberg Hamiltonian[30]. This would be particularly interesting for fcc Ni, where similar calculations based on the LDA yield incorrect results for Curie temperature and for spin stiffness. Finally, the present scheme can serve as a basic building block for a future implementation of the LDA+DMFT method and its applications to systems with observable deviations from the Fermi-liquid behavior either due to strong electron correlations or disorder.

Acknowledgements

Financial support for this work was provided by the Grant Agency of the Czech Republic (Project 202/01/0764), and the Grant Agency of the Academy of Sciences of the Czech Republic (Project A 1010829).

REFERENCES

1. P. Hohenberg and W. Kohn, Phys. Rev. **136**, B864 (1964).
2. W. Kohn, Rev. Mod. Phys. **71**, 1253 (1999).

3. M. Petersilka, U.J. Gossmann, and E.K.U. Gross, Phys. Rev. Lett. **76**, 1212 (1996).
4. L. Lægsgaard and A. Svane, Phys. Rev. B **59**, 3450 (1999).
5. L. Hedin, Phys. Rev. **139**, A796 (1965).
6. M. Springer, F. Aryasetiawan, and K. Karlsson, Phys. Rev. Lett. **80**, 2389 (1998).
7. V. Anisimov, J. Zaanen, and O.K. Andersen, Phys. Rev. B **44**, 943 (1991).
8. F. Aryasetiawan and O. Gunnarsson, Phys. Rev. Lett. **74**, 3221 (1995).
9. S. Hüfner, G.K Wertheim, N.V. Smith, and M.M Traum, Solid State Commun. **11**, 323 (1972).
10. M. Springer, F. Aryasetiawan, and K. Karlsson, Phys. Rev. Lett. **76**, 1212 (1996).
11. J. Igarashi, P. Unger, K. Hirai, and P. Fulde, Phys. Rev. B **49**, 16181 (1994).
12. F. Manghi, V. Bellini, and C. Arcangeli, Phys. Rev. B **56**, 7149 (1997).
13. A. Georges, G. Kotliar, W. Krauth, and M. Rozenberg, Rev. Mod. Phys. **68**, 13 (1996).
14. S.Y. Savrasov and G. Kotliar, Phys. Rev. Lett. **84**, 3670 (2000).
15. A.I. Lichtenstein and M.I. Katsnelson, Phys. Rev. B **57**, 6884 (1998).
16. M.I. Katsnelson and A.I. Lichtenstein, J. Phys.: Cond. Matt. **11**, 1037 (1999).
17. J. Hubbard, Proc. Roy. Soc. London, Ser. A **276**, 238 (1963).
18. N.E. Bickers and D.J. Scalapino, Ann. Phys. (N.Y.) **193**, 206 (1989).
19. K. Held, I.A. Nekrasov, N. Blümer, V.I. Anisimov, and D. Vollhardt, arXiv:cond-mat/0010395.
20. K. Held, G. Keller, V. Eyert, D. Vollhardt, and V.I. Anisimov, Phys. Rev. **86**, 5345 (2001).
21. A.I. Lichtenstein and M.I. Katsnelson, arXiv:cond-mat/0102297.
22. V. Drchal, V. Janiš, and J. Kudrnovský, in *Electron Correlations and Materials Properties* edited by A. Gonis et al., (Kluwer Academic/Plenum Publishers, NY, 1999), p. 273.
23. V. Drchal, V. Janiš, and J. Kudrnovský, Phys. Rev. B **60**, 15664 (1999).
24. I. Turek, V. Drchal, J. Kudrnovský, M. Šob, and P. Weinberger, *Electronic Structure of Disordered Alloys, Surfaces and Interfaces* (Kluwer, Boston, 1997)
25. M.M. Steiner, R.C. Albers, and L.J. Sham, Phys. Rev. B **45**, 13272 (1992).
26. B. Velický, H. Ehrenreich, and S. Kirkpatrick, Phys. Rev. **175**, 747 (1968).
27. V. Drchal, J. Phys. Chem. Solids **40**, 393 (1979).
28. J. Kudrnovský and V. Drchal, Phys. Rev. B **41**, 7515 (1990).
29. Landolt-Börnstein, *Numerical Data and Functional Relationships in Science and Technology, New Series, Volume 19a, Magnetic Properties of Metals*, editor H.P.J. Wijn (Springer, Berlin, 1986).
30. M.I. Katsnelson and A.I. Lichtenstein, Phys. Rev. B **61**, 8906 (2000).

QUANTAL DENSITY FUNCTIONAL THEORY

Viraht Sahni*

1. INTRODUCTION

Traditional time-dependent (TD) density functional theory (DFT) is based on two postulates: the Hohenberg-Kohn theorem [1,2] that the Schrödinger wavefunction for a system of electrons in an external field is a unique functional of the electronic density $\rho(\mathbf{r}t)$ to within a TD phase, and on the stationary action principle. The density is thereby obtained via the Euler equation. The theorem is proved for the case when the potential energy $v(\mathbf{r}t)$ of an electron in an external field is Taylor expandable about some initial time. However, the explicit functional dependence of the wavefunction on the density is not defined. The fact that the wavefunction is a functional of the density is explicitly employed in the Kohn-Sham (KS) version [1,2] of the theory. The basic idea of KS DFT is to transform the physical system as described by the Schrödinger equation to one of noninteracting Fermions with equivalent density. We refer to these model Fermions as the S system, S being a mnemonic for 'single Slater' determinant. The existence of such an S system, referred to as noninteracting v-representability, has recently been proved [3,4] for Taylor-expandable external potential energies. In the TD KS description of the S system, all the many-body effects are incorporated in an electron-interaction action functional $\mathcal{A}_{ee}^{KS}[\rho]$. The corresponding effective electron-interaction potential energy $v_{ee}(\mathbf{r}t)$ of the S system Fermions is defined as the functional derivative $\delta \mathcal{A}_{ee}^{KS}[\rho]/\delta\rho(\mathbf{r}\,t)$. How the different electron correlations present within the S system

* Department of Physics, Brooklyn College and The Graduate School of the City University of New York, 2900 Bedford Avenue, Brooklyn, New York 11210

are incorporated in the action functional $\mathcal{A}_{ee}^{KS}[\rho]$ or its functional derivative is not described by the theory.

Time-independent (TI) KS DFT [5] is based on the Hohenberg-Kohn theorems [6] that the stationary-state Schrödinger wavefunction for electrons in an arbitrary external field, is a functional of the density $\rho(\mathbf{r})$, and that the density may be obtained by the variational principle for the energy with respect to arbitrary variations of the density. Once again the existence of a corresponding S system is assumed. All the many-body effects are incorporated in an KS electron-interaction energy functional $E_{ee}^{KS}[\rho]$, and the electron-interaction potential energy of the S system Fermions is defined as the functional derivative $\delta E_{ee}^{KS}[\rho]/\delta\rho(\mathbf{r})$. Since the explicit functional dependence of the wavefunction on the density is unknown, the manner by which the various electron correlations are incorporated in the functional $E_{ee}^{KS}[\rho]$ or its functional derivative is again not described by the theory.

KS DFT is one way to describe the S system. There is, however, another distinct way by which the model system of noninteracting Fermions with equivalent density may be described. This description, which is independent of traditional DFT, and valid for ***arbitrary*** external field, is referred to as Quantal density functional theory (Q-DFT). Q-DFT therefore constitutes another method for the determination of electronic structure within the framework of local effective potential energy theory. In Q-DFT, the fact that the wavefunction is a functional of the density is not explicitly employed. Hence, there are no functionals or functional derivatives in Q-DFT. The functional dependence of the wavefunction on the density is used only to the degree that it explains why a system with density equivalent to that of Schrodinger theory is being constructed.

Q-DFT is a description of the S system in terms of classical fields that pervade all space, the sources of these fields being quantum-mechanical expectations of Hermitian operators taken with respect to the wavefunction. The sources, and thus the fields, are each separately representative of the different electron correlations that must be accounted for in the S system. These correlations are those due to the Pauli exclusion principle and Coulomb repulsion, and Correlation-Kinetic and Correlation-Current-Density effects, the latter two being a consequence of the difference in kinetic energy and current density between the Schrodinger and S systems. The sources and fields are explicitly defined, and thus so are the contributions of the different electron correlations to the total energy and local electron-interaction potential energy $v_{ee}(\mathbf{r}t)$ of the S system Fermions. Q-DFT thus provides the physics of the mapping from Schrodinger theory to that of the S system. As a consequence, it provides a rigorous physical interpretation of the KS DFT action and energy functionals of the density and of their functional derivatives.

We describe TD Q-DFT [7-9] in the following section. TI Q-DFT then constitutes a special case [10-13]. In the section that follows, we describe the physical interpretation of TD KS DFT. In that section we also describe an analysis [9] performed via TD Q-DFT and adiabatic coupling constant perturbation theory to interpret and thereby to explain the various electron correlations that contribute to the KS DFT 'exchange' and 'correlation' potential energies. These potential energies are defined, respectively, as the functional derivatives of the KS exchange and correlation action functionals of the density. The interpretations of the energy functionals and functional derivatives of TI KS theory then constitute a special case of those of TD theory [14].

2. QUANTAL DENSITY FUNCTIONAL THEORY

In order to understand TD Q-DFT one must first understand TD Schrödinger theory from the perspective of fields and quantal sources. We thus begin with this description [15] of Schrödinger theory.

2.1. Time-Dependent Schrödinger Theory

Consider a system of N electrons in an *arbitrary* external field such that the potential energy of an electron at positron **r** and time t is the real function v(**r**,t). The Schrödinger equation for this system is

$$\hat{H}(t)\Psi(\mathbf{X}t) = i\frac{\partial\Psi(\mathbf{X}t)}{\partial t} , \tag{1}$$

where the Hamiltonian $\hat{H} = \hat{T} + \hat{V}(t) + \hat{U}$ with

$$\hat{T} = -\frac{1}{2}\sum_i \nabla_i^2, \quad \hat{V}(t) = \sum_i v(\mathbf{r}_i\, t), \quad \hat{U} = \frac{1}{2}\sum_{i,j}{}' \frac{1}{|\mathbf{r}_i - \mathbf{r}_j|} , \tag{2}$$

$\mathbf{X} = \mathbf{x}_1, \ldots. \mathbf{x}_N$, $\mathbf{x} = \mathbf{r}\sigma$, σ the spin coordinate, and $\Psi(\mathbf{X}t)$ the wavefunction.

We next define various quantal sources which give rise to fields, and which are expectations of Hermitian operators taken with respect to the wavefunction. These sources of fields are the density $\rho(\mathbf{r}\,t)$, the spinless single particle density matrix $\gamma(\mathbf{r}\mathbf{r}'t)$, the pair-correlation density $g(\mathbf{r}\mathbf{r}'t)$, and the current density $\mathbf{j}(\mathbf{r}t)$.

2.1.1. *Quantal Sources*

The electron density $\rho(\mathbf{r}\,t)$ is N times the probability of an electron being at **r**:

$$\rho(\mathbf{r}\,t) = N\sum_\sigma \int \Psi^*(\mathbf{r}\,\sigma, \mathbf{X}^{N-1}, t)\,\Psi(\mathbf{r}\,\sigma, \mathbf{X}^{N-1}, t)\,d\mathbf{X}^{N-1} \tag{3}$$

$$= \langle \Psi(t) | \,\hat{\rho}\, | \Psi(t) \rangle , \tag{4}$$

where $\mathbf{X}^{N-1} = \mathbf{x}_2,\ldots\mathbf{x}_N,$, $d\mathbf{X}^{N-1} = d\mathbf{x}_2,\ldots d\mathbf{x}_N, \int d\mathbf{x} = \Sigma_\sigma \int d\mathbf{r}$, and $\hat{\rho}$ is the Hermitian operator

$$\hat{\rho}(\mathbf{r}) = \sum_i \delta(\mathbf{r}_i - \mathbf{r}) \ . \tag{5}$$

The spinless single particle density matrix $\gamma(\mathbf{r}\mathbf{r}'t)$ is defined as

$$\gamma(\mathbf{r}\mathbf{r}'t) = N\sum_\sigma \int \Psi^*(\mathbf{r}\sigma, \mathbf{X}^{N-1}, t)\Psi(\mathbf{r}'\sigma, \mathbf{X}^{N-1}, t)d\mathbf{X}^{N-1} \tag{6}$$

$$= \langle \Psi(t)| \ \hat{\gamma}(\mathbf{r}\mathbf{r}')|\Psi(t)\rangle \ , \tag{7}$$

where the Hermitian operator

$$\hat{\gamma}(\mathbf{r}\mathbf{r}') = \hat{A} + i\hat{B} \tag{8}$$

$$A = \frac{1}{2}\sum_j [\delta(\mathbf{r}_j - \mathbf{r})T_j(\mathbf{a}) + \delta(\mathbf{r}_j - \mathbf{r}')T_j(-\mathbf{a})]$$

$$B = -\frac{i}{2}\sum_j [\delta(\mathbf{r}_j - \mathbf{r})T_j(\mathbf{a}) - \delta(\mathbf{r}_j - \mathbf{r}')T_j(-\mathbf{a})] \ ,$$

$T_j(\mathbf{a})$ is a translation operator such that $T_j(\mathbf{a})\Psi(\ldots.\mathbf{r}_j\ldots,t) = \Psi(\ldots\mathbf{r}_j + \mathbf{a},\ldots,t)$, and $\mathbf{a} = \mathbf{r}' - \mathbf{r}$.

The pair-correlation density $g(\mathbf{r}\mathbf{r}'t)$ is the density at $\mathbf{r}'$ for an electron at $\mathbf{r}$:

$$g(\mathbf{r}\mathbf{r}'t) = \langle \Psi(t)| \ \hat{P}(\mathbf{r}\mathbf{r}')|\Psi(t)\rangle / \rho(\mathbf{r}t) \ , \tag{9}$$

where the Hermitian pair-correlation operator $\hat{P}(\mathbf{r}\mathbf{r}')$ is

$$\hat{P}(\mathbf{r}\mathbf{r}') = \sum_{i,j}{}' \delta(\mathbf{r}_i - \mathbf{r})\delta(\mathbf{r}_j - \mathbf{r}') \ . \tag{10}$$

This source is representative of electron correlations due to the Pauli exclusion principle and Coulomb repulsion. The pair-correlation density may be further partitioned into its local and nonlocal components as

$$g(\mathbf{r}\mathbf{r}'t) = \rho(\mathbf{r}'t) + \rho_{xc}(\mathbf{r}\mathbf{r}'t) \ , \tag{11}$$

where $\rho_{xc}(\mathbf{r}\mathbf{r}'t)$ is the Fermi-Coulomb hole charge at $\mathbf{r}'$ for an electron at $\mathbf{r}$.

The current density $\mathbf{j}(\mathbf{r}t)$ may be expressed in terms of the single particle density matrix source $\gamma(\mathbf{r}\mathbf{r}'t)$ or as the expectation of the current density operator $\hat{\mathbf{j}}(\mathbf{r})$:

$$\mathbf{j}(\mathbf{r}\,t)=\frac{i}{2}N\sum_{\sigma}\int\left[\Psi(t)\nabla\Psi^{*}(t)-\Psi^{*}(t)\nabla\Psi(t)\right]d\,\mathbf{X}^{N-1} \tag{12}$$

$$=\frac{i}{2}\sum_{\sigma}\left[\nabla'-\nabla''\right]\gamma(\mathbf{x}'\mathbf{x}''t)\big|_{\mathbf{x}'=\mathbf{x}''=\mathbf{x}} \tag{13}$$

$$=\langle\Psi(t)|\hat{\mathbf{j}}(\mathbf{r})|\Psi(t)\rangle\ , \tag{14}$$

where

$$\hat{\mathbf{j}}(\mathbf{r})=\frac{1}{2i}\sum_{j}\left[\nabla_{\mathbf{r}_j}\delta(\mathbf{r}-\mathbf{r}_j)+\delta(\mathbf{r}-\mathbf{r}_j)\nabla_{\mathbf{r}_j}\right]\ . \tag{15}$$

These sources have the properties:

$$\int\rho(\mathbf{r}\,t)d\,\mathbf{r}=N\ ,\ \gamma(\mathbf{r}\mathbf{r}\,t)=\rho(\mathbf{r}\,t)\ ,\ \int g(\mathbf{r}\mathbf{r}'t)d\,\mathbf{r}'=N-1\ , \tag{16}$$

and $\int\rho_{xc}(\mathbf{r}\mathbf{r}'t)d\,\mathbf{r}'=-1$.

The sources give rise to the following fields.

2.1.2 Fields

The electron-interaction field $\mathcal{E}_{ee}(\mathbf{r}t)$ is obtained via Coulomb's law from the pair-correlation density:

$$\mathcal{E}_{ee}(\mathbf{r}\,t)=\int\frac{g(\mathbf{r}\mathbf{r}'t)(\mathbf{r}-\mathbf{r}')}{|\mathbf{r}-\mathbf{r}'|^{3}}d\,\mathbf{r}' \tag{17}$$

$$=\mathcal{E}_{H}(\mathbf{r}\,t)+\mathcal{E}_{xc}(\mathbf{r}\,t)\ , \tag{18}$$

where the Hartree $\mathcal{E}_{H}(\mathbf{r}t)$ and Pauli-Coulomb $\mathcal{E}_{xc}(\mathbf{r}t)$ components are defined, respectively, in terms of their sources as

$$\mathcal{E}_{H}(\mathbf{r}\,t)=\int\frac{\rho(\mathbf{r}')(\mathbf{r}-\mathbf{r}')}{|\mathbf{r}-\mathbf{r}'|^{3}}d\,\mathbf{r}'\ \text{ and }\ \mathcal{E}_{xc}(\mathbf{r}\,t)=\int\frac{\rho_{xc}(\mathbf{r}\mathbf{r}'t)(\mathbf{r}-\mathbf{r}')}{|\mathbf{r}-\mathbf{r}'|^{3}}d\,\mathbf{r}'\ . \tag{19}$$

The differential density field **D**(**r**t) whose source is the density is defined as

$$\mathbf{D}(\mathbf{r}\,t)=\mathbf{d}(\mathbf{r}\,t)/\rho(\mathbf{r}\,t)\ , \tag{20}$$

where

$$\mathbf{d}(\mathbf{r}\,t)=-\frac{1}{4}\nabla\nabla^2\rho(\mathbf{r}\,t)\ . \tag{21}$$

The kinetic field **Z**(**r**t) whose source is the spinless single particle density matrix is defined as

$$\mathbf{Z}(\mathbf{r}\,t)=\mathbf{z}(\mathbf{r}\,t;[\gamma])/\rho(\mathbf{r}\,t)\ , \tag{22}$$

where the field **z**(**r**t) is defined by its component z_α(**r**t) as

$$z_\alpha(\mathbf{r}\,t)=2\sum_\beta\frac{\partial}{\partial r_\beta}t_{\alpha\beta}(\mathbf{r}\,t)\ , \tag{23}$$

and where $t_{\alpha\beta}$(**r**t) is the kinetic-energy-density tensor defined in turn as

$$t_{\alpha\beta}(\mathbf{r}\,t)=\frac{1}{4}\left[\frac{\partial^2}{\partial r'_\alpha\,\partial r''_\beta}+\frac{\partial^2}{\partial r'_\beta\,\partial r''_\beta}\right]\gamma(\mathbf{r}'\mathbf{r}''t)\Big|_{\mathbf{r}'=\mathbf{r}''=\mathbf{r}}\ . \tag{24}$$

The current density field **J**(**r**t) is obtained from its source, the current density or equivalently the spinless single particle density matrix, as the temporal derivative

$$\mathbf{J}(\mathbf{r}\,t)=\frac{1}{\rho(\mathbf{r}\,t)}\frac{\partial}{\partial t}\mathbf{j}(\mathbf{r}\,t)\ . \tag{25}$$

The Hartree $\mathcal{E}_H$(**r**t) and differential density **D**(**r**t) fields are conservative so that $\nabla\times\mathcal{E}_H(\mathbf{r}\,t)=\nabla\times\mathbf{D}(\mathbf{r}\,t)=0$. The fields $\mathcal{E}_{ee}$(**r**t) [or $\mathcal{E}_{xc}$(**r**t)], **Z**(**r**t) and **J**(**r**t) are in general not conservative. Their sum, however, always is, so that $\nabla\times[\mathcal{E}_{ee}(\mathbf{r}\,t)+\mathbf{Z}(\mathbf{r}\,t)+\mathbf{J}(\mathbf{r}\,t)]=0$. For systems of certain symmetry or systems that have been symmetrized, these fields too may be separately conservative, in which case $\nabla\times\mathcal{E}_{ee}(\mathbf{r}\,t)=\nabla\times\mathbf{Z}(\mathbf{r}\,t)=\nabla\times\mathbf{J}(\mathbf{r}\,t)=0$.

2.1.3 *Pure State Differential Virial Theorem*

That the quantum system of N electrons in the presence of an arbitrary external field may be described entirely in terms of the various fields defined above follows from the pure state differential virial theorem [7,8] according to which

$$\nabla v(\mathbf{r}\,t) = -\mathbf{F}(\mathbf{r}\,t)\,, \tag{26}$$

where

$$\mathbf{F}(\mathbf{r}\,t) = -\boldsymbol{\mathcal{E}}_{ee}(\mathbf{r}\,t) + \mathbf{D}(\mathbf{r}\,t) + \mathbf{Z}(\mathbf{r}\,t) + \mathbf{J}(\mathbf{r}\,t)\,. \tag{27}$$

The evolution of the system with time is described by the fields in this theorem which can also be shown to be equivalent to Ehrenfest's theorem. The (nonconserved) total energy E(t) of the system, however, is expressed in integral virial form in terms of these fields, and its evolution with time is described by these expressions.

2.1.4 Total Energy and its Components

The total energy E(t), which is the expectation of the Hamiltonian, is the sum of the kinetic T(t), external $E_{ext}(t)$, and electron-interaction $E_{ee}(t)$ components:

$$E(t) = \langle \Psi(t) |\, \hat{H}(t)\, | \Psi(t) \rangle$$

$$= T(t) + E_{ext}(t) + E_{ee}(t)\;. \tag{28}$$

These components may be written in terms of their respective fields as follows:

The kinetic energy T(t) is

$$T(t) = \langle \Psi(t) |\; \hat{T}\; | \Psi(t) \rangle$$

$$= -\frac{1}{2}\int d\mathbf{r}\,\rho(\mathbf{r}\,t)\,\mathbf{r}\cdot\mathbf{Z}(\mathbf{r}\,t)\,. \tag{29}$$

The electron-interaction potential energy $E_{ee}(t)$ is

$$E_{ee}(t) = \langle \Psi(t) |\; \hat{U}\; | \Psi(t) \rangle$$

$$= \int d\mathbf{r}\,\rho(\mathbf{r}\,t)\,\mathbf{r}\cdot\boldsymbol{\mathcal{E}}_{ee}(\mathbf{r}\,t) \tag{30}$$

$$= E_H(t) + E_{xc}(t)\;, \tag{31}$$

where the Hartree or Coulomb self-energy $E_H(t)$ and the quantum-mechanical exchange-correlation energy $E_{xc}(t)$ are defined, respectively, as

$$E_H(t) = \int d\mathbf{r}\,\rho(\mathbf{r}\,t)\,\mathbf{r}\cdot\boldsymbol{\mathcal{E}}_H(\mathbf{r}\,t) \tag{32}$$

and

$$E_{XC}(t) = \int d\mathbf{r}\rho(\mathbf{r}t)\mathbf{r}\cdot \mathcal{E}_H(\mathbf{r}t) \; . \tag{33}$$

The external potential energy $E_{ext}(t)$ which is the expectation

$$E_{ext}(t) = \langle \Psi(t) | \hat{V}(t) | \Psi(t) \rangle$$

$$= \int d\mathbf{r}\rho(\mathbf{r}t)v(\mathbf{r}t) \; , \tag{34}$$

depends upon all the internal fields present within the quantum system. From the differential virial theorem Eq.(26), it follows that the potential energy v(**rt**) of an electron at each instant of time is the work done to move the electron from some reference point at infinity to its position at **r** in the force of the conservative field **F**(**rt**):

$$v(\mathbf{r}t) = -\int_{\infty}^{\mathbf{r}} \mathbf{F}(\mathbf{r}'t)\cdot d\mathbf{l}' \; . \tag{35}$$

This work done is path independent since $\nabla \times \mathbf{F}(\mathbf{r}t) = 0$. Thus, the external potential energy may be expressed in terms of the internal fields.

The expressions for the energy components are independent of whether or not the fields are conservative.

2.2 Time-Independent Schrödinger Theory

The description of stationary state Schrödinger theory in terms of quantal sources and fields is a special case of the TD version. For TI external fields such that the potential energy of an electron is v(**r**), the Schrödinger equation is

$$H\Psi_n(\mathbf{X}t) = i\frac{\partial\Psi(\mathbf{X}t)}{\partial t} = E_n\,\Psi_n, \tag{36}$$

where the $\Psi_n(\mathbf{X}t)$ are the wavefunctions and E_n the energy eigenvalues. The solutions of Eq. (36) are of the form

$$\Psi_n(\mathbf{X}t) = \psi_n(\mathbf{X})e^{-iE_n t} \; , \tag{37}$$

where the functions $\psi_n(\mathbf{X})$ and eigenvalues E_n are determined by the TI Schrödinger equation

$$H\psi_n(\mathbf{X}) = E_n\psi_n(\mathbf{X}) \; . \tag{38}$$

The various quantal sources and fields are determined by the functions $\psi_n(\mathbf{X})$, the phase factor vanishing as a consequence of the definitions of the sources as expectation values. Further, in the TI case, there is no current density field **J**(**rt**). The differential virial theorem is the same, but with this term absent. The components of the total energy, and the potential energy v(r) due to the external field, are defined as before but by the static fields $\mathcal{E}_{ee}(\mathbf{r})$, $\mathbf{D}(\mathbf{r})$, and $\mathbf{Z}(\mathbf{r})$. These remarks are valid for both ground and excited states of Schrödinger theory. We refer the reader to Ref. 15 for an example of the structure of these various fields for an exactly solvable model.

2.3 Time Dependent S System

With the assumption of noninteracting v representability, the Schrödinger equation for the model S system of noninteracting Fermions with density equivalent to that of the interacting system is

$$\hat{H}_s(t)\varphi_i(\mathbf{x}\,t) = i\frac{\partial \varphi_i(\mathbf{x}\,t)}{\partial t} \,, \tag{39}$$

where

$$H_s(t) = -\frac{1}{2}\nabla^2 + v_s(\mathbf{r}\,t) \,, \tag{40}$$

$$v_s(\mathbf{r}\,t) = v(\mathbf{r}\,t) + v_{ee}(\mathbf{r}\,t) \,, \tag{41}$$

and the wavefunction is the single Slater determinant $\Phi\{\varphi_i\}$ of the orbitals $\varphi_i(\mathbf{x}t)$. Here $v_{ee}(\mathbf{r}t)$ is the electron interaction potential energy of the model Fermions representative of the various electron correlations present in the S system. Since $v_{ee}(\mathbf{r}t)$ is the potential energy of such a Fermion, there must exist a corresponding force field $\mathcal{F}(\mathbf{r}t)$. Furthermore, this force field must represent all the different electron correlations present. Certainly, it must incorporate correlations due to the Pauli exclusion principle and Coulomb repulsion as these are the electron correlations in Schrödinger theory. But it must also account for the fact that the kinetic energy and current density of the Schrödinger and S systems differ. Thus, there are contributions to this force field from Correlation-Kinetic and Correlation-Current-Density effects. However, prior to deriving the force field $\mathcal{F}(\mathbf{r}t)$, let us define the quantal sources and state the differential virial theorem for the S system.

2.3.1 Quantal Sources

The electron density $\rho(\mathbf{r}t)$ which is the same as in Schrödinger theory is expressed in terms of the orbitals $\varphi_i(\mathbf{x}t)$ as

$$\rho(\mathbf{r}\,t) = \langle \Phi\{\varphi_i\} | \hat{\rho}(\mathbf{r}) | \Phi\{\varphi_i\} \rangle = \sum_i \sum_\sigma |\varphi_i(\mathbf{x}\,t)|^2 \,. \tag{42}$$

The Dirac density matrix $\gamma_s(\mathbf{r}\mathbf{r}'t)$ as expressed in terms of the orbitals is

$$\gamma_s(\mathbf{rr}'t) = \langle \Phi\{\varphi_i\} | \hat{\gamma}(\mathbf{rr}') | \Phi\{\varphi_i\} \rangle = \sum_i \sum_\sigma \varphi_i^*(\mathbf{r}\sigma, t)\varphi_i(\mathbf{r}'\sigma, t) . \tag{43}$$

The pair-correlation density $g_s(\mathbf{rr}'t)$ is

$$g_s(\mathbf{rr}'t) = \langle \Phi\{\varphi_i\} | \hat{P}(\mathbf{rr}') | \Phi\{\varphi_i\} \rangle / \rho(\mathbf{r}t) \tag{44}$$

$$= \rho(\mathbf{r}'t) + \rho_x(\mathbf{rr}'t), \tag{45}$$

where the Fermi hole $\rho_x(\mathbf{rr}'t)$ is defined in the standard manner as

$$\rho_x(\mathbf{rr}'t) = -|\gamma_s(\mathbf{rr}'t)|^2 / 2\rho(\mathbf{r}t) . \tag{46}$$

The current density $\mathbf{j}_s(\mathbf{r}t)$ is defined as

$$\mathbf{j}_s(\mathbf{r}t) = \frac{i}{2} \sum_i \sum_\sigma [\varphi_i(\mathbf{x}t)\nabla\varphi_i^*(\mathbf{x}t) - \varphi_i^*(\mathbf{x}t)\nabla\varphi_i(\mathbf{x}t) \tag{47}$$

$$= \frac{i}{2}[\nabla' - \nabla'']\gamma_s(\mathbf{r}'\mathbf{r}''t)|_{\mathbf{r}'=\mathbf{r}''=\mathbf{r}} \tag{48}$$

$$= \langle \Phi\{\varphi_i\} | \hat{\mathbf{j}}(\mathbf{r}) | \Phi\{\varphi_i\} \rangle . \tag{49}$$

These sources have the properties:

$\gamma_s(\mathbf{rr}t) = \rho(\mathbf{r}t)$; $\int g_s(\mathbf{rr}'t)d\mathbf{r}' = N-1$; $\int \rho_x(\mathbf{rr}'t)d\mathbf{r} = -1$, $\rho_x(\mathbf{rr}t) = -\rho(\mathbf{r}t)/2$; $\rho_x(\mathbf{rr}'t) \leq 0$.

2.3.2 *Differential Virial Theorem for S system*

The differential virial theorem satisfied by the orbitals of the S system is [7,8]

$$\nabla v(\mathbf{r}t) = -\mathbf{F}_s(\mathbf{r}t) , \tag{50}$$

where

$$\mathbf{F}_s(\mathbf{r}t) = \nabla v_{ee}(\mathbf{r}t) + \mathbf{D}(\mathbf{r}t) + \mathbf{Z}_s(\mathbf{r}t) + \mathbf{J}_s(\mathbf{r}t) . \tag{51}$$

Here the kinetic field $\mathbf{Z}_s(\mathbf{r}t)$ is defined in terms of its source as

$$\mathbf{Z}_s(\mathbf{r}t) = \mathbf{z}_s(\mathbf{r}t; [\gamma_s]) / \rho(\mathbf{r}t), \tag{52}$$

$$z_{s,\alpha}(\mathbf{r}\,t)=2\sum_{\beta}\frac{\partial}{\partial r_{\beta}}t_{s,\alpha\beta}(\mathbf{r}\,t)\ , \tag{53}$$

and where $t_{s,\alpha\beta}(\mathbf{r}t)$ is the S system kinetic energy density:

$$t_{s,\alpha\beta}(\mathbf{r}\,t)=\frac{1}{4}\left[\frac{\partial^2}{\partial r'_{\alpha}\,\partial r''_{\beta}}+\frac{\partial^2}{\partial r'_{\beta}\,\partial r''_{\alpha}}\right]\gamma_s(\mathbf{r}'\mathbf{r}''\,t)\Big|_{\mathbf{r}'=\mathbf{r}''=\mathbf{r}}\ . \tag{54}$$

The current density field $\mathbf{J}_s(\mathbf{r}t)$ is the temporal derivative

$$\mathbf{J}_s(\mathbf{r}\,t)=\frac{1}{\rho(\mathbf{r}\,t)}\frac{\partial}{\partial t}\mathbf{j}_s(\mathbf{r}\,t)\ . \tag{55}$$

2.4 Time-Independent S System

In this article, for the TI case, we restrict ourselves to transformations from the ground state of Schrödinger theory to an S system in its ground state. Then, once again, with the assumption of existence, the differential equation for the S system of noninteracting Fermions with density equivalent to that of stationary state Schrödinger theory is

$$\left[-\frac{1}{2}\nabla^2+v_s(\mathbf{r})\right]\varphi_i(\mathbf{x})=\epsilon_i\,\varphi_i(\mathbf{x})\ ,\quad i=1,\ldots N \tag{56}$$

where

$$v_s(\mathbf{r})=v(\mathbf{r})+v_{ee}(\mathbf{r})\ , \tag{57}$$

ϵ_i are the eigenvalues, and the wavefunction is the single Slater determinant $\Phi\{\varphi_i\}$. The electron-interaction potential energy $v_{ee}(\mathbf{r})$ in this case is representative of only electron correlations due to the Pauli principle, Coulomb repulsion, and Correlation-Kinetic effects. Again, there must exist an effective force field $\mathcal{F}(\mathbf{r})$ such that $v_{ee}(\mathbf{r})$ is the corresponding potential energy of the model system Fermion.

The definitions of the quantal sources, the fields, and the resulting differential virial theorem are the same as for the TD S system with the difference that the time parameter t and current density field $\mathbf{J}_s(\mathbf{r}t)$ are absent.

2.5 Quantal Density Functional Theory Definitions

It is now possible by employing the differential virial theorems for the Schrödinger and S systems, to determine the conservative field $\mathcal{F}(\mathbf{r}t)$ and thereby the Q-DFT definitions of the electron-interaction potential energy $v_{ee}(\mathbf{r}t)$ and total energy $E(t)$.

2.5.1 Electron-Interaction Potential Energy

Equating the differential virial theorems of Eqs. (26) and (50), we obtain

$$\nabla \mathrm{v}_{ee}(\mathbf{r}\,t) = -\mathcal{F}(\mathbf{r}\,t) \ , \tag{58}$$

where the effective internal field $\mathcal{F}(\mathbf{r}t)$ is

$$\mathcal{F}(\mathbf{r}\,t) = \mathcal{E}_{ee}(\mathbf{r}\,t) + \mathcal{Z}_{t_c}(\mathbf{r}\,t) + \mathcal{J}_c(\mathbf{r}\,t) \ . \tag{59}$$

The electron interaction field $\mathcal{E}_{ee}(\mathbf{r}t)$ is as defined before by Eqs. 17-19. The Correlation-Kinetic $\mathcal{Z}_{t_c}(\mathbf{r}\,t)$ and Correlation-Current-Density $\mathcal{J}_c(\mathbf{r}t)$ fields are defined, respectively, as

$$\mathcal{Z}_{t_c}(\mathbf{r}\,t) = \mathbf{Z}_s(\mathbf{r}\,t) - \mathbf{Z}(\mathbf{r}\,t) \ , \tag{60}$$

and

$$\mathcal{J}_c(\mathbf{r}\,t) = \mathbf{J}_s(\mathbf{r}\,t) - \mathbf{J}(\mathbf{r}\,t) \ . \tag{61}$$

It is evident that these latter fields represent the correlation contributions of the kinetic and current density effects to the electron interaction potential energy $\mathrm{v}_{ee}(\mathbf{r}t)$. From Eq. (58) it then follows that the potential energy $\mathrm{v}_{ee}(\mathbf{r}t)$ is the work done at each instant of time to move the model system Fermion from some reference point at infinity to its position at **r** in the force of the conservative field $\mathcal{F}(\mathbf{r}t)$:

$$\mathrm{v}_{ee}(\mathbf{r}\,t) = -\int_{\infty}^{\mathbf{r}} \mathcal{F}(\mathbf{r}'t)\cdot d\mathbf{l}' \ . \tag{62}$$

This work done is path independent since $\nabla \times \mathcal{F}(\mathbf{r}\,t) = 0$. The fields $\mathcal{E}_{ee}(\mathbf{r}t)$, $\mathcal{Z}_{t_c}(\mathbf{r}\,t)$, $\mathcal{J}_c(\mathbf{r}t)$ are in general not conservative. Their sum always is.

The effective internal field $\mathcal{F}(\mathbf{r}t)$ may be further decomposed by writing the Fermi-Coulomb hole $\rho_{xc}(\mathbf{r}\mathbf{r}'t)$ as a sum of its Fermi $\rho_x(\mathbf{r}\mathbf{r}'t)$ and Coulomb $\rho_c(\mathbf{r}\mathbf{r}'t)$ holes:

$$\rho_x(\mathbf{r}\mathbf{r}'t) = \rho_x(\mathbf{r}\mathbf{r}'t) + \rho_c(\mathbf{r}\mathbf{r}'t). \tag{63}$$

This defines the Coulomb hole and its sum rule

$$\int \rho_c(\mathbf{r}\mathbf{r}'t)\, d\mathbf{r}' = 0 \ . \tag{64}$$

The Fermi and Coulomb hole charge distributions in turn give rise to the Pauli $\mathcal{E}_x(\mathbf{r}t)$ and Coulomb $\mathcal{E}_c(\mathbf{r}t)$ fields, respectively:

$$\mathcal{E}_x(\mathbf{r}\,t)=\int\frac{\rho_x(\mathbf{r}\mathbf{r}'t)(\mathbf{r}-\mathbf{r}')}{|\mathbf{r}-\mathbf{r}'|^3}d\mathbf{r}' \quad \text{and} \quad \mathcal{E}_c(\mathbf{r}\,t)=\int\frac{\rho_c(\mathbf{r}\mathbf{r}'t)(\mathbf{r}-\mathbf{r}')}{|\mathbf{r}-\mathbf{r}'|^3}d\mathbf{r}' \ . \tag{65}$$

Thus, the field $\mathcal{F}(\mathbf{r}t)$ may be written as a sum of its Hartree $\mathcal{E}_H(\mathbf{r}t)$, Pauli $\mathcal{E}_x(\mathbf{r}t)$, Coulomb $\mathcal{E}_c(\mathbf{r}t)$, Correlation-Kinetic $\mathcal{Z}_{t_c}(\mathbf{r}\,t)$, and Correlation-Current-Density $\mathcal{J}_c(\mathbf{r}t)$ components:

$$\mathcal{F}(\mathbf{r}\,t)=\mathcal{E}_H(\mathbf{r}\,t)+\mathcal{E}_x(\mathbf{r}\,t)+\mathcal{E}_c(\mathbf{r}\,t)+\mathcal{Z}_{t_c}(\mathbf{r}\,t)+\mathcal{J}_c(\mathbf{r}\,t) \ . \tag{66}$$

In this manner, it is possible to study a system as a function of the electron correlations. Including just the first two terms corresponds to the Pauli correlated approximation. One could subsequently include Coulomb correlations via the field $\mathcal{E}_c(\mathbf{r}t)$, and then Correlation-Kinetic and Correlation-Current-Density contributions via the fields $\mathcal{Z}_{t_c}(\mathbf{r}\,t)$ and $\mathcal{J}_c(\mathbf{r}t)$.

Since the Hartree field $\mathcal{E}_H(\mathbf{r}t)$is conservative, it may be written as $\mathcal{E}_H(\mathbf{r}t) = -\nabla W_H(\mathbf{r}t)$ where

$$W_H(\mathbf{r}\,t)=-\int_\infty^{\mathbf{r}}\mathcal{E}_H(\mathbf{r}'t)\cdot d\mathbf{l}'=\int\frac{\rho(\mathbf{r}'t)}{|\mathbf{r}-\mathbf{r}'|}d\mathbf{r}' \ . \tag{67}$$

Thus, the electron interaction potential energy $v_{ee}(\mathbf{r}t)$ may be expressed as

$$v_{ee}(\mathbf{r}\,t)=W_H(\mathbf{r}\,t)+\left(-\int_\infty^{\mathbf{r}}\left[\mathcal{E}_x(\mathbf{r}'t)+\mathcal{E}_c(\mathbf{r}'t)+\mathcal{Z}_{t_c}(\mathbf{r}'t)+\mathcal{J}_c(\mathbf{r}'t)\right]\cdot d\mathbf{l}'\right) \ . \tag{68}$$

For systems with symmetry such that the curl of each component of $\mathcal{F}(\mathbf{r}t)$ vanishes separately we have

$$v_{ee}(\mathbf{r}\,t)=W_H(\mathbf{r}\,t)+W_X(\mathbf{r}\,t)+W_C(\mathbf{r}\,t)+W_{t_c}(\mathbf{r}\,t) \ , \tag{69}$$

where $W_x(\mathbf{r}t)$, $W_c(\mathbf{r}t)$, and $W_{t_c}(\mathbf{r}\,t)$are, respectively, the work done in the fields $\mathcal{E}_x(\mathbf{r}t)$, $\mathcal{E}_c(\mathbf{r}t)$, and $\mathcal{Z}_{t_c}(\mathbf{r}\,t)$. For systems of such symmetry the field $\mathcal{J}_c(\mathbf{r}t) = 0$, and does not contribute to the potential energy. The field $\mathcal{J}_c(\mathbf{r}t)$ vanishes because both $\nabla\times\mathcal{J}_c(\mathbf{r}\,t)=0$ and $\nabla\cdot\mathcal{J}_c(\mathbf{r}\,t)=0$, the latter following from the continuity equations

$$\nabla\cdot\mathbf{j}(\mathbf{r}\,t)=\frac{\partial\rho(\mathbf{r}\,t)}{\partial t} \quad \text{and} \quad \nabla\cdot\mathbf{j}_s(\mathbf{r}\,t)=\frac{\partial\rho(\mathbf{r}\,t)}{\partial t} \ . \tag{70}$$

2.5.2 *Total Energy and its Components*

The kinetic energy $T_s(t)$ of the S system Fermions may be expressed in terms of the kinetic field $\mathbf{Z}_s(\mathbf{r}t)$ so that

$$T_s(t) = \langle \Phi\{\varphi_i\} | \hat{T} | \Phi\{\varphi_i\} \rangle$$

$$= -\frac{1}{2} \int d\mathbf{r} \rho(\mathbf{r}t) \mathbf{r} \cdot \mathbf{Z}_s(\mathbf{r}t) \ . \tag{71}$$

Since $T_s(t)$ is the kinetic energy of noninteracting Fermions, we must add the correlation contribution to the kinetic energy $T_c(t)$ – the Correlation-Kinetic energy – if the total energy E(t) is to be the same as that of Schrödinger theory. Furthermore, Correlation-Current-Density effects do not contribute explicitly to the total energy since it can be shown [9] that

$$\int d\mathbf{r} \rho(\mathbf{r}t) \mathbf{r} \cdot \mathbf{J}_c(\mathbf{r}t) = 0 \ . \tag{72}$$

However, their contribution is implicit via the orbitals $\varphi_i(\mathbf{x}t)$ generated by the potential energy $v_{ee}(\mathbf{r}t)$. As noted previously, for systems of special symmetry $\mathcal{J}_c(\mathbf{r}t)$ vanishes, and thus for such symmetries these effects contribute to neither the potential nor the total energy. Therefore, the total energy may be expressed as

$$E(t) = T_s(t) + \int d\mathbf{r} \rho(\mathbf{r}t) v(\mathbf{r}t) + E_{ee}(t) + T_c(t) \ , \tag{73}$$

where the Correlation-Kinetic energy $T_c(t)$ is defined as

$$T_c(t) = \frac{1}{2} \int d\mathbf{r} \rho(\mathbf{r}t) \mathbf{r} \cdot \mathcal{Z}_{t_c}(\mathbf{r}t) \ , \tag{74}$$

and where the electron-interaction energy is defined in terms of the electron-interaction field $\mathcal{E}_{ee}(t)$ as given by Eq. (30).

This then is the TD Q-DFT description of the potential and total energy of noninteracting Fermions with electronic density and total energy equivalent to that of Schrödinger theory. The description, as derived above, is in terms of classical fields and their quantal sources. The evolution of the S system in time is described by the differential virial theorem of Eq. (58). For this evolution to be equivalent to that of Ehrenfest's theorem, the average of the effective internal force $\mathcal{F}(\mathbf{r}t)$ must vanish.

Time-independent Q-DFT constitutes a special case of the TD version. The expressions for the total energy E and its components, and for the effective force field $\mathcal{F}(\mathbf{r})$ and potential energy $v_{ee}(\mathbf{r})$, are the same as in TD Q-DFT with the difference that the time parameter t and the Correlation-Current-Density field $\mathcal{J}_c(\mathbf{r}t)$ are absent. Further, as is well known [16] from previous work on traditional density functional theory, the highest occupied eigenvalue of the S system differential equation Eq. (56) has the rigorous physical interpretation of being the negative of the ionization potential or electron affinity as the case may be. We refer the reader to Ref. [12] for an exactly solvable model example of the structure of the various sources and fields within TI Q-DFT.

In the following section we present the basic tenets of the Kohn-Sham DFT description of the S system. We then explain the physical interpretation of the action /energy functionals and of the functional derivatives of this theory as arrived at by Q-DFT.

3. KOHN-SHAM DENSITY FUNCTIONAL THEORY AND ITS PHYSICAL INTERPRETATION.

TD DFT is based on the Runge-Gross [1] extension of the first Hohenberg-Kohn [6] theorem, and on the stationary-action principle. Accordingly, for electrons in an external field such that their potential energy v(rt) is Taylor expandable about some initial time t_o, the Schrödinger wavefunction is a functional of the TD density to within a purely TD phase. The solution of the TD Schrödinger equation Eq. (1), with the system in some initial state $\Psi(t_o) = \Psi_o$, corresponds to a stationary point of the quantum-mechanical action integral

$$\mathcal{A} = \int_{t_o}^{t_1} dt \langle \Psi(t) | i\frac{\partial}{\partial t} - \hat{H}(t) | \Psi(t) \rangle \ . \tag{75}$$

Since the wavefunction is a functional of the density, the action $\mathcal{A}$ is such a functional:

$$\mathcal{A}[\rho] = \int_{t_o}^{t_1} dt \langle \Psi[\rho](t) | i\frac{\partial}{\partial t} - \hat{H}(t) | \Psi[\rho](t) \rangle \ . \tag{76}$$

The functional $\mathcal{A}[\rho]$ will have a stationary point at the correct TD density, so that the density may be obtained via the Euler equation

$$\frac{\delta \mathcal{A}[\rho]}{\delta \rho(\mathbf{r}\, t)} = 0 \ , \tag{77}$$

with the appropriate boundary conditions. The functional $\mathcal{A}[\rho]$ is well-defined only for v-representable densities [1,2].

3.1 Time-Dependent Kohn-Sham Theory

In the Kohn-Sham (KS) description of the S system defined by Eqs. 39-41, all the many-body effects – those due to Pauli and Coulomb correlations, and Correlation-Kinetic and Correlation-Current-Density effects – are incorporated into an electron-interaction action functional $\mathcal{A}_{ee}^{KS}[\rho]$. The electron-interaction potential energy $v_{ee}(\mathbf{r}t)$ of the model Fermions of Eq. (41) is defined on the basis of the Euler equation as the functional derivative

$$v_{ee}(\mathbf{r}t)=\frac{\delta\mathcal{A}_{ee}^{KS}[\rho]}{\delta\rho(\mathbf{r}t)} . \tag{78}$$

The functional $\mathcal{A}_{ee}^{KS}[\rho]$ may be decomposed into its Hartree $\mathcal{A}_{H}[\rho]$ and 'KS exchange-correlation' $\mathcal{A}_{xc}^{KS}[\rho]$ action functionals, and the latter may be further decomposed into its 'KS exchange' $\mathcal{A}_{x}[\rho]$ and 'KS correlation' $\mathcal{A}_{c}^{KS}[\rho]$ components. Thus,

$$\mathcal{A}_{ee}^{KS}[\rho]=\mathcal{A}_{H}[\rho]+\mathcal{A}_{xc}^{KS}[\rho] \tag{79}$$

$$=\mathcal{A}_{H}[\rho]+\mathcal{A}_{x}^{KS}[\rho]+\mathcal{A}_{c}^{KS}[\rho] , \tag{80}$$

where

$$\mathcal{A}_{H}[\rho]=\frac{1}{2}\int_{t_0}^{t_1}dt\int d\mathbf{r}\,d\mathbf{r}'\frac{\rho(\mathbf{r}t)\rho(\mathbf{r}'t)}{|\mathbf{r}-\mathbf{r}'|} , \tag{81}$$

$$\mathcal{A}_{x}[\rho]=\frac{1}{2}\int_{t_0}^{t_1}dt\int d\mathbf{r}\,d\mathbf{r}'\frac{\rho(\mathbf{r}t)\rho_{x}(\mathbf{r}\mathbf{r}'t)}{|\mathbf{r}-\mathbf{r}'|} , \tag{82}$$

and $\mathcal{A}_{c}^{KS}[\rho]$ is defined by Eq. (80). The potential energy $v_{ee}(\mathbf{r}t)$ is then

$$v_{ee}(\mathbf{r}t)=v_{H}(\mathbf{r}t)+v_{xc}(\mathbf{r}t) \tag{83}$$

$$=v_{H}(\mathbf{r}t)+v_{x}(\mathbf{r}t)+v_{c}(\mathbf{r}t) , \tag{84}$$

where

$$v_{H}(\mathbf{r}t)=\frac{\delta\mathcal{A}_{H}[\rho]}{\delta\rho(\mathbf{r}t)}=\int\frac{\rho(\mathbf{r}'t)}{|\mathbf{r}-\mathbf{r}'|}d\mathbf{r}' , \tag{85}$$

and

$$v_{xc}(\mathbf{r}t)=\frac{\delta\mathcal{A}_{xc}^{KS}[\rho]}{\delta\rho(\mathbf{r}t)} , \quad v_{x}(\mathbf{r}t)=\frac{\delta\mathcal{A}_{x}^{KS}[\rho]}{\delta\rho(\mathbf{r}t)} , \quad v_{c}(\mathbf{r}t)=\frac{\delta\mathcal{A}_{c}^{KS}[\rho]}{\delta\rho(\mathbf{r}t)} . \tag{86}$$

The total (nonconserved) energy E(t) is defined within TD KS DFT as

$$E(t)=T_s[\rho](t)+\int d\mathbf{r}\rho(\mathbf{r}t)v(\mathbf{r}t)+E_{ee}^{KS}[\rho](t) \ , \tag{87}$$

where $E_{ee}^{KS}[\rho](t)$ is the KS electron interaction energy functional.

Note that the action functionals $\mathcal{A}_{ee}^{KS}[\rho]$, $\mathcal{A}_{xc}^{KS}[\rho]$, $\mathcal{A}_{x}^{KS}[\rho]$, and $\mathcal{A}_{c}^{KS}[\rho]$ are at present unknown. However, these action functionals, their derivatives, and the functional $E_{ee}^{KS}[\rho](t)$ can be provided a rigorous physical interpretation via TD Q-DFT which we next present.

3.2 Physical Interpretation of TD Kohn-Sham Theory via Q-DFT

We focus here on the interpretation of the functional derivative of the action $\mathcal{A}_{ee}^{KS}[\rho]$ and its components, and of the KS electron-interaction energy $E_{ee}^{KS}[\rho](t)$. The reason for focusing on these properties is that it makes the corresponding interpretation for the TI case more accessible.

It is evident from the Q-DFT equation Eq. (62) that the functional derivative $\delta\mathcal{A}_{ee}^{KS}[\rho]/\delta\rho(\mathbf{r}t)$ of Eq. (78) is the work done, at each instant of time, to move an electron from its reference position at infinity to its position at **r** in the force of the conservative field $\mathcal{F}(\mathbf{r}t)$:

$$v_{ee}(\mathbf{r}t)=\frac{\delta\mathcal{A}_{ee}^{KS}[\rho]}{\delta\rho(\mathbf{r}t)}=-\int_{\infty}^{\mathbf{r}}\mathcal{F}(\mathbf{r}'t)\cdot d\mathbf{l}' \ , \tag{88}$$

where

$$\mathcal{F}(\mathbf{r}t)=\mathcal{E}_{ee}(\mathbf{r}t)+\mathcal{Z}_{t_c}(\mathbf{r}t)+\mathcal{J}_c(\mathbf{r}t) \ . \tag{89}$$

The various components of the field $\mathcal{F}(\mathbf{r}t)$ are as defined previously by Eqs. 17, 60, and 61. By equating the Q-DFT energy expression Eq. (73) to the KS DFT equation Eq. (87) we see that the KS electron-interaction energy $E_{ee}^{KS}[\rho](t)$ is the sum of the quantum-mechanical electron-interaction energy $E_{ee}(t)$ and the Correlation-Kinetic $T_c(t)$ energy:

$$E_{ee}^{KS}[\rho](t)=E_{ee}(t)+T_c(t) \ , \tag{90}$$

where the energy components are defined in terms of the respective fields as

$$E_{ee}(t)=\int d\mathbf{r}\rho(\mathbf{r}t)\mathbf{r}\cdot\mathcal{E}_{ee}(\mathbf{r}t) \ , \tag{91}$$

and

$$T_c(t)=\frac{1}{2}\int d\mathbf{r}\rho(\mathbf{r}t)\mathbf{r}\cdot\mathcal{Z}_{t_c}(\mathbf{r}t) \quad . \tag{92}$$

We reiterate that these energy expressions are valid whether or not the individual fields are conservative.

3.2.1 Physical Interpretation of Kohn-Sham 'Exchange' and 'Correlation' Potential Energies

It is also possible [9] via TD Q-DFT and adiabatic coupling constant perturbation theory [17] to provide a rigorous physical interpretation for the TD KS DFT exchange and correlation potential energies in terms of the different electron correlations which contribute to them.

In adiabatic coupling constant (λ) theory, the Hamiltonian $\hat{H}^{\lambda}(t)$ is defined as $\hat{H}^{\lambda}(t)=\hat{T}+\hat{V}_{\lambda}(t)+\lambda\hat{U}\,(0\le\lambda\le 1)$, where the operator $\hat{V}_{\lambda}=\sum_i v_{\lambda}(\mathbf{r}_i\, t)$ is constrained such that its addition to $\lambda\hat{U}$ leads to the density for the fully interacting system ($\lambda = 1$). Thus, the wavefunction $\Psi^{\lambda}(\mathbf{X}t)$ is such that the expectation $\langle\Psi^{\lambda}(t)|\hat{\rho}|\Psi^{\lambda}(t)\rangle=\rho^{\lambda=1}(\mathbf{r}t)=\rho(\mathbf{r}t)$. The $\lambda = 0$ case corresponds to the S system. In this theory the density is independent of λ. Thus, we may also write the S system potential energy as $v_s(\mathbf{r}t)=v_{\lambda}(\mathbf{r}t)+v_{ee}^{\lambda}(\mathbf{r}t)$. The corresponding Q-DFT equations are the same except that they are written with a superscript λ whenever appropriate.

The wavefunction $\Psi^{\lambda}(\mathbf{X}t)$ is next expanded in powers of the coupling constant:

$$\Psi^{\lambda}(\mathbf{X}t)=\Phi\{\varphi_i\}(\mathbf{X}t)+\lambda\Psi_1(\mathbf{X}t)+\lambda^2\Psi_2(\mathbf{X}t)+\ \ldots \tag{93}$$

where $\Phi\{\varphi_i\}$ is the Slater determinant of S system. The resulting pair-correlation density $g^{\lambda}(\mathbf{rr'}t)$ and single particle density matrix $\gamma^{\lambda}(\mathbf{rr'}t)$ sources are

$$g^{\lambda}(\mathbf{rr'}t)=g_s(\mathbf{rr'}t)+\lambda g_1^c(\mathbf{rr'}t)+\lambda^2 g_2^c(\mathbf{rr'}t)+\ \ldots \tag{94}$$

and

$$\gamma^{\lambda}(\mathbf{rr'}t)=\gamma_s(\mathbf{rr'}t)+\lambda\gamma_1^c(\mathbf{rr'}t)+\lambda^2\gamma_2^c(\mathbf{rr'}t)+\ \ldots, \tag{95}$$

where $g_s(\mathbf{rr'}t)$ and $\gamma_s(\mathbf{rr'}t)$ are the corresponding S system sources defined previously by Eqs. (43) and (44), and $O_1^c(\mathbf{rr'}t)=\langle\Psi_1(t)|\hat{O}|\Phi(t)\rangle+\langle\Phi(t)|\hat{O}|\Psi_1(t)\rangle$, etc.

The fields $\mathcal{E}^{\lambda}_{ee}(\mathbf{r}\,t)$, $\mathcal{Z}^{\lambda}_{t_c}(\mathbf{r}\,t)$, and $\mathcal{J}^{\lambda}_{c}(\mathbf{r}\,t)$ which arise from these sources are

$$\mathcal{E}^{\lambda}_{ee}(\mathbf{r}\,t) = \lambda\mathcal{E}_{H}(\mathbf{r}\,t) + \lambda\mathcal{E}_{x}(\mathbf{r}\,t) + \lambda^{2}\mathcal{E}_{c,1}(\mathbf{r}\,t) + \lambda^{3}\mathcal{E}_{c,2}(\mathbf{r}\,t) + \ldots \quad (96)$$

$$\mathcal{Z}^{\lambda}_{t_c}(\mathbf{r}\,t) = -\lambda\mathcal{Z}_{t_c,1}(\mathbf{r}\,t) - \lambda^{2}\mathcal{Z}_{t_c,2}(\mathbf{r}\,t) - \lambda^{3}\mathcal{Z}_{t_c,3}(\mathbf{r}\,t) - \ldots . \quad (97)$$

$$\mathcal{J}^{\lambda}_{c}(\mathbf{r}\,t) = -\lambda\mathcal{J}_{c,1}(\mathbf{r}\,t) - \lambda^{2}\mathcal{J}_{c,2}(\mathbf{r}\,t) - \lambda^{3}\mathcal{J}_{c,3}(\mathbf{r}\,t) - \ldots.., \quad (98)$$

where $\mathcal{E}_{c,1}(\mathbf{r}\,t) = \int d\mathbf{r}'\, g^{c}_{1}(\mathbf{r}\mathbf{r}'\,t)(\mathbf{r}-\mathbf{r}')/|\mathbf{r}-\mathbf{r}'|^{3}$, $\mathcal{Z}_{t_c,1}(\mathbf{r}\,t) = \mathcal{Z}(\mathbf{r}\,t;\gamma^{c}_{1})/\rho(\mathbf{r}\,t)$, and $\mathcal{J}_{c,1}(\mathbf{r}\,t) = (\partial \mathbf{j}(\mathbf{r}\,t;\gamma^{c}_{1})/\partial t)/\rho(\mathbf{r}\,t)$, etc.

The KS theory Hartree $\mathcal{A}_{H}[\rho]$ and exchange $\mathcal{A}_{x}[\rho]$ actions as defined by Eqs.(81) and (82) scale linearly in $\lambda: \mathcal{A}^{\lambda}_{H} = \lambda\mathcal{A}_{H}$, $\mathcal{A}^{\lambda}_{x} = \lambda\mathcal{A}_{x}$. Thus, the corresponding functional derivatives $v_{H}(\mathbf{r}t)$ and $v_{x}(\mathbf{r}t)$ also scale linearly:

$$v^{\lambda}_{H}(\mathbf{r}\,t) = \lambda\, v_{H}(\mathbf{r}\,t) \ , \quad (99)$$

and

$$v^{\lambda}_{x}(\mathbf{r}\,t) = \lambda\, v_{x}(\mathbf{r}\,t) \, . \quad (100)$$

Now since $\nabla v^{\lambda}_{ee}(\mathbf{r}\,t) = -\mathcal{F}^{\lambda}(\mathbf{r}\,t)$ with $\mathcal{F}^{\lambda}(\mathbf{r}\,t) = \mathcal{E}^{\lambda}_{ee}(\mathbf{r}\,t) + \mathcal{Z}^{\lambda}_{t_c}(\mathbf{r}\,t) + \mathcal{J}^{\lambda}_{c}(\mathbf{r}\,t)$, we have on substituting the KS definition of $v^{\lambda}_{ee}(\mathbf{r}\,t)$ that

$$\nabla\left[v^{\lambda}_{H}(\mathbf{r}\,t) + v^{\lambda}_{x}(\mathbf{r}\,t) + v^{\lambda}_{c}(\mathbf{r}\,t)\right] = -\left[\mathcal{E}^{\lambda}_{ee}(\mathbf{r}\,t) + \mathcal{Z}^{\lambda}_{t_c}(\mathbf{r}\,t) + \mathcal{J}^{\lambda}_{c}(\mathbf{r}\,t)\right] . \quad (101)$$

On substitution of the expansions for the fields and the scaling relations, and on equating terms of equal order, we have the components of the KS potential energy in terms of the fields:

$$\nabla v_{H}(\mathbf{r}\,t) = -\mathcal{E}_{H}(\mathbf{r}\,t) \quad (102)$$

$$\nabla v_{x}(\mathbf{r}\,t) = -\left[\mathcal{E}_{x}(\mathbf{r}\,t) - \mathcal{Z}_{t_c,1}(\mathbf{r}\,t) - \mathcal{J}_{c,1}(\mathbf{r}\,t)\right] \quad (103)$$

$$\nabla v_{c,2}(\mathbf{r}\,t) = -\left[\mathcal{E}_{c,1}(\mathbf{r}\,t) - \mathcal{Z}_{t_c,2}(\mathbf{r}\,t) - \mathcal{J}_{c,2}(\mathbf{r}\,t)\right] \quad (104)$$

where $v_{c,2}(\mathbf{r}t)$ etc, are the KS correlation potential energy component to $O(\lambda^{2})$, etc. Note that the KS correlation potential energy commences in second order.

From Eq. (102) it is evident that the functional derivative $v_H(\mathbf{r}t)$ is the work done $W_H(\mathbf{r}t)$ in the Hartree field $\mathcal{E}_H(\mathbf{r}t)$. (See Eq. 67). However, as the Hartree field is conservative, $v_H(\mathbf{r}t)$ may be expressed as in Eqs. (67) or (85).

From Eq. (103) we see that the functional derivative $v_x(\mathbf{r}t)$ – the KS exchange potential energy – is the work done to move an electron in a conservative field $\mathbf{R}(\mathbf{r}t)$:

$$v_x(\mathbf{r}t) = -\int_{\infty}^{\mathbf{r}} \mathbf{R}(\mathbf{r}'t)\cdot d\mathbf{l}' \;, \tag{105}$$

where $\mathbf{R}(\mathbf{r}t) = \mathcal{E}_x(\mathbf{r}t) - \mathcal{Z}_{t_c,1}(\mathbf{r}t) - \mathcal{J}_{c,1}(\mathbf{r}t)$. Since $\nabla \times \mathbf{R}(\mathbf{r}t) = 0$, this work done is path independent. We also see that the KS exchange potential $v_x(\mathbf{r}t)$ is representative not only of electron correlations due to the Pauli principle, but also of lowest order $O(\lambda)$ contributions due to Correlation-Kinetic and Correlation-Current-Density effects through the fields $\mathcal{Z}_{t_c,1}(\mathbf{r}t)$ and $J_{c,1}(\mathbf{r}t)$, respectively.

From Eq. (104), the lowest order $O(\lambda^2)$ KS Correlation potential energy component $v_{c,2}(\mathbf{r}t)$ is the work done in a conservative field:

$$v_{c,2}(\mathbf{r}t) = -\int_{\infty}^{\mathbf{r}} \left[\mathcal{E}_{c,1}(\mathbf{r}'t) - \mathcal{Z}_{t_c,2}(\mathbf{r}'t) - \mathcal{J}_{c,2}(\mathbf{r}'t)\right]\cdot d\mathbf{l}' \;. \tag{106}$$

This field and thus $v_{c,2}(\mathbf{r}t)$, is representative of lowest order Coulomb correlations, and second order Correlation-Kinetic and Correlation-Current-Density effects. At each order the work done is path independent.

For systems of special symmetry, such as spherically symmetrized ones etc., there are no Correlation-Current-Density contributions as noted previously. Furthermore, the KS potential energy components may then be determined as the sum of the work done in the remaining individual fields.

The various action functionals of KS theory may also be expressed in terms of the fields. This is achieved by first deriving the integral virial theorem for ***arbitrary*** coupling constant strength. Integrating this expression over time for the fully interacting ($\lambda = 1$) and the noninteracting S system ($\lambda = 0$) cases then leads to expressions for the actions in terms of the fields. For further details we refer the reader to Ref. [9].

3.3 Time-Independent Kohn-Sham Theory and its Physical Interpretation via Q-DFT.

TI ground state density functional theory is based on the two Hohenberg-Kohn theorems: the stationary state Schrödinger wavefunction is a functional of the density $\rho(\mathbf{r})$ so that the ground state energy is such a functional, and that the density may be obtained by the variational principle for the energy for arbitrary variations of the density. In the TI case, the densities must satisfy the weaker condition of N representability [18].

The fundamental premise of TI Kohn-Sham theory is the assumption of noninteracting N representability. The many body correlations due to the Pauli principle, Coulomb repulsion, and Correlation-Kinetic effects are incorporated in the KS electro-interaction energy functional $E_{ee}^{KS}[\rho]$, and the energy is expressed as

$$E[\rho] = T_s[\rho] + \int d\mathbf{r}\rho(\mathbf{r})\, v(\mathbf{r}) + E_{ee}^{KS}[\rho] \ , \tag{107}$$

where T_s is the kinetic energy of the noninteracting Fermions. The corresponding S system electron interaction potential energy $v_{ee}(\mathbf{r})$, as derived via the variational principle, is then the functional derivative

$$v_{ee}(\mathbf{r}) = \frac{\delta E_{ee}^{KS}[\rho]}{\delta\rho(\mathbf{r})} \ . \tag{108}$$

The physical interpretation of the functional $E_{ee}^{KS}[\rho]$ and of its Hartree $E_H[\rho]$, KS exchange $E_x^{KS}[\rho]$, and KS correlation $E_c^{KS}[\rho]$ components, and of the functional derivative $v_{ee}(\mathbf{r})$ and its components $v_H(\mathbf{r}) = \delta E_H[\rho]/\delta\rho(\mathbf{r})$, $v_x(\mathbf{r}) = \delta E_x^{KS}[\rho]/\delta\rho(\mathbf{r})$, and $v_c(\mathbf{r}) = \delta E_c^{KS}[\rho]/\delta\rho(\mathbf{r})$ in terms of fields and quantal sources is the same as in the TD case with the exception that the Correlation-Current-Density field is absent from the effective force field $\mathcal{F}(\mathbf{r})$. (See Eqs. 88-92, 105, 106). Thus, the functional derivative $v_{ee}(\mathbf{r})$ of Eq. (108) is the work done to move an electron in the force field $\mathcal{F}(\mathbf{r}) = \mathcal{E}_{ee}(\mathbf{r}) + \mathcal{Z}_{t_c}(\mathbf{r})$. The functional $E_{ee}^{KS}[\rho] = E_{ee} + T_c$, where E_{ee} is expressed in integral virial form in terms of the field $\mathcal{E}_{ee}(\mathbf{r})$ and T_c in terms of the field $\mathcal{Z}_{t_c}(\mathbf{r})$.

A Q-DFT and adiabatic coupling constant perturbation theory analysis has been performed for the KS exchange $E_x^{KS}[\rho]$ and Correlation $E_c^{KS}[\rho]$ energy functionals with interesting conclusions. For example, we know that the KS correlation energy commences in $O(\lambda^2)$. However, it can be proved that there are no Coulomb correlation contributions to the lowest order term $E_{c,2}^{KS}[\rho]$, and that this term is due entirely to Correlation-Kinetic effects to second order. We refer the reader to the original literature [14] for further details.

4. CONCLUDING REMARKS

TD Quantal density functional theory is a description of the physics of mapping from TD Schrödinger theory to that of the S system of noninteracting Fermions with equivalent density and energy. The local electron-interaction potential energy $v_{ee}(\mathbf{r}t)$ of the S system Fermions in which all the many-body effects are incorporated is the work done, at each instant of time, to move the electron in the force of a conservative field $\mathcal{F}(\mathbf{r}t)$. The components of this field are the electron interaction $\mathcal{E}_{ee}(\mathbf{r}t)$, Correlation-

Kinetic $\mathcal{Z}_{t_c}(\mathbf{r}\,t)$, and Correlation-Current-Density $\mathcal{J}_c(\mathbf{r}t)$ fields. These fields are respectively representative of electron correlations due to the Pauli exclusion principle and Coulomb repulsion, Correlation-Kinetic, and Correlation-Current-Density effects. In turn, the electron interaction E_{ee} and Correlation-Kinetic T_c energy components of the total energy are expressed in integral virial form in terms of the $\mathcal{E}_{ee}(\mathbf{r}t)$ and $\mathcal{Z}_{t_c}(\mathbf{r}\,t)$ fields, respectively.

TI Quantal density functional theory constitutes a special case of TD theory. The TI Q-DFT described in this article is that of the mapping from a ***ground*** state of Schrödinger theory to an S system which is also in its ***ground*** state. The physical description of the electron interaction potential energy $v_{ee}(\mathbf{r})$ is the same: it is the work done to move an electron in the force of a conservative field $\mathcal{F}(\mathbf{r})$, whose components are the electron interaction $\mathcal{E}_{ee}(\mathbf{r})$ and Correlation-Kinetic $\mathcal{Z}_{t_c}(\mathbf{r})$ fields. The expressions for the electron interaction E_{ee} and Correlation-Kinetic T_c energy components of the total energy in terms of these fields are again the same as in the TD case.

TI Quantal density functional theory of excited states [19,20] is a description of the mapping from an ***excited*** state of Schrödinger theory to that of an S system. Once again, the description of the physics of transformation is the same as in the case of the ground state. However, the choice of the S system is ***arbitrary***, in that the S system may ***either*** be in an ***excited*** state with the same configuration as in Schrödinger theory, or it may be in its ***ground*** state with a different configuration. In each case the corresponding local electron interaction potential energy $v_{ee}(\mathbf{r})$ is the work done in a conservative field, the difference in the potential energies arising entirely from the Correlation-Kinetic field component.

Quantal density functional theory also provides a rigorous physical interpretation, in terms of electron correlations, of the action and energy functionals and derivatives of traditional Kohn-Sham DFT. For example, the Kohn-Sham 'exchange' potential energy $v_x(\mathbf{r})$ is shown to be representative not ***only*** of correlations due to the Pauli exclusion principle as commonly assumed, but also to incorporate the lowest-order Correlation-Kinetic and Correlation-Current-Density effects. The Kohn-Sham correlation potential energy $v_c(\mathbf{r})$ in turn is representative of Coulomb correlations and all the higher-order Correlation-Kinetic and Correlation-Current-Density effects. There are other insights on Kohn-Sham DFT derived via Q-DFT that are not touched upon in this article. An important one concerns the physical origin of the discontinuity of the functional derivative $\delta E_{ee}^{KS}[\rho]/\delta\rho(\mathbf{r})$ as the electron number passes through an integer value. The discontinuity is of particular significance in the dissociation of molecules and band structure calculations. It can be analytically proved [13] via Q-DFT that Pauli and Coulomb correlations do not contribute to this discontinuity, and that the discontinuity is entirely a consequence of Correlation-Kinetic effects.

This article has focused primarily on the basic tenets of Q-DFT, and the physical interpretation of Kohn-Sham DFT that it has provided. The applications, as the theory has evolved, have thus far principally been to the few-electron nonuniform density system in atoms, and the many-electron inhomogeneity at metal surfaces. The reader is referred to fairly recent review articles describing the results in atoms [10] and at metal surfaces [21,22].

5. ACKNOWLEDGEMENT

I wish to thank Tony Gonis, Nick Kioussis, and Mike Ciftan for the invitation to attend and present the above work at the 2nd International Workshop on Electron Correlations and Materials Properties.

REFERENCES

1. E. Runge and E. K. U. Gross, Phys. Rev. Lett. **52**, 997 (1984).
2. E. K. U. Gross, J. F. Dobson, and M. Petersilka, Top. Curr. Chem. **181**, 81 (1996).
3. R. van Leeuwen, Phys. Rev. Lett. **82**, 3863 (1999).
4. N. T. Maitra and K. Burke, Phys. Rev. A **63**, 042501 (2001).
5. W. Kohn and L. J. Sham, Phys. Rev. **140**, A1133 (1965).
6. P. Hohenberg and W. Kohn, Phys. Rev. **136**, B864 (1964).
7. Z. Qian and V. Sahni, Phys. Lett A **247**, 303 (1998).
8. Z. Qian and V. Sahni, Int. J. Quantum Chem. **78**, 341 (2000).
9. Z. Qian and V. Sahni, Phys. Rev. A **63**, 042508 (2001).
10. V. Sahni, Top. Curr. Chem. **182**, 1 (1996).
11. V. Sahni, Phys. Rev. A **55**, 1846 (1997).
12. Z. Qian and V. Sahni, Phys. Rev. A **57**, 2537 (1998).
13. Z. Qian and V. Sahni, Phys. Rev. B **62**, 16364 (2000).
14. Z. Qian and V. Sahni, Int. J. Quantum Chem. **80**, 555 (2000).
15. V. Sahni, J. Mol. Struct. (Theochem) **501**, 91 (2000).
16. J. P. Perdew, R. G. Parr, M. Levy, and J. L. Balduz, Phys. Rev. Lett. **49**, 1691 (1982); M. Levy, J. P. Perdew, and V. Sahni, Phys. Rev. A **30**, 2745 (1984); C. O. Almbladh and U. von. Barth, Phys. Rev. B **31**, 3231 (1985).
17. J. Harris and R. O. Jones, J. Phys. F: Met. Phys. **4**, 1170 (1974); D. C. Langreth and J. P. Perdew, Solid State Commun. **17**, 1425 (1975); Phys. Rev. B **15**, 2884 (1977); O. Gunnarsson and B. Lundqvist, Phys. Rev. **B** 13, 4274 (1976).
18. M. Levy, Proc. Natl. Acad. Sci. USA **76**, 6062 (1979); Phys. Rev. A **26**, 1200 (1982); E. H. Lieb, in *Density Functional methods in Physics*, Vol. 123 of NATO *Advanced Study Institute Series B: Physics*, edited by R. M. Dreizler and J. da Providencia (Plenum, New York, 1985).
19. V. Sahni, L. Massa, R. Singh, and M. Slamet, Phys. Rev. Lett. (2001)(to be published).
20. M. Slamet and V. Sahni, Int. J. Quantum Chem. (2001) (to be published).
21. V. Sahni, Prog. Surf. Sci. **54**, 115 (1997).
22. V. Sahni and A. Solomatin, Adv. Quantum Chem. **33**, 241 (1999).

UNIFORM ELECTRON GAS FROM TWO-PARTICLE WAVEFUNCTIONS

Paola Gori-Giorgi *

1. INTRODUCTION

A common approach to the many-electron problem (atoms, molecules and solids) is its transformation into a fewer-particle problem. In Density Functional Theory[1,2] (DFT) one rewrites the ground-state energy of a many-electron system as a functional of just the one-electron density $n(\mathbf{r})$, the diagonal part of the one-particle reduced density matrix. In recent years, much attention has been devoted to approaches which rewrite the system energy as a functional of the one- or the two-particle reduced density matrix.[3-9] With respect to DFT, these approaches have the advantage that the kinetic energy functional is known and that they provide more information about the many-body wavefunction. In this contest, the relevance of two-electron wavefunctions (geminals) has been pointed out: the use of an antisymmetrized geminal power wavefunction, i.e., a many-body wavefunction built in terms of geminals with the correct symmetry under particle permutation, provides the basis for a formally correct one-particle reduced density matrix functional theory.[3]

With these concepts in mind, it is of great interest to start with a simple many-body system, the uniform electron gas, a limit which should be recovered by any approximate approach to the many-electron problem of nonuniform density. A more detailed discussion about reduced-density matrix functionals applied to the uniform electron gas can be found in Refs. 7,10–12. In this paper, an approximation for the unknown two-electron wavefunctions (geminals) of the uniform electron gas is found, starting from the effective screened Coulomb potential proposed by Overhauser.[13] The short-range part (small electron-electron distances) of the corresponding pair density is found to be in accurate agreement with the newest Quantum Monte

*Paola Gori-Giorgi, Unità INFM and Department of Physics, University of Rome "La Sapienza", Piazzale A. Moro 2, I-00185 Rome, Italy. E-mail: gp.giorgi@caspur.it

Electron Correlations and Materials Properties 2
Edited by Gonis *et al.*, Kluwer Academic/Plenum Publishers, 2002

Carlo data.[16] This means that the short-range part of these geminals is reliable and could serve as a useful term of comparison for other two-electron approaches to the uniform electron gas.[7,10-12]

2. ELECTRON PAIR-DENSITIES, ETC.

Given an N-electron wavefunction $\Psi(\mathbf{r}_1\sigma_1, ...\mathbf{r}_N\sigma_N)$, we define the pair density

$$\rho_2(\mathbf{r}, \mathbf{r}') = N(N-1) \sum_{\sigma_i=1}^{N} \int \prod_{j=3}^{N} d\mathbf{r}_j |\Psi(\mathbf{r}\sigma_1, \mathbf{r}'\sigma_2, ...\mathbf{r}_N\sigma_N)|^2, \tag{1}$$

the one-particle density matrix

$$\rho_1(\mathbf{r}, \mathbf{r}') = N \sum_{\sigma_i=1}^{N} \int \Psi^*(\mathbf{r}\sigma_1, \mathbf{r}_2\sigma_2, ...\mathbf{r}_N\sigma_N)\Psi(\mathbf{r}'\sigma_1, \mathbf{r}_2\sigma_2, ...\mathbf{r}_N\sigma_N) d\mathbf{r}_2...d\mathbf{r}_N, \tag{2}$$

and the one-electron density

$$n(\mathbf{r}) = \rho_1(\mathbf{r}, \mathbf{r}) = \frac{1}{N-1} \int d\mathbf{r}' \rho_2(\mathbf{r}, \mathbf{r}'). \tag{3}$$

While $n(\mathbf{r})d\mathbf{r}$ is the probability of finding an electron in $d\mathbf{r}$, $\rho_2(\mathbf{r}, \mathbf{r}')d\mathbf{r}d\mathbf{r}'$ is the probability of finding one electron in $d\mathbf{r}$ and another in $d\mathbf{r}'$. We also define the pair-distribution function $g(\mathbf{r}, \mathbf{r}')$:

$$\rho_2(\mathbf{r}, \mathbf{r}') = n(\mathbf{r})n(\mathbf{r}')g(\mathbf{r}, \mathbf{r}'). \tag{4}$$

By integrating Eq. (1) over $\mathbf{r}'$, we find that

$$\int d\mathbf{r}' n(\mathbf{r}') [g(\mathbf{r}, \mathbf{r}') - 1] = -1. \tag{5}$$

In other words, the density $n(\mathbf{r}')[g(\mathbf{r}, \mathbf{r}') - 1]$ of the exchange-correlation hole around an electron at $\mathbf{r}$ represents a deficit of one electron.

3. UNIFORM ELECTRON GAS (JELLIUM)

The three-dimensional jellium model consists of N electrons enclosed in a box of volume V (periodically repeated in space) in the presence of a neutralizing background of uniform positive charge density $n^+ = N/V$. The non relativistic jellium is thus governed by the hamiltonian (in Hartree atomic units):

$$H = \sum_{i=1}^{N} \frac{\mathbf{p}_i^2}{2} + \frac{1}{2} \sum_{i \neq j=1}^{N} \frac{1}{|\mathbf{r}_i - \mathbf{r}_j|} + \Lambda, \tag{6}$$

where Λ represents the effect of the background. The hamiltonian of Eq. (6) describes a model solid whose positive ionic charges are smeared throughout the whole crystal volume to yield a shapeless, uniform positive background (whence the nickname of jellium). When studying this model, one is usually interested in its macroscopic properties, i.e., the thermodynamic limit ($N, V \to \infty$ keeping $n = N/V$ constant) of its extensive physical quantities per particle or per volume. Two numbers are enough to describe its zero-temperature phase diagram, namely, the one-electron density $n = N/V$ and the spin polarization $\zeta = |N_\uparrow - N_\downarrow|/N$, where $N_{\uparrow(\downarrow)}$ is the number of spin-up (down) electrons ($N = N_\uparrow + N_\downarrow$). Instead of the particle density n, it is often convenient to use the Wigner-Seitz radius r_s (in units of the Bohr radius) given by $r_s = \left(\frac{4\pi}{3}n\right)^{-1/3}$; it is also useful to define the Fermi wavevector k_F, simply related to the r_s parameter by $k_F = \left(3\pi^2 n\right)^{1/3} = (\frac{9\pi}{4})^{1/3}\frac{1}{r_s}$.

Beeing the system homogeneous and isotropic, the pair-distribution function only depends on $r = |\mathbf{r}_1 - \mathbf{r}_2|$, and parametrically on r_s and ζ.

4. JELLIUM FROM TWO-ELECTRON WAVEFUNCTIONS

The pair-distribution function of jellium can be built starting from two-electron wavefunctions.[13,14] We first rewrite the non-interacting gas (ideal Fermi gas) in terms of two-electron wavefunctions. This first step can seem redundant, since the ideal gas can be written in terms of one-electron wavefunctions, as a Slater determinant of plane waves. However, treating first the ideal gas is essential to set the proper normalization. The two-electron wavefunctions for the interacting gas, in fact, are found by solving a scattering problem in an effective potential, with the normalization condition that the ideal gas is recovered when the potential is set to zero. In this paper, only the $\zeta = 0$ case is analysed. For the generalization to the $\zeta \neq 0$ gas see Ref. 14.

4.1. Ideal Fermi gas

If we select a pair of electrons at random in the spin-unpolarized uniform gas, there is one chance in four that they will be in the singlet state, $\uparrow\downarrow - \downarrow\uparrow$, and three chances in four that they will be in one of the triplet states, $\uparrow\uparrow$, $\downarrow\downarrow$, $\uparrow\downarrow + \downarrow\uparrow$. In the case of no electron-electron interaction, the corresponding two-electron spatial wavefunctions can be rewritten in the center-of-mass reference system as

$$\Psi(\mathbf{r}, \mathbf{R}) = \frac{1}{\sqrt{2}} e^{i\mathbf{K}\cdot\mathbf{R}} \left(e^{i\,\mathbf{k}\cdot\mathbf{r}} \pm e^{-i\,\mathbf{k}\cdot\mathbf{r}}\right), \tag{7}$$

where "+"is for the singlet state and "−" is for the triplet state, and

$$\mathbf{R} = \tfrac{1}{2}(\mathbf{r}_1 + \mathbf{r}_2), \qquad \mathbf{r} = \mathbf{r}_2 - \mathbf{r}_1, \qquad \mathbf{K} = \mathbf{k}_1 + \mathbf{k}_2, \qquad \mathbf{k} = \tfrac{1}{2}(\mathbf{k}_2 - \mathbf{k}_1). \tag{8}$$

Beeing the system isotropic, it is convenient to expand the plane waves into spherical harmonics

$$e^{i\,\mathbf{k}\cdot\mathbf{r}} = \sum_{\ell=0}^{\infty}(2\ell+1)\, i^\ell\, P_\ell(\cos\theta)\, j_\ell(k\,r), \tag{9}$$

where P_ℓ are Legendre polynomials and j_ℓ are spherical Bessel functions. Then

$$\Psi_{\text{singlet}}(\mathbf{r},\mathbf{R}) = \sqrt{2}\,e^{i\mathbf{K}\cdot\mathbf{R}} \sum_{\substack{\ell=0\\ \text{even }\ell}}^{\infty} (2\ell+1)\, i^\ell\, P_\ell(\cos\theta)\, j_\ell(k\,r) \tag{10}$$

$$\Psi_{\text{triplet}}(\mathbf{r},\mathbf{R}) = \sqrt{2}\,e^{i\mathbf{K}\cdot\mathbf{R}} \sum_{\substack{\ell=1\\ \text{odd }\ell}}^{\infty} (2\ell+1)\, i^\ell\, P_\ell(\cos\theta)\, j_\ell(k\,r). \tag{11}$$

The pair-distribution function can be obtained by Ψ_{singlet} and Ψ_{triplet} by giving them the proper weight,

$$g(r) = \tfrac{1}{4}\langle|\Psi_{\text{singlet}}(\mathbf{r})|^2\rangle + \tfrac{3}{4}\langle|\Psi_{\text{triplet}}(\mathbf{r})|^2\rangle, \tag{12}$$

and by considering that in the uniform electron gas there is a probability $p(k)$ that two electrons have a given relative momentum $k = \frac{1}{2}|\mathbf{k}_1 - \mathbf{k}_2|$. In Eq. (12) the symbol $\langle\rangle$ means, in fact, that an average over $p(k)$ and over the solid angle has to be performed. If one is interested in the spin-resolved pair-distribution functions, $g_{\uparrow\uparrow}(r)$ and $g_{\uparrow\downarrow}(r)$, corresponding to parallel- and antiparallel-spin interactions, and such that for the unpolarized gas

$$g = \tfrac{1}{2}(g_{\uparrow\uparrow} + g_{\uparrow\downarrow}), \tag{13}$$

one has to consider that $\frac{1}{3}$ of the triplet state ($\uparrow\downarrow + \downarrow\uparrow$) contributes to the antiparallel-spin correlations and $\frac{2}{3}$ of it ($\uparrow\uparrow$ and $\downarrow\downarrow$) to the parallel-spin correlations. So, we have

$$g_{\uparrow\downarrow}(r) = \tfrac{1}{2}\langle|\Psi_{\text{singlet}}(\mathbf{r})|^2\rangle + \tfrac{1}{2}\langle|\Psi_{\text{triplet}}(\mathbf{r})|^2\rangle \tag{14}$$

$$g_{\uparrow\uparrow}(r) = \langle|\Psi_{\text{triplet}}(\mathbf{r})|^2\rangle, \tag{15}$$

where $\langle\rangle$ denotes again average over $p(k)$ and over the solid angle. Performing the spherical average over the solid angle, we obtain:

$$g_{\uparrow\downarrow}(r) = \sum_{\ell=0}^{\infty}(2\ell+1)\langle j_\ell^2(k\,r)\rangle \tag{16}$$

$$g_{\uparrow\uparrow}(r) = 2\sum_{\substack{\ell=1\\ \text{odd }\ell}}^{\infty}(2\ell+1)\langle j_\ell^2(k\,r)\rangle. \tag{17}$$

Equation (16) immediately gives the exact result for a noninteracting gas, i.e., $g_{\uparrow\downarrow}(r) = 1$ for each r. To obtain the noninteracting $g_{\uparrow\uparrow}(r)$ from Eq. (17), we need to average over k. In the noninteracting electron gas, the probability distribution $p(k)$ for $k = \frac{1}{2}|\mathbf{k}_2 - \mathbf{k}_1|$ can be obtained geometrically by considering two three-dimensional vectors $\mathbf{k}_1$ and $\mathbf{k}_2$ with $0 \le |\mathbf{k}_{1(2)}| \le k_F$, where k_F is the Fermi wavevector. The probability $p(k)$ is then

$$p(k) = 24\frac{k^2}{k_F^3} - 36\frac{k^3}{k_F^4} + 12\frac{k^5}{k_F^6}, \tag{18}$$

with k ranging from 0 to k_F (see Fig. 1). By using two known series which involve

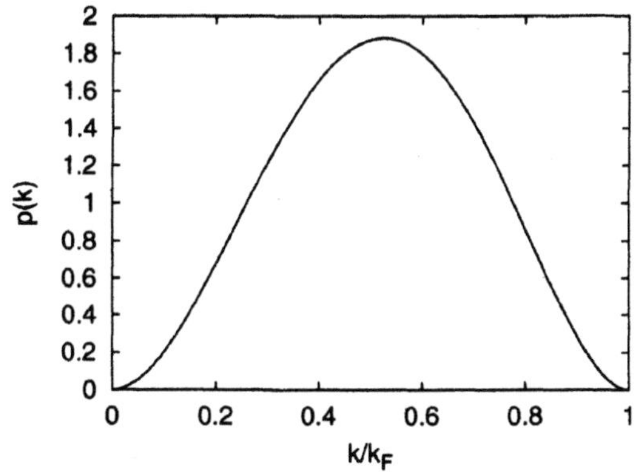

Figure 1: Relative-momentum ($k = \frac{1}{2}|\mathbf{k}_2 - \mathbf{k}_1|$) probability distribution $p(k)$ for the noninteracting gas.

j_ℓ^2,

$$\sum_{\ell=0}^{\infty}(2\ell+1)j_\ell^2(x) = 1 \qquad \sum_{\ell=0}^{\infty}(-1)^\ell(2\ell+1)j_\ell^2(x) = \frac{\sin(2x)}{2x}, \tag{19}$$

we can rewrite $g_{\uparrow\uparrow}$ as

$$g_{\uparrow\uparrow}(r) = \sum_{\ell=0}^{\infty}(2\ell+1)\langle j_\ell^2(kr)\rangle - \sum_{\ell=0}^{\infty}(-1)^\ell(2\ell+1)\langle j_\ell^2(kr)\rangle = \left\langle 1 - \frac{\sin(2kr)}{2kr}\right\rangle. \tag{20}$$

This means:

$$g_{\uparrow\uparrow}(r) = \int_0^{k_F}\left[1 - \frac{\sin(2kr)}{2kr}\right]\left(24\frac{k^2}{k_F^3} - 36\frac{k^3}{k_F^4} + 12\frac{k^5}{k_F^6}\right) dk. \tag{21}$$

Performing this integral gives the known Hartree-Fock (noninteracting) $g_{\uparrow\uparrow}$,

$$g_{\uparrow\uparrow}^{\mathrm{HF}} = 1 - 9\left[\frac{\sin(k_F r) - k_F r\cos(k_F r)}{(k_F r)^3}\right]^2. \tag{22}$$

Numerically, in the range $0 \leq k_F r \leq 6$ and with a truncation of the infinite sum over ℓ at $\ell_{\max} = 7$, Eq. (17) reproduces the exact Eq. (22) within an accuracy of 10^{-6}. When the scaled variable $k_F r$ (or equivalently r/r_s) is used, the pair-distribution function of the noninteracting gas does not depend on r_s. The explicit dependence on r_s only appears when Coulomb repulsion is taken into account in the wavefunction.

4.2. Interacting electron gas

The interacting case can now be treated by introducing an effective potential which describes the electron-electron interactions in a uniform electron gas. The spherical Bessel functions j_ℓ which appear in Eqs. (16) and (17) are solution of the noninteracting radial Schrödinger equation

$$\left[\frac{d^2}{dr^2} - \frac{\ell(\ell+1)}{r^2} + k^2\right] u_\ell = 0 \qquad u_\ell = kr\, j_\ell(kr). \tag{23}$$

If we introduce an effective radial potential $V(r, r_s)$ which depends on the electron-electron distance r and parametrically on the electron density r_s, we can solve the corresponding interacting radial Schrödinger equation,

$$\left[\frac{d^2}{dr^2} - \frac{\ell(\ell+1)}{r^2} - V(r, r_s) + k^2\right] u_\ell = 0 \qquad u_\ell = kr\, R_\ell(r, k, r_s). \tag{24}$$

The radial functions R_ℓ will depend parametrically on k and on r_s. We can insert them into Eqs. (16) and (17), and find the corresponding $g_{\uparrow\downarrow}$ and $g_{\uparrow\uparrow}$:

$$g_{\uparrow\downarrow}(r, r_s) = \sum_{\ell=0}^{\infty}(2\ell+1)\langle R_\ell^2(r, k, r_s)\rangle \tag{25}$$

$$g_{\uparrow\uparrow}(r, r_s) = 2\sum_{\substack{\ell=1 \\ \text{odd } \ell}}^{\infty}(2\ell+1)\langle R_\ell^2(r, k, r_s)\rangle. \tag{26}$$

To compute the average over all the possible relative k (represented again by the symbol $\langle\rangle$) one should in principle use the interacting momentum distribution, which deviates from the Fermi step function of the ideal gas. This would slightly change $p(k)$ of Fig. 1 by adding a "tail" for $k > k_F$ and lowering the maximum. For practical purposes, the use of the noninteracting $p(k)$ is enough to give good results.[14] Notice that unless the potential $V(r, r_s)$ is very sophisticated, the treatment just described will fail to recover the long-range correlations, which are mainly governed by collective modes, and will fail to satisfy the particle-conservation sum rule of Eq. (5).

5. SOLUTION OF THE OVERHAUSER MODEL

In this section, we compute an interacting pair-distribution function following the procedure just described by using the simple model potential $V(r, r_s)$ proposed by Overhauser.[13] This simple model gives very good results for the short-range ($r < r_s$) part of $g(r)$.[14]

5.1. The Overhauser potential

Overhauser[13] proposed a simple and reasonable model for the screened Coulomb repulsion $V(r, r_s)$ in the uniform electron gas: he took the sphere of volume n^{-1} as the boundary within which the screening charge density is ne and outside of which it is zero. We thus have the electrostatic potential due to a point charge $-e$ in the origin plus a sphere around it of radius r_s and of uniform positive charge density

ne. In Hartree atomic units:

$$
\begin{aligned}
V(r,r_s) &= \frac{1}{r_s}\left(\frac{r_s}{r}+\frac{r^2}{2r_s^2}-\frac{3}{2}\right) && r \le r_s \\
V(r,r_s) &= 0 && r > r_s.
\end{aligned} \tag{27}
$$

This is equivalent to assuming that in the interacting gas the probability of finding three electrons in a sphere of radius r_s is exactly zero, an assumption which is nearly true. In fact, numerical estimates of this probability for the electron gas show that it is indeed small.[15] (At $r_s = 5$ the ratio between the probabilities of finding three and two electrons in the same sphere of radius r_s is about 1/11; for larger r_s this ratio is lower, and for smaller r_s it is higher, being about 1/7 at $r_s = 0$.) Thus, for interelectronic distances $r < r_s$ we expect the Overhauser potential to be close to the true potential felt by an electron moving in a uniform electron gas when another electron is fixed at the origin. In the region $r > r_s$ the potential is set to zero, and so is not expected to be reliable. We also expect to have results that become more accurate as the density decreases, since the probability of having three electrons in the same sphere of radius r_s becomes lower and lower. Finally, at high and intermediate densities our results will be much closer to the true $g(r)$ for antiparallel-spin correlations than for parallel-spin ones. When two electrons of opposite spins are in the same sphere of radius r_s, a third electron is excluded from the sphere because of both the Pauli principle and the Coulomb repulsion. For a pair of parallel-spin electrons, only the Coulomb repulsion prevents a third electron of opposite spin from entering the sphere of radius r_s, a mechanism which becomes less efficient as the density (and thus the kinetic energy) increases.

5.2. Solution of the model

The radial Schrödinger equation corresponding to the Overhauser potential can be solved analytically (but not in closed-form),[14] and the functions $R_\ell(s,q,r_s)$ (where $s = r/r_s$ and $q = kr_s$ are scaled variables) are found. As an example, in the left panel of Fig. 2, the radial functions corresponding to $r_s = 4$, $\ell = 0$, and three different values of the scaled relative momentum $q = kr_s$ are reported, together with the effective potential. For an unpolarized uniform gas, the probability distribution for q can be obtained from Eq. (18) by setting $k_F = \frac{1}{r_s}(\frac{9\pi}{4})^{1/3}$,

$$p(q) = \frac{16}{3\pi}q^2\left[2-\left(\frac{12}{\pi}\right)^{1/3}q+\frac{4}{9\pi}q^3\right], \tag{28}$$

with q ranging from 0 to $(9\pi/4)^{1/3} \approx 1.91916$. This $p(q)$ is exact for a noninteracting gas. As said, $p(q)$ for the interacting gas slightly deviates from Eq. (28), and depends explicitly on r_s. However, in the region where the potential is reliable, $r/r_s \le 1$, we see from Fig. 2 that the q-dependence of R_ℓ is very weak, as already pointed out by Overhauser.[13] We thus expect to have no significant change in the short-range ($r < r_s$) part of g if we use an interacting momentum distribution instead of Eq. (28).

In the right panel of Fig. 2 the $g(r)$ obtained by solving the Overhauser model are reported, and compared with the newest diffusion Quantum Monte Carlo (QMC) data of Ortiz, Harris, and Ballone.[16] In the range $0 \le r/r_s \le 2$, a truncation of

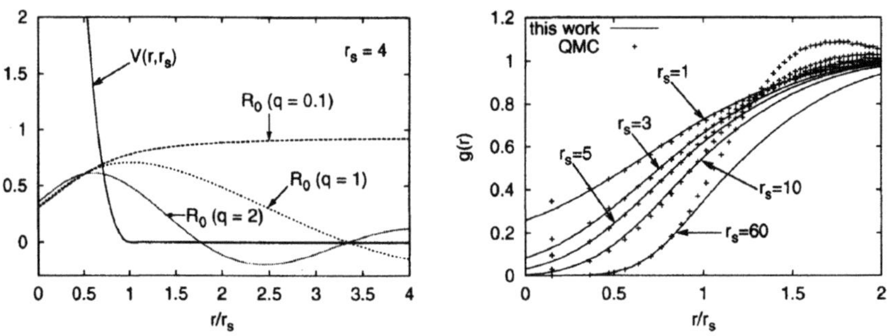

Figure 2: Left panel: the Overhauser potential $V(r, r_s)$ for $r_s = 4$, together with the radial wavefunctions $R_{\ell=0}$ for different values of the scaled relative momentum $q = kr_s$. Right panel: pair-distribution functions $g(r)$ of jellium computed with the Overhauser potential compared to the Quantum Monte Carlo data of Ortiz, Harris and Ballone.[16]

the infinite sum over ℓ in Eqs. (25) and (26) at $\ell_{\max} = 7$ is enough to reach good convergence.[14] We see that there is accurate agreement with the QMC data for $0.5 \lesssim r/r_s \lesssim 1$ for a wide range of electron densities. In the shortest-range region, $r/r_s \lesssim 0.5$, the QMC data are known to suffer large errors and are not so reliable. (In this region there is in fact a significant discrepancy between the data from Ref. 16 and those from Ref. 17.) Thus, for $r/r_s \lesssim 0.5$ the present treatment should provide results much closer to the true $g(r)$. As said, for $r/r_s > 1$ the results obtained with the Overhauser potential are not reliable.

6. SUMMARY AND CONCLUSIONS

Two-electron wavefunctions (geminals) for the uniform electron gas of density $n = 3/4\pi r_s^3$ are found using the simple screened Coulomb potential proposed by Overhauser. These wavefunctions give pair-densities in agreement with the QMC simulations in the short-range ($r < r_s$) region. They should thus be accurate for $r < r_s$, and could be used as a comparison for other two-electron approaches to the jellium model, and for testing reduced-density matrix energy functionals. In particular, it would be of great interest to compare them with the geminals which will be computed following the approach described by P. Ziesche in this book.

The main lack of the present treatment is the violation of the particle-conservation sum rule. A possible solution to this problem is to write down a self-consistent set of equations in which the effective potential is unknwon, but the "exact" pair-density of jellium[18,19] is used to generate it.

ACKNOWLEDGMENTS

Financial support from MURST (the Italian Ministry for University, Research and Technology) through COFIN99 is acknowledged.

References

[1] P. Hohenberg and W. Kohn, *Phys. Rev.* **136**, B864 (1964).

[2] W. Kohn and L. J. Sham, *Phys. Rev.* **140**, A1133 (1965).

[3] D. Mazziotti, *J. Chem. Phys.* **112**, 10125 (2000) and the references cited therein.

[4] K. Yasuda, *Phys. Rev. A* **63**, 32517 (2001).

[5] A. Gonis and T. C. Schulthess, *J. Phys.: Condens. Matt.* **10**, 3535 (1998).

[6] S. Goedecker and C. J. Umrigar, *Phys. Rev. Lett.* **81**, 866 (1998).

[7] P. Ziesche, in: *Electron Correlations and Material Properties*, edited by A. Gonis *et al.* (Kluwer Academic/Plenum Publishers, New York 1999).

[8] P. Ziesche, in: *Many-Electron Densities and Reduced Density Matrices*, edited by J. Cioslowski (Kluwer Academic/Plenum Publishers, New York 2000).

[9] J. Cioslowski and K. Pernal, *J. Chem. Phys.* **111**, 3396 (1999).

[10] J. Cioslowski and K. Pernal, *Phys. Rev. A* **61**, 34503 (2000).

[11] J. Cioslowski, P. Ziesche, and K. Pernal, *Phys. Rev. B* **63**, 205105 (2001).

[12] P. Ziesche, in this book.

[13] A. W. Overhauser, *Can. J. Phys.* **73**, 683 (1995).

[14] P. Gori-Giorgi and J. P. Perdew, *Phys. Rev. B* **64**, 155102 (2001).

[15] P. Ziesche, J. Tao, M. Seidl, and J. P. Perdew, *Int. J. Quantum Chem.* **77**, 819 (2000).

[16] G. Ortiz, M. Harris, and P. Ballone, *Phys. Rev. Lett.* **82**, 5317 (1999).

[17] D. M. Ceperley and B. J. Alder, *Phys. Rev. Lett.* **45**, 566 (1980).

[18] J. P. Perdew and Y. Wang, *Phys. Rev. B* **46**, 12947 (1992); **56**, 7018 (1997).

[19] P. Gori-Giorgi, F. Sacchetti, and G. B. Bachelet, *Phys. Rev. B* **61**, 7353 (2000).

DENSITY FUNCTIONAL THEORY FROM THE EXTREME LIMITS OF CORRELATION

Michael Seidl *

1. INTRODUCTION

The difficulty of the quantum-mechanical many-body problem of atoms, molecules, solids, etc. is mainly due to the mutual Coulomb repulsion between the electrons. In Hartree-Fock theory, this two-particle interaction is approximated by an effective single-particle potential. In terms of the rest interaction, the perturbation theory by Møller and Plesset[1] provides a weak-interaction expansion for the ground state energy. If carried to high orders, however, this expansion seems to diverge even in cases as promising as the neon atom.[2]

The extreme opposite to the familiar weak-interaction limit are hypothetic electrons that repel each other infinitely strongly but have at the same time a given finite density distribution $\rho(\mathbf{r})$. This strong-interaction limit can be mathematically simple although it is not accessible by perturbation theory. In the following sections we shall see how these two extreme limits can be combined to predict properties of realistic electron systems.

The introductory section 2 illustrates the contrast between weak and strong interactions in the many-electron wave function. Section 3 summarizes a systematic analysis of the strong interaction limit for spherically symmetric two-electron systems. A simple model for general systems is also available. A natural bridge connecting the concepts of weak and strong interaction is provided by the coupling-constant integral of density functional theory (DFT, section 4). The interaction-strength interpolation of section 5 yields a compact density functional for the exchange-correlation energy, based only on the limits of extreme (weak and strong) interactions.

*Michael Seidl, Institute of Theoretical Physics, University of Regensburg, D-93040 Regensburg, Germany.

2. WEAK AND STRONG CORRELATIONS

The simplicity of the strong-interaction limit results from extremely strong correlations between the electrons. In the present work, "correlation" refers to the tendency of the particles in a quantum state Ψ to avoid each other in space.

An example for a wave function without any correlations is the ground state of N non-interacting identical bosons in some attractive external potential,

$$\Psi(\mathbf{r}_1, ..., \mathbf{r}_N) = \psi(\mathbf{r}_1) \cdots \psi(\mathbf{r}_N). \tag{1}$$

This symmetric product of N equivalent factors $\psi(\mathbf{r}_i)$ has the particle density

$$\rho(\mathbf{r}) = N|\psi(\mathbf{r})|^2. \tag{2}$$

The positions of N particles in a quantum state Ψ can always be measured simultaneously, resulting in a 'snapshot' that shows N discrete points in space. For the uncorrelated state (1), such a measurement can be simulated by a Metropolis algorithm, selecting random positions according to the probability distribution $\frac{1}{N}\rho(\mathbf{r})$. The typical result, shown in Fig. 1a, can qualitatively serve as a model also for non-interacting electrons. In contrast to the product (1), the electronic wave function is a Slater determinant, but the weak correlations due to Pauli exclusion (exchange effects) cannot produce a distinctive visible effect in a distribution of discrete positions such as in Fig. 1a.

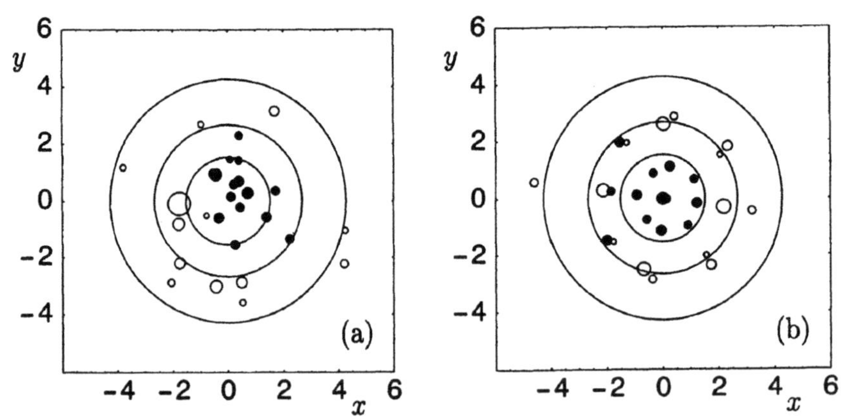

Figure 1: (a) Model for a "snapshot" of $N = 28$ electron positions in a non-interacting state with the exponential particle density $\rho(\mathbf{r}) = \frac{N}{8\pi}a^{-3}e^{-r/a}$. (b) The same for 28 electrons in the given density $\rho(\mathbf{r})$ but with infinitely strong repulsion. Shown are projections on the x, y-plane (lengths are in units of the parameter a). The z-coordinates are indicated by the sizes of the symbols, making distant particles appear small. The three large circles indicate the concentric spheres that contain on the average 20%, 50%, and 80%, respectively, of the particles. Positions spatially outside the 50%-sphere are plotted as open circles.

To have also a model for the opposite limit of infinitely strong repulsion in a given density $\rho(\mathbf{r})$, we may consider a system of N *classical* point electrons, exposed to a continuous rigid background of positive charge that has a properly chosen charge density $\rho_+(\mathbf{r}) \approx \rho(\mathbf{r})$. To minimize the total electrostatic energy, the classical point electrons take certain static equilibrium positions, trying to neutralize the continuous positive background as much as this is possible by a system of discrete negative charges.

The numerical result of such a minimization of the classical electrostatic energy is plotted in Fig. 1b. To account approximately for the self-interaction error in the positive background, its charge density was chosen to be $\rho_+(\mathbf{r}) = \frac{N-1}{N}\rho(\mathbf{r})$. In striking contrast to Fig. 1a, the positions in Fig. 1b are distributed in the density $\rho(\mathbf{r})$ in a regular arrangement without any accidental clustering. Such finite "crystals" of electrons are reminiscent of "Wigner molecules", predicted for 2D (low-density) quantum dots.[3,4] We consider Fig. 1b as a typical "snapshot" of the positions of N strongly interacting electrons in a quantum state with smooth density $\rho(\mathbf{r})$.

While this is only an illustrative model, the true strong-interaction limit is analysed in the following section for spherical two-electron systems ($N = 2$), revealing that quantum-mechanical kinetic-energy effects are in fact suppressed in that limit.

3. STRICTLY CORRELATED ELECTONS

To study strongly repulsive electrons with a given quantum-mechanical density distribution $\rho(\mathbf{r})$, we consider the Hamiltonian

$$\hat{H}^\alpha[\rho] = \hat{T} + \alpha\hat{V}_{ee} + \hat{V}^\alpha_{ext}[\rho]. \tag{3}$$

Here, $\hat{T}$ is the operator of the kinetic energy. $\hat{V}_{ee} = \sum_{i<j} |\hat{\mathbf{r}}_i - \hat{\mathbf{r}}_j|^{-1}$ is (in atomic units where $e^2 = \hbar = m_e = 1$) the two-particle Coulomb repulsion between realistic electrons which is scaled here by a parameter (coupling constant) $\alpha \geq 0$. To enforce a given density distribution $\rho(\mathbf{r})$ on the ground state of this Hamiltonian, we introduce a proper α-dependent attractive external potential $\hat{V}^\alpha_{ext}[\rho] = \sum_i v^\alpha_{ext}([\rho]; \hat{\mathbf{r}}_i)$. For the rest of this section, we confine ourselves to spherically symmetric two-electrom systems ($N = 2$).

3.1. The Limit $\alpha \to \infty$

In the limit $\alpha \to \infty$, the two electrons in the ground state of the Hamiltonian (3) are expected to be always exactly on opposite sides of the center at distances r_1 and r_2 that are "strictly correlated" in terms of a "co-motion" function $f(r)$,[5,6]

$$r_2 = f(r_1). \tag{4}$$

This behavior is illustrated in Fig. 2a. $f(r)$ must be a monotonically decreasing function of the variable r, driving one electron away from the center as the other one approaches from the opposite side. In particular,

$$f(r_0) = r_0 \tag{5}$$

where r_0 is the 50%-radius of the density ρ, fixed by $\int_0^{r_0} dr\,(4\pi r^2)\,\rho(r) = \frac{N}{2} = 1$. Solving the differential equation[6]

$$f'(r) = -\frac{r^2\,\rho(r)}{f(r)^2\,\rho(f(r))} \tag{6}$$

to the boundary condition (5), the function $f(r)$ is uniquely determined by the spherical density $\rho(r)$. An example is plotted in Fig. 2b.

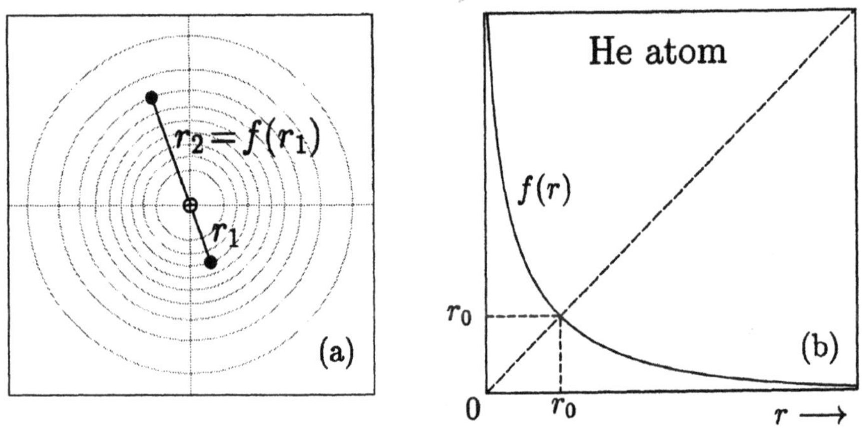

Figure 2: (a) A "snapshot" of two strictly correlated electrons in a spherical density that is indicated by its contours, cf. Fig. 1b. (b) The co-motion function $f(r)$ for an accurate density $\rho(r)$ of the He atom,[7] obtained from the differential equation (6). The length scale is defined by the 50%-radius $r_0 = 0.81$ a.u.; note that $f(r_0) = r_0$, due to Eq. (5).

The correlation structure in this SCE state is completely charcterized by the simple co-motion function $f(r)$. In particular, the ground-state expectation value $V_{ee}^{\alpha}[\rho]$ of the operator $\hat{V}_{ee}$ takes in the limit $\alpha \to \infty$ the explicit form[6]

$$V_{ee}^{SCE}[\rho] \equiv \lim_{\alpha\to\infty} V_{ee}^{\alpha}[\rho] = 2\pi \int_0^{\infty} dr \frac{r^2\,\rho(r)}{r + f(r)}. \tag{7}$$

The limit of the external potential $\hat{V}_{ext}^{\alpha}[\rho]$ in the Hamiltonian (3) is given by

$$v_{ext}^{SCE}([\rho];r) \equiv \lim_{\alpha\to\infty} \frac{v_{ext}^{\alpha}([\rho];r)}{\alpha} = w(r) \tag{8}$$

where the potential function $w(r)$ is the solution of the differential equation[6]

$$w'(r) = \frac{1}{[r + f(r)]^2} \tag{9}$$

to the boundary condition $w(\infty) = 0$.

3.2. Large, but finite $\alpha \gg 1$

To study small deviations from the SCE behavior that occur at large but finite $\alpha \gg 1$, we model the true Hamiltonian (3) by

$$\hat{H}_0^\alpha[\rho] = \hat{T} + \alpha\hat{V}_{ee} + \alpha\hat{V}_{ext}^{SCE}[\rho] \tag{10}$$

where $\hat{V}_{ext}^{SCE}[\rho] = \sum_i v_{ext}^{SCE}([\rho]; \hat{r}_i)$. Note that, unlike $\hat{H}^\alpha[\rho]$, the Hamiltonian $\hat{H}_0^\alpha[\rho]$ has a mildly α-dependent ground-state density ρ_0^α that coincides exactly with the given density ρ only in the limit $\alpha \to \infty$. Although the difference between ρ_0^α and ρ might be small at large α, we cannot expect that the following results for finite $\alpha \gg 1$ are exact.

It can be shown[6] that the potential-energy function

$$\alpha V_{pot}(\mathbf{r}_1, \mathbf{r}_2) \;\equiv\; \frac{\alpha}{|\mathbf{r}_1 - \mathbf{r}_2|} + \alpha\left[w(r_1) + w(r_2)\right] \tag{11}$$

in the Hamiltonian $\hat{H}_0^\alpha[\rho]$ is equal to the constant $\alpha w(0)$, provided that the coordinates $\mathbf{r}_1$ and $\mathbf{r}_2$ form a pair of strictly correlated positions, $\{\mathbf{r}_{10} = r\mathbf{e}, \mathbf{r}_{20} = -f(r)\mathbf{e}\}$, with some radius r and unit vector $\mathbf{e}$.

In the general case, we write $\mathbf{r}_1 = \mathbf{r}_{10} + \mathbf{u}_1$ and $\mathbf{r}_2 = \mathbf{r}_{20} + \mathbf{u}_2$ where the independent displacement vectors $\mathbf{u}_1$ and $\mathbf{u}_2$ can describe deviations from the strictly correlated motion. Taylor expansion of the function (11) with respect to $\mathbf{u}_1$ and $\mathbf{u}_2$ around any given pair $\{\mathbf{r}_{10}(r, \mathbf{e}), \mathbf{r}_{20}(r, \mathbf{e})\}$ of SCE positions reveals that such a deviating motion is confined by a strong attractive oscillator-type effective potential $\propto \alpha$ that tends to drive the electrons into the SCE state.[6] Correspondingly, as in the Born-Oppenheimer approximation to the motion of the atomic nuclei in a molecule, we treat at large $\alpha \gg 1$ the stricly-correlated motion of the two electrons at constant potential energy $\alpha w(0)$ as infinitely slow, superimposed by fast zero-point oscillations driven by a strong restoring force of the order $O(\alpha)$.

In fact, there exists a coordinate transformation $\{\mathbf{u}_1, \mathbf{u}_2\} \to \{\mathbf{S}, \mathbf{s}\}$ (s. Ref.[6]) such that $\hat{H}_0^\alpha[\rho]$ can be expanded around any pair $\{\mathbf{r}_{10}(r, \mathbf{e}), \mathbf{r}_{20}(r, \mathbf{e})\}$ of SCE positions,

$$\begin{aligned}\hat{H}_0^\alpha([\rho]; r) \;&=\; \frac{m}{2}\dot{\mathbf{S}}^2 \;+\; \alpha\, w(0) \\ &+\; \frac{m}{2}\dot{\mathbf{s}}^2 \;+\; \alpha\left[\frac{k_1(r)}{2}(x^2 + y^2) + \frac{k_3(r)}{2}z^2\right] \;+\; O(\{\mathbf{S}, \mathbf{s}\}^3).\end{aligned} \tag{12}$$

Obviously, the coordinate $\mathbf{S}$ describes an effective free particle with constant potential energy $\alpha\, w(0)$ representing strictly correlated motion of the two electrons. In contrast, the coordinate $\mathbf{s} = \{x, y, z\}$ describes an effective particle in an oscillator-type external potential with spring constants $\alpha k_1(r)$ and $\alpha k_3(r)$, where

$$k_1(r) = \frac{2}{[r + f(r)]^3}\left(\frac{r}{f(r)} + \frac{f(r)}{r}\right), \quad k_3(r) = \frac{-4}{[r + f(r)]^3}\left(\frac{1}{f'(r)} + f'(r)\right). \tag{13}$$

The ground-state energy of the Hamiltonian (12) is the constant potential energy of the free particle plus the zero-point energy of the harmonic oscillator,

$$E_0^\alpha([\rho]; r) \;=\; \alpha\, w(0) \;+\; \alpha^{1/2}\left[\tfrac{2}{2}\hbar\omega_1(r) + \tfrac{1}{2}\hbar\omega_3(r)\right], \tag{14}$$

with local frequencies $\omega_i(r) = k_i(r)^{1/2}$. Correspondingly, the ground state energy of the Hamiltonian $\hat{H}_0^\alpha[\rho]$ at large $\alpha \gg 1$ is the average of $E_0^\alpha([\rho];r)$ in the probability density $\frac{1}{2}\rho(r)$,

$$E_0^\alpha[\rho] = \left\{V_{ee}^{SCE}[\rho] + V_{ext}^{SCE}[\rho]\right\}\alpha + C[\rho]\,\alpha^{1/2} \quad (\alpha \gg 1). \tag{15}$$

Here, $V_{ext}^{SCE}[\rho] \equiv \int d^3r \frac{1}{2}\rho(r) v_{ext}^{SCE}(r)$ and the coefficient in the zero-point energy is

$$C[\rho] = 2\pi \int_0^\infty dr\, r^2\, \rho(r)\, \left[\hbar\omega_1(r) + \tfrac{1}{2}\hbar\omega_3(r)\right]. \tag{16}$$

The kinetic-energy contribution $T^\alpha[\rho]$ to Eq. (15) must be in the zero-point term $C[\rho]\,\alpha^{1/2}$. Consequently, due to the virial theorem for a harmonic oscillator,

$$T^\alpha[\rho] \to \tfrac{1}{2}C[\rho]\,\alpha^{1/2} \qquad (\alpha \to \infty). \tag{17}$$

This result indicates that in the limit $\alpha \to \infty$ the full kinetic energy $T^\alpha[\rho]$ overwhelmingly dominates the non-interacting one, $T_s[\rho]$ (section 4), but is at the same time suppressed by the potential energy which is of the order $O(\alpha)$.

A generalization of the SCE concept to systems with $N > 2$ electrons is not straight-forward. However, an accurate approximation to SCE is the PC model of Eqs. (28) and (29) which applies to any N-electron density ρ.

4. THE COUPLING-CONSTANT INTEGRAL OF DFT

The complicated physics of realistic electron systems can be found somewhere between the weak-interaction limit, represented by the perturbation expansion, and the strong-interaction limit, discussed in the preceding section 3. A natural bridge joining these relatively simple extreme limits is provided by the coupling-constant integral of density functional theory (DFT), Eq. (19) below.

Due to DFT,[8,9] the ground-state energy E of a realistic electron system in an external potential $v_{ext}(\mathbf{r})$ is a unique functional of the ground-state density $\rho(\mathbf{r})$,

$$E[\rho] = T_s[\rho] + \int d^3r\; v_{ext}(\mathbf{r})\; \rho(\mathbf{r}) + U[\rho] + E_{xc}[\rho]. \tag{18}$$

Here, $T_s[\rho]$ is the kinetic energy of a system of non-interacting electrons with ground-state density ρ. (This is the ground-state kinetic energy of the Hamiltonian $\hat{H}^0[\rho]$). $U[\rho] \equiv \frac{1}{2}\int d^3r \int d^3r'\; \rho(\mathbf{r})\rho(\mathbf{r}')/|\mathbf{r}-\mathbf{r}'|$ is the Hartree energy which treats the interacting electrons as a continuous classical charge distribution with density ρ. The only contribution to expression (18) that must be approximated in practice is the so-called exchange-correlation energy $E_{xc}[\rho]$ which includes the extra kinetic energy missing in the non-interacting functional $T_s[\rho]$ and accounts for the quantum-mechanical correlation effects ignored by the classical continuum functional $U[\rho]$.

An exact expression for $E_{xc}[\rho]$ is the coupling-constant integral,[10,11]

$$E_{xc}[\rho] = \int_0^1 d\alpha\; W_\alpha[\rho]. \tag{19}$$

The integrand is given by $W_\alpha[\rho] \equiv V_{ee}^\alpha[\rho] - U[\rho]$ where $V_{ee}^\alpha[\rho]$ is the expectation value of the operator $\hat{V}_{ee}$ in the ground state of the Hamiltonian $\hat{H}^\alpha[\rho]$, Eq. (3) (which by construction has the density ρ). Fig. 3 shows the ISI model of section 5 for the integrand $W_\alpha[\rho]$, evaluated with the density ρ of the He atom.

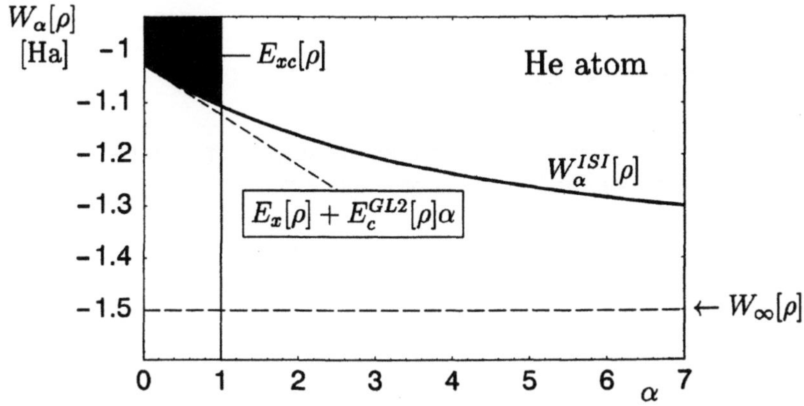

Figure 3: The model (23) for the coupling-constant integrand $W_\alpha[\rho]$ (solid curve, in hartree units), evaluated with an accurate density ρ[7] of the He atom. The integral (19) is indicated by shading. The weak-interaction limit (20) to linear order is plotted as an inclined dashed line while the horizontal line marks the asymptotic limit $W_\infty[\rho]$.

In general, $W_\alpha[\rho]$ is a monotonically decreasing function of α that approaches its lower bound $W_\infty[\rho]$ as $\alpha \to \infty$. The weak-interaction ($\alpha \to 0$) limit is controlled by DFT perturbation theory,[12]

$$W_\alpha[\rho] = E_x[\rho] + 2E_c^{GL(2)}[\rho]\,\alpha \qquad (\alpha \to 0), \tag{20}$$

which is essentially the Taylor expansion of $W_\alpha[\rho]$ around $\alpha = 0$. Both the DFT exchange energy $E_x[\rho]$ and the 2nd-order DFT correlation energy $E_c^{GL(2)}[\rho]$ are known explicitly in terms of the Kohn-Sham (KS) single-particle orbitals.[12]

In the opposite strong-interaction ($\alpha \to \infty$) limit, $W_\alpha[\rho]$ has the asymptotic behavior

$$W_\alpha[\rho] \;\to\; W_\infty[\rho] + W'_\infty[\rho]\alpha^{-1/2} \qquad (\alpha \to \infty). \tag{21}$$

For spherical two-electron systems, the coefficients can be constructed from the SCE concept of section 3. Obviously, $W_\infty[\rho] = V_{ee}^{SCE}[\rho] - U[\rho]$ or, explicitly,

$$W_\infty[\rho] \;=\; 2\pi \int_0^\infty dr \frac{r^2\,\rho(r)}{r + f(r)} - U[\rho]. \tag{22}$$

Using the expansion (12), we can evaluate, in terms of the simple functions $\rho(r)$ and $f(r)$, the expectation value of the operator $\hat{V}_{ee}$ in the ground state of the

Hamiltonian $\hat{H}_0^\alpha[\rho]$. The result is[6,13] $V_{ee}^{SCE}[\rho] + \widetilde{W}'_\infty[\rho]\alpha^{-1/2} + O(\alpha^{-1})$, where $\widetilde{W}'_\infty[\rho]$ is a model for the coefficient $W'_\infty[\rho]$ in Eq. (21). For an accurate density ρ of the He atom,[7] we have $W_\infty[\rho] = -1.5006$ and $\widetilde{W}'_\infty[\rho] = 0.1975$ (in hartree units). Since the Hamiltonians $\hat{H}_0^\alpha[\rho]$ and $\hat{H}^\alpha[\rho]$ are not the same at finite $\alpha < \infty$, we cannot claim that the functional $\widetilde{W}'_\infty[\rho]$ is identical with the true coefficient $W'_\infty[\rho]$ of the expansion (21).

5. INTERACTION-STRENGTH INTERPOLATION

The simplest analytical function that shares the asymptotes (20) and (21) with the unknown exact integrand $W_\alpha[\rho]$ is probably given by[14]

$$W_\alpha^{ISI}[\rho] = W_\infty[\rho] + \frac{X[\rho]}{\sqrt{1+Y[\rho]\,\alpha} + Z[\rho]}. \tag{23}$$

The parameters are fixed by

$$X[\rho] = \frac{xy^2}{z^2}, \qquad Y[\rho] = \frac{x^2y^2}{z^4}, \qquad Z[\rho] = \frac{xy^2}{z^3} - 1, \tag{24}$$

in terms of the quantities $x = -4E_c^{GL\,2}[\rho]$, $y = W'_\infty[\rho]$, and $z = E_x[\rho] - W_\infty[\rho]$. Using only the coefficients of the weak- and the strong-interaction limits (20) and (21), the function (23) provides an "interaction-strength interpolation" (ISI) between these limits. It can be integrated analytically in Eq. (19) to yield the ISI exchange-correlation energy,[14]

$$E_{xc}^{ISI}[\rho] = W_\infty + \frac{2X}{Y}\left[(1+Y)^{1/2} - 1 - Z\ln\left(\frac{(1+Y)^{1/2}+Z}{1+Z}\right)\right]. \tag{25}$$

The model $W_\alpha^{ISI}[\rho]$ has the correct scaling behavior of the exact integrand,[14]

$$W_\alpha^{ISI}[\rho] = \alpha W_{\alpha=1}^{ISI}[\rho_{1/\alpha}], \tag{26}$$

where $\rho_\lambda(\mathbf{r}) \equiv \lambda^3\rho(\lambda\mathbf{r})$ is a scaled density.[15] Therefore,[16] the ISI functional (25) has the correct high-density limit,[14]

$$\lim_{\lambda\to\infty} \tfrac{1}{\lambda} E_{xc}^{ISI}[\rho_\lambda] = E_x[\rho]. \tag{27}$$

For the He atom, $E_x[\rho] = -1.0246$[7] and $E_c^{GL2}[\rho] = -0.0483$.[17] Using these data along with the values for $W_\infty[\rho]$ and $\widetilde{W}'_\infty[\rho]$ given in section 4, the functional (25) predicts $E_c^{ISI}[\rho] \equiv E_{xc}^{ISI}[\rho] - E_x[\rho] = -0.0424$ for the correlation energy $E_c[\rho]$ of the He atom (exact value: -0.042[18]).

While a generalization of the SCE concept of section 3 to systems with $N > 2$ electrons is not straight-forward, the PC model of Ref.[19] provides a simple and accurate approximation to the strong-interaction limit. Treating one electron as a point charge inside an empty sphere that is surrounded by a continuous fluid representing the other $N-1$ electrons (cf. Fig. 1b), this "point-charge-plus-continuum" (PC) model leads to simple second-order gradient approximations,

$$W_\infty[\rho] \approx W_{PC}[\rho] = \int d^3r \left[A\,\rho(\mathbf{r})^{4/3} + B\,\frac{|\nabla\rho(\mathbf{r})|^2}{\rho(\mathbf{r})^{4/3}}\right], \tag{28}$$

$$W'_\infty[\rho] \approx W'_{PC}[\rho] = \int d^3r \left[C\,\rho(\mathbf{r})^{3/2} + D\,\frac{|\nabla\rho(\mathbf{r})|^2}{\rho(\mathbf{r})^{7/6}}\right]. \tag{29}$$

The predicted coefficients $A = -\frac{9}{10}(\frac{4\pi}{3})^{1/3} = -1.451$, $B = \frac{3}{350}(\frac{3}{4\pi})^{1/3} = 5.317 \times 10^{-3}$, and $C = \frac{1}{2}(3\pi)^{1/2} = 1.535$ are realistic while the value of D has the wrong sign. Therefore, D has been adjusted to the meta-GGA of Ref.[20] which is self-correlation free and therefore expected to work well in the strong-interaction limit. This procedure yields the value $D = -0.02558$.

Employing the PC models (28) and (29) for $W_\infty[\rho]$ and $W'_\infty[\rho]$, the ISI functional (25) predicts with competitive accuracy atomization energies of molecules,[14] correlation energies of atoms,[19] and the correlation energy $e_c(r_s)$ of the 2D uniform electron gas for all values of the density parameter $r_s \equiv (\pi\rho_{2D})^{-1/2}$.[5] This is achieved by using only the relatively simple extreme limits of weak and strong interactions.

The success of ISI is particularly remarkable, since it makes use of the exact exchange energy, see Eq. (27). Almost all the other accurate density functionals available are not compatible with exact exchange, but rely on a fortunate error cancellation between approximate exchange and approximate correlation. In other words, the ISI correlation energy $E_c^{ISI}[\rho] \equiv E_{xc}^{ISI}[\rho] - E_x[\rho]$ must be very accurate.

6. SUMMARY AND CONCLUSIONS

This article focusses on the strong-interaction limit of DFT as the counterpart of the perturbation expansion for weak interactions. Although it is far from reality, the strong-interaction limit appears to provide just that information on correlations which is not accessible by the perturbation expansion. Useful for electronic-structure calculations on real electronic systems, this limit is at the same time mathematically simple, as is illustrated here in section 3 by the SCE concept for spherically symmetric two-electron systems.

The SCE concept is a candidate for the exact quantum-mechanical solution to that limit. In the SCE state, the two electrons are dynamically connected in terms of a co-motion function $f(r)$ which is determined uniquely by the given spherical density $\rho(r)$. The attractive external potential $\alpha w(r)$, required to make two strongly repulsive electrons maintain a given smooth and finite quantum-mechanical density distribution, can be constructed explicitly. In particular, the SCE concept yields a simple functional for the asymptotic limit $W_\infty[\rho]$ of the coupling-constant integrand. A simple and accurate approximation to SCE, applicable to any N-electron system, is the PC model.

The interaction-strength interpolation (ISI) of section 5 combines the PC model (employing the gradient functionals $W_{PC}[\rho]$ and $W'_{PC}[\rho]$) with the leading terms of the perturbation expansion (the exchange energy $E_x[\rho]$ and the second-order correlation energy $E_c^{GL2}[\rho]$) in an analytical model $W_\alpha^{ISI}[\rho]$ for the coupling-constant integrand $W_\alpha[\rho]$. By construction, the resulting approximation $E_{xc}^{ISI}[\rho]$ for the exchange-correlation energy shares many fundamental properties with the unknown exact functional. In particular, unlike all the other approximate correlation functionals available, the ISI functional appears to be compatible with exact exchange.

REFERENCES

1. C. Møller and M. S. Plesset, *Phys. Rev.* **46**, 618 (1934).
2. J. Olsen, O. Christiansen, H. Koch, and P. Jørgensen, *J. Chem. Phys.* **105**, 5082 (1996).
3. S.M. Reimann, M. Koskinen, and M. Manninen, *Phys. Rev. B* **62**, 8108 (2000).
4. R. Egger, W. Häusler, C.H. Mak, H. Grabert, *Phys. Rev. Lett.* **82**, 3320 (1999).
5. M. Seidl, J. P. Perdew, and M. Levy, *Phys. Rev. A* **59**, 51 (1999).
6. M. Seidl, *Phys. Rev. A* **60**, 4387 (1999).
7. E. Clementi and C. Roetti, *At. data nucl. Data Tables* **14**, 177 (1974).
8. R. G. Parr and W. Yang, *Density-Functional Theory of Atoms and Molecules* (Oxford University Press, New York, 1989).
9. R.M. Dreizler and E.K.U. Gross , *Density Functional Theory* (Springer, Berlin, 1990)
10. D. C. Langreth and J. P. Perdew, *Solid State Commun.* **17**, 1425 (1975).
11. O. Gunnarsson and B. I. Lundqvist, *Phys. Rev. B* **13**, 4274 (1976).
12. A. Görling and M. Levy, *Phys. Rev. B* **47**, 13 105 (1993); Phys. Rev. A **52**, 4493 (1995).
13. M. Seidl, (unpublished).
14. M. Seidl, J. P. Perdew, and S. Kurth, *Phys. Rev. Lett.* **84**, 5070 (2000).
15. M. Levy, in *The single-Particle Density in Physics and Chemistry*, eds. N.H. March and B.M. Deb, Academic Press, London (1987).
16. M. Levy, *Phys. Rev. A* **43**, 4637 (1991).
17. M. Pindl, *Diploma thesis*, University of Regensburg, October 2000 (unpublished).
18. C. J. Umrigar and X. Gonze, *Phys. Rev. A* **50**, 3827, (1994); C. J. Umrigar, private communication.
19. M. Seidl, J. P. Perdew, and S. Kurth, *Phys. Rev. A* **62**, 012502 (2000).
20. J. P. Perdew, S. Kurth, A. Zupan, and P. Blaha, *Phys. Rev. Lett.* **82**, 2544 (1999); *ibid.* **82**, 5179 (1999) (E); and references therein.

AB INITIO STUDY OF IRON MAGNETISM ALONG BAIN'S PATH

Applications to overlayers

Martin Friák, Mojmír Šob, and Václav Vitek*

A detailed theoretical study of magnetic behavior of iron along the bcc-fcc (Bain's) transformation paths at various atomic volumes is presented. The total energies and magnetic moments are calculated by spin-polarized full-potential LAPW method and are displayed in contour plots as functions of tetragonal distortion *c/a* and volume; borderlines between various magnetic phases are shown. Stability of tetragonal magnetic phases of γ-Fe is discussed. The calculated phase boundaries are used to predict the lattice parameters and magnetic states of iron overlayers on various (001) substrates. Detailed comparison of the LDA and GGA results is performed.

1. INTRODUCTION

Iron thin films and overlayers on various metallic substrates have attracted a lot of interest in recent years[1-14]. They exhibit a rich variety of magnetic phases. For example, on the (001) surface of Ag, Pd and Cu_3Au, they are ferromagnetic (FM). On $Cu_{84}Al_{16}$ and on Cu, both ferromagnetic and antiferromagnetic (AFM) ordering was observed. In addition to that, considerable differences in magnetic properties of pulsed-laser-deposited and thermally deposited iron films on Cu(001) were reported[10]. This is connected with the fact that the density of the iron films on Cu and $Cu_{84}Al_{16}$ substrates is close to the range for which a longitudinal spin-density instability was predicted in bulk γ-Fe[15]. At these densities, ferromagnetic high-spin and low-spin states and antiferromagnetic states differ only very little in energy and, therefore, the magnetic ground state of iron the films on those substrates will depend critically on volume and symmetry. Pseudomorphic iron

* Martin Friák, Institute of Physics of Materials, Academy of Sciences of the Czech Republic, Žižkova 22, CZ-616 62 Brno, Czech Republic, and Institute of Condensed Matter Physics, Faculty of Science, Masaryk University, Kotlářská 2, CZ-611 37 Brno, Czech Republic. Mojmír Šob, Institute of Physics of Materials, Academy of Sciences of the Czech Republic, Žižkova 22, CZ-616 62 Brno, Czech Republic. Václav Vitek, Department of Materials Science and Engineering, University of Pennsylvania, Philadelphia, 3231 Walnut St., PA 19104-6272, U.S.A.

Electron Correlations and Materials Properties 2
Edited by Gonis *et al.*, Kluwer Academic/Plenum Publishers, 2002

overlayers adopt the lattice dimensions of the substrate in the (001) plane and relax the interlayer distance.

There is a stress in the (001) plane keeping the structure of the film and of the substrate coherent, and the stress perpendicular to this plane vanishes due to relaxation. A tetragonal phase arises, which may be stable or metastable[16]. Iron films and overlayers on (001) substrates have, therefore, a tetragonal structure (Fig. 1).

To advance our fundamental understanding of energetics of tetragonal iron, we performed ab initio calculations of electronic structure (ES) and total energy along tetragonal (Bain's) deformation paths at various volumes[17]. It turns out that the results of these ab initio calculations may be used to understand and predict the structure and magnetic ordering of iron overlayers on various (001) substrates. Here we discuss this application in more detail.

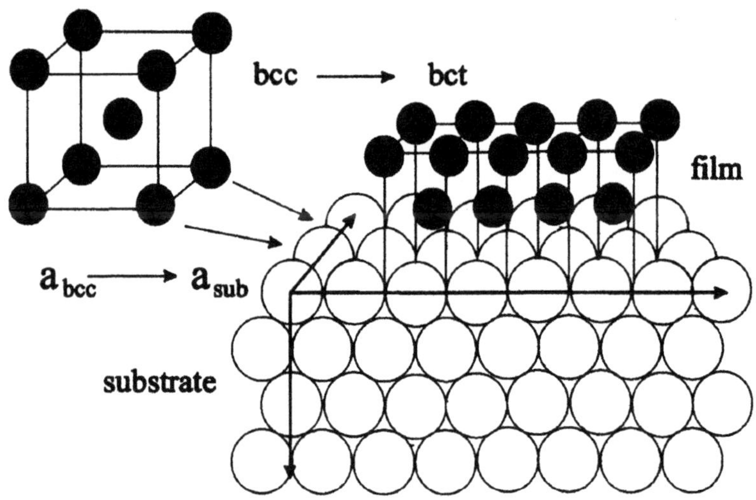

Figure 1. Epitaxial growth of overlayers on a (001) substrate.

2. METHODOLOGY

Craievich *et al.*[18] have shown that some energy extrema on constant-volume transformation paths are dictated by the symmetry. Namely, most of the structures encountered along the transformation paths between some higher-symmetry structures, say between bcc and fcc at the Bain's path, have a symmetry that is lower than cubic. At those points of the transformation path where the symmetry of the structure is higher the derivative of the total energy with respect to the parameter describing the path must be zero. These are the so-called symmetry-dictated extrema. However, other extrema may occur that are not dictated by symmetry and reflect properties of the specific material. Configurations corresponding to energy minima at the transformation paths represent stable or metastable structures and may mimic atomic arrangements that could be encountered when investigating thin films[16] and extended defects (e.g. dislocations[19, 20]).

We start with the bcc structure and consider it as a tetragonal one with the *c/a* ratio equal to 1. Subsequently, we perform a tetragonal deformation (uniaxial deformation along the [001] axis), i.e. we change the *c/a* ratio and the structure becomes tetragonal indeed. However, at $c/a = \sqrt{2}$, we arrive to the fcc structure, which has again cubic symmetry (Fig. 2). The points $c/a = 1$ and $c/a = \sqrt{2}$ correspond to the only higher-symmetry structures along the tetragonal deformation path and, therefore, symmetry-dictated extrema of the total energy may be expected here[18, 19].

Let us note that many papers define the *c/a* such that the fcc structure is considered as a tetragonal one with $(c/a)^* = 1$; then $(c/a)^* = (c/a)/\sqrt{2}$ and the bcc structure corresponds to the ratio $(c/a)^* = (\sqrt{2})/2$.

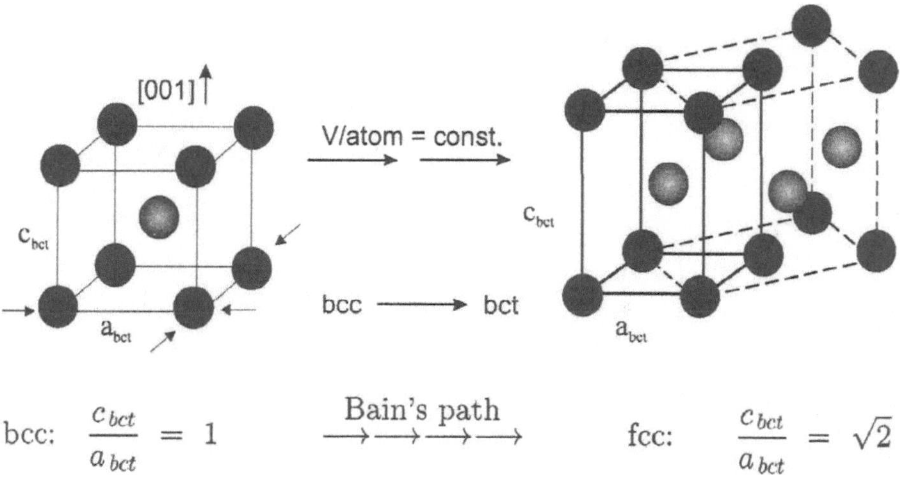

Figure 2. The bcc and fcc structures along the tetragonal deformation path.

We calculate the total energy and magnetic moment of iron along the tetragonal deformation paths keeping the atomic volume constant; the region of atomic volumes studied extends from $V/V_{exp} = 0.84$ till $V/V_{exp} = 1.05$. We include nonmagnetic (NM), ferromagnetic (FM) and two antiferromagnetic states (Fig. 3), namely the single-layer antiferromagnetic one (AFM1), in which the (001) planes have alternating magnetic moments (↑↓↑↓ ...), and double-layer one (AFMD), where the pairs of (001) planes have alternating magnetic moments (↑↑↓↓ ...). Table 1 summarises the symmetry of all calculated iron phases along the tetragonal transformation path.

For the electronic structure calculations, we utilized the full-potential linearized augmented plane waves (FLAPW) WIEN97 code[21]. The FLAPW method is one of the most accurate ab initio (first-principles) methods for calculating electronic structure of solids. Here the electron wave functions are expressed as a linear combination of convenient basis functions and the total energy of the system is minimized by means of the variational principle. This approach leads to the system of Kohn-Sham equations that must be solved selfconsistently, and this is the primary computational task. The total energy itself consists of a few parts; a special attention is paid to the exchange-correlation part which emerges from the density functional theory (DFT) of Hohenberg, Kohn and Sham.

The FLAPW method employs a combined system of basis functions. Inside of non-overlapping so-called muffin-tin spheres constructed around each nucleus, radial Schrödinger equation is solved and a linear combination of its solutions (radial functions) multiplied by spherical harmonics is used to express the electronic wave functions in this region. In the remaining space (interstitial region) the electronic wave functions are expressed by a linear combination of plane waves. The condition that basis functions inside the muffin-tins must match the corresponding basis functions (plane waves) in the interstitial region on the sphere boundary in value and slope determines certain coefficients in the expression for the basis functions and such basis functions are called "augmented plane waves" (APW).

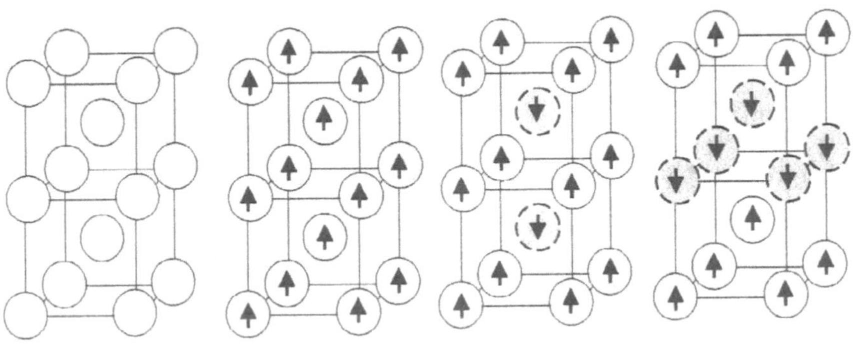

Figure 3. Non-magnetic (NM), ferromagnetic (FM), antiferromagnetic single-layer (AFM1) and double-layer (AFMD) states.

For the above mentioned exchange-correlation part of the total energy, the local density approximation (LDA) or its spin-polarized version is relevant for most problems. Unfortunately, it fails in case of some important materials, e.g. it predicts an incorrect ground state of iron. One way how to go beyond the LDA is so-called generalized gradient approximation (GGA); it includes gradients of the electronic density and fixes some problems as e.g. the ground state of iron mentioned above.

Table 1. Symmetry of calculated iron phases along Bain's transformation path: lt stands for layered tetragonal.

	$c/a < 1$	$c/a = 1$	$1 < c/a < \sqrt{2}$	$c/a = \sqrt{2}$	$c/a > \sqrt{2}$
NM	A6	A2	A6	A1	A6
FM	A6	A2	A6	A1	A6
AFM1	$L1_0$	B2	$L1_0$	$L1_0$	$L1_0$
AFMD	lt	lt	lt	lt	lt

In the present paper, all electronic structure calculations were performed self-consistently within both LDA[22] and GGA[23] and the total energy and magnetic moment using both approximations were compared in detail. The muffin-tin radius of iron atoms of 1.90 au was kept constant for all calculations, the product of the number of k-points and number of non-equivalent atoms in the basis was equal to 6000, and the product of the muffin-tin radius and the maximum reciprocal space vector, $R_{MT}k_{max}$, was equal to 10. The maximum l value for the waves inside the atomic spheres, l_{max}, and the largest reciprocal vector, G, in the charge Fourier expansion, G_{max}, was set to 12 and 15, respectively.

3. RESULTS AND DISCUSSION

Fig. 4 displays the variation of total energies of iron along the tetragonal deformation path at the experimental lattice volume of the FM bcc iron of 11.72 Å^3 and at the atomic volume of 10.21 Å^3. The NM and FM states exhibit energy extrema at $c/a = 1$ and also at $c/a = \sqrt{2}$ corresponding to higher-symmetry structures (bcc and fcc). However, the AFM1 iron keeps its cubic symmetry only for $c/a = 1$, i.e. for the bcc structure. At the point of $c/a = \sqrt{2}$ the atoms occupy the fcc lattice positions, but as the atoms with spins up and down are not equivalent, the resulting symmetry is tetragonal, exactly the same as the symmetry of the structures in the neighborhood of $c/a = \sqrt{2}$, and no higher-symmetry structure occurs here. Therefore, no symmetry-dictated extremum of total energy at the point $c/a = \sqrt{2}$ corresponding to the fcc structure (γ-Fe) is to be expected. And, indeed, the total energy curves of AFM1 states exhibit, in general, a nonzero derivative at $c/a = \sqrt{2}$ (Fig. 4); this was also found by Qiu *et al.*[24]. Consequently, γ-Fe with AFM1 ordering is unstable with respect to tetragonal deformation. The other minimum of the AFM1 curves is also not dictated by symmetry. As to the AFMD iron, it is never cubic (see Table 1) and no symmetry-dictated extrema occur along the tetragonal deformation path.

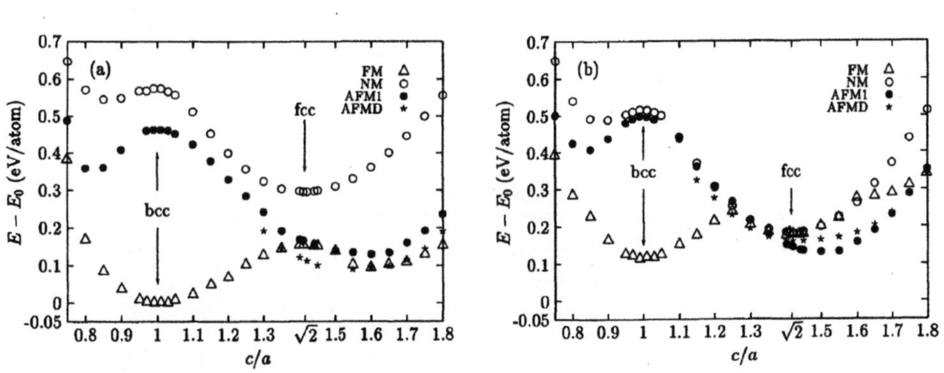

Figure 4. Variations of total energies (per atom) of iron along the constant-volume bcc-fcc transformation path at the experimental atomic volume V_{exp} = 11.72 Å^3 (a) and at the atomic volume V=10.21 Å^3 (b) relative to the equilibrium energy E_0 of FM bcc iron calculated within the GGA.

It may be seen from Fig. 4a that the FM bcc iron has the lowest energy at the experimental atomic volume. In the region from $c/a \approx 1.4$ till $c/a \approx 1.7$, the AFMD ordering is most favorable. This situation is somewhat changed at sufficiently lower atomic volumes. At V=10.21 Å^3, the energy of the FM bcc state is still the lowest one. However, for $1.4 < c/a < 1.8$, the AFM1 states are most favorable (Fig. 4b).

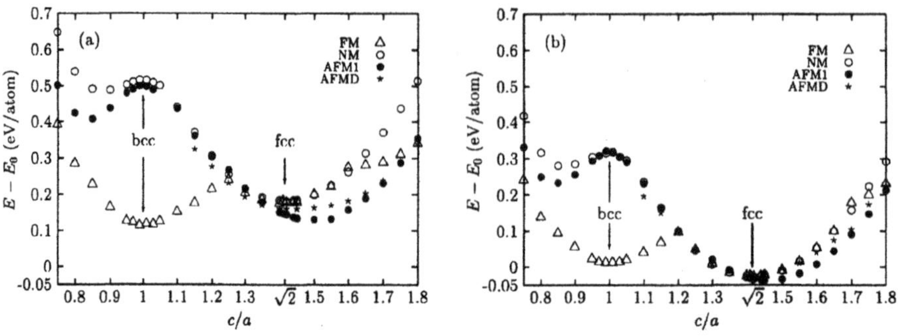

Figure 5. The same as in Fig. 4 but calculated using LDA.

Fig. 5 displays the total energy dependence for the same states as Fig. 4 but calculated within the LDA. All symmetry-dictated extrema are well pronounced and also positions of the other extrema not dictated by symmetry for all phases are nearly identical as found within the GGA. Special attention should be paid to the states in the neighborhood of the fcc structure at the lower volume (Fig. 5b) that are lower in energy than the equilibrium ferromagnetic state of bcc iron.

In Table 2, the intervals of *c/a* corresponding to the states lowest in energy calculated within GGA and LDA are displayed. These intervals are nearly identical for both approximations. The shape of the energy curves corresponding to the same type of magnetic order is remarkably similar in the GGA and LDA, but their mutual shift is somewhat different.

Table 2. Magnetic order corresponding to the lowest energy states along the tetragonal deformation path for two atomic volumes, V_{exp} = 11.72 Å^3 and V=10.21 Å^3 (87 % of V_{exp}) calculated within the GGA and LDA. The equilibrium FM bcc state within the LDA was found at the volume V/V_{exp} = 0.90.

V_{exp}	GGA		
phase	FM	AFMD	FM
c/a	(0.75, 1.35)	(1.35, 1.70)	(1.70, 1.80)
$V = 0.87\ V_{exp}$	GGA		
phase	FM	AFMD	AFM1
c/a	(0.75, 1.25)	(1.25, 1.35)	(1.35, 1.80)
V_{exp}	LDA		
phase	FM	AFMD	FM
c/a	(0.75, 1.30)	(1.30, 1.75)	(1.75, 1.80)
$V = 0.87\ V_{exp}$	LDA		
phase	FM	AFMD	AFM1
c/a	(0.75, 1.20)	(1.20, 1.25)	(1.25, 1.80)

To see the differences between the LDA and GGA more clearly, we performed the total energy calculations along the Bain's path for nine volumes ranging from 84 % to 105 % of experimental volume. Below we present the results in the form of contour plots.

The total energy of non-magnetic (NM) phase is plotted in Figure 6. For GGA the positions of the symmetry-dictated global minimum corresponding to the fcc structure, the saddle point for the bcc structure and another minimum (not dictated by symmetry) are denoted by a cross, star and triangle. Within the LDA, they are shifted to lower volumes and are out of range of our present calculation. Let us note that in case of the two-dimensional representation, some extrema encountered in one-dimensional total energy profiles manifest themselves as saddle points.

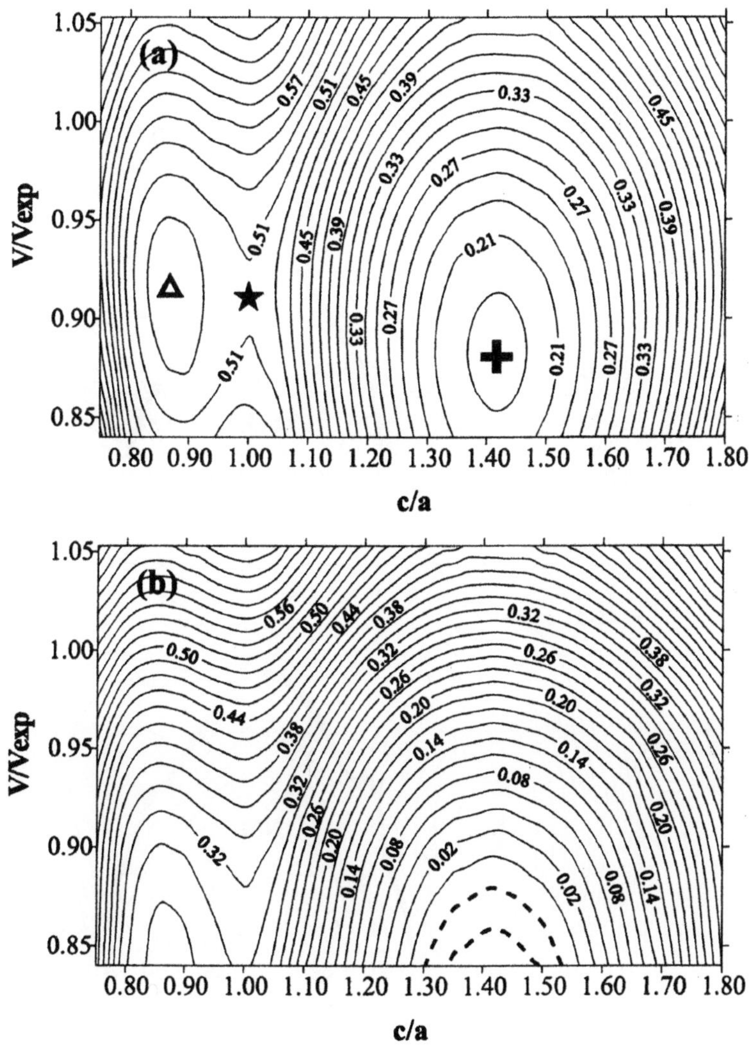

Figure 6. Contour plots of total energy (per atom) of NM iron as a function of *c/a* ratio and volume calculated within the GGA (a) and LDA (b) relative to the energy of the equilibrium FM bcc state. The contour interval is 20 meV/atom. Thick dashed lines stand for the states lower in energy than the equilibrium FM bcc state. The cross corresponds to the global symmetry-dictated minimum, the star to the symmetry-dictated saddle point and the triangle to another minimum not dictated by symmetry.

Fig. 7 shows the total energy of ferromagnetic (FM) states in the same coordinates. The global minimum (iron ground state) is marked by a cross. Again, it is symmetry-dictated. Within the GGA, it is located at 98.5 % of experimental volume. Therefore, the calculated equilibrium volume is about 1.5 % lower than the experimental value, which may be considered as a very good agreement. The LDA yields the equilibrium FM bcc state at 90 % of experimental volume. A shallow minimum, not dictated by symmetry, was found within the GGA at nearly experimental volume (it is denoted by a triangle in Fig. 7a). There is a region where, within the LDA, ferromagnetic solution of Kohn-Sham equations was not stable and the spin-polarized calculations ended up as non-magnetic.

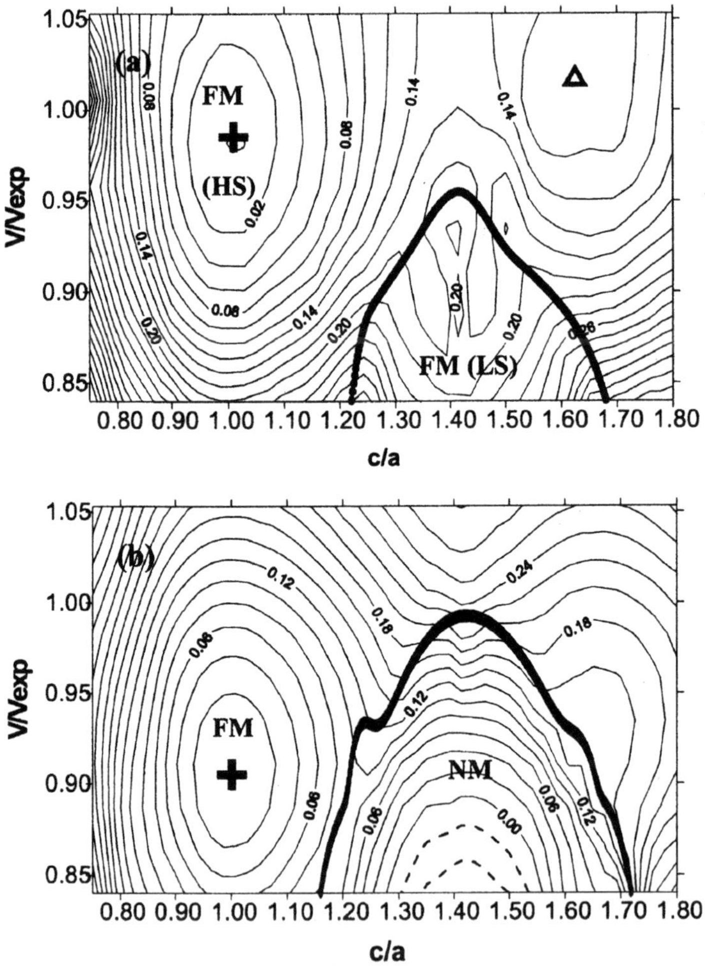

Figure 7. Contour plots of total energy (per atom) of FM iron in the same coordinates as in Fig. 6 calculated within the GGA (a) and LDA (b) relative to the energy of the equilibrium FM bcc state. The contour interval is 20 meV/atom. Thick dashed lines show the states lower in energy than the equilibrium FM bcc state. A border-line between FM (HS) and FM (LS) states in case of GGA (a) and FM and NM states in case of LDA (b) is shown. The crosses correspond to the minima, the triangle shows a minimum of the total energy not dictated by symmetry.

Thus, within the LDA, there is a borderline between the ferromagnetic (FM) and non-magnetic (NM) states. In case of GGA a similar borderline separates the region of the states with magnetic moments of about 0.70 – 0.90 μ_B (low-spin states, LS) from the states with magnetic moments of about 2.0 – 2.5 μ_B (high-spin states, HS). Here the difference between the LDA and GGA is not only quantitative but also qualitative. The contour plots of magnetic moments per atom are shown in Fig. 8 with the same borderlines as in Fig. 7.

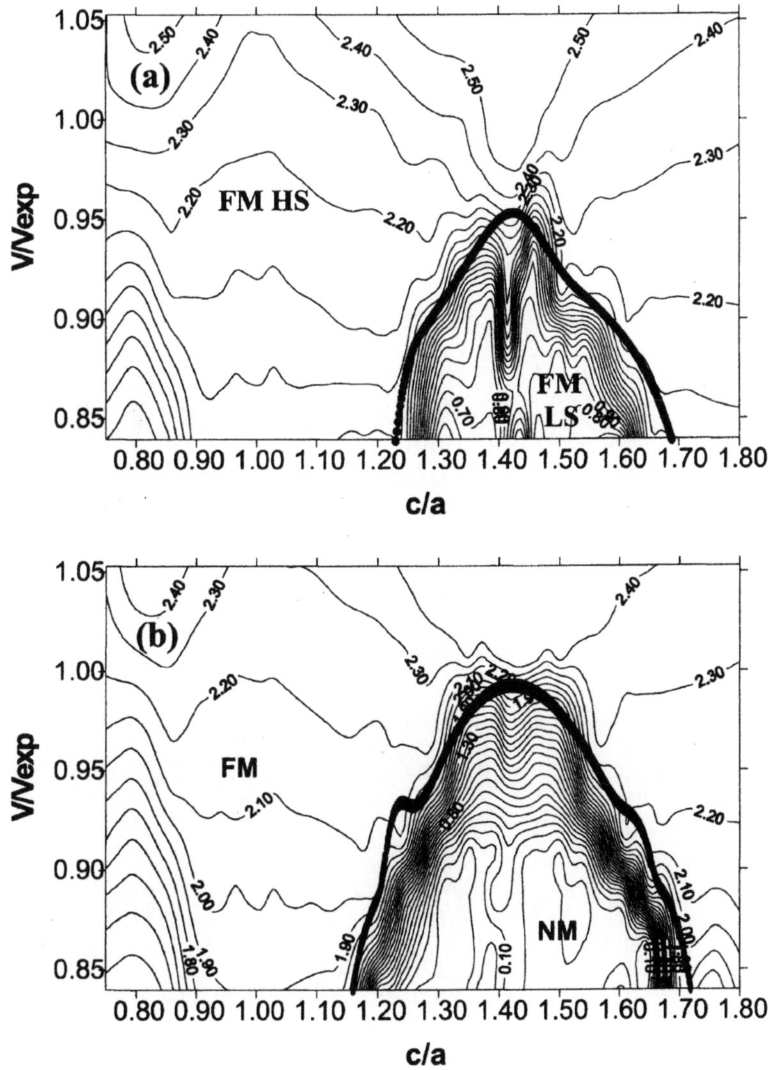

Figure 8. Contour plots of magnetic moments per atom in FM iron in the same coordinates as in Fig. 6 calculated within the GGA (a) and LDA (b). The contour interval is 0.1 μ_B. A borderline between the FM (HS) and FM (LS) states in case of GGA (a) and between the FM and NM states in case of LDA (b) is shown.

As far as the total energy of antiferromagnetic states in single-layer arrangement is concerned, the situation is very similar to that of non-magnetic states with one important exception, which is the position of the global energy minimum. Symmetry-dictated saddle point is again situated exactly at the bcc structure but the global minimum of AFM1 states marked by the cross is not dictated by symmetry because these states do not form a higher-symmetry structure in this region. We may see again that γ-Fe (fcc structure, $c/a = \sqrt{2}$) with AFM1 ordering is unstable with respect to tetragonal deformation (cf. also Figs. 4 and 5). A local minimum was found in the neighborhood of the bcc structure. It may be seen in both Figs. 9a and 9b; all the other extrema within the LDA are out of range of Fig. 9b.

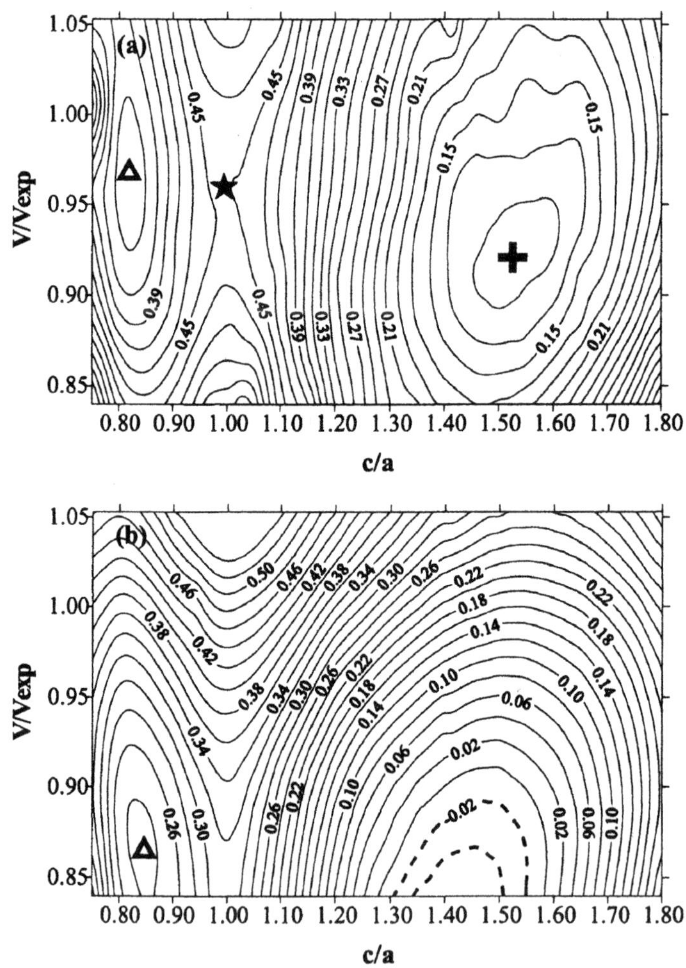

Figure 9. Contour plots of total energy (per atom) of AFM1 iron in the same coordinates as in Fig. 6 calculated within the GGA (a) and LDA (b) relative to the energy of the equilibrium FM bcc state. The contour interval is 20 meV/atom. Thick dashed lines show the states lower in energy than the equilibrium FM bcc state. The cross corresponds to the global minimum, the triangle to a minimum not dictated by symmetry and the star to the saddle point.

Fig. 10 displays behavior of magnetic moments of AFM1 iron as a function of volume and tetragonal deformation calculated within the GGA and LDA. Only magnetic moments corresponding to the atoms with one spin orientation are shown. As magnetic moment is not a variational quantity within the DFT, the contours are not as smooth as in case of total energy. Similarly as in Fig. 8, the magnetic moments calculated within GGA are somewhat higher that those obtained using the LDA.

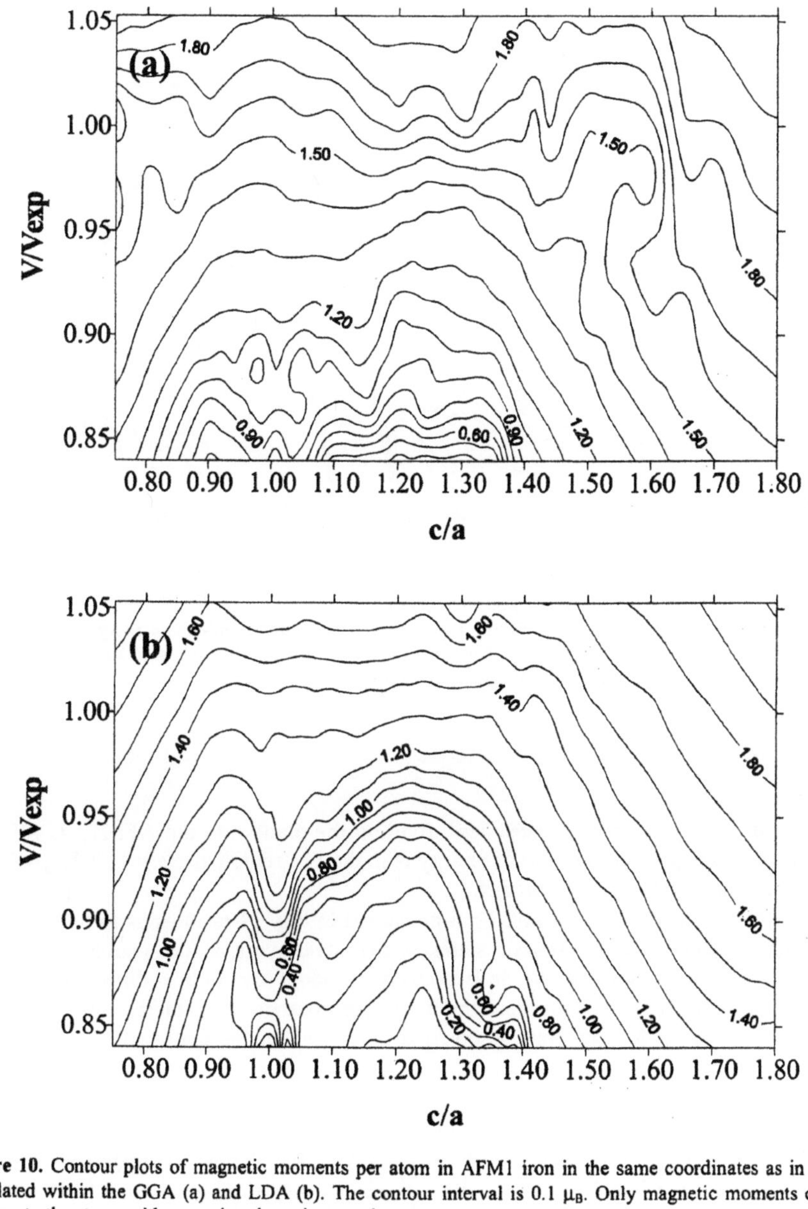

Figure 10. Contour plots of magnetic moments per atom in AFM1 iron in the same coordinates as in Fig. 8 calculated within the GGA (a) and LDA (b). The contour interval is 0.1 μ_B. Only magnetic moments corresponding to the atoms with one spin orientation are shown.

The antiferromagnetic states with double-layer arrangement (AFMD) have the lowest energy only in a limited region of contour plots presented here. We show them as a part of a complex contour plot representing the states that exhibit the lowest value of the total energy at each point of the c/a and V/V_{exp} plane. Such a contour plot for GGA is displayed in Fig. 11. It contains two phase boundaries as two transition regions occur, namely FM/AFMD and AFMD/AFM1.

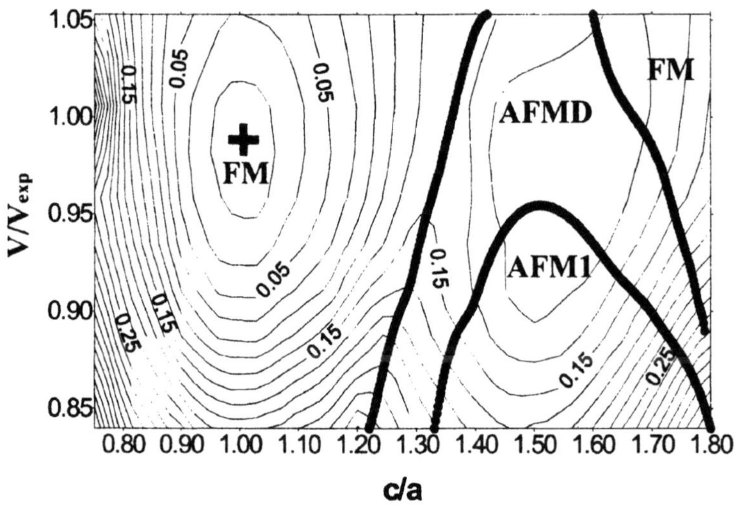

Figure 11. Total energy (per atom) of iron as a function of c/a ratio and volume relative to the FM bcc equilibrium state energy calculated within the GGA. Only states with the minimum energy are shown. The contour interval is equal to 20 meV/atom. Thick lines show the FM/AFMD and AFMD/AFM1 phase boundaries. The cross corresponds to the global symmetry-dictated minimum.

We can clearly see the "horseshoes" dividing the plane into the AFM1, AFMD and FM regions. The global minimum of energy is in the FM region at $c/a = 1$, $V/V_{exp} = 0.985$, which corresponds to the bcc structure.

The total energy profile of bulk iron presented in Fig. 11 enables us to predict easily the lattice parameters and magnetic states of iron overlayers at (001) substrates. Let us suppose that the pseudomorphic iron overlayers adopt the lattice dimensions of the substrate in the (001) plane and relax the interlayer distance (characterized by c/a). If the lattice constant of a fcc substrate is equal to a_{sub}, then in the coordinates $x = c/a$, $y = V/V_{exp}$ and $z = E-E_0$, the surfaces corresponding to a fixed a_{sub} in the (001) planes are the planes $y = kx$, where $k = (\sqrt{2}/8)(a^3_{sub} / V_{exp})$. The configuration and magnetic state of iron overlayers on a (001) substrate should correspond to the energy minimum constrained to this plane provided that the effect of the substrate/overlayer interface is not very strong. In Fig. 12, these planes for different values of a_{sub} are displayed by straight lines together with available experimental results and our theoretical predictions.

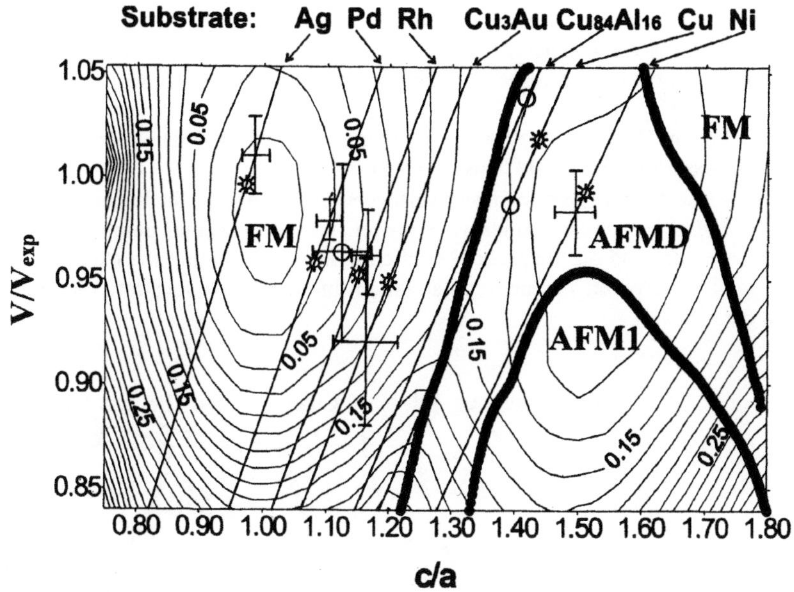

Figure 12. The same as in Fig.11, but with the added straight lines corresponding to constant lateral lattice parameters of various (001) substrates, as described in the text. The crosses composed from the vertical and horizontal error bars centered at those straight lines represent the structures of Fe films on the corresponding substrates found experimentally. One of these crosses is out of the line; it stands for the experimental structure of Fe on $Cu_3Au(001)$ films[6] and its center is denoted by an open circle. The other two open circles show the experimental structures where no error bars were given. Other smaller open circles combined with asterisks represent our theoretical results[17].

The experimental point[1] for Fe films on Ag(001) corresponds to a slightly distorted bcc structure and lies in the FM region, in agreement with experimental findings[2]. Our theoretical point is within the experimental limits. A similar situation takes place for iron films on Pd(001)[3] and Rh(001)[4]. The films exhibit body-centered tetragonal structure and the theoretical results are not very far from the experimental ones (for Rh, our result is again within experimental limits). The points for Fe films on Pd and Rh substrates still lie in the FM region. The FM ordering of films on Pd was also confirmed by experiment[3]. As to the Rh substrate, the magnetic structure of the films was not investigated experimentally yet[4].

As for $Cu_3Au(001)$ substrate, there are two experimental findings for thin films. One of them[5] lies at the straight line for the Cu_3Au substrate in Fig. 6, and our theoretical point is within the experimental limits. Another one[6] is somewhat shifted from the Cu_3Au line. The structure of the films corresponds again to a tetragonally strained FM bcc phase, in agreement with experiment[6].

Iron films on $Cu_{84}Al_{16}(001)$ substrate are reported[7] to be fcc with the atomic volume equal to 12.15 $Å^3$. The corresponding point lies very close to the phase boundary between the FM and AFMD ordering in the fcc region. This is also in accordance with experimental findings: up to four ML, high-spin FM state is reported, for higher thicknesses a low-spin and/or AFM phase was found[7]. Here we do not have enough calculated results to find out the theoretical point, but it will not be probably too much higher than the point with coordinates (c/a, V/V_{exp}) =(1.44, 1.05).

Another system close to the FM/AFMD phase boundary are the Fe films on Cu(001). This fact confirms a conclusion[25] that FM and AFMD phases are energetically almost degenerate along the line corresponding to the lattice constant of Cu and somewhat favored over the AFM1 phase, and is closely connected with a variety of magnetic states found in the Fe films on Cu(001). Our theoretical point lies in the AFMD region and is not very far from the experimental result[26] (see Fig. 12).

Finally, the straight line for iron overlayers on Ni(001) is, for lower volumes, close to the AFMD/AFM1 phase boundary. However, the experimental point[13] is distinctly in the AFMD region, and our theoretical prediction lies again within the experimental error bars. In reality, however, magnetic polarization due to FM Ni substrate may induce FM order in the film[14].

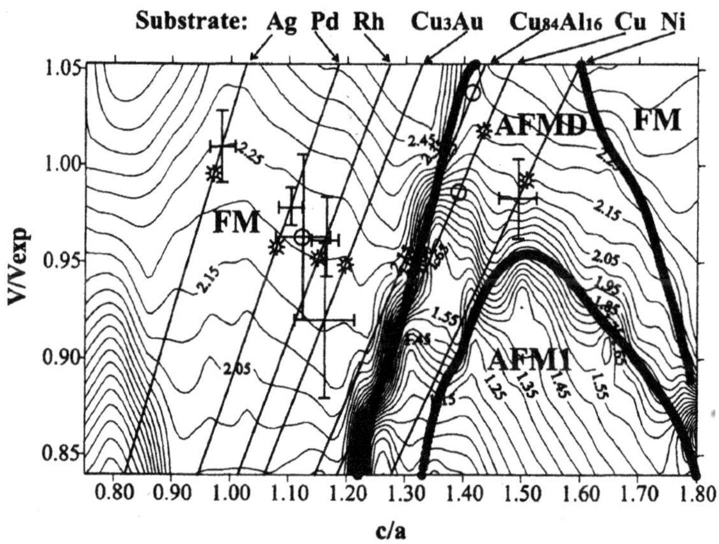

Figure 13. Magnetic moment per atom of states with the lowest energy at each configuration as a function of *c/a* and volume calculated within the GGA. The contour interval is 0.05 μ_B per atom. Thick lines show the FM/AFMD and AFMD/AFM1 phase boundaries. The meaning of the added straight lines and other symbols is the same as in Fig. 11.

Fig. 13 displays behavior of magnetic moments of iron as a function of volume and tetragonal deformation. Only magnetic moments corresponding to the states with the lowest energies for a given configuration are shown. This figure enables us to predict theoretically the magnetic moments of iron overlayers at various (001) substrates. The contour plot demonstrates that the magnetic moment decreases with decreasing atomic volume. The phase boundaries (FM/AFMD and AFMD/AFM1) are accompanied by a dramatic reduction in the value of magnetic moment when changing from one phase to another. The theoretical and experimental results for the (001) iron thin films and overlayers on various substrates are collected in Table 3, where the substrates are given in decreasing order of lattice constants. Due to the lack of experimental magnetic moment data in most cases, the last column of Table 3 may be considered as theoretical predictions.

Table 3. Type of magnetic ordering of iron overlayers on (001) substrates and the value of the *c/a* ratio in comparison with theoretically predicted results within the GGA and calculated magnetic moments of atoms in iron overlayers. AFM denotes an antiferromagnetic ordering without further specification. The experimental *c/a* ratio expressed as a single number means that no experimental error bars were given.

substrate	a_{sub} (Å) (exp.)	magnetic ordering (exp.)	c/a (exp.)	magnetic ordering (theory)	c/a (theory)	μ [μ_B] (theory)
Ag[1, 2]	4.086	FM	0.98 ± 0.02	FM	0.97	2.23
Pd[3]	3.891	FM	1.10 ± 0.01	FM	1.08	2.18
Rh[4]	3.804	—	1.16 ± 0.02	FM	1.15	2.20
Cu_3Au[5, 6]	3.745	FM	1.125 ± 0.045	FM	1.17	2.20
$Cu_{84}Al_{16}$[7]	3.649	FM/AFM	1.41	AFMD	1.44	2.35
Cu[8, 9, 10]	3.615	FM	1.39	AFMD	1.45	2.20
Cu[11, 12, 10]	3.615	AFM	1.39	AFMD	1.45	2.20
Ni[13, 14]	3.524	—	1.48 ± 0.02	AFMD	1.50	2.15

The difference between the total energy profiles obtained within the GGA and LDA may be seen by comparing Figs. 12 and 14. Whereas the GGA reproduces the FM bcc ground state of iron fairly well, the global minimum of energy within the LDA is, with the volume and *c/a* as free variables, nonmagnetic fcc, at the volume V/V_{exp} =0.82 and about 63 meV/atom below the FM bcc local minimum (this global minimum is out of the range of Fig. 14).

Surprisingly, the positions of borderlines between individual magnetic phases are not very different within the GGA and LDA. This is due to two mutually compensating effects. First, the whole "landscape" of the LDA is, roughly speaking, shifted to lower volumes with respect to the GGA one (phase boundaries are shifted in the same manner). Second, the local minima and the energy surface in the neighborhood of the fcc structure

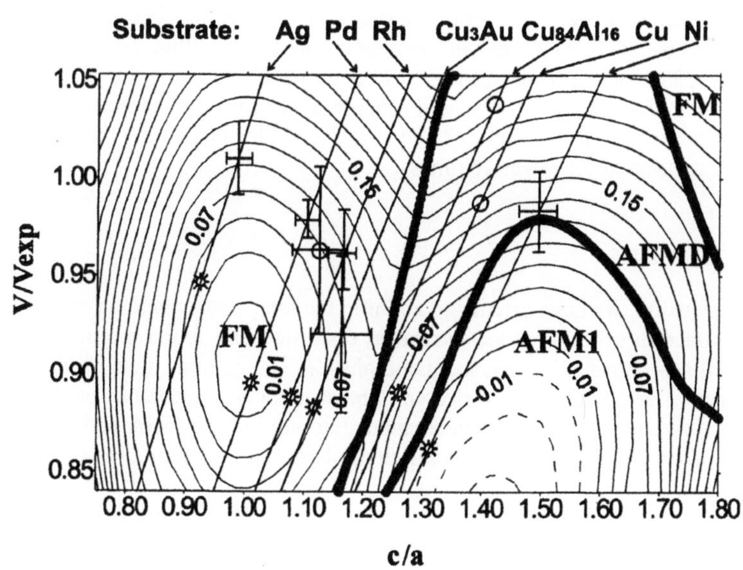

Figure 14. The same as in Fig.11 but calculated within the LDA. Smaller open circles combined with asterisks represent our theoretical results[17].

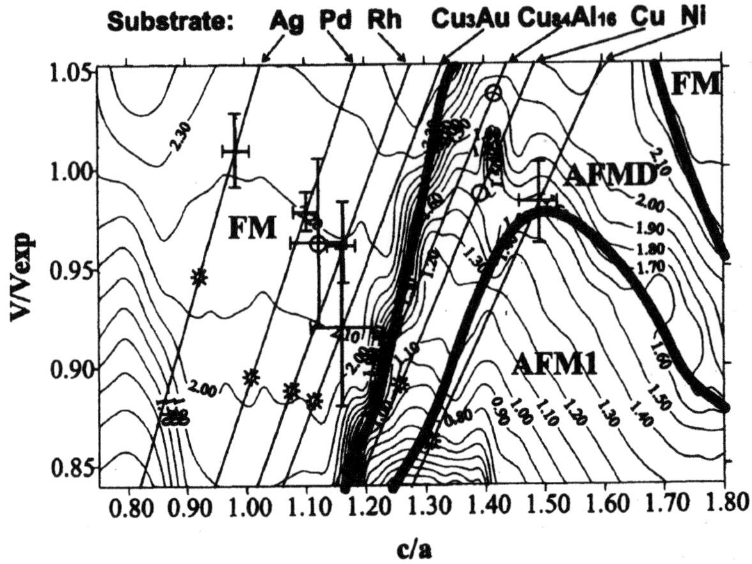

Figure 15. Magnetic moment per atom of states that minimize the total energy as a function of *c/a* and volume calculated within the LDA. The contour interval is 0.1 μ_B per atom. Thick lines show the FM/AFMD and AFMD/AFM1 phase boundaries. The meaning of the added straight lines and other symbols is the same as in Fig. 13.

are shifted to lower energies with respect to the energy surface in the neighborhood of the bcc structure (this shifts the phase boundaries back to higher volumes). The second effect seems to be stronger and, as a result, the phase boundaries in LDA occur even at higher volumes than in GGA. The magnetic moments calculated within the LDA are displayed in Figure 15. They are somewhat lower than those obtained within the GGA. However, as may be seen in Fig. 14, the LDA would provide rather wrong theoretical prediction of the lattice parameters of Fe films on metallic substrates. This is due to a considerable shift of the energy minima to the lower volumes.

4. CONCLUSIONS

In conclusion, we have calculated ab initio the total energy and magnetic moment of iron as a function of volume and tetragonal distortion and found the minimum energies and borderlines between various magnetic phases. The contour plots of total energies of non-magnetic, ferromagnetic and two antiferomagnetic arrangements, namely the single-layered and double-layered one, were calculated within both GGA and LDA, compared in detail and used for understanding and prediction of lattice parameters, magnetic states and magnetic moments of Fe films on various metallic (001) substrates; our theoretical results obtained within the generalized gradient approximation are in a very good agreement with available experimental data. Further, we have confirmed that the antiferromagnetic fcc γ-Fe is not stable with respect to tetragonal distortion.

5. ACKNOWLEDGEMENTS

This research was supported by the Grant Agency of the Academy of Sciences of the Czech Republic (Project No. A1010817), by the Grant Agency of the Czech Republic (Project No. 106/99/1178), by the Ministry of Education of the Czech Republic (Project No. ME-264 and 577/2001), by the National Science Foundation-International Programs (Grant No. INT-96-05232), and by the U.S. Department of Energy, Basic Energy Sciences (Grant No. DE-FG02-98ER45702). A part of this study has been performed in the framework of the COST Action P3 (Project No. OC P3.10). The use of the computer facilities at the MetaCenter of the Masaryk University, Brno, and at the Boston University Scientific Computing and Visualization Center is acknowledged.

REFERENCES

1. H. Li, Y. S. Li, J. Quinn, D. Tian, J. Sokolov, F. Jona, and P. M. Marcus, *Phys. Rev. B* **42**, 9195-9198 (1990).
2. M. Stampanoni, A.Vaterlaus, M. Aesschliman, and F. Meier, *Phys. Rev. Lett.* **59**, 2483-2485 (1997).
3. H. Li, Y. S. Li, J. Quinn, D. Tian, F. Jona, and P. M. Marcus, *Phys. Rev. B* **43**, 3959-3968 (1991).
4. A. M. Begley, S. K. Kim, F. Jona, and P. M. Marcus, *Phys. Rev. B* **48**, 1786-1793 (1993).
5. B. Feldmann, B. Schirmer, A. Sokoll, and M. Wuttig, *Phys. Rev. B* **57**, 1014-1023 (1998).
6. B. Schirmer, B. Feldmann, and M. Wuttig, *Phys. Rev. B* **58**, 4984-4991 (1998).
7. W. A. Macedo, F. Sirotti, G. Panaccione, A. Schatz, W. Keune, W. N. Rodrigues, and G. Rossi, *Phys. Rev. B* **58**, 11534-11538 (1998).
8. J. Thomassen, F. May, B. Feldmann, M. Wuttig, and H. Ibach, *Phys. Rev. Lett.* **69**, 3831-3834 (1992).
9. Th. Detzel, M. Vonbank, M. Donath, and V. Dose, *J. Magn. Magn. Mat.* **147**, L1 (1995).
10. H. Jenniches, J. Shen, Ch. V. Mohan, S. S. Manoharan, J. Barthel, P. Ohresser, M. Klaua, and J. Kirschner, *Phys. Rev. B* **59**, 1196-1208 (1999).
11. R. D. Ellebrock, A. Fuest, A. Schatz, W. Keune, and R. A. Brand, *Phys. Rev. Lett.* **74**, 3053-3056 (1999).
12. D. J. Keavney, D. F. Storm, W. J. Freeland, I. L. Grigorov, and C. Walker, *Phys. Rev. Lett.* **74**, 4531-4534 (1995).
13. G. C. Gazzadi, P. Luches, A. di Bona, L. Marassi, L. Pasquali, S. Valeri, and S. Nannarone, *Phys. Rev. B* **61**, 2246-2253 (2000).
14. B. Schirmer, and M. Wuttig, *Phys. Rev. B* **60**, 12945-12949 (1999).
15. V. L. Moruzzi, P. M. Marcus, and J. Kübler, *Phys. Rev. B* **39**, 6957-6961 (1989).
16. P. Alippi, P. M. Marcus, and M. Scheffler, *Phys. Rev. Lett.* **78**, 3892-3895 (1997).
17. M. Friák, M. Šob, and V. Vitek, *Phys. Rev. B* **63**, 052405-1-052405-4 (2001).
18. P. J. Craievich, M. Weinert, J. M. Sanchez, and R. E. Watson, *Phys. Rev. Lett.* **72**, 3076-3079 (1994).
19. M. Šob, L. G. Wang, and V. Vitek, *Comput. Mat. Sci.* **8**, 100-106 (1997).
20. V. Paidar, L. G. Wang L.G., M. Šob, and V. Vitek, *Modelling and Simulation Mat. Sci. Eng.* **7**, 369-381 (1999).
21. P. Blaha, K. Schwarz, and J. Luitz, WIEN97, Technical University of Vienna 1997 (improved and updated Unix version of the original copyrighted WIEN-code, which was published by Blaha, P., Schwarz, K., and Sorantin., P., *Comput. Phys. Commun.* **59**, 399-415 (1990)).
22. J. P. Perdew and Y. Wang, *Phys. Rev.* B **45**, 13 244–13 249 (1992)
23. J. P. Perdew, S. Burke, and M. Ernzerhof, *Phys. Rev. Lett.* **67**, 3865-3868 (1996).
24. S. L. Qiu, P. M. Marcus, and H. Ma, *J. Appl. Phys.* **87**, 5932-5934 (2000).
25. D. Spišák, and J. Hafner, *Phys. Rev. B* **61**, 16129-16136 (2000).
26. S. H. Lu, J. Quinn, D. Tian, F. Jona, and P. M. Marcus, *Surf. Sci.* **209**, 364-378 (1989).

INDEX

MIX
Papier aus verantwortungsvollen Quellen
Paper from responsible sources
FSC® C105338

If you have any concerns about our products,
you can contact us on
ProductSafety@springernature.com

In case Publisher is established outside the EU,
the EU authorized representative is:
Springer Nature Customer Service Center GmbH
Europaplatz 3, 69115 Heidelberg, Germany

Printed by Libri Plureos GmbH
in Hamburg, Germany